AF252005

Selected Papers of

Alan Hoffman

With Commentary

Selected Papers of

Alan Hoffman

With Commentary

edited by

Charles A. Micchelli
State University of New York, Albany, USA

World Scientific
New Jersey • London • Singapore • Hong Kong

Published by

World Scientific Publishing Co. Pte. Ltd.

5 Toh Tuck Link, Singapore 596224

USA office: Suite 202, 1060 Main Street, River Edge, NJ 07661

UK office: 57 Shelton Street, Covent Garden, London WC2H 9HE

Library of Congress Cataloging-in-Publication Data
Hoffman, A. J. (Alan Jerome), 1924–
    [Selections. 2003]
    Selected papers of Alan Hoffman with commentary / edited by Charles Micchelli.
     p. cm.
    Includes bibliographical references.
    ISBN 981-02-4198-4 (alk. paper)
     1. Combinatorial analysis 2. Programming (Mathematics) I. Micchelli, Charles, A. II.
   Title.

QA164 .H64 2003   2003053547
511'.60--dc21

British Library Cataloguing-in-Publication Data
A catalogue record for this book is available from the British Library.

The editor and publisher would like to thank the following organisations and publishers of the various journals and books for their assistance and permission to reproduce the selected reprints found in this volume:

Accad. Naz. Dei Lincei
Academic Press
American Mathematical Society
Canadian Mathematical Society
CNRS
Duke University Press
Elsevier Science Publishing Co., Inc.
Gordon and Breach Science Publishers Ltd.

Institute of Mathematical Statistics, USA
International Business Machines Corporation
National Institute of Standards and Technology
Office of Naval Research
Princeton University Press
Society for Industrial and Applied Mathematics
Springer-Verlag
Taylor and Francis Group Company

While every effort has been made to contact the publishers of reprinted papers prior to publication, we have not been successful in some cases. Where we could not contact the publishers, we have acknowledged the source of the material. Proper credit will be accorded to these publishers in future editions of this work after permission is granted.

Printed by FuIsland Offset Printing (S) Pte Ltd, Singapore

This book is dedicated to Elinor,
who gives me love, laughter, joy and youth

Alan Hoffman
Greenwich, Connecticut
May 26, 2003

Preface

Alan Hoffman was my mentor, colleague, and friend for nearly thirty years at the IBM T. J. Watson Research Center Yorktown Heights, New York. I often think of Mathematics as work, indeed hard work, albeit glorious work. Certainly, it is not easy work! Alan made me remember the fun of doing Mathematics and I am grateful to him for that. Editing this selection of his collected work was a joy. I am privileged to have been granted the opportunity to do it.

Charles A. Micchelli
Mohegan Lake, New York
June 12, 2003

Contents

Biography

Born May 30, 1924 in New York City. Educated at Columbia (AB, 1947, PhD, 1950). Served in U.S. Army, 1943–46.

1950–51   Member, Institute for Advanced Study, Princeton
1951–56   Mathematician, National Bureau of Standards, Washington
1956–57   Scientific Liason Officer, Office of Naval Research, London, U.K.
1957–61   Consultant, Management Consultation Services, General Electric Company, New York
1961–2002   Research Staff Member, T. J. Watson Research Center, IBM, Yorktown Heights
2002–   IBM Fellow Emeritus, T. J. Watson Research Center, IBM, Yorktown Heights

Adjunct or Visiting Professor at:
Technion, Israel Institute of Technology, 1965
City University of New York, 1965–1976
Yale University, 1975–1985 and 1991
Stanford University, 1980–1991
Rutgers University, 1990–1996
Georgia Institute of Technology, 1992–93

Students:

Fred Buckley, City University of New York,	1978
Michael Doob, City University of New York,	1969
Michael Gargano, City University of New York	1975
Allan Gewirtz, City University of New York	1967
Refael Hassin, Yale University	1977
Leonard Howes, City University of New York	1970
Basharat Jamil, City University of New York	1976
Sidney Jacobs, City University of New York	1971
Deborah Kornblum, City University of New York	1978
S. Thomas McCormick, Stanford University	1983
Louis Quintas, City University of New York	1967
Peter Rolland, City University of New York	1976
Howard Samowitz, City Univeristy of New York	1979
Robert Singleton, Princeton University	1962
Lennox Superville, City University of New York	1978

Present or past service on editorial boards of:
Linear Algebra and its Applications (founding editor)
Mathematics of Operations Research
Discrete Mathematics
Discrete Applied Mathematics
Naval Research Logistics Quarterly
Journal of Combinatorial Theory
Combinatorica
SIAM Journal of Discrete Mathematics
SIAM Journal of Applied Mathematics
Mathematics of Computation
International Computing Center Bulletin

Honors:
Member, National Academy of Sciences, 1982–
IBM Fellow, 1978–
DSc (hon.) Technion, 1986
Fellow, American Academy of Arts and Sciences, 1987–
Fellow, New York Academy of Sciences, 1975–
Phi Beta Kappa Lecturer, 1989–90
Special issue of Lin. Alg. Appl., 1989, for 65th birthday
von Neumann Prize (Operations Research Society &
Institute of Management Science), 1992
Founder's Award, Mathematical Programming Society, 2000
Fellow, Institute for Operations Research and Management Science, 2002–

List of Publications

1. On the foundations of inversion geometry, *Trans. Amer. Math. Soc.* **17**, 218–242 (1951).
2. A note on cross ratio, *Amer. Math. Monthly* **58**, 613–614 (1951).
3. Chains in the projective line, *Duke Math. J.* **18**, 827–830 (1951).
4. Cyclic affine planes, *Can. J. Math.* **4**, 295–301 (1952).
5. On approximate solutions of systems of linear inequalities, *J. Res. Natl. Bureau Stds.* **49**, 263–265 (1952).
6. (with H. W. Wielandt) The variation of the spectrum of a normal matrix, *Duke Math. J.* **20**, 37–39 (1953).
7. (with M. Mannos, D. Sokolowsky and N. Wiegmann) Computational experience in solving linear programs, *J. Soc. Ind. Appl. Math.* **1**, 17–33 (1953).
8. On a combinatorial theorem, *Natl. Bureau Stds. Rep.* 2377 (1953).
9. On an inequality of Hardy, Littlewood and Polya, *Natl. Bureau Stds. Rep.* 2974 (1953).
10. Cycling in the simplex algorithm, *Natl. Bureau Stds. Rep.* 2974 (1953).
11. (with O. Taussky) A characterization of normal matrices, *J. Res. Natl. Bureau Stds.* **52**, 17–19 (1954).
12. (with R. Bellman) On a theorem of Ostrowski and Taussky, *Arch. Math.* **5**, 123–127 (1954).
13. (with K. Fan) Lower bounds for the rank and location of the eigenvalues of a matrix, in *Contributions to the Solution of Systems of Linear Equations and the Determination of Eigenvalues, Appl. Math. Series No. 39*, 117–130, Washington (1954).
14. (with J. Gaddum and D. Sokolowsky) On the solution of the caterer problem, *Nav. Res. Logist. Quart.* **1**, 223–229 (1954).
15. (with W. Jacobs) Smooth patterns of production, *Management Sci.* **1**, 86–91 (1954).
16. (with H. Antosiewicz) A remark on the smoothing problem, *Management Sci.* **1**, 92–95 (1954).
17. (with K. Fan) Some metric inequalities in the space of matrices, *Proc. Amer. Math. Soc.* **6**, 111–116 (1955).
18. How to solve a linear programming problem, *Proc. Second Linear Programming Symposium*, 397–424, Washington (1955).
19. SEAC determines low bidders, *Res. Rev.* 11–12 (April, 1955).
20. (with G. Voegeli) Linear programming, *Res. Rev.* 23–27 (August, 1955).
21. Linear programming, *Appl. Mech. Rev.* **9**, 185–187 (1956).
22. (with M. Newman, E. Straus and O. Taussky) On the number of absolute points of a correlation, *Pac. J. Math.* **6**, 83–96 (1956).
23. (with H. Kuhn) On systems of distinct representatives and linear programming, *Amer. Math. Monthly* **63**, 455–460 (1956).

24. (with J. Kruskal) Integral boundary points of convex polyhedra, *Ann. Math. Studies* **38**, 223–246, Princeton (1956).

25. (with G. Dantzig) Dilworth's theorem on partially ordered sets, *Ann. Math. Study* **38**, 207–214, Princeton (1956).

26. (with H. Kuhn) On systems of distinct representatives, *Ann. Math. Study* **38**, 199–202, Princeton (1956).

27. Generalization of a theorem of Konig, *J. Wash. Acad. Sci.* **46**, 211–212 (1956).

28. (with K. Fan and I. Glicksberg) Systems of inequalities involving convex functions, *Proc. Amer. Math. Soc.* **8**, 617–622 (1957).

29. Geometry, Chap. 8, 97–102, *Handbook of Physics*, McGraw-Hill (1958).

30. Linear programming, *McGraw-Hill Encyclopedia of Science and Technology* **7**, 522–523 (1960).

31. Some recent applications of the theory of linear inequalities to extremal combinatorial analysis, *Proc. Symp. in Applied Mathematics*, Amer. Math. Soc., 113–127 (1960).

32. On the uniqueness of the triangular association scheme, *Ann. Math. Statist.* **31**, 492–497 (1960).

33. On the exceptional case in a characterization of the arcs of a complete graph, *IBM J. Res. Dev.* **4**, 487–496 (1960).

34. (with R. Singleton) On Moore graphs with diameters 2 and 3, *IBM J. Res. Dev.* **4**, 497–504 (1960).

35. (with W. Hirsch) Extreme varieties, concave functions and the fixed charge problem, *Commun. Pure Appl. Math.* **14**, 355–369 (1961).

36. (with M. Richardson) Block design games, *Can. J. Math.* **13**, 110–128 (1961).

37. (with I. Heller) On unimodular matrices, *Pac. J. Math.* **12**, 1321–1327 (1962).

38. (with R. Gomory) Finding optimal combinations, Science and Technology, 26–33 (July, 1962).

39. On the polynomial of a graph, *Amer. Math. Monthly* **70**, 30–36 (1963).

40. (with R. Gomory) On the convergency of an integer-programming process, *Naval Res. Logistics Quart.* **10**, 121–123 (1963).

41. On simple linear programming problems, *Proc. Symp. in Pure Mathematics* *VII*, 317–327, Amer. Math. Soc. (1963).

42. Dynamic programming, *5th IBM Medical Symposium*, Endicott (1963).

43. On the duals of symmetric partially balanced incomplete block designs, *Ann. Math. Statist.* **34**, 528–531 (1963).

44. (with J. Griesmer and A. Robinson) On symmetric bimatrix games, *IBM Res. Rep.* RC959 (1963).

45. On abstract dual linear programs, *Naval Res. Logistics Quart.* **10**, 369–373 (1963).

46. (with H. Markowitz) Shortest path, assignment and transportation problems, *Naval Res. Logistics Quart.* **10**, 375–379 (1963).

47. Large linear programs, *Proceedings IFIP Congress* 1962, 173–176.

48. (with P. Gilmore) A characterization of comparability graphs and of interval graphs, *Can. J. Math.* **16**, 539–548 (1964).

49. (with M. McAndrew) Linear inequalities and analysis, *Amer. Math. Monthly* **71**, 416–418 (1964).

50. (with R. Gomory and N. Hsu) Some properties of the rank and invariant factors of matrices, *Can. Math. Bull.* **7**, 85–96 (1964).

51. On the line graph of the complete bipartite graph, *Ann. Math. Statist.* **35**, 883–885 (1964).
52. (with D. Fulkerson and M. McAndrew) Some properties of graphs with multiple edges, *Can. J. Math.* **17**, 166–177 (1965).
53. (with M. McAndrew) The polynomial of a directed graph, *Proc. Amer. Math. Soc.* **16**, 303–309 (1965).
54. On the line graph of a projective plane, *Proc. Amer. Math. Soc.* **16**, 297–392 (1965).
55. (with D. Ray-Chaudhuri) On the line graph of a finite affine plane, *Can. J. Math.* **17**, 687–694.
56. (with D. Ray-Chaudhuri) On the line graph of a symmetric balanced incomplete block design, *Trans. Amer. Math. Soc.* **116**, 238–252 (1965).
57. On the nonsingularity of real matrices, *Math. Comput.* **XIX**, No. 89, 56–61 (1965).
58. On the nonsingularity of real partitioned matrices, *Int. J. Comput. Appl. Math.* **4**, 7–17 (1965).
59. (with P. Camion) On the nonsingularity of complex matrices, *Pac. J. Math.* **17**, 211–214 (1966).
60. (with R. Karp) On nonterminating stochastic games, *Management Sci.* **12**, 359–370 (1966).
61. Ranks of matrices and families of cones, *Trans. New York Acad. Sci.* **29**, 375–377 (1967).
62. Three observations on nonnegative matrices, *J. Res. Nat. Bureau Stds.* **71b**, 39–41 (1967).
63. The eigenvalues of the adjacency matrix of a graph, *Proc. Symp. on Combinatorial Mathematics*, 578–584, Univ. North Carolina (1967).
64. Some recent results on spectral properties of graphs, Beitrage zur Graphentheorie (Proceedings of an International Colloquium), 75–80, B. G. Teubner, Leipzig (1968).
65. Estimation of eigenvalues of a matrix and the theory of linear inequalities, *Proc. AMS Lectures in Applied Mathematics*, II, part 1, Mathematics of the Decision Sciences, 295–300 (1968).
66. Bounds for the rank and eigenvalues of a matrix, *Proc. IFIP Congress* 1968, 111–113.
67. A special class of doubly stochastic matrices, *Aequationes Math.* **2**, 319–326 (1969).
68. The change in the least eigenvalue of the adjacency matrix of a graph under imbedding, *SIAM J. Appl. Math.* **17**, 664–671 (1969).
69. On the covering of polyhedra by polyhedra, *Proc. Amer. Math. Soc.* **23**, 123–126 (1969).
70. On unions and intersections of cones, *Recent Progress in Combinatorics*, 103–109, Academic Press, New York (1969).
71. (with E. Haynsworth) Two remarks on copositive matrices, *Linear Algebra Appl.* **2**, 387–392 (1969).
72. Generalizations of Gersgorin's theorem, Lectures given at Univ. California, Santa Barbara (1969).
73. (with T. Rivlin) When is a team "mathematically" eliminated?, *Proc. Princeton Symposium on Math. Programming*, 1966, 391–401, Princeton (1970).

74. On the variation of coordinates in subspaces, *Ann. Mat. Pura Appl.* (4) **86**, 53–59 (1970).

75. (with R. Varga) Patterns of dependence in generalizations of Gersgorin's theorem, *SIAM J. Numer. Anal.* **7**, 571–574 (1970).

76. $-1-\sqrt{2}$?, *Combinatorial Structures and Their Applications*, 173–176, Gordon and Breach, New York (1970).

77. On eigenvalues and colorings of graphs, *Graph Theory and its Applications*, 79–91, Academic Press, New York (1970).

78. (with L. Howes) On eigenvalues and colorings of graphs II, *Proc. International Conference for Combinatorial Mathematics, Ann. New York Acad. Sci.* **175**, 238–242 (1970).

79. Combinatorial aspects of Gerschgorin's theorem, *Recent Trends in Graph Theory*, 173–179, Springer-Verlag, New York (1970).

80. On vertices near a given vertex of a graph, *Studies in Pure Mathematics*, 131–136, Academic Press, New York (1971).

81. Eigenvalues and partitionings of the edges of a graph, *Linear Algebra Appl.* **5**, 137–146 (1972).

82. (with S. Winograd) On finding all shortest distances in a directed network, *IBM J. Res. Dev.* **16**, 412–414 (1972).

83. On limit points of spectral radii of nonnegative symmetric integral matrices, *Graph Theory and its Applications*, 165–172, Springer-Verlag (1972).

84. Some applications of graph theory, *Proc. Third S.E. Conference on Combinatorics and Computing*, 9–14 (1972).

85. Sparse matrices, *Proc. Second Manitoba Conference on Numerical Mathematics* (Univ. Manitoba, Winnipeg, 1972), pp. 19–26, *Congr. Numer.*, No. VII, Utilitas Math., Winnipeg, 1973.

86. On spectrally bounded graphs, in *A Survey of Combinatorial Theory*, 277–283, North-Holland (1973).

87. (with M. Martin and D. Rose) Complexity bounds for regular finite difference and finite element grids, *SIAM J. Numer. Anal.* **10**, 364–369 (1973).

88. (with F. Pereira) On copositive matrices with -1, 0, 1 entries, *J. Comb. Theory* **A14**, 302–309 (1973).

89. (with W. Donath) Lower bounds for the partitioning of graphs, *IBM J. Res. Dev.* **17**, 420–425 (1973).

90. (with R. Brayton and D. Coppersmith) Self-orthogonal latin squares of all orders $n = 2$, 3, 6, *Bull. Amer. Math. Soc.* **80**, 116–118 (1974).

91. On eigenvalues of symmetric $(+1, -1)$ matrices, *Israel J. Math.* **17**, 69–75 (1974).

92. A generalization of max flow-min cut, *Math. Programming* **6**, 352–359 (1974).

93. (with D. Fulkerson and R. Oppenheim) On balanced matrices, *Math. Programming Study* **1**, 120–132 (1974).

94. Eigenvalues of graphs, in *Studies in Graph Theory*, Part II, 225–245, Mathematical Association of America (1975).

95. Linear G-functions, *Linear and Multilinear Algebra* **3**, 45–52 (1975).

96. On the spectral radii of topologically equivalent graphs, in *Recent Advances in Graph Theory*, 273–282, Czechoslovak Academy of Sciences (1975).

97. Applications of Ramsey style theorems to eigenvalues of graphs, in *Combinatorics*, 245–260, D. Reidel Publishing Company, Dordrecht (1975).

98. Spectral functions of graphs, *Proc. of International Congress of Mathematics* 1974, Vol. 2, 461–464, Canadian Mathematical Congress (1975).

99. On convex cones in C^n, *Bull. of the Institute of Mathematics*, Academia Sinica **3**, 1–6 (1975).

100. On spectrally bounded signed graphs, *Proc. 21st Conference of Army Mathematicians*, 1–6, El Paso (1975).

101. (with R. Brayton and D. Coppersmith) Self-orthogonal latin squares, Colloquio Internationale sulle Teorie Combinatorie, Tomo II, 509–517, Accademia Nazionale Dei Lincei (1976).

102. Total unimodularity and combinatorial theorems, *Linear Algebra Appl.* **13**, 103–108 (1976).

103. On graphs whose least eigenvalue exceeds $-1 - \sqrt{2}$, *Linear Algebra Appl.* **16**, 153–166 (1977).

104. (with B. Jamil) On the line graph of the complete tripartite graph, *Linear Multilinear and Algebra* **7**, 10–25 (1977).

105. (with R. Brayton and T. Scott) A theorem on inverses of convex sets of real matrices, with application to the worst-case DC problem, *IEEE Trans. Circuits Systems* CAS-24, 409–415 (1977).

106. On signed graphs and grammians, *Geometrie Dedicata* **6**, 455–470 (1977).

107. (with D. Schwartz) On partitions of a partially ordered set, *J. Comb. Theory* **B23**, 3–13 (1977).

108. On limit points of the least eigenvalue of a graph, *Ars Comb.* **3**, 3–14 (1977).

109. Linear programming and combinatorial problems, *Proc. of a Conference on Discrete Mathematics and its Applications*, 65–92, Indiana University (1976).

110. (with P. Joffe) Nearest S-matrices of given rank and the Ramsey problem for eigenvalues of bipartite S-graphs, Colloques Internationaux C.N.R.S. 260, Problèmes Combinatoire et Théorie des Graphes, 237–240 (1977).

111. (with C. Berge) Multicoloration dans les hypergraphes unimodulaires et matrices dont les coefficients sont des ensembles, Colloques Internationaux C.N.R.S. 260, Problemes Combinatoire et Theorie des Graphes, 27–30 (1977).

112. (with R. Graham and H. Hosoya) On the distance matrix of a graph, *J. Graph Theory* **1**, 85–88 (1977).

113. (with R. Oppenheim) Local unimodularity in the matching polytope, *Ann. Discrete Math.* **2**, 201–209 (1978).

114. (with D. E. Schwartz) On lattice polyhedra, *Proc. 5th Hungarian Colloquium on Combinatorics*, 593–598 (1978).

115. (co-editor, M. Balinski) Polyhedral Combinatorics, Mathematical Programming Study 8, North-Holland (1978).

116. D. R. Fulkerson's contributions to polyhedral combinatorics, *Math. Programming Study* **8**, 17–23 (1978).

117. On lattice polyhedra III: Blockers and anti-blockers of lattice clutters, *Math. Programming Study* **8**, 197–207 (1978).

118. Helly numbers of some sets in R^n, *IBM Res. Rep.* RC7319 (1978).

119. Some greedy ideas, *IBM Res. Rep.* RC7279 (1978).

120. The role of unimodularity in applying linear inequalities to combinatorial theorems, *Ann. Discrete Math.* **4**, 73–84 (1979).

121. (with F. Buckley) On the mixed achromatic number and other functions of graphs, *Graph Theory and Related Topics*, 105–119, Academic Press (1979).

122. Linear programming and combinatorics, *Proc. Symposia in Pure Mathematics* **34**, 245–253, American Mathematical Society (1979).

123. Binding constraints and Helly numbers, *Ann. New York Acad. Sci.* **319**, 284–288 (1979).

124. (with F. Granot) Polyhedral aspects of discrete optimization, *Ann. Discrete Math.* **4**, 183–190 (1979).

125. (with P. Erdös and S. Fajtlowicz) Maximum degree in graphs of diameter 2, *Networks* **10**, 87–96 (1980).

126. (With H. Gröflin) Matroid intersections, *Combinatorica* **1**, 43–47 (1981).

127. Ordered sets and linear programming, *Ordered Sets*, 619–654, D. Reidel Publishing Company (1981).

128. (with E. Barnes) On bounds for eigenvalues of real symmetric matrices, *Linear Algebra Appl.* **40**, 217–223 (1981).

129. (with H. Gröflin) Lattice polyhedra II: Construction and examples, *Ann. Discrete Math.* **15**, 189–203 (1982).

130. (with D. Gale) Two remarks on the Mendelsohn–Dulmage theorem, *Ann. Discrete Math.* **15**, 171–177 (1982).

131. Extending Greene's theorem to directed graphs, *J. Comb. Theory* **A34**, 102–107 (1983).

132. (with J. Goffin) On the relationship between the Hausdorff distances and matrix distances of ellipsoids, *Linear Algebra Appl.* **52/53**, 301–313 (1983).

133. On greedy algorithms in linear programming, *Proc. 4th Japanese Mathematical Programming Symposium*, 1–13, Kobe (1983).

134. (with E. Barnes) Partitioning, spectra and linear programming, in *Progress in Combinatorial Optimization*, 13–26, Academic Press (1984).

135. (with S. McCormick) A fast algorithm that makes matrices optimally sparse, in *Progress in Combinatorial Optimization*, 185–196, Academic Press (1984).

136. (with M. Held, E. Johnson and P. Wolfe) Aspects of the Traveling Salesman problem, *IBM J. Res. Dev.* **28**, 476–486 (1984).

137. (with R. Brualdi) On the spectral radius of $(0,1)$ matrices, *Linear Algebra Appl.* **65**, 133–146 (1985).

138. (with R. Aharoni and I. Hartman) Path partitions and packs of acyclic digraphs, *Pac. J. Math.* **118**, 249–259 (1985).

139. (with G. Dantzig and T. Hu) Triangulations (tilings) and certain block triangular matrices, *Math. Programming* **31**, 1–14 (1985).

140. (with E. Barnes) On transportation problems with upper bounds on leading rectangles, *SIAM J. Algebraic Discrete Methods* **6**, 487–496 (1985).

141. (with A. Kolen and M. Sakarovitch) Totally balanced and greedy matrices, *SIAM J. Algebraic Discrete Methods* **6**, 721–730 (1985).

142. (with B. Eaves, U. Rothblum and H. Schneider) Line sum symmetric scaling of square nonnegative matrices, *Math. Programming Study* **25**, 124–141 (1985).

143. (with P. Wolfe) Minimizing a unimodal function of two integer variables, *Math. Programming Study* **25**, 76–87 (1985).

144. (with P. Wolfe) History of Traveling Salesman problem, in *The Traveling Salesman*, 1–15, John Wiley (1985).

145. On a conjecture of Kojima and Saigal, *Linear Algebra Appl.* **71**, 159–160 (1985).

146. On greedy algorithms that succeed, *Surv. Comb.*, 97–112, Cambridge University Press (1985).

147. (with C. Lee) On the cone of nonnegative circuits, *Discrete Comput. Geom.* **1**, 229–239 (1986).

148. (with G. Golub and G. Stewart) A generalization of the Eckart–Young–Mirsky matrix approximation theorem, *Linear Algebra Appl.* **88/89**, 317–327 (1987).

149. (with A. Tucker) Greedy packing and series-parallel graphs, *J. Comb. Theory* **A47**, 6–15 (1988).

150. On greedy algorithms for series parallel graphs, *Math. Programming* **40**, 197–204 (1988).

151. (with B. C. Eaves and H. Hu) Linear programming with spheres and hemispheres of objective vectors, *Math. Programming* **51**, 1–16 (1991).

152. Linear Programming at the National Bureau of Standards, in *History of Mathematical Programming, Collection of Personal Reminiscences*, edited by J. K. Lenstra, A. Rinooy-Kan and A. Schrijver, 62–64, Elsevier Science Publishers, Amsterdam (1991).

153. (with E. R. Barnes and U. Rothblum) Optimal partitions having disjoint conic and convex hulls, *Math. Programming* **54**, 69–86 (1992).

154. On simple combinatorial optimization problems, *Discrete Math.* **106/107**, 285–289 (1992).

155. (with Ilan Adler and Ron Shamir) Monge and feasibility sequences in general flow problems, *Discrete Appl. Math.* **44**, 21–38 (1993).

156. (co-editors R. W. Cottle and D. Goldfarb) Festschrift in honor of Philip Wolfe, *Math. Programming* **B62** (1993).

157. (with W. W. Bein and P. Brucker) Series parallel composition of greedy linear programming problems, *Math. Programming* **B62**, 1–14 (1993).

158. (with A. F. Veinott, Jr.) Staircase transportation problems with superadditive rewards and cumulative capacities, *Math. Programming* **B62**, 199–214 (1993).

159. (with U. Rothblum) A proof of the convexity of the range of a nonatomic vector measure using linear inequalities, *Linear Algebra Appl.* **199**, 373–379 (1994).

160. (with E. Barnes) Bounds for the spectrum of normal matrices, *Linear Algebra Appl.* **201**, 79–90 (1994).

161. Special Issue, Mathematical Sciences Department, T. J. Watson Research Center, *IBM J. Res. Dev.* **38**, number 3 (1994), guest editor.

162. (with E. V. Denardo, T. Mackenzie and W. R. Pulleyblank) A nonlinear allocation problem, *IBM J. Res. Dev.* **38**, 301–306 (1994).

163. (with O. Güler and U. G. Rothblum) Approximations to solutions to systems of linear inequalities, *SIAM J. Matrix Anal. Appl.* **16**, 688–696 (1995).

164. (with M. Hofmeister and P. Wolfe) A note on almost regular matrices, *Linear Algebra Appl.* **226–228**, 105–108 (1995).

165. (co-editors A. Blokhuis and W. Haemers) Special Issue honoring J. J. Seidel, *Linear Algebra Appl.* **226–228** (1995).

166. (with U. Faigle and W. Kern) A characterization of nonnegative box greedy matrices, *SIAM J. Discrete Math.* **9**, 1–6 (1996).

167. (with D. De Werra, N. V. R. Mahadev and U. N. Peled) Restrictions and preassignments in preemptive open shop scheduling, *Discrete Appl. Math.* **68**, 169–188 (1996).

168. (with D. Hershkowitz and H. Schneider) On the existence of sequences and matrices with prescribed partial sums of elements, *Linear Algebra Appl.* **265**, 71–92 (1997).

169. (with D. Coppersmith and U. G. Rothblum) Inequalities of Rayleigh quotients and Bounds on the spectral radius of nonnegative symmetric matrices, *Linear Algebra Appl.* **263**, 201–220 (1997).

170. (with D. Cao, V. Chvátal and A. Vince) Variations on a theorem of Ryser, *Linear Algebra Appl.* **260**, 215–222 (1997).

171. (with W. Pulleyblank and J. Tomlin) On computing Ax and $\pi^T A$, when A is sparse, *Ann. Numer. Math.* **4**, 359–368 (1996).

172. (with C. Micchelli) On a measure of dissimilarity between positive definite matrices, *Ann. Numer. Math.* **6**, 351–358 (1996).

173. What Olga did for me, *Linear Algebra Appl.* **280**, 13 (1998).

174. (co-editors M. Minoux and A. Vanelli) Special issue on VLSI, *Discrete Appl. Math.* **90** (1999).

175. (with B. Schieber) The edge versus path incidence matrix of series parallel graphs and greedy packing, *Discrete Appl. Math.* **113**, 275–284 (2001).

176. Gersgorin variations I: On a theme of Pupkov and Solov'ev, *Linear Algebra Appl.* **304**, 173–177 (2000).

177. Polyhedral combinatorics and totally ordered abelian groups, to appear in *Math Programming, Series A*.

178. On a problem of Zaks, *J. Comb. Theory* **A93**, 371–377 (2001).

179. (with B. L. Dietrich) On greedy algorithms, partially ordered sets and submodular functions, *IBM J. Res. Dev.* **47**, 25–30 (2003).

180. (with J. Lee and J. Williams) New upper bounds for maximum-entropy sampling, mODA6 — Advances in Model-Oriented Design and Analysis, edited by A. C. Atkinson, P. Hackel, W. Muller, Physica Verlag, 143–153 (2001).

181. (with H. Gröflin, A. Gaillard and W. R. Pulleyblank) On the submodular matrix representation of a digraph, *Theor. Comput. Sci.* **287**, 563–570 (2002).

182. (with K. Jenkins and T. Roughgarden) On a game in directed graphs, *Information Processing Letters* **83**, 13–16 (2002).

Patent

Processing System and Method for Performing Sparse Matrix Multiplication by Reordering Vector Blocks, U.S. Patent 5,905,666, May 18, 1999, with W. Pulleylank and J. Tomlin.

Autobiographical Notes

At a cocktail party, I am asked by a stranger, "What do you DO?".
"I do mathematics."
"Are you an accountant? Are you a programmer?"
"No. I prove theorems."
"Oh." The stranger searches the room wildly, hoping for rescue.

Who among us, my fellow mathematicians, has not had this approximate experience, not once but several times? Members of the general public, along with our spouses, siblings, parents and children, have not the foggiest idea of what we do and why we do it; but they must be curious about the mores of one of the world's oldest professions, and of its professors.

Why did we decide to become mathematicians, rather than lawyers or writers or entrepreneurs? What are the perils and pleasures of mathematical research?

Some of you may also have my taste for reminiscences. I learned from Ky Fan how Fréchet's students queued outside the professor's office one day a week waiting their turns to see him and report on their work. Alexander Ostrowski told me of the fierce competition among the students at Göttingen; Richard Courant told me anecdotes about its faculty. I heard from Magnus Hestenes and Adrian Albert about the old days at the University of Chicago, from Helmut Wielandt about moving from group theory to numerical analysis, from R. C. Bose how circumstances changed him from a geometer at a small women's college to the creator of much of the theory of experimental designs.

I love this mix of history and gossip. I also try to learn how and why certain topics started (why did George Dantzig create linear programming?) and what were the inspirations for particular theorems. In many respects I think of mathematics as a magnificent sport, of which I am a devoted player and fan.

Over the years I have been asked about the origin of some of my theorems and topics, and recently sat for a long video interview devoted to my activities in the early days of Optimization. These experiences led me to propose the inclusion, in this volume of Selected Papers, of comments about the origin or after-life of the papers included; I always seek such information from other mathematicians about their work. The major features of my education and career are sketched in these Autobiographical Notes. Let me say that to be a mathematician in the second half of the twentieth century was a spectacular opportunity. I hope this description of my small part in that great adventure will interest contemporary and future readers.

Early Years

Until I was almost 13, our family lived in Bensonhurst, a neighborhood in the borough of Brooklyn, New York City, not far from Gravesend Bay and Coney Island.

Boys played games after school: I remember being pretty good at boxball, marbles, stoopball and mumbletypeg, terrible at stickball and touch football. My older sister, Mildred, was vivacious and talented; our father, Jesse, was a manufacturer of women's dresses and subwayed daily to Manhattan; our mother, Muriel, managed the home. I was a good student in all subjects not requiring neatness or dexterity, and thought in a vague way that I would make my career in one of the learned professions. The idea that there was a profession called mathematician and an activity called mathematical research reached me before I was 12, because I had a mathematician cousin, George Comenetz, who had received his PhD from Columbia University in 1932.

Shortly before my thirteenth birthday, our family moved to the upper West Side of Manhattan, four blocks south of Columbia University. In the apartment next door was a retired actuary and passionate lover of mathematics, S. A. Joffe, who at this time was helping R. C. Archibald edit the journal *Mathematical Tables and Other Aids to Computation*. Mr. Joffe encouraged me to be a mathematician (and NOT to be an actuary) all the years I lived there.

I went to the wonderful George Washington High School, where my classmates and I studied geometry the old-fashioned way: we learned it, from a modernized Euclid, as a system of postulates and axioms from which we could deduce theorems. Sixty-five years later, I still get chills recalling that course. The concept that concrete information about tangible objects (triangles, trapezoids, etc.) could be found (and MUST be verified) by rigorous deduction from assumptions dazzled me like nothing I had ever seen before, and very little that I have seen since. Learning the concept of rigor was a stunning epiphany: what intellectual encounter is comparable? And you could make up proofs which differed from those in the text, and still be right; in fact, you could guess and try to prove your own theorems! I felt like King Arthur in Camelot! The teacher was Mr. Parelhoff (that's how I remember the spelling), and I will revere him forever.

Intermediate algebra was much inferior. I recall mild amusement at the introduction of strange Greek words like "polynomial", which I thought were designed to inhibit rather than advance learning. Trigonometry had some fun (proving miscellaneous identities), but wasn't in the same league as plane geometry. But the trigonometry course did have a distinctive and provocative feature: elaborate work with tables of functions. This forced me to confront the concept that mathematics could be useful as well as beautiful, and I was not so comfortable with that revelation. From Eric Temple Bell's "Men of Mathematics", I had concluded that useless mathematics was in principle nobler than useful mathematics. Pure mathematicians were monks in service to a higher religion whose secrets were known only to the residents of the monastery.

I graduated in January of 1940, and planned to start college (Columbia if I won a scholarship, City College of New York if I didn't) in the fall. Now was the time to choose a career. I loved mathematics, but I was also swept by other passions: the sonorities of poetry (large swaths of Kipling, Keats, Poe, Housman, even Alfred Noyes were committed to memory), the glamor of journalism (foreign correspondents interviewed diplomats and met beautiful women), indeed any activity closely or remotely connected to writing. I loved mathematics, but I think I loved literature more. Given the choice to be Shakespeare or Gauss, I would have instantly chosen Shakespeare; but knowing that I was neither Shakespeare nor Gauss, I felt that I could pretend to be Gauss more deftly. I was also confident that, if I were unable to make my way in mathematics, I would have a reasonable chance in any scholarly profession.

So, when I began college, my first (but not only) choice of career was to be a professor of mathematics. I believed I had better than average talent for it, and I was not dissuaded because at that time the prospects for gainful employment were dismal. The one conceivable alternative was physics. Einstein was a physicist; ergo, physics was an important subject. But the elementary physics course at Columbia was intoned in uninspiring lectures in a huge auditorium. The smaller recitation sections, led by the brilliant but sleepy Willis Lamb and Martin Schwarzschild, didn't compensate.

Here is how I spent my time at Columbia before going off to the Army in February of 1943. I had the great good fortune to meet some upperclassmen (Fred Bagemihl, Bernie Gelbaum, ...) and graduate students (Leonard Gillman, Fritz Steinhardt, ... and especially Ernst Straus) who took seriously my boyish enthusiasm for mathematics. I also had the good fortune to meet Alex Heller and Max Rosenlicht, who were respectively one and two years behind me, and far more gifted. Yet, even though Ernst, Alex and Max had much more firepower, I did not feel deficient in imagination; and I knew that mathematics is a subject in which it is just as meritorious to raise questions as to answer them, maybe even more so. Laymen tend to think mathematicians are just quintessential problem-solving wizards, but we members of the profession know they are wrong.

Also, a mathematics professor not only proves theorems (more accurately, conceives and proves theorems), but also lectures. I joined the Debate Council to overcome (or, at least, learn to live with) the fear of public speaking I had known in high school. I also became active in movements to support American aid for the Allies in the war against the Axis, and movements to get America to enter the war directly. I found the classes in the history of governments, philosophy and literature unbelievably stirring; these courses are the core of the Columbia College educational experience, and their memory a bond among alumni. (Amazing but true: at an evening meeting of the National Academy of Sciences where awards were made for different scientific achievements, I heard five of the honorees, all of them scientists, call this curriculum part of the foundation of their research careers.)

My teachers in mathematics were, principally, the incomparable J. F. Ritt, whose wit and pedagogy were legendary, and G. A. Pfeiffer. Most of what I know about teaching I learned from Ritt: how to demand the concentration of the class when necessary, how to break the tension of concentration with a witty aside, how to give students some whiff of the history behind mathematical ideas. I spent years trying to learn how to do these things gracefully.

Pfeiffer, on the other hand, was considered such a poor teacher by some of the students that, even in those ancient times when faculty received tremendous deference, a group of them had persuaded Dean Hawkes that Pfeiffer was unfit to teach calculus. But Pfeiffer was a marvelous teacher for serious students. Through a flaw in the course selection system, I took a challenging course in set theory (with a text by Hausdorff, in German), for which I was woefully unprepared mathematically and linguistically. Pfeiffer taught the course by reading the book and discussing whether in various places the author's arguments were flawed! To be invited to consider such questions, in a small class of less than 10 serious students, made me feel that I had been invited to be a pledge by the august fraternity of mathematicians. Then I took from Pfeiffer a course in foundations of geometry, where the material was easier for me. And I liked this course so much that my thesis grew from it, as I explain in my comments a bit further down in this section, and in the first chapter of this collection of papers. Unfortunately my education in calculus was a little skimpy; and even to this day, though I use calculus occasionally, and sometimes with panache, I don't feel totally confident or trust my instincts.

I stayed multiply involved with clubs and organizations and studies even though (or maybe because) a terrible war was in progress and my own military service was imminent. In fact, I intentionally hastened that military service by not registering for the fall semester of my junior year. I kept going to classes and participating in other activities, but I just didn't register. Nobody seemed to notice except the Army, which called me to service in February 1943.

How can I describe my three years in the Army? It was the climactic event of my life, with adventure magnified by the sensibilities of youth, and I cannot recreate that experience in words. But I will describe how these three years of service intersected my mathematical career. In the early weeks of basic training at Fort Eustis, the anti-aircraft artillery school, deep in the Tidewater country of Virginia, I got an idea for what I hoped would be an interesting project in the foundations of geometry: to develop axioms for a geometry of circles. I was lucky, and found phenomena in the geometry of circles analogous to phenomena in the geometry of lines. It turned out that the analogies were better than I could have imagined. They were there, but they weren't exact: like a painting perfectly planned but slightly, and in accordance with the artist's intention, a little unbalanced to provoke aesthetic interest. Of course, pursuing this project took some effort. At one time, I carried in my head a swirling vision of a configuration in space containing ten points and nine circles arranged on seven spheres; and I had to construct this configuration gradually, because the construction was the fulfillment of a mathematical argument created in fits and starts, with errors that had to be discovered and corrected at every step. And I had to carry this in my head, because I can't draw. Fortunately, even in the midst of intense basic training, there was so much "hurry up and wait" that I could cultivate and memorize the diagram.

(I also began with this experience the practice of not writing down partial results until I am convinced I know enough to write a full paper about the topic I am studying. It is a ridiculous practice, which I recommend to no one. It has, however, one virtue: if I return to this topic, after an interval of months or years, I am not handicapped by having a record of the way I thought about it before. In *The Psychology of Invention in the Mathematical Field*, Hadamard argues, with supporting quotations from Poincaré and other savants, that the ability (sometimes) to solve a problem that had been intractable when first put aside months or years earlier is because the mind has been unconsciously working on it. I think it is equally (maybe more) plausible that returning to the problem after an interval allows you to pursue it with a smaller chance of following false paths that you are lucky you forgot.)

After a month, I was sent to the anti-aircraft artillery metorology school, and after graduation became an instructor at the school. What I taught were some rudiments of trigonometry that were used in tracking ballons to plot winds aloft. I did not feel comfortable teaching, and I think I did it poorly. Later, as the anti-aircraft school was closing, I was sent to the University of Maine to study electrical engineering. Spofford Kimball gave the mathematics courses I attended. I was astounded by the clarity of his lectures, and any potential interest in studying electrical engineering was blown outside to the cold winds of Orono and the Penobscot River. Besides, I had already in basic training done some work I considered interesting mathematics. How could the study of electrical engineering compete for my enthusiasm? I now had among my belongings (which I kept until they were stolen from me in June 1945 when I went on a weekend trip from Augsburg to Paris) the slim volume *Introduction to the Theory of Numbers* by Dickson, which I hardly ever opened, and the heavy two-volume *Projective Geometry*, by Veblen and Young, which I did open from time to time. Besides these stimuli, I received mail from Straus, and each letter contained about ten mathematical questions, research questions appropriate for my limited knowledge, not problems suitable for or taken from a student magazine. I made limited progress on only two or three per letter, and there were only a few letters, but what a joy to receive them.

After Maine, I moved to the Signal Corps: training at Fort Monmouth, NJ, in the rudiments of long-lines telephony; on to C Company, 3186th Signal Service Battalion; arrived Liverpool in early December 1944, eventually (with the war almost over) to southern France and southern Germany (I was in Nesselwang on the day the war in Europe ended), traveled from Marseilles to Manila (Hiroshima's bombing occurred when we were already west of the Panama Canal) to Yokohama, returned home in February 1946 almost exactly three years from the day I entered. I thought a little bit about mathematics during this year and a half, but I didn't DO mathematics. I did some teaching: when the battalion reassembled in Augsburg after various individuals and teams had operated division-to-corps (and higher echelon) radiotelephone communications all over Europe, some of us set up a little university and conducted short courses where we taught each other what we knew (the 3186th was a very unusual battalion). We later did the same in Manila. I wrote down my forays into circle geometry in little blue examination booklets, and thought about how I would show this work to my friends and to Prof. Pfeiffer.

A sudden case of pneumonia kept me from enrolling for the spring semester. But when I recovered, I discovered that Pfeiffer was no longer at Columbia and Prof. E. R. Lorch was now teaching the courses in foundations of geometry. I showed him my blue booklets, confident that he would find the material suitable for a master's essay. But he surprised me by suggesting that it could be the basis for a dissertation. That was great news. More great news was that I was asked to teach a mathematical survey course at the Columbia College of Pharmacy, even though I was still an undergraduate. Here was a chance to practice pedagogy. My short teaching career at Fort Eustis had not been a rousing success; I did slightly better in Augsburg and Manila; but I was very uncertain that I could function adequately in my chosen profession of teaching.

On the opening day of the fall semester, I wrote my name on the board, I described how homework would be assigned and exams scheduled, gave an outline of the material to be covered, answered questions like "do you grade on the curve?", answered a few mathematical questions at my desk after the bell had rung and most of the class had left the room; and realized that my pulse was racing at 120 beats a minute. I had held the attention of the class for the full 50 minutes, I hadn't panicked or felt fear of any kind, my sentences had been reasonably grammatical, I had given some clear answers to a few technical questions: in short, I had felt comfortable as a teacher, and was totally confident that, if asked, I could do it forever. And, apart from my maladroit handling of a case of cheating, I thought that year at Pharmacy went very well. Also, before the end of the academic year, which means before I had graduated from College, I had worked over the research I had done in the army sufficiently to be very confident that I could get a thesis from it. So my career was on track!

During the year, I had become very fond of Esther Walker, the sister of my army buddy Alex. Buoyed by my success in research and teaching, I proposed marriage. She accepted, and we were married on the same day that I graduated from college. And, in the fall, I began graduate studies at Columbia brimming with confidence.

But what I did not realize at first was that having the thesis more or less in hand before leaving college had deplorable consequences. In the first place, it gave me an incentive to stay on at Columbia rather than go elsewhere, where, I believe, I would have been forced to learn much more. Second, anxiety about a dissertation is a wonderful incentive to study; but I didn't have such an incentive, and I loafed. I loafed with such devotion and consistency that I failed my algebra examination for the doctorate on the first attempt.

Candor compels me to report another personal flaw that first appeared during my years as a graduate student: an almost childlike impatience. I found sitting in a chair for an hour and a half lecture almost unendurable. I found reading technical material for that long very hard. Over the years I have developed various strategies for coping with this impatience, but I am nevertheless amazed that I have been able to function as a scholar with this handicap.

Eventually I passed all my exams, defended my thesis and departed for a postdoctoral fellowship, sponsored by the Office of Naval Research, at the Institute for Advanced Study in Princeton. In those years, the Institute was the summit of the mathematical world (Morse, Weyl, Siegel, von Neumann, ..., none of whom I ever engaged in any significant conversation. I did, however, spend much time with Oswald Veblen, who wrote the book I carried with me in the army). When we arrived at the Institute in midsummer, 1950, I realized it was now time, actually way past time, for me to start working diligently on my career. Besides my thesis, I had proved a few small theorems in the theory of complex variables which were perhaps publishable, but I felt they were probably a little too little. Now was the time to really get to work: to get up in the morning, have breakfast, walk from our miner's shack on Springdale Road to my office (next to Einstein) in a corner on the first floor of Fuld Hall, and work. As a student, the only significant time I had spent concentrating on mathematics were in occasional all-night sessions, lying in bed and worrying some problem like a dog with a bone. But I now realized that, whatever be the level of my talent and knowledge, my accomplishments would be proportional to the amount of time I devoted to my profession. And I wrote a few papers, one of which I have included in the first chapter; what was more important was that I established a rhythm of work. You are a mathematician, you do mathematics. Since that time, with one exception that I will describe later, I have been UNABLE (not unwilling, but unable) to enjoy two weeks free of mathematics; it has become an addiction.

Because this essay is pledged to my professional work, I refrain from detailing the host of experiences during that fabulous year 1950–51. I shall not describe how the intellectuals living in Institute housing left poisoned meat under the bottom floor to persuade an odiferous skunk to prowl elsewhere. The skunk ate the meat and promptly died, forcing a lengthy evacuation of the building. But I should mention our friendships with the Estrins and Bigelows, who were then building the IAS computer, an activity in which I took not the slightest interest; how Marston Morse would comment at seminars how much the speaker's research was related to Morse's work many years earlier, which behavior, regrettably, I now find myself occasionally imitating; attending Artin's lectures at Princeton University (he was fabulously dramatic); appreciating the other attractions at the Institute besides science (Panofsky on the history of art and Kennan on government, for example, were fantastically brilliant; Scandinavian meteorologists and their spouses incredibly handsome).

National Bureau of Standards, Office of Naval Research (London), General Electric Company

I needed a job when the fellowship ended. I was not especially choosy, but I did not have any academic offer in a location where I wanted to live. Feeling desperate, and contrary to almost everyone's advice that working for the government was inappropriate and would mean abandoning research for money, I wrote to the National Bureau of Standards (NBS, now the National Institutes of Science and Technology, NIST), and two days later received a telephone call offering me a job! Even though I knew nothing, as I readily acknowledged, of applied mathematics! I was so impressed that they were so impressed by my résumé (and the letters of recommendation that had brought me to the Institute) that I accepted instantly. Nor have I ever regretted the decision. The entire arc of my career is based on the experiences of the five years I spent in Washington at NBS. Initially, I thought that, after some time at the Bureau, I could in some future year find the academic position for which I had prepared. I would wear tweed jackets with leather patches, smoke pipes, deliver witty lectures and, from time to time, discover and nurture or possibly romance some brilliant student. But that never happened.

The Applied Mathematics Division at the Bureau had a contract with the Office of the Air Controller of the United States Air Force to pursue a program of research and computing in a subject of which I had never heard, called "linear programming", and I was hired to help fulfill that contract. Since linear programming was very new, my ignorance of traditional applied mathematics was not an insuperable obstacle; and I found the new subject a delicious combination of challenge and fun. It was also a marvelous opportunity to contribute to the early development of a part of mathematics that has thrived in many contexts: practical operations research, applications to engineering and many other branches of science, applications to combinatorial theory and to diverse topics in numerical analysis. When a subject is new, any simple insight is potentially a fundamental theorem, and you may be lucky enough to have the theorem identified by your name. Another virtue of early entry into a priesthood is friendship with other acolytes and the joy of exchanging ideas and problems.

But these contacts with George Dantzig, the father of linear programming, and his colleagues at the Air Force, gave me much more than a mathematical playground. George and his buddies believed that they were developing a technique for helping to run an organization (at least part of an organization) through the use of mathematical models. I am always skeptical, as a point of honor, of claims about the applicability of mathematics. But in this instance I was totally wrong, as I realized after a few months. And my experiences using linear programming in business contexts, while never my main interest, introduced me to the worlds of operations research, management consulting, manufacturing research, finance... a whole constellation of cultures I would otherwise have never encountered. I never felt exactly at home in these cultures, but I enjoyed the exposures.

At that time, most of the Bureau's laboratories were in a campus-like setting, with gorgeous azaleas and other flora, west of Connecticut Avenue in northwest Washington. From our apartment in Silver Spring, Maryland, a scenic drive through Rock Creek Park, including a ford over a small stream, brought me to the laboratory. Several of us, including the linear programming group, worked in a large room, bullpen style, so the atmosphere was very social. I loved the place. I loved the people and the work. I had infinite energy: I was writing papers, supervising calculations, working jointly with others on calculations and on formulating models, and I felt powerful. I collaborated with marvelous mathematicians (such as Morris Newman and Olga Taussky, who were also employees; visitors Ky Fan and Helmut Wielandt; colleagues elsewhere such as Dick Bellman, Joe Kruskal and Harold Kuhn). I learned fascinating concepts in experimental designs from colleagues in statistics, particularly Bill Connor and Marvin Zelen. (Of course, I wasn't becoming a professor, although I did some adjunct lecturing at American University, and George Washington University, polishing my skills in anticipation of the time when, in due course, I would leave the government for full-time academic employment.)

I also learned there were things I did not do well. I had no special skill in programming: I wrote a code in 1951 which just didn't run. It didn't make errors, it just never succeeded in starting the computer (our machine, SEAC, examined the first eight words of a code to find, among other things, instructions about continuing. But SEAC refused to accept my eight words). Though I believe to this day that the fault lay in the hardware, not my program, the experience was disheartening and I never wrote another program. For some calculations, I had a knack for choosing what to put in the computer and what to put on the magnetic tape; but I realized that I had no special skill or talent in numerical work generally. This was also obvious from conversations with John Todd, my boss, and with Jim Wilkinson, the world's expert on numerical linear algebra: that they had instincts and insights that I would never attain.

I owe a special debt to John Todd and to his wife Olga Taussky. Jack was fair, firm, friendly, and shielded our group from intermittent crises as best he could. Olga was employed as a consultant (probably because of nepotism rules), and she was counselor, den mother, mentor to me and to all the young mathematicians. She taught us the culture of the profession: how to conceive problems and questions, how to write papers, indeed the whole publication process (how to referee papers, how to respond to editors' comments). She was delighted (I am sure it was genuine, not feigned for effect) when one of our mathematical investigations went well, and sympathetic when it went poorly. I loved Olga. In later years, I have tried to do for others what she did for me.

By 1956 I felt I had done yeoman work, and I coveted what appeared to me to be the most glamorous job in mathematics: scientific liason officer (mathematics) at the London branch of the Office of Naval Research. This office had been created shortly after the end of World War II to reestablish connections between European and American scholars. The job of scientific liason officer entailed travel around Britain and the continent, visiting laboratories and universities, attending meetings, writing reports of what you learned, and helping Europeans and Americans with similar interests get to know each other. The big attraction was the expatriate adventure, not as exotic as in the 1920's, but far less common than it is now. And living in England, the linguistic and historic homeland of Americans! I persuaded Joe Weyl at ONR headquarters that I was ideally suited for the position, and also was assured by Jack Todd that I would be welcomed back at NBS.

Our year and a half in London had its share of hardships, particularly since our daughter Eleanor was only two years old and Elizabeth less than 6 months when we arrived, but I think that on the whole the experience was positive for Esther and me. We had nice holidays in Devon, Cornwall and the Lake District; in Paris and in southern France; and in Rome, Venice, Naples and Taormina. The year and a half was a perfect sabbatical. A sabbatical (which was, essentially, what the ONR job was for me) is more satisfying than travelling to meetings. Mathematical conferences, the "leisure of the theory class", are delightful venues for listening and learning, for renewing and establishing friendships, for preaching (this is Paul Erdös' language for lecturing) and preening, even for a little (actually, remarkably little) lust. But on a sabbatical, you learn to think like, and live like, and really be a local. And in England! Where (for example), thanks to the courtesy of John Hammersley, I could feast with the Fellows of Trinity College, Oxford, in the Tower Room of the Senior Common Room, on wine, fruit and conversation!

I found that, although I was determined to concentrate on the job, and NOT do mathematics, I was unable to refrain. In fact, when we were crowded into a room of an apartment hotel near Marble Arch before finding our permanent place on Kensington Square, I started doing mathematics to drown out the wails of my crying baby and the screams of a warring couple across the courtyard: it is well-known in the profession that doing mathematics eases pain and sorrow. (That Marble Arch research, closely related to interests of Al Tucker at Princeton, went well. When I returned to the States and told my results to him, he suggested to the organizers of a symposium on combinatorial mathematics, to which he had already been invited, that I should speak in his place. I kept the title "Some recent applications of the theory of linear inequalities to extremal combinatorial analysis" which Al had already chosen, and the paper appears in Chap. 5 of this book. Tucker was amazingly generous.)

I also did mathematics in hotel rooms and bars and sleeping cars. Half asleep on a train to Frankfurt, I discovered a beautiful theorem connecting a topic in algebra to the geometry of circles; and I lectured on this work at Frankfurt and at Mainz. I thought this theorem is maybe too beautiful to be true, and this sentiment was reinforced when I got back to London and realized my proof was flawed. The theorem is true, however, as Jeff Kahn showed in his thesis many years later. The one place where I couldn't do mathematics was at the office. Julian Cole, the distinguished aerodynamicist, and I faced each other across identical abutting desks and it was impossible to concentrate.

Instead of returning to Washington, or looking for an academic position, I investigated two job opportunities in New York. One was with a fledgling mathematical research group in IBM at a beautiful location (Lamb Estate) in Northern

Westchester. But the group was tiny and I did not think it had much future; also, at NBS, we were always a little snooty about IBM for no good reason that I can remember. The other was at the headquarters of General Electric, teaching operations research to people in the various departments of the company and helping them in any appropriate way when they returned to their respective departments. This was an established operation, the location in midtown Manhattan very exciting, the salary very attractive, and the chance to see if I could (and if operations research could) succeed in business was intriguing. So I accepted.

The job was fascinating in three respects: (1) since we were, organizationally, close to the Chairman, I learned something of life at court, where the mood of the monarch is constantly assessed from clues offered by tone of voice, tilt of eyebrows and the like; (2) because GE was a very diverse company, I became friends with people making jet engines, steam irons, military electronics (light and heavy), steam turbines (large and medium), lamps, plastics, and so on; (3) our group's location within Management Consulting Service gave me a chance (really, an obligation) to observe the culture of Management, which was very different from any I had known. Peter Drucker was a frequent visitor and probably our intellectual leader. I found him wonderfully entertaining in conversations between adjacent urinals, but the discussions of management philosphy seemed (I want to say this respectfully) less profound than their reputation.

The job was very satisfying in many respects, although some part of me kept telling myself I belonged somewhere else. My boss Mel Hurni said it was OK for me to do mathematics if it didn't interfere with my assigned duties. But it was clear that he wasn't thrilled by this research, almost all of which was orthogonal to the mission of our group. I was, nevertheless, very active mathematically at this time even though it seemed incongruous to do such work high in an elegant office building in the heart of Manhattan at 57th St. and Park Avenue.

In the summer of 1960 the mathematics department at IBM Research, still at the Lamb Estate in Westchester, but now large and active, invited me to participate in a summer workshop on combinatorics. I was only able to come for a couple of days, but I was dazzled by the atmosphere. The campus reminded me of NBS, except that the Lamb Estate was prettier. There was a large lawn for frisbee and other games, and various small cottages, no two alike, with fireplaces and attics adjacent to offices. And people all around doing mathematics! It was time, I concluded, to quit GE. I had been invited to join IBM by Herman Goldstine, Herb Greenberg and Ralph Gomory several times over the years, and in 1961 I accepted. In the back of my mind I thought: this seems like a great place to work, but it probably won't last. So I will stay here a couple of years and get myself a proper position as professor in some nice place when the atmosphere sours. But that time never came.

Alan Hoffman at Columbia, New York, 1948

Ernst Straus, Cambridge, 1950

Alan, Esther, Liz and Eleanor Hoffman at Brandeis, Waltham, 1977

Emilie Haynsworth, Alan, Olga Taussky and Gene Golub, Gatlinburg, 1964

Alan lecturing at conference celebrating George Dantzig's 85th birthday, Philadelphia, 1999

Alan with Elinor Hershaft, Greenwich, 2002

Alan, Phil Wolfe and Jack Edmonds at Georgia Tech, Atlanta, 2000

Founder's Award ceremony, Mathematical Programming Society, Atlanta, 2000. (From left) Phil Wolfe, Harold Kuhn, Harry Markowitz, Ralph Gomory, George Dantzig, Alan, Guus Zoutendijk and William Davidon.

Don Coppersmith and Alan at IBM, Yorktown Heights, 2003

IBM, CUNY, LAA, Stanford

My first act on joining IBM was to leave for a summer workshop on combinatorics at the Rand Corporation in Santa Monica. Tucker was chairman of the workshop, and I was delighted to discover that his working definition of combinatorial mathematics, so far as I could tell, was all (well, almost all) things of interest to me. So I felt very much at home; I knew most of the participants from other places, and I understood almost everything that was discussed (mathematics has become so specialized that it is possible to attend a conference even in one's specialty and not understand some parts of what other people are discussing). I also met the young Jack Edmonds, who was at the start of his brilliant career; and Claude Berge (sculptor, writer, discoverer and collector of primitive art, mathematician, . . .), incredibly versatile but at that point in his life very lonely. After the summer, I came back to Westchester and to a windowless Saarinen building which, several miles from the bucolic Lamb Estate, was the new home of IBM Research.

At NBS, ONR London and GE, I had always been among the youngest. When I joined the mathematics department at IBM Research, I was immediately among the oldest. Most of my colleagues were recent Ph.D.'s, in the act of establishing themselves and their careers; I had received my degree eleven years earlier, I had some reputation as a leader in my fields of interest, and I felt like a senior and acted like a senior. For several years I made it a practice to get to know all the mathematicians in the department and learn what they were doing. And what began as an emulation of what Olga did for me was great fun mathematically as well: my younger colleagues were very smart and were doing very interesting work.

I stopped the practice of entering stranger's offices invitationless a few years later, after a nine-month period as acting director of the Mathematical Sciences Department while Shmuel Winograd was away, when I thought my visits could be regarded as (and occasionally really were) management inquiries. This was not the only consequence of my acting directorship. Because I felt the position required full time attention, I vowed to do no mathematics whatever during my tenure. My explicit objective as director, which raised a few eyebrows, was to maximize the euphoria of members of the department, and that required all my attention. I abstained from research from September to the Christmas holiday, when a febrile convulsion from encephalitis put me in a coma for two days. I awoke in what was obviously a hospital room with a neurologist asking questions clearly intended to test my alertness and cognition; obviously, something bad had happened to me. Although I thought my answers coherent, I asked myself after the neurologist left if my brain could function on an adult level. And was I still capable of serious thinking? So I decided to test myself, did some mental arithmetic with two-digit numbers (I add, therefore I think), and concluded that maybe I could still function in my chosen profession. Then I asked myself: what was that intriguing mathematical question I had put aside four months earlier when I began as acting director, and I remembered, and the addiction to research resumed. For the rest of my tenure, research consumed a much larger fraction of my time and energy than I thought (and think) was appropriate.

My term ended with an annual report in verse, most of which was atrocious. But I was proud of the preamble, an adaptation of the opening of Coleridge's Kubla Khan. And the experience as director taught me to sympathize with all the incumbents who had that responsibility, even when some of their actions strained my geniality.

I will devote the next section to miscellaneous reminiscences of IBM and my research. What happened to my intention of becoming a professor? I didn't make it, except as a visitor. But that turned out well for me in the following sense. When I was invited to teach at various places, the courses I taught were always in subjects where I had a special interest (why ask a visitor to teach elementary calculus?), so I have never had the fatiguing experience of teaching material I found boring. Pretty lucky for me. But I have always thought of myself as both mathematician and educator, and used both descriptors in Who's Who in America. And I have spent many hours lecturing in the classroom and advising Ph.D. students.

While at General Electric, and in my first years with IBM, I taught classes at New York University, Columbia, the New School, Yeshiva University and the Technion (Israel Institute of Technology). I had been a Zionist since high school, and teaching at the Technion, located in the beautiful northern Israeli city of Haifa, fulfilled a lifetime dream. After an initial hurricane, my family and I enjoyed wonderful weather and hospitality and touring. The course I taught there was on the basic facts of linear programming. I was proud of the notes I prepared, which I think treated that material better than any text I have seen, but I never published them.

When I returned from Israel in 1965, I began teaching at the Graduate Center of City University of New York. All the teaching I had done in the preceding ten years had been in linear programming, but my CUNY courses were in combinatorics; and they were, for me, as fine a teaching experience as I could have imagined. My first year there was the best. I had about seven students (all of whom became professionals) and two professors, the graduate mathematics program was very new, there was an air of excitement in the classroom. And, starting that year, I worked with a corps of students for a long enough time that, even though I was an adjunct professor, I supervised theses. That is a precious obligation and opportunity: getting a degree in mathematics can (and generally does) change a person's life, and guiding a student towards that degree is, for the teacher as well, a moving experience. I owe a debt to my students for the satisfaction they have given me, and I am proud of their character as well as their scholarship. I am especially proud of my first CUNY student, Allan Gewirtz (after whom is named the Gewirtz graph), who, among other achievements, has an unbelievable record of community service at Brooklyn College, where he taught mathematics and was also dean of general studies; in the ambulance corps and as president of a hospital in Monmouth County, New Jersey; and as a teacher in drug rehabilitation centers in upstate New York; among other activities.

Teaching and research are intertwined: a sizable fraction of what I teach comes from, or is closely related to my research; and a sizable fraction of my research is a consequence of some kind of academic encounter. I am sure that these facts were arguments used by Ralph Gomory when he shepherded through the bureaucracy permission for me and others to teach part-time even while employed full-time by IBM.

Eventually I stopped teaching at CUNY because New York City had financial troubles, and taught elsewhere. I had an amazingly large number of students at CUNY for an adjunct professor, which of course I attributed to my superior skills.

I never duplicated that success at Yale, Stanford, Georgia Tech or Rutgers, but I have many plausible explanations for this failure.

A few words about my experience as founding editor of *Linear Algebra and its Applications* (LAA). Alexander Ostrowski spent a sabbatical at IBM in 1967 and asked if I would consider becoming the editor-in-chief of a new mathematical journal specializing in linear algebra. He had already been in contact with the publishing house American Elsevier about the possibility of inaugurating such a journal, but did not wish to assume that editorial responsibility himself. Of course, I was thrilled to be asked and accepted without hesitation. I had been managing editor of the *Naval Research Logistics Quarterly* twelve years earlier, but starting a journal was a bigger undertaking.

Looking back, I think that Lore Henlein, the editor at Elsevier, and I did a splendid job in getting LAA started. My part was establishing a scope not too narrow and not too wide, recruiting a capable board, soliciting the first papers, and establishing a rhythm for the editorial activities. The first issue appeared in 1968. "Lore, Lore, hallelujah!", I wrote in a congratulatory note to Ms. Henlein. The journal still flourishes, but I ceased being editor in 1972. I was not able to maintain the routine of paying systematic attention to the fates of the papers that were submitted, or the responsiveness of the referees; and I was unable to recruit administrative help that could have done it for me. I would also agonize indecisively about various editorial disputes. My board urged me to quit, which I did eventually, although I should have done it much sooner. Hans Schneider, to whom I am eternally grateful, came to my rescue and was appointed the next editor-in-chief. He was later joined by Richard Brualdi, and most recently by Volker Mehrmann, as joint editors-in-chief.

Besides humility, I learned a lot from this experience about my own limitations. I also learned that friends were as willing to forgive as I was eager to apologize.

In 1980, hoping to escape from the cold New York winters which aggravated the miseries I was suffering from asthma, I started teaching mathematical topics at the Stanford University operations research department during the winter quarter, and continued for almost ten years. At Stanford, I experienced for the first time "student evaluations", and learned what they thought of my teaching. In general, they thought my handwriting illegible, my organization of the material rather sloppy, and my enthusiasm great; I could not have agreed more with these evaluations. (J. F. Ritt would have ranked tops in all criteria.) My energy was spent on conveying the joy of mathematics: how to think of questions to investigate, questions which are natural, or ingenious, or simply fun. And I did not mind if occasionally, but not too often, I lost my way and had difficulty reconstructing an argument. It is good to (1) show students how to identify and challenge an obstacle, which I would do aloud and invite audience participation, and (2) in this process, let them "hear" how your mind works, which is something they will never get from textbooks. So even your mistakes as a teacher are helpful. But I never was as shamelessly unprepared as Professor Paul Smith was at Columbia. Smith was constantly getting lost, and turning sheepishly to Alex Heller to rescue him. Students learned nothing from Smith's lectures except the lesson that even a brilliant mathematician can have a bad memory and think on his feet very slowly. In an odd way, that was good for our morale.

More about lecturing. Students generally give you clues about how well you are doing. If the material is difficult, they look puzzled, implicitly pleading for further or alternative expositions of the material. If they understand the material, they nod affirmatively and vigorously, encouraging you to move on to the next topic. The only time this system of indicators failed me was when I was an instructor at Barnard College, the women's college of Columbia Univesity, in 1949–1950. The women's faces showed nothing, neither understanding nor confusion. It took me months to recognize tiny signals, like the twist of a neck or the adjustment of spectacles, for clues about the effect of a lecture. There was no such problem at Stanford: the reactions of the students were very easy to read.

These winters became the best part of each year. Our friends (Cottles, Dantzigs, Bermans, Mannes, Veinotts, Jacobsteins, Curtis Eaves, Gene Golub...) were incredibly hospitable, the University atmosphere exciting, the outdoors inviting (Foothill Park, Point Lobos, Half Moon Bay...).

In 1986, Esther contracted a fatal blood disease. We continued to visit Stanford, which she loved as much as I did, but her energy level gradually decreased. At Stanford, she typically would spend morning attending a class or visiting with friends, and the afternoons resting. At home she bravely continued her activities as much as she could. We even travelled together to some meetings. She died in the summer of 1988. I am comforted that she did not seem to be in physical pain; indeed, when last I saw her alive, at the hospital where she seemed to be recovering from a sudden strange shock to her blood chemistry, she was joking with the nurses. Her courage in the last years was incredible.

A year after Esther died, I met the young, beautiful and incandescent Elinor Hershaft, an accomplished interior designer. We married in September 1990, and have lived happily ever after in Greenwich, Connecticut. Ellie wasn't able to go away for the winter with me, and after a while I stopped going to warm places like Stanford and Atlanta (my IBM colleagues Earl Barnes and Ellis Johnson had become professors at Georgia Tech). I also began, at the invitation of my friend Peter Hammer, teaching one semester each year at Rutgers in New Jersey, but stopped about six years ago. Now that I am retired, I hope to resume teaching. Ellie has "youthanized" me, and I love the classroom dynamics.

Reminiscences

Finally, let me close these Notes with further reminiscences about my years at the IBM Research Center, and further comments about my research habits. First, some miscellaneous memories.

For about 15 years, a group of about eight of us played The Coin Game at lunch. The ostensible purpose was, via an elaborate set of rules involving holding coins in fists and guessing the sum of the number of coins collectively held, to decide who would fetch and pay for dessert for the group. But the real purpose of The Coin Game was the exchange of taunts and insults about the intellectual prowess of the other players, each taunt an expression of affection for the taunted. The Game faded away not long after Alan Konheim (who shared with Roy Adler the leadership of The Game) moved to University of Califoria at Santa Barbara.

Courtesy of an old friend from the Bureau of Standards, Leon Nemerever, I worked at the election night programs for CBS in 1962 and 1964, representing

IBM on a committee which reviewed the computer output for sanity before passing its predictions to the correspondents on TV. Very late in the night in '64, our committee, giddy with weariness and boredom, decided on a victor in a Senate race for a state in the Southwest, even though the computer said the race was too close to call. And we were wrong! Fortunately, the correspondents were at that hour equally tired and never broadcast our error.

I remember coming into the office early one morning and seeing Bryant Tuckerman jump four feet into the air when he looked at the results of the previous night's computation and saw that he had found the largest number known to be prime. Although the mathematics of the algorithm used was standard, Bryant's code, designed to exploit every feature of the computer, was outstanding. Bryant also had a flair for cryptography. Incidentally, I knew his father, L. B. Tuckerman, one of the world's experts on metrology at the Bureau of Standards; and I remember an eerie feeling watching Bryant once write on a blackboard an equation identical to one I had seen his father write two decades earlier (Bryant's equation dealt with cryptography, his father's with the measurement of gravity!).

I remember being at the Johnson Space Center near Houston, Texas, where Gerry Rubin and I worked on a proposal for scheduling the experiments to be carried by the space shuttle (they were then thinking of flights occurring as often as once a month; by the way, we didn't get the contract). Suddenly Bob Brayton called me to say that a summer student, Don Coppersmith, had completed the proof of our paper on self-orthogonal latin squares (in Chap. 2 of this book). I remember H. V. Smith demonstrating that checkers was a really really deep game during social moments when he and I were working together on a financial planning project for IBM that turned out to be hugely successful. I remember a discussion with an engineer from our plant at Poughkeepsie who tried to convince me that his method for scheduling the manufacture of a certain part was valid because of Hoffman's circulation theorem; I told him I thought he was wrong, but I never told him I was Hoffman.

John Cocke, a leader in the development of RISC architecture, compiler optimization and other innovations in computing software and hardware, was the glue that held the different parts of the laboratory together. I remember seeing him shuffle down our aisle, cigarette ashes dropping like snow in his path, to pick Alan Konheim's brain; and Alan responding with an ingenious generating function approach to the problem Cocke proposed. David Sayre, who had a dual career both in programming (he was, for example, in the original Fortran team), and in crystallography (where he had an international reputation) gave me wonderful private lectures on the art and science of imaging small objects. Similarly, Dick Toupin, whose knowledge of mechanics, electricity and magnetism was awesome, taught me how magnetic inks could be used in printing, the rigorous basis of Saint Venant's principle in mechanics, and other wonders.

Probably my favorite recollection of other people's work was that of Roy Adler and his friends (Coppersmith, Bruce Kitchens, Martin Hassner) creating new sliding block codes for magnetic recording, based on what Roy and Brian Marcus had done years earlier on symbolic dynamical systems. It is a prime example showing that the distinctions between pure and applied mathematics (which had captivated me when I was in high school) are very difficult to sustain in contemporary mathematics. Furthermore, it's a losing game to preach that there is an essential difference, to split hairs on the distinctions between applied mathematics and applicable mathematics, and so on.

Next, some technical comments about my work. I will try to answer questions about why I do what I do, and how I do it. When I left the Bureau of Standards in 1956, I knew well the vocabulary and issues then current in the use of computers. When I joined IBM in 1961, almost all of that knowledge was obsolete. I would go to meetings where acronyms were confidently exchanged around the table, and I grew weary constantly asking for definitions. This made me a little uncertain about how useful I could be in computing work involving linear programming, and certainly had an impact (I don't know whether for better or worse) on my contribution to the company. Five years earlier, I probably shared with the economist Ragnar Frisch the distinction of having supervised more calculations in that subject than anyone else; and I had presumed that this experience was a principal reason I was hired. Fortunately, Ralph Gomory had such a spectacular grasp of all issues involving the practice and theory of linear programming that I did not feel guilty about my deficiencies. In fact, as Phil Wolfe, Ellis Johnson, and others joined our organization, I always felt that work was in hands more capable than mine, so I did not feel at all guilty as I rotated among other research interests.

What were these other interests? When I began my career, I was happy to swing at any problem anyone threw at me. Maybe I was too gregarious. In fact, when I was acting director of the department, I discovered in my own personnel file(!) a letter of recommendation, written about me before I was hired, which cited my willingness to work on other people's problems and failure to specialize as a possibly negative attribute to complement the compliments in the rest of the letter. But over the years, the combined influences of Straus, Ryser, Dantzig and Taussky made me principally interested in the interplay among ideas in three mathematical subjects: combinatorial mathematics, linear programming and matrix theory. So I have belonged to those three mathematical communities, felt welcome in each, but for many reasons feel most at home with linear programmers.

Did I think of myself as a member of the community of mathematicians, or the community of IBM employees? My locus of loyalty was unequivocally the Department of Mathematical Sciences at the Research Center. I had feelings of kinship with other people in IBM, particularly mathematicians; and I certainly considered mathematicians around the world, particularly in my areas of research, as brothers. But mainly I identified with the Department. I have known some of my colleagues for forty years (Phil Wolfe more than fifty years), many for twenty to thirty years. They have tolerated my idiopathic humming and punning with only the mildest complaints. I have the strongest sentimental attachments to my Department colleagues, past and present, and even to the Department as an institution. We had such a good time! I wish the Department's history were better known and appropriately celebrated.

How do I do my research? Well, each question or problem has its own scenario. Most of my attempts, say 90%, to do something end in failure. The big attraction of mathematics is that it IS hard, and I have learned to live with the experience of making errors and following trails that lead nowhere, and consequently really enjoy the rare occasions when I succeed. Also, I like to keep several problems juggling more or less simultaneously, so I do not linger too long depressed by failure to progress in one of them.

One magic September, all the mathematics on my mind at the time was resolved! I had the ideas for about seven papers on a variety of topics suddenly succeed. Of course I didn't write them up at once. At other times, I have gone for as long as eight months with no success.

Of course, what is most delicious for me is to spend months or years intermittently on a mathematical question, occasionally adding more to my knowledge of what is characteristic of the situation, and finally reach a point where I think I know, or I almost know, the theorem or theorems I want to prove and think I can prove. Then I have a choice of working very very hard to finish the project, or sitting back and savoring the pleasure of anticipating the joy I will have when I finally prove what I believe to be a correct description of the mathematical question and its answers. And accompanying this anticipation, planning already how I will introduce the mathematical issue when I give a lecture about it, and what jokes and anecdotes will enliven the talk. Writing is not as big a pleasure, and I have a backlog of about a dozen papers essentially completed but not yet written in satisfactory form. But I will: doing mathematics and writing papers is what I do. I am also committed to preparing with Uri Rothblum a monograph on aspects of matrix theory inspired by Gersgorin's theorem.

All of my collaborators are my friends, each and every one; you will find their names in my list of publications. But I want to make special mention of: George Dantzig and Olga Taussky, responsible respectively for inspiring my interests in linear programming and in matrix theory; Helmut Wielandt and Ky Fan: when I was a rookie, they invited me to partcipate in interesting aspects of linear algebra they were creating; Harold Kuhn and David Gale, in fond recollection of the early '50s, when we taught each other to use the ostensibly practical subject of linear programming to prove aesthetic combinatorial theorems that were ostentatiously useless; Dijen Ray-Chaudhuri and Heinz Gröflin, because each developed one of my ideas into much more than I had thought possible; Earl Barnes and Uri Rothblum, because with each I had several times the fun arising when one of us has an insight in some mathematical situation and the other made it much clearer and more general; Ralph Gomory (not only for his creative mathematics: he has the unusual gift of thinking about issues of the practical world with the same creative intensity); and Don Coppersmith, about whom the least I can say is that his talent is incredible, so is his thoughtfulness, and I have come to depend on both.

Two other mathematicians influenced me enormously. Herb Ryser's work showed me what a powerful weapon matrices were for proving combinatorial theorems; in particular, the Bruck–Ryser–Chowla theorem blew my mind. Jaap Seidel and I were simpatico in taste and outlook, and boosted each other in the early days of research on the spectrum of graphs.

I tried to make a more complete list of the mathematicians who influenced my scholarship even if we never published anything (or published very little) together. There were about seventy names already on the list when I was only halfway through my memory bank, so I stopped the effort. The world is not burdened with an excess of mathematicians, and I have known many of them: I started doing mathematics when I was 18, I will soon be 79, and I have had wonderful colleagues. There is no way I can find words to describe my gratitude and feelings of kinship.

This project was created and nurtured by Professor Charles Micchelli of the State University of New York, The University at Albany, who was my colleague for thirty years at IBM, and by Ms. E. H. Chionh of World Scientific Publishing Company, Singapore. I thank them for their initiative, wisdom and patience.

Greenwich, Connecticut
May 26, 2003

Geometry

1. On the foundations of inversion geometry

In February 1943, just after taking a course on the foundations of geometry from Professor George Pfeiffer at Columbia, I joined the U.S. Army. During idle moments in basic training, I speculated about what could be the ingredients of an axiomatic foundation for a geometry of circles. I assumed that such a foundation already existed, so I tried to imagine a variation in which tangency did not exist: i.e. if two circles on a sphere had one point in common, they had another common point. My notion was that the usual geometry of circles was "affine". Tangency was analogous to parallelism, and I was going to do something "projective". I succeeded in constructing a finite example. I also created an incidence theorem that could not be proved from incidence axioms for circles on a sphere, but could be proved if there were a world of spheres in a 3-space. This was analogous to Desargues' theorem in projective geometry. Also (discovered later) an analogue of Pappus' theorem, and of the relation betwen the two.

I thought all this was pretty nifty, and decent work for a college junior. Eventually, because I showed there could only be a finite number of finite examples, I lost interest in the "projective" geometry of circles, but not in the "affine" case. I returned to Columbia in 1946 thinking that what I had done might be suitable, eventually, for an M.A. thesis, but E.R. Lorch thought it might be suitable, with some additional work, for a dissertation (especially since it turned out that the geometry of circles had not been so thoroughly axiomatized after all). I eventually wrote such a dissertation, with much help from Hing Tong and Donald Coxeter on the exposition, and this is it.

There are axioms in this treatment which make the underlying field have the property that it is ordered, and every positive number is a square. While such a property is perfectly proper, I never learned how to dispense with the axioms which compel this property.

2. Cyclic affine planes

This paper is a souvenir of the postdoctoral year I spent at the Institute for Advanced Study. I want to take this occasion to apologize to Gerry Estrin who proved the principal theorem in this paper and who, instead of being merely thanked in an acknowledgment, should have been a joint author, at least.

There were three earlier papers on difference sets I knew about (by Singer, Bose and Shrikhande, and Marshall Hall). Singer proved the existence of difference sets for projective planes; Hall proved some fascinating theorems (I especially liked the multiplier theorem) about such sets; Bose and Shrikhande proved the existence of difference sets for affine planes. So I "completed the square": my paper is to Hall's as Bose–Shrikhande is to Singer; or you could say mine is to Bose–Shrikhande as Hall is to Singer.

The best thing that happened to me professionally during that year was meeting Herbert Ryser. The Bruck–Chowla–Ryser theorem about projective planes and Ryser's theorem about duality for symmetric block designs influenced my work enormously. They showed you could use matrix theory to prove combinatorial theorems, even though the theorems never mentioned matrices in either the hypotheses or conclusions. What a dazzling idea!

3. On the number of absolute points of a correlation

This is the first paper where I used eigenvalues to prove a combinatorial theorem, so it has sentimental value for that reason alone. It was undertaken when Morris Newman and I were invited to UCLA to participate in a summer workshop of the National Security Agency. Unfortunately, our clearances were not completed in time to enable us to participate in the classified work of the conference, so we frolicked in unclassified work. In this paper, we revisited work of Baer and Ball to reprove their theorems using an approach through matrix theory and some skill in algebraic number theory the latter being the contribution of my fabulous coauthors. Besides Newman, these were Ernst Straus and Olga Taussky-Todd. Ernst, one of the most brilliant and most principled persons I have ever met, did me the honor of taking me seriously when I was a freshman at Columbia and he was a graduate student. Olga was my quasi-supervisor at the National Bureau of Standards, and (among many other lessons) taught me that matrix theory was a beautiful subject, even when it wasn't applied to combinatorial theorems.

4. On unions and intersections of cones

It is probably hard to recognize from the text of this paper, but it began by first considering how "diagonal dominance" conditions (which imply nonsingularity of complex matrices) could be weakened if the matrices were real. The weakened condition asserted a property of the unions of a certain family of cones. I then discovered completely by accident that Ky Fan, some years earlier, had generalized a topological theorem of Al Tucker, in such a way as to highlight a property of the intersections of cones. I was able in this paper to connect the condition on unions with the condition on intersections, with the key argument being the trivial (and, at first blush, irrelevant) statement that a matrix is nonsingular if and only if its transpose is nonsingular!

5. Binding constraints and Helly numbers

Herbert Scarf sent me a copy of his theorem (also proved by David Bell and J. P. Doignon) that solving an integer linear program in n dimensions could require as many as $2^n - 1$ of the inequalities to specify the answer, but no more. I did not understand his proof ... and initially did not know of the others ... so I concocted my own, which proceeded on the basis of axioms for a certain abstract system of convex sets (I had returned to "axiomland" from the Army and my student days, and I learned from a wonderful survey of Helly's theorem and its relatives by Danzer, Grunbaum and Klee about the concept of abstract convexity. Without the good fortune of attending the symposium where their survey was first presented, I doubt that I would have succeeded). I proved a theorem about my abstract system that implied the theorem of Bell, Doignon and Scarf, but only mentioned integer programming in the last paragraph. This theorem is, I think, relevant to what have come to be called antimatroids and/or convex geometries, but did not, so far as I know now, have any influence.

Reprinted from the
Trans. Amer. Math. Soc., Vol. 71, No. 2 (1951), pp. 218–242

ON THE FOUNDATIONS OF INVERSION GEOMETRY

BY

ALAN J. HOFFMAN

Introduction. A 3-dimensional inversion geometry over an ordered field V in which every nonnegative number is a square may be defined as a partially ordered set Π of objects called points, circles, spheres, and inversion space with the properties:

(i) if p is any point, then there is an affine geometry whose "points," "lines," "planes," and "3-space" are, respectively, the points of Π other than p, the circles containing p, the spheres containing p, and the inversion space;

(ii) the underlying field of this affine geometry is V;

(iii) this affine geometry can be made a Euclidean geometry in such a way that the "circles" and "spheres" of the Euclidean geometry are, respectively, the circles of Π not containing p and the spheres of Π not containing p.

The purpose of this paper is to give axioms for Π that will be sufficient to establish (i), (ii), and (iii). The only undefined relation is the ordering relation $\leq$, which means, geometrically, that all our axioms are incidence axioms. There does not seem to be any particular interest in finding alternative statements of (i), so (i) is simply assumed (1.4). Additional assumptions are added (2.11 and 2.12), and the remainder of the paper is devoted to proving that these axioms are sufficient for (ii) and (iii).

The extension of this work to higher dimensions is straightforward, and we have concentrated on the 3-dimensional case for the sake of simplicity. The 2-dimensional case, however, is different in many ways[1], and will be treated in a future paper.

It is rather surprising that the literature contains so few investigations of the foundations of inversion geometry as an autonomous subject[2]. Certainly much less is known about the postulates for inversion geometry than for other geometries. The present paper is an effort to remedy this deficiency.

We wish to thank H. S. M. Coxeter, Tong Hing, and E. R. Lorch for their invaluable advice at various stages in the preparation of this manuscript.

1. The first set of postulates. In this section, we postulate that our set

Presented to the Society, December 28, 1948; received by the editors November 5, 1950 and, in revised form, January 20, 1951.

[1] For the most notable difference, see footnote 7.

[2] In [7] (numbers in brackets refer to the bibliography at the end of the paper), Pieri has treated the 3-dimensional case over the real numbers, and [10] contains a discussion by van der Waerden of the 2-dimensional case over a general field. The principal ideas of these papers are given in footnotes 11 and 15. More recently, Petkantschin [6] has discussed the 2-dimensional case over the real numbers, and it is easy to reformulate his postulates so that the only undefined relation is incidence.

Π has the property (i) of the introduction[3]. Π is then imbedded in a lattice Λ, which is, for present purposes, more easily manageable.

1.1 AXIOM. Π is a set with a binary relation $\leq$ defined on it.

An element $p \in \Pi$ with the property $x \leq p$ implies $x = p$ is called a point. We reserve the letters p, q, r, $\cdots$, z for points.

1.2 AXIOM. If $a \in \Pi$, then there is a point p such that $p \leq a$.

1.3 AXIOM. If $p \leq a$ and $a \leq b$, then $p \leq b$.

1.4 AXIOM. If $p \in \Pi$, then the following subsystem of Π, under the relation $\leq$, is a 3-dimensional affine geometry from which the zero element has been deleted: all points other than p, and all elements a such that $p \leq a$, $p \neq a$.

We note some immediate consequences.

(a) Under $\leq$, Π is a partially ordered set. This follows at once from the properties of points and the fact that an affine geometry is a partially ordered set.

(b) There is an element $I \in \Pi$ such that $a \in \Pi$ implies $a \leq I$. Let p be a point of Π. The affine geometry corresponding to p of 1.4 contains a greatest element, which we denote by I. We show that I has the required property; that is, I is the greatest element of Π. If $a \in \Pi$, let $q \leq a$ (1.2). Let J be the greatest element in the affine geometry corresponding to q. By 1.4, we have $p \leq J \leq I$ and $q \leq J \leq I$. Thus $I = J$ and $a \leq I$.

(c) We proceed to the imbedding of Π in a lattice. For each pair of distinct p, q we adjoin to Π the symbol $P_{p,q}$, obtaining a new set Π', $\Pi \subset \Pi'$. We now extend the relation $\leq$ to Π' by the following rules: $p \leq P_{p,q}$, $q \leq P_{p,q}$; if $a \in \Pi$ and p, $q \leq a$, then $P_{p,q} \leq a$; $P_{p,q} \leq P_{p,q}$. It is easy to see that Π' is a partially ordered set under the extended definition of $\leq$. Further, if we denote by $\Pi'(p)$ the subsystem of Π' consisting of all elements $a \in \Pi'$ such that $p \leq a$, then $\Pi'(p)$ is an affine geometry in which p is the zero element, for $\Pi'(p)$ is clearly isomorphic to the affine geometry described in 1.4.

$\Pi'(p)$ has a unique extension to a projective geometry of the same dimension. We now adjoin to Π' the "elements at infinity" of the projective extension of $\Pi'(p)$, for each $p \in \Pi$, obtaining a set $\Pi'' \supset \Pi'$. We extend the relation $\leq$ to Π'' by the following rule: if a, $b \in \Pi''$ and there is a point p such that a and b are elements in the projective extension of $\Pi'(p)$, and if in that projective geometry a is contained (properly or improperly) in b, then we say $a \leq b$. It is clear that Π'' is partially ordered set. Finally, we adjoin to Π'' an element 0, obtaining a set Λ, and extend $\leq$ by the conditions: $0 \leq 0$; $0 \leq a$ for all $a \in \Pi''$. Λ is of course a partially ordered set under $\leq$, and indeed a lattice.

---

[3] There are many ways to effect this. See [5] and also [1, p. 109, ex. 12]. We shall for the most part follow [1, chap. I] for the general terminology of ordered sets. We assume familiarity with the lattice-theoretic formulation of projective geometry of Birkhoff and Menger.

Proof. We show that if a, $b \in \Lambda$, then Λ contains $a \cup b$ and $a \cap b$. If at least one of a, b is 0, the result is immediate, so we assume the contrary. Hence, there exist p and q such that $p \leq a$, $q \leq b$. Let us assume first that p and q can be chosen so that $p = q$. Then the existence of $a \cup b$ and $a \cap b$ follows from the fact that a projective geometry is a lattice. The other possibility is that for every choice of $p \leq a$ and $q \leq b$, we have $p \neq q$. In this case, it is immediate that $a \cap b = 0$, and what remains to be shown is the existence of $a \cup b$. First, $p \cup q$ exists, and $p \cup q = P_{p,q}$; for l.u.b. (p, q) in Π' is $P_{p,q}$, by its definition, and the successive imbedding of Π' in Π'' and Λ preserves l.u.b.'s. Next, $((p \cup q) \cup a)$ exists, since $p \leq p \cup q$, $p \leq a$. Similarly, $(((p \cup q) \cup a) \cup b)$ exists, and clearly is $a \cup b$.

Henceforth, unless otherwise specified, an "element" is an element of Λ.

1.5 *Some notations and definitions.* The expression "a is contained in b" or "b contains a" means $a \leq b$. "a is properly contained in b" means $a < b$; that is, $a \leq b$, $a \neq b$. $a \cup b$ is called the join of a and b, $a \cap b$ is called the intersection of a and b.

Λ is clearly a lattice of dimension 5. Using $d(\ )$ for the dimension function, we have $d(0) = 0$, $d(p) = 1$ for all points p, $d(a) = 1$ implies a is a point. An element of dimension 2 is called a pair. We reserve the letters P, Q, R, $\cdots$, Z for pairs. An element of dimension 3 is called a circle. The symbol $[\ ]$ will be used to designate a circle in various ways; for example, $[a\ b] =$ "the circle containing the elements a, b"; or $[a\ b] =$ "the elements a and b are contained in a circle." The context will clarify the usage. An element of dimension 4 is called a sphere. The symbol $\{\ \}$ will be used for spheres in the same manner that $[\ ]$ is used for circles. I, the unique element of dimension 5, is called the inversion space.

0 and all elements of Π' are called ordinary elements of Λ. All other elements of Λ are called singular. Thus, the only singular elements are certain pairs, circles, and spheres; namely, those elements of Π'' that are not elements of Π'. Note that the ordinary pairs are precisely those elements of Π' that are not elements of Π.

We shall specify that an element is singular by attaching the unique point it contains as a subscript; for example, a_p is a singular element containing p. Observe that according to the construction of Λ there is one and only one singular sphere containing p. Singular spheres will be denoted by capital letters at the beginning of the alphabet. Thus, A_p is the unique singular sphere containing p. Two ordinary elements of Λ are said to be tangent if their intersection is a singular element([4]).

The following statements are obvious:

1.5.1 If a is ordinary, and $d(a) = n \neq 0$, then there exist p_1, $\cdots$, p_n such that $p_1 \cup \cdots \cup p_n = a$.

([4]) This is precisely the reason for introducing singular elements. Otherwise, we would be bothered in various places to consider tangency as a special case when, in fact, it is not.

1.5.2 (a) If a is ordinary, $b < a$, then there exists $p \leqq a$, $p \not< b$, $p \neq b$.

(b) For any $a \in \Lambda$, if $p \not< a$, $p \neq a$, then $d(p \cup a) = 1 + d(a)$.

1.5.3 If $a \cap b \neq 0$, then $d(a) + d(b) = d(a \cup b) + d(a \cap b)$.

1.5.4 Every singular element containing p is contained in A_p. Conversely, if $a \leqq A_p$, $a \neq 0$, p, then a is singular.

1.5.5 If a is an ordinary circle (sphere), $p < a$, $q \not< a$, then there exists one and only one circle (sphere) containing p and q and tangent to a. Such a circle (sphere) is said to be tangent to a at p.

1.6.1 LEMMA. *Every ordinary circle contains an infinite number of points.*

Proof. If we use 1.4, any two ordinary circles contain the same number of points (which may be infinite). If this number is finite, say $n+1$, then the number of points on each ordinary sphere is n^2+1, the number of points of Λ is n^3+1, the number of ordinary spheres containing a given point is $n^3+n^2 +n$. If D is the number of ordinary spheres, then by counting the number of incidences of points with spheres, we have

$$(n^3 + n^2 + n)(n^3 + 1) = (n^2 + 1)D.$$

From 1.5.1, we know $n > 1$; but for $n > 1$, this equation is satisfied by no integer D.

This lemma will be useful in assuring that we have "enough" points with which to operate. The following theorem, which we shall use a great deal from §4 on, is an easy consequence of the fact that every ordinary circle contains at least four points. We omit its proof.

1.6.2 THEOREM. *Let τ be any 1-1 transformation of the set of points of Λ onto itself such that $[p_1p_2p_3p_4]$ implies $[\tau p_1 \tau p_2 \tau p_3 \tau p_4]$. Then τ can be extended uniquely to an automorphism $\bar{\tau}$ of Λ.*

2. **The definition of inversion and the second set of postulates.** We now prove a sequence of theorems corresponding to the construction of the ideal point in a four-dimensional incidence geometry[5]. These lay the foundation for the definition of inversion.

2.1 THEOREM[6]. *Let P_1, P_2, P_3 be distinct pairs, not all contained in a circle, such that $[P_1P_2]$, $[P_1P_3]$, $[P_2P_3]$. Let p be any point not contained in any of these three circles. By 1.5.2(b) and 1.5.3, $[p\,P_1] \cap [p\,P_2]$ is a pair, say P. Then $[P\,P_3]$.*

Proof. We first note that $\{P_1P_2P_3\}$, since by 1.5.3, $d(P_1 \cup P_2 \cup P_3)$

---

[5] Our lattice Λ and the semi-lattice of incidence geometry considered by Gorn in [3] are sufficiently similar that the work of [3] is applicable here. (This remark was made in [5]. Its meaning is that an inversion geometry is an example of an incidence geometry.) 2.2, 2.6, and 2.7 are restatements, for present use, of theorems of [3].

[6] This is condition E of [3]. The proof of this theorem has been known, although it does not seem to be in the literature.

$= d((P_1 \cup P_2) \cup (P_1 \cup P_3)) = d(P_1 \cup P_2) + d(P_1 \cup P_3) - d(P_1) = 3 + 3 - 2 = 4.$ We now consider two cases:

(i) $p \not< \{P_1 P_2 P_3\}$. Then $P < \{p\ P_1 P_3\} \cap \{p\ P_2 P_3\} = [p\ P_3]$. Hence, $[P\ P_3]$.

(ii) $p < \{P_1 P_2 P_3\}$ [7]. By 1.5.2(a), there exists a point $q \not< \{P_1 P_2 P_3\}$. By case (i), $q < Q = [q\ P_1] \cap [q\ P_2] \cap [q\ P_3]$. Now $p \not< Q$, for otherwise, $p = Q \cap \{P_1 P_2 P_3\} < [q\ P_1] \cap \{P_1 P_2 P_3\} = P_1$, contrary to hypothesis. Therefore, $[p\ Q]$. Further, by 1.5.3, $[p\ Q] \cap \{P_1 P_2 P_3\}$ is a pair, which we denote by P. We show that P has the required property. We have $P < \{P_1 P_2 P_3\} \cap \{p\ [Q\ P_i]\}$ and $P_i < \{P_1 P_2 P_3\} \cap \{p\ [Q\ P_i]\}$. Hence, $[P\ P_i]$ for $i = 1, 2, 3$, which is the result sought.

2.2 COROLLARY. *If P_1 and P_2 are distinct pairs such that $[P_1 P_2]$, $q \not< [P_1 P_2]$, $r \not< [P_1 P_2]$, $Q = [q\ P_1] \cap [q,\ P_2]$, $R = [r\ P_1] \cap [r\ P_2]$, $p < [P_1 P_2]$, then $[p\ Q] \cap [p\ R] \cap [P_1 P_2]$ is a pair.*

Proof. If $Q = R$, the result is immediate, so let us assume the contrary· We show first that $[Q\ R]$. If $r < [q\ P_1]$ or $r < [q\ P_2]$, this follows at once from the definition of Q and R; if $r \not< [q\ P_1]$ and $r \not< [q\ P_2]$, this follows from 2.1. Applying 2.1 now to Q, R, P_1 and p, we have the corollary, provided $[Q\ R\ P_1]$ is false. If $[Q\ R\ P_1]$, then apply 2.1 to Q, R, P_2 and p.

2.3 DEFINITION. We say that P_1, P_2 are anallagmatic pairs (of a fundamental involution) if

(i) $[P_1 P_2]$, and

(ii) $P_1 \cap P_2 = 0$.

The terminology will be shown to be appropriate in 2.8.

2.4 LEMMA. *If P_1, P_2 are anallagmatic pairs, $p \not< [P_1 P_2]$, $P_3 = [p\ P_1] \cap [p\ P_2]$, then P_1, P_3 are anallagmatic pairs.*

Proof. Condition (i) of 2.3 is satisfied at once. Assume (ii) is not satisfied· Then $P_3 \cap P_1 \neq 0$ implies that $[p\ P_1] \cap [p\ P_2] \cap P_1 = P_1 \cap [p\ P_2] \neq 0$. Since $p \not< [P_1 P_2]$, it follows that $P_1 \cap [p\ P_2]$ is a point, say q. Therefore, $[p\ P_2] = [q\ P_2] = [P_1 P_2]$, so $p < [P_1 P_2]$, which violates the hypothesis.

2.5 THEOREM. *Let P_1, P_2 be anallagmatic pairs. Then there exists a function F with the following properties:*

(a) *F is mapping of the set of points of Λ into the set of pairs of Λ.*

(b) *$p < F(p)$.*

(c) *P_1 and P_2 are in the image of F.*

(d) *If $Q \neq R$ are in the image of F, then Q, R are anallagmatic pairs.*

---

[7] Although this case involves only objects contained in a sphere, an example due to Hjelmslev (see [2, p. 229], where the example is obviously intended to apply to 2.1 (ii)) shows that its proof requires the use of a point not on the sphere. This striking analogy to the Desargues situation in projective geometry was pointed out in [4], where it was strengthened by showing that Miquel's theorem (see footnote 15) implies 2.1 (ii), as Pappus implies Desargues.

Proof. If $p < P_1$ (or $p < P_2$), we define $F(p) = P_1$ (or P_2). If $p < [P_1P_2]$, then we define $F(p) = [p\ P_1] \cap [p\ P_2]$. If $p < [P_1P_2]$, but $p < P_1$, $p < P_2$, then we define $F(p)$ to be the pair discussed in the conclusion of Corollary 2.2.

(a), (b), and (c) are satisfied immediately. (d) is a consequence of Theorem 2.1 and is essentially a summation of all we have done so far in §2. We omit a formal proof.

2.5.1 REMARK. F is uniquely determined by any two distinct pairs in its image.

2.6 We now generalize 2.1 in such a way as to lead to a definition and discussion of coaxal circles.

THEOREM. *Let a_1, a_2, a_3 be distinct circles not all contained in a sphere such that $\{a_1a_2\}$, $\{a_1a_3\}$, $\{a_2a_3\}$. Let p be a point not contained in any of these three spheres. Then there is a unique circle a containing p such that $\{a\ a_1\}$, $\{a\ a_2\}$, $\{a\ a_3\}$.*

Proof. If $a_1 \cap a_2 \neq 0$, then $a_1 \cap a_2$ is a pair P. $P < \{a_1a_3\} \cap \{a_2a_3\} = a_3$. The a of the theorem is then $[p\ P]$.

If $a_1 \cap a_2 = 0$, then by the above, $a_1 \cap a_3 = a_2 \cap a_3 = 0$. Choose $p_1 < a_1$, $p_2 < a_2$, $p_3 < a_3$. Further, let $P_1 \neq Q_1$ be chosen so that $p_1 < P_1$, $p_1 < Q_1$, $[P_1Q_1] = a_1$. Let $P_2 = [p_2P_1] \cap a_2$, $Q_2 = [p_2Q_1] \cap a_2$. Then the P's and Q's determine respectively a function F_P and a function F_Q in accordance with 2.5. We note first that $F_P(p_3) < a_3$. For $F_P(p_3) = [p_3P_1] \cap [p_3P_2] < \{p_3a_1\} \cap \{p_3a_2\} = \{a_3a_1\} \cap \{a_3a_2\} = a_3$. Similarly, $F_Q(p_3) < a_3$. Further, if x is any point of Λ, then $F_P(x) \neq F_Q(x)$. For if $x \not< a_1$, then since $[P_1F_P(x)]$, we have $P_1 = [p_1\ F_P(x)] \cap a_1$. Similarly, $Q_1 = [p_1\ F_Q(x)] \cap a_1$. If $F_P(x) = F_Q(x)$, then $P_1 = Q_1$, which is a contradiction. If $x < a_1$, the same argument would establish $P_2 = Q_2$, and this would violate what we have just shown in the case $x = p_2$. In particular, $F_P(p) \neq F_Q(p)$, $F_P(p_i) \neq F_Q(p_i)$ ($i = 1, 2, 3$). Let $a = [F_P(p)\ F_Q(p)]$. We shall show that $\{a\ a_i\}$ for $i = 1, 2, 3$; that is, $d(a \cup a_i) = 4$.

$$d(a \cup a_i) = d([F_Q(p)\ F_P(p)] \cup [F_Q(p_i)\ F_P(p_i)])$$

$$= d([F_P(p)\ F_P(p_i)] \cup [F_Q(p)\ F_Q(p_i)]) = 4,$$

by 1.5.3 and 2.5 (b). The uniqueness of a is obvious.

2.7 DEFINITION. Let k be an ordinary sphere, a_1 and a_2 circles contained in k, $p < k$, $p \not< a_1 \cap a_2$, $q \not< k$. The circle $a = k \cap \{p[\ \{q\ a_1\} \cap \{q\ a_2\}]\}$ is called the circle containing p coaxal with a_1 and a_2.

It must be shown, of course, that a does not depend on the choice of q. That this is indeed the case can be proven from 2.6 in a manner precisely analogous to the derivation of 2.2 from 2.1, and we omit the details. It is clear that if $r < a$, then a is also the circle containing r coaxal with a_1 and a_2; also, if $a_1 \cap a_2 \neq 0$, then the circle containing p coaxal with a_1 and a_2 is $p \cup (a_1 \cap a_2)$.

The set of all circles coaxal with two given circles is called a coaxal set of circles. It is clear that any two circles of a coaxal set determine the coaxal set.

2.8 *Definitions.* Returning to the function F of 2.5, we see that F induces in a natural way a 1-1 transformation τ of the set of points of Λ onto itself. τ is defined as follows:

(i) if $F(p)$ is a singular pair, then $\tau p = p$;

(ii) if $F(p)$ is an ordinary pair, $F(p) = p \cup q$, then $\tau p = q$.

As a transformation, τ is an involution, and we call τ a fundamental involution. A fundamental involution, then, is a point-point transformation associated with a function F in the prescribed way. Any pair in the image of F is said to be anallagmatic with respect to τ (or anallagmatic under τ), which justifies the terminology of 2.3. It follows from 2.5.1 that a fundamental involution is completely determined by any two distinct anallagmatic pairs.

2.8.1 DEFINITION. If τ is the fundamental involution associated with a function F, then a is said to be anallagmatic with respect to τ (or anallagmatic under τ) if a contains a pair P in the image of F.

2.9 *Corollaries.* We leave the proofs to the reader.

2.9.1 If a and b are anallagmatic with respect to τ, and if $a \cap b \neq 0$, then $a \cap b$ is anallagmatic with respect to τ.

2.9.2 If $k = \{a\ b\}$ is an ordinary sphere, and if a and b are circles anallagmatic with respect to τ, then every circle coaxal with a and b is also anallagmatic under τ.

2.9.3 If τ, ρ are distinct fundamental involutions, and k is an ordinary sphere anallagmatic under both τ and ρ, then the circles on k anallagmatic under both τ and ρ are a coaxal set of circles.

2.9.4 If a is anallagmatic with respect to τ, and $p < a$, then $\tau p < a$.

2.9.5 τ and ρ will have a common anallagmatic pair if and only if the coaxal circles of 2.9.3 have a common pair (which, of course, will be the common anallagmatic pair).

2.9.6 If $a \neq 0$ is ordinary and not anallagmatic under τ, $d(a) = n$, $a = p_1 \cup \cdots \cup p_m$, and P_i is the anallagmatic pair containing p_i, then $d(P_1 \cup \cdots \cup P_m) = n + 1$.

2.10 *Definitions.* If τ is a fundamental involution, a point p such that $\tau p = p$ is called a fixed point or double point of τ. A fundamental involution τ is called a negative inversion if for a and b ordinary circles anallagmatic under τ, $\{a\ b\}$, then $a \cap b$ is an ordinary pair. Any other fundamental involution is called a positive inversion, or briefly, an inversion.

The axioms previously given are insufficient, and we add two more. Specifically, we shall require that (positive) inversions possess further properties. Now, inversions have been defined in Λ, and our axioms should be given as properties of Π. It is not difficult, however, to define inversion exclusively in terms of Π, and we leave this to the reader. The axioms of this section are then to be considered as axioms of Π; but their content is precisely the same as if they were given as properties of Λ, and it is as such that we shall use them.

2.11 AXIOM[8]. If τ is an inversion, then every anallagmatic ordinary circle contains at least one fixed point of τ.

It follows that a fundamental involution is an inversion if and only if it admits a double point.

2.12 AXIOM. If τ is an inversion, and if $[p_1 p_2 p_3 p_4]$, then $[\tau p_1\, \tau p_2\, \tau p_3\, \tau p_4]$; that is, any inversion satisfies the hypothesis of 1.6.2.

This completes our list of assumptions. It is possible to replace 2.12 by either of the following axioms:

(i) If ρ and τ are inversions, then $\rho\tau\rho$ is a fundamental involution. (Hence, $\rho\tau\rho$ is an inversion, for if x is any fixed point of τ, then ρx is a fixed point of $\rho\tau\rho$.)

(ii) If $\tau_1, \cdots, \tau_n$ are inversions, if $\rho = \tau_n \cdots \tau_1$, and if x_1, x_2, and x_3 are distinct points such that $\rho x_i = x_i$ ($i = 1, 2, 3$), then $x < [x_1 x_2 x_3]$ implies $\rho x = x$[9].

By virtue of 1.6.2, Axiom 2.12 implies that any inversion τ, or any composition of inversions, has a unique extension to an automorphism of Λ. No confusion will arise if we use the same symbol for the automorphism that we have hitherto used for the point transformation, and henceforth we shall do so. It is convenient to note here a few useful facts about automorphisms of Λ.

If ϕ is an automorphism of Λ, and

2.12.1 if a, b, c are coaxal circles, then ϕa, ϕb, ϕc are coaxal circles;

2.12.2 if a_p is a singular element, then ϕa_p is also singular;

2.12.3 if P, Q, R are anallagmatic pairs of a fundamental involution, then ϕP, ϕQ, ϕR are anallagmatic pairs of a fundamental involution;

2.12.4 if $P = p \cup q$, and if $\phi P = P$, then either $\phi p = p$ or $\phi p = q$; if $\phi p = q$, then $\phi q = p$; if $\phi p = p$, then $\phi q = q$.

2.13 THEOREM. *Let k be an ordinary sphere anallagmatic under an inversion τ. Then k contains a circle c such that* (i) $p < k$, $\tau p = p$ *imply* $p < c$; (ii) $p < c$ *implies* $\tau p = p$; (iii) c *is not anallagmatic under* τ; (iv) *if a is an ordinary anallagmatic circle on k, then $a \cap c$ is an ordinary pair. We call c the circle of inversion of τ on k.*

Proof. 2.11 implies that there are at least three fixed points of τ on k. We shall show that all fixed points lie on a circle. Assume the contrary, and

<hr>

[8] This is our only "order" axiom, in contrast with the variety of order axioms in [4] and [5]. This assumption compels our field V to have the property described in the introduction. It is quite clear that, conversely, this assumption is satisfied in any inversion geometry over V.

[9] (i) is a special case of Miquel's theorem, as stated in 4.17. (ii) is reminiscent of axiom P in projective geometry [9, vol. I, p. 95].

let x_1, x_2, x_3, x_4 be fixed but not contained in a circle. Then $[x_1\ x_2\ x_3]$ and $[x_1\ x_2\ x_4]$ cannot both be anallagmatic, for that would imply $\tau : x_1 \rightarrow x_2$; so at least one of them, say $[x_1\ x_2\ x_3]$, is not anallagmatic. By 2.12, $p < [x_1\ x_2\ x_3]$ implies $\tau p < [x_1\ x_2\ x_3]$, so $\tau p = p$, for otherwise $[x_1\ x_2\ x_3]$ would be anallagmatic. Let $r \neq x_4$, $r \not< [x_1\ x_2\ x_3]$ be any point of k, and let $c_1 = [r\ x_4\ x_1]$, $c_2 = [r\ x_4\ x_2]$. Then c_i contains two points of $[x_1\ x_2\ x_3]$ or is tangent to $[x_1\ x_2\ x_3]$, so by 2.12 or 2.12.2, $\tau c_i = c_i$. Hence, 2.12.4 implies $\tau r = r$. Let $[r\ s\ t]$ be any circle of k containing r and anallagmatic under τ, and let ρ be the inversion defined by $\rho : s \rightarrow t$, $r \rightarrow r$[10]. If $u \neq r$ is a fixed point of ρ on k, and a is the unique circle (2.9.6) containing r and u anallagmatic under ρ, then $a \cap A_r$ and $a \cap A_u$ are anallagmatic under both ρ and τ. Therefore, $\rho = \tau$ (2.8), which is impossible. Hence, if c is a circle of k containing three fixed points of τ, say $c = [x_1\ x_2\ x_3]$, (i) is proven. (iii) follows from the fact that we can certainly find a point $x \in \Lambda$ such that $x \not< c$, $\tau x = x$. If c were anallagmatic, $\{x\ c\}$ would be an anallagmatic sphere whose fixed points were not contained in a circle. (ii) has already been proven, under the temporary (but now verified assumption) that $[x_1\ x_2\ x_3] = c$ is not anallagmatic. By 2.11, an ordinary anallagmatic circle on k contains at least one point of c, hence it contains two points, or c would be anallagmatic. This proves (iv).

2.14 LEMMA. *Assume that we are given a set of coaxal circles on an ordinary sphere k such that if a and b are distinct circles of the set, then $a \cap b = 0$. Then there exist exactly two points p and q such that $A_p \cap k$ and $A_q \cap k$ are circles of the coaxal set.*

Proof. There certainly exist no more than two points p and q with the given property. For if $A_r \cap k$ is another singular circle of the coaxal set, let σ be the inversion under which $A_p \cap [p\ q\ r]$ and $A_q \cap [p\ q\ r]$ are anallagmatic pairs. Then by 2.9.2, $A_r \cap k$ is anallagmatic under σ, so $\sigma r = r$. Therefore, $[p\ q\ r]$ would be anallagmatic under σ and contain three double points. This violates 2.13 (iii).

Now to show that there are at least two points with the specified property. Let a and b be two ordinary circles of the given set of coaxal circles (if such ordinary circles did not exist, then the assertion to be proven would be immediately true). Let μ be any fundamental involution under which a and b are anallagmatic (μ clearly exists). Since $a \cap b = 0$, it follows from 2.10 that μ is an inversion. By 2.13, k contains a circle c which is the circle of inversion of μ. c intersects a in two points, say s_1, s_2; c intersects b in two points, say t_1, t_2. Let ν be the fundamental involution determined by $\nu : s_1 \rightarrow s_2$, $t_1 \rightarrow t_2$. By 2.10, ν is an inversion, and c is anallagmatic under ν. Hence, c contains two points, p and q, which are double points of ν. Therefore, p and q are double

---

([10]) The final sentence of 2.8 shows that ρ is determined by these stipulations, for $s \cup t$ and $[r\ s\ t] \cap A_r$ are pairs anallagmatic under ρ.

points of both μ and ν, so $A_p \cap k$, $A_q \cap k$, a and b are anallagmatic under both μ and ν. By 2.9.3, p and q are the desired points.

2.15 Theorem. *Given any ordinary circle d on a sphere k, then there exists a unique inversion under which k is anallagmatic and d is the circle of inversion.*

Proof. There cannot be more than one such inversion, for then the singular circles containing the points of d and contained in k would be coaxal, violating 2.14. To show that there is at least one such inversion, let p, q, and r be distinct points contained in d, and s a point not contained in k. Let

$$S = (s \cup (A_p \cap k)) \cap (s \cup (A_q \cap k)) \cap (s \cup (A_r \cap k)).$$

Since $S \cap k = 0$, there exists a fundamental involution τ under which S and k are anallagmatic. By 2.8.1, $(s \cup (A_p \cap k))$ is anallagmatic, hence by 2.9.1 $A_p \cap k$ is anallagmatic, so $\tau p = p$, and τ is an inversion. Similarly, $\tau q = q$, $\tau r = r$. Hence, d is the circle of inversion on k of τ.

2.15.1 **Definition.** We shall find it occasionally convenient, when the anallagmatic ordinary sphere under discussion, say k, is well understood, to designate an inversion τ, whose circle of inversion on k is c, by τ_c.

2.16. Theorem. *Let a be an ordinary circle contained in a sphere k, $r < k$, $r \not< a$, $s = \tau_a r$. Then $A_r \cap k$, $A_s \cap k$ and a are coaxal.*

Proof. Let $x < a$. By 2.12, 2.12.1, and 2.12.2, if b is the unique circle containing x coaxal with $A_r \cap k$ and $A_s \cap k$, then $\tau_a b = b$. To prove $b = a$, it is sufficient to show that b is not anallagmatic with respect to τ_a. Assume the contrary. Then $[r\ s\ x] \cap b$ is singular. Let σ be the inversion which has $A_r \cap [r\ s\ x]$ and $A_s \cap [r\ s\ x]$ as anallagmatic pairs. Then by 2.9.2, b is anallagmatic under σ, so by 2.9.1 $[r\ s\ x] \cap b$ is also anallagmatic under σ. Hence, r, s, and x are double points of σ, violating 2.13.

3. **Harmonic sets and orthogonal circles**[11].

3.1 **Definition.** If p, q, r, and s are distinct, $[p\ q\ r\ s]$, and there is an inversion τ such that $\tau p = p$, $\tau q = q$, $\tau r = s$, we say $H(p\ q,\ r\ s)$ (read "$p\ q,\ r\ s$ are a harmonic set"). It is obvious that if p, q, r are given, s is uniquely determined, $s \neq r$ by 2.13, and we therefore sometimes say: "s is the harmonic conjugate of r with respect to p and q."

3.1.1 **Corollary.** *If $H(p\ q,\ r\ s)$, then $H(r\ s,\ p\ q)$.*

Proof. Let k be any sphere containing $[p\ q\ r\ s]$, and let $a_r = A_r \cap k$, $a_s = A_s \cap k$. By 3.1 and 2.15.1, k contains a circle a such that $\phi_a r = s$, and $a \cap [p\ q\ r\ s] = p \cup q$. Let τ be the inversion given by $\tau: p \to q$, $r \to r$. Then a_r and a are anallagmatic under τ, so by 2.16 and 2.9.2, a_s is also anallagmatic under τ. Hence $\tau s = s$. By the definition of τ, this implies $H(r\ s,\ p\ q)$.

---

[11] The axioms of inversion geometry given by Pieri in [7] place principal emphasis on the notion of harmonic sets and arrive at inversions through them.

 A. J. HOFFMAN [September

3.1.2 COROLLARY. *If ϕ is any automorphism of Λ, and if $H(p\,q,\,r\,s)$, then $H(\phi\,p\,\phi\,q,\,\phi\,r\,\phi\,s)$.*

This follows easily from 2.12.2 and 2.12.3.

3.2 DEFINITION. If a and b are ordinary circles contained in a sphere k, we say $a\perp b$ (read "a is orthogonal to b") if b is anallagmatic under ϕ_a.

The following statements are immediate:

3.2.1 If $a\perp b$ and τ is an automorphism of Λ, then $\tau a\perp\tau b$.

3.2.2 $a\perp b$ implies that $a\cap b$ is an ordinary pair.

3.2.3 If a is an ordinary circle contained in a sphere k, and if p, q are distinct points contained in k with $\phi_a p\neq q$, then there exists one and only one circle b such that p, $q<b<k$ and $a\perp b$.

3.2.4 If a and b are circles tangent at a point p, and if c is a circle such that $p<c<\{a\,b\}$ and $c\perp a$, then $c\perp b$.

3.2.5 THEOREM. *$a\perp b$ implies $b\perp a$.*

Proof. Let $b\cap a=p\cup q$ (3.2.2). Let r be any other point contained in a. Since $a\perp b$, it follows that $[r\cup(A_p\cap b)]$ is tangent to $[r\cup(A_q\cap b)]$. Let $\phi_b r=s$. It is clear $r\neq s$. By 2.12.2, $\phi_b A_p=A_p$ and $\phi_b A_q=A_q$, so that $[s\cup(A_p\cap b)]$ is tangent to $[s\cup(A_q\cap b)]$; that is, s is a fixed point of ϕ_a, or $s<a$. It follows that a is anallagmatic under ϕ_b, since a contains the anallagmatic pair $r\cup s$; hence, $b\perp a$.

3.3 THEOREM. *If a and b are ordinary circles on a sphere k and $a\perp b$, then $\tau_a\,\tau_b=\tau_b\,\tau_a$.*

Proof. Let $\rho=\tau_b\,\tau_a\,\tau_b\,\tau_a$. By 1.6.2, it is sufficient to show that ρ takes every point of Λ into itself. Let $a\cap b=p\cup q$. It is clear that $\rho p=p$, $\rho q=q$. Let x be any other point of Λ, and let X be the unique pair containing x that is anallagmatic under τ_b. We then have by 3.2.5 that X, $A_p\cap a$, $A_q\cap a$ are distinct pairs anallagmatic with respect to τ_b. Since $A_p\cap a$ and $A_q\cap a$ are taken into themselves by τ_a, it follows from 2.12.3 that $\tau_a X$, $A_p\cap a$ and $A_q\cap a$ are anallagmatic pairs of a fundamental involution, which clearly is τ_b. Hence $\rho X=\tau_b\,\tau_a\,\tau_b(\tau_a X)=\tau_b\,\tau_a(\tau_a X)=\tau_b X=X$. This shows that if X is singular, then $\rho x=x$. If X is ordinary, $X=x\cup y$, say, then $\tau_a X=\tau_a x\cup\tau_a y$. Hence $\rho x=\tau_b\,\tau_a(\tau_b\,\tau_a x)=\tau_b\,\tau_a(\tau_a y)=\tau_b y=x$.

3.4 COROLLARY. *Under the hypothesis of 3.3, if $a\cap b=p\cup q$, and if P_p is a singular pair such that $p<P_p<k$, then $\tau_b\,\tau_a P_p=P_p$.*

Proof. Assume, temporarily, that if $x<k$, $x\neq p$, q, then $\tau_b\,\tau_a x\neq x$. Let c be any ordinary circle contained in k such that $q\not<c$, $P_p<c$. Let $d=\tau_b\,\tau_a c$, so that by 3.3 we have $c=\tau_b\,\tau_a d$. If $c=d$ (actually, this cannot occur), then since p is a fixed point of both τ_a and τ_b, we have $\tau_b\,\tau_a P_p=\tau_b\,\tau_a(A_p\cap c)=\tau_b\,\tau_a A_p\cap\tau_b\,\tau_a c=A_p\cap c=P_p$. If $c\neq d$, then consider the pair $c\cap d$. It is easy to see that this pair is taken into itself by $\tau_b\,\tau_a$. Consequently, if $c\cap d$ is ordinary, say

$c \cap d = p \cup x$, then by 2.12.4, $\tau_b \, \tau_a x = x$, which contradicts our assumption. Therefore, $c \cap d$ is singular, so that $c \cap d = A_p \cap c = P_p$, and we have $\tau_b \, \tau_a P_p = P_p$.

It remains, then, to prove our assumption. We first note that $A_p \cap k$ and $A_q \cap k$ are anallagmatic with respect to both τ_a and τ_b. Hence, by 2.9.5, τ_a and τ_b do not have a common anallagmatic pair. If $\tau_a x = \tau_b x$, then x is either a fixed point of both inversions (impossible, since $x \nless a \cap b$) or x is contained in an ordinary pair anallagmatic under both inversions, which violates the preceding sentence. Hence, $\tau_a x \neq \tau_b x$, which is equivalent to our assumption.

3.5 COROLLARY. *Under the hypothesis of 3.3, if c is any circle such that $a \cap b = p \cup q < c < k$, then $\tau_b \, \tau_a c = c$.*

Proof. Let $P_p = A_p \cap c$, so that $c = q \cup P_p$. By 3.4, $\tau_b \, \tau_a c = \tau_b \, \tau_a (q \cup P_p)$ $= \tau_b \, \tau_a q \cup \tau_b \, \tau_a P_p = q \cup P_p = c$.

3.6 COROLLARY. *Under the hypothesis of 3.3, if x is a point contained in k but $x \neq p$, q, then $H(p \, q, \, x \, \tau_b \, \tau_a x)$.*

Proof. That x is distinct from $\tau_b \, \tau_a x$ was shown in 3.4 and $[p \, q \, x \, \tau_b \, \tau_a x]$ follows from 3.5. By 2.9.2, if c is the circle containing x coaxal with $A_p \cap k$ and $A_q \cap k$, then $\tau_b \, \tau_a x < c$. If σ is the inversion under which $A_p \cap [p \, q \, x]$ and $A_q \cap [p \, q \, x]$ are anallagmatic pairs, then by 2.13 and 2.12.4, we have $\sigma : p \rightarrow p, \, q \rightarrow q, \, x \rightarrow \tau_b \, \tau_a x$. This is equivalent to the statement to be proven.

3.7 COROLLARY. *If a and b are ordinary circles tangent at a point p, and if P_p is a singular pair such that $p < P_p < \{a \, b\}$, then $\tau_b \, \tau_a P_p = P_p$.*

Proof. Let $c < \{a \, b\}$ be a circle orthogonal to a and containing p. By 3.2.4, $c \perp b$. Note that $\tau_b \, \tau_a = \tau_b \, \tau_a (\tau_c \, \tau_c) = (\tau_b \, \tau_c)(\tau_a \, \tau_c)$, by 3.3. An application of 3.4 completes the proof.

3.8 COROLLARY. *On an ordinary sphere k, there do not exist more than three mutually orthogonal circles.*

3.9 THEOREM. *If k is a sphere anallagmatic under a negative inversion τ, then the restriction to k of τ is the composition of (positive) inversions.*

The theorem is actually true throughout Λ, not merely on k. But the stronger result, which would require some preparation, is not needed for what follows.

Proof. By 2.11, k is an ordinary sphere, and if a is any circle contained in k anallagmatic with respect to τ, then a is also ordinary. Consider ϕ_a. The circles on k anallagmatic under both τ and ϕ_a form a coaxal set of circles which by 2.10 have an ordinary pair, say $p \cup q$, in common. By 2.9.5, $p \cup q$ is anallagmatic with respect to both τ and ϕ_a.

In order to prove the theorem, it will suffice to prove that for every $x < k$,

$x \neq p$, q, the point τx is the harmonic conjugate of $\phi_a x$ with respect to p and q. For assume that this has been shown. Let c and d be two orthogonal circles on k, each containing p and q. Then by 3.6, we have $\tau x = \phi_d \phi_c \phi_a x$. Further, p and q will be double points of both ϕ_d and ϕ_c, so that $\tau = \phi_d \phi_c \phi_a$ for every point of k.

Now to prove our statement, let b be the circle containing p and q orthogonal to $[p\ q\ x]$ (3.2.3). Since $p \cup q < b$, and $p \cup q$ is anallagmatic under ϕ_a, it follows that $a \perp b$. Therefore, $a \cap b$ is an ordinary pair, say $s \cup t$; and since a and b are anallagmatic under τ, so is $s \cup t$. The harmonic conjugate of $\phi_a x$ with respect to p and q is clearly $\phi_b(\phi_a x)$. Because $a \perp b$, we have (3.5) $[x\ \phi_b\ \phi_a x\ s\ t]$. Because $[p\ q\ x]$ is anallagmatic under both ϕ_a and ϕ_b, we have $[x\ \phi_b\ \phi_a x\ p\ q]$. Therefore, since $p \cup q$ and $s \cup t$ are anallagmatic with respect to τ, it follows that $x \cup \phi_b\ \phi_a x$ is anallagmatic with respect to τ. This, however, is equivalent to the statement to be proved.

An obvious and important consequence is that τ has a unique extension to an automorphism of the sublattice of Λ consisting of all elements contained in k, which can be further extended to an automorphism of Λ. No confusion will arise when we use the same symbol for the automorphism as for the point transformation.

3.10 Lemma. *Let k be an ordinary sphere, a an ordinary circle contained in k. Let ϕ be an automorphism of Λ such that $\phi k = k$, and $\phi x = x$ for all $x < a$. Then, restricted to k, ϕ is either the identity mapping or ϕ_a.*

Proof. It may be that k contains a point p such that $p \not< a$ and $\phi p = p$. Let q be any other point of k not contained in a. Let b and c be distinct circles, each containing p, q and a point of a. It is clear that $\phi b = b$, $\phi c = c$, so that $\phi(b \cap c) = b \cap c$. By 2.12.4, we have $\phi q = q$, so that in this case ϕ, restricted to k, is the identity map.

On the other hand, it may be that if $p < k$, $p \not< a$, then $\phi p \neq p$. Let b_1, b_2 be circles such that $p < b_i \perp a$ ($i = 1$, 2). Let $b_i \cap a = x_i \cup y_i$. By 3.2.1, $\phi\ b_i$ is a circle contained in k orthogonal to a, and containing x_i and y_i, so that by 3.2.3, $\phi\ b_i = b_i$. Since $b_1 \cap b_2 = p \cup \phi_a p$, an application of 2.12.4 shows that in this case ϕ, restricted to k, is ϕ_a.

3.11 Theorem. *If a pair P is anallagmatic under three (not necessarily distinct) fundamental involutions τ_1, τ_2, τ_3, then $\tau_3 \tau_2 \tau_1$ is a fundamental involution.*

Proof. *Case* (i). P is an ordinary pair, say $P = p \cup q$. Let r be any point of Λ other than p or q. It is clear that $[p\ q\ r\ \tau_3 \tau_2 \tau_1 r]$. Let ρ be the fundamental involution defined by $\rho: p \rightarrow q$, $r \rightarrow \tau_3 \tau_2 \tau_1 r$, and let $\phi = \rho\ \tau_3 \tau_2 \tau_1$. It is clearly sufficient to show that if s is any point of Λ, then $\phi s = s$. Let k be any sphere containing p, q, r, s. Let b be the circle on k containing p and r and orthogonal to $[p\ q\ r]$ (3.2.3). Since $\phi\ p = p$, $\phi\ r = r$ and $\phi[p\ q\ r] = [p\ q\ r]$, it follows from 2.12, 3.9, and 3.2.1 that $\phi\, b = b$. Let t be any point contained in b other than p. Since $\phi[t\ p\ q] = [t\ p\ q]$ and $b \cap [t\ p\ q] = p \cup t$, it follows from 2.12.4 that $\phi\, t = t$. Further, since $\phi\, q = q$ and $q \not< b$, it follows from 3.10 that $\phi\, s = s$.

Case (ii). P is a singular pair, say $P = P_p$. Let r be any point of Λ other than p, and let ρ be the inversion with respect to which P_p is anallagmatic, and which takes r into $\tau_3\, \tau_2\, \tau_1 r$. If we let ϕ designate the automorphism $\rho\, \tau_3\, \tau_2\, \tau_1$, it is sufficient as in Case (i) to show that if s is any point of Λ, then $\phi\, s = s$. Let k be any sphere containing P_p, r, and s. As in Case (i), if b is the circle on k that contains r, and p is orthogonal to $[r\ P_p]$, then ϕ takes every point of b into itself. We need only consider, then, the case that $s \not< b$. Note that the circles of inversion of τ_1, τ_2, τ_3, and ρ have a singular pair containing p in common, since each of these circles contains p and is orthogonal to $[r\ P_p]$. Hence by 3.7, if Q_p is any singular pair such that $p < Q_p < k$, then $\phi\, Q_p = Q_p$.

Let t and u be distinct points contained in b other than p. Then $[t\ s\ p] = t \cup Q_p$, where $Q_p = A_p \cap [t\ s\ p]$. By the preceding paragraph, $\phi[t\ s\ p] = Q(t \cup Q_p) = \phi t \cup \phi\, Q_p = t \cup Q_p = [t\ s\ p]$. Similarly, $\phi[u\ s\ p] = [u\ s\ p]$. An application of 2.12.4 proves that $\phi s = s$.

4. **Coordinates on a circle.** We begin the process of introducing coordinates, by establishing a field of points contained in any ordinary circle. This is closely analogous to the well known field of points of a conic[12], with 3.11 playing the role of Pascal's theorem. Then we show that this field is an ordered field in which every positive number is a square.

Let c be any ordinary circle, and let three distinct points contained in c be identified by the labels 0, 1, ∞. (The context will always enable us to distinguish the point 0 from the 0 element of Λ.)

4.1 DEFINITION. Let x, $y < c$; x, $y \neq 0$, ∞. Let ϕ be the fundamental involution given by $\phi\colon 0 \to \infty$, $x \to y$, and let $z = \phi 1$. We say that $z = xy$ (z is x multiplied by y). Note that we do not require that x and y be distinct.

4.2 DEFINITION. Let x, $y < c$; $y \neq \infty$. Let ϕ be the inversion under which c is anallagmatic and $\phi\colon x \to y$, $\infty \to \infty$. Let $z = \phi 0$. We say that $z = x + y$ (z is x added to y). Note that x and y need not be distinct.

4.3 THEOREM. *The points of c other than 0 and ∞ constitute an abelian group under multiplication. The points of c other than ∞ constitute an abelian group under addition.*

Proof. We shall prove the statement about multiplication; the proof for addition is essentially the same. It is clear that all we need show is associativity, since all the other postulates for an abelian group are clearly satisfied. Let x, y, z be arbitrary (not necessarily distinct) points contained in c other than 0 and ∞. Let

------

[12] See [9, vol. I, p. 231].

$$\phi_1: \ x \rightarrow y, \quad 0 \rightarrow \infty, \ 1 \rightarrow xy,$$
$$\phi_2: \ y \rightarrow z, \quad 0 \rightarrow \infty, \ 1 \rightarrow yz,$$
$$\phi_3: \ z \rightarrow xy, \ 0 \rightarrow \infty.$$

We shall show that $\phi_3 yz = x$, which is equivalent to $(xy)z = x(yz)$.

By 3.11, $\phi_3 \phi_2 \phi_1$ is an involution. Since $\phi_3 \phi_2 \phi_1 x = xy$, it follows that $x = \phi_3 \phi_2 \phi_1 xy = \phi_3 \phi_2 1 = \phi_3 yz$.

4.4 DEFINITION. If $x \neq \infty$, we say $0x = x0 = 0$; $\infty + x = x + \infty = \infty$. If $x \neq 0$, we say $\infty \, x = x \infty = \infty$.

4.5 THEOREM. *If $x \neq 0$, ∞, then there exist two fundamental involutions ϕ_1, ϕ_2 such that for all $p < c$, we have $px = \phi_2 \phi_1 x$. In other words, multiplication by a point other than 0 or ∞ is the restriction to c of a lattice automorphism. Similarly, addition by a point other than ∞ is the restriction to c of a lattice automorphism.*

Proof. For the same reason as in 4.3, we confine our attention to the statement about multiplication. Let $\phi_1: 0 \rightarrow \infty$, $1 \rightarrow 1$, and let $\phi_2: 0 \rightarrow \infty$, $1 \rightarrow x$. Assume first that $p \neq 0$, ∞. Then $\phi_2 \phi_1 p = \phi_2 p^{-1} = y$, where $p^{-1} y = x$; that is, $y = px$, which was to be proven. If $p = 0$, we have $\phi_2 \phi_1 0 = 0 = 0x$, by 4.4. The case $p = \infty$ is handled in the same way.

4.6 THEOREM. *The points of c other than ∞ constitute a field under the given definitions of addition and multiplication.*

Proof. In view of 4.3, all we need show is: if x, y, $z < c$, and x, y, $z \neq \infty$, then $x(y+z) = xy + xz$. If at least one of x, y, z is 0, the result is immediate, so we assume the contrary. Further, assume temporarily that $y \neq z$ and that $0 \neq y + z$. By 4.2, the pairs $A_\infty \cap c$, $y \cup z$ and $0 \cup (y+z)$ are anallagmatic pairs of an inversion. By 4.5 and 2.12.3, it follows that $A_\infty \cap c$, $xy \cup xz$, and $0 \cup x(y+z)$ are anallagmatic pairs of an inversion; that is, $x(y+z) = xy + xz$.

If $y = z$, then replace the pair $y \cup z$ in the preceding discussion by $A_y \cap c$. If $0 = y + z$, then replace the pair $0 \cup (y+z)$ by $A_0 \cap c$. In each case, the remainder of the proof is essentially the same.

4.7 DEFINITION[13]. Let p, q, r, s be distinct and $[p \ q \ r \ s]$. Let ϕ be the fundamental involution given by $\phi: p \rightarrow q$, $r \rightarrow s$. If ϕ is a negative inversion, we say $p \ q \,|\, r \ s$ (p and q separate r and s). If ϕ is an inversion, we say $p \ q \nmid r \ s$ (p and q do not separate r and s).

Observe that separation (or nonseparation) of pairs of points is preserved by any automorphism of Λ. Further, observe that the field of points on a circles furnishes an easy criterion for separation. Let us set $p = 0$, $q = \infty$, $r = 1$. Then $p \ q \,|\, r \ s$ if and only if $s \neq p$, q, r and s is not a square in this field. $p \ q \nmid r \ s$ if and only if $s \neq p$, q, r and s is a square in this field.

---

[13] This definition comes from [7].

4.8 THEOREM. *If $p\,q|r\,s$, then $p\,r{\nmid}q\,s$.*

Proof. Let $a = [p\,q\,r\,s]$, and k be any sphere containing a. Let b and c be circles contained in k such that $p\cup q < b\perp a$, $r\cup s < c\perp a$. By hypothesis, $b\frown c$ is an ordinary pair, say $b\cap c = t\cup u$.

Now $p\,r{\nmid}q\,s$ if and only if the fundamental involution determined by $\tau\colon p{\to}r$, $q{\to}s$ has a double point. But τ clearly interchanges b and c by 3.2.1, so that $\tau\colon t\cup u = t\cup u$. Further, $t\cup u$ is not anallagmatic under τ, for that would imply that b is anallagmatic under τ, which is impossible. Hence $\tau t \neq u$, which by 2.12.4 implies $\tau t = t$. Hence $p\,r{\nmid}q\,s$.

4.9 THEOREM. *If $p\,q{\nmid}x\,y$, $q\,r{\nmid}x\,y$, and $p\neq r$, then $p\,r{\nmid}x\,y$.*

Proof. It is clear that $[p\,q\,r\,x\,y]$. We designate this circle by c. Set $x = 0$, $y = \infty$, $q = 1$, which determines a field of points contained in c. We work in this field. $p\,q{\nmid}x\,y$ implies that there exists $t < c$ such that $t^2 = p$. $q\,r{\nmid}x\,y$ implies that there exists $u < c$ such that $u^2 = r$. Then $(tu)^2 = pr$. Let ϕ be the fundamental involution determined by $\phi\colon p{\to}r$, $0{\to}\infty$. Then $\phi\colon 1{\to}pr$, $tu{\to}tu$. Hence $p\,r{\nmid}0\,\infty$; that is, $p\,r{\nmid}x\,y$.

4.10 COROLLARY. *If $p\,q|x\,y$ and $q\,r{\nmid}x\,y$, then $p\,r|x\,y$.*

4.11 COROLLARY. *If $p\,q|x\,y$, $q\,r|x\,y$, and $p\neq r$, then $p\,r{\nmid}x\,y$.*

Proof. $p\,q|x\,y$, so by 4.8, $p\,x{\nmid}q\,y$ and $p\,y{\nmid}q\,x$. Secondly, $q\,r|x\,y$, so by 4.8, $r\,x{\nmid}q\,y$ and $r\,y{\nmid}q\,x$.

By 4.9, we have $p\,r{\nmid}q\,y$ and $p\,r{\nmid}q\,x$. Applying 4.9 again, we have $p\,r{\nmid}x\,y$.

4.12 THEOREM. *If $H(p\,q,\,r\,s)$, then $p\,q|r\,s$.*

Proof. Let $a = [p\,q\,r\,s]$, k be any sphere containing a. Since $H(p\,q,\,r\,s)$, k contains a circle b such that $p\cup q < b$ and $\phi_b r = s$. Next, consider the fundamental involution $\tau\colon p{\to}q$, $r{\to}s$. b is anallagmatic under τ, so if τ were an inversion, then b would contain a double point of τ, say t. But $r\cup s$ is also anallagmatic under τ, so $[r\,s\,t]$ would be tangent to b. Since $[r\,s\,t]$ is also anallagmatic under ϕ_b, we would have a violation of 2.13. Hence τ is a negative inversion, which proves the theorem.

4.13 THEOREM. *Let a be any ordinary circle, 0, 1, ∞ three distinct points contained in a, and let F be the field of points determined by 0, 1, ∞. Then F is an ordered field in which $x \geqq 0$ if and only if x is a square.*

Proof. It is easy to see that it is sufficient to show that F satisfies:
(i) -1 is not a square, and
(ii) if x is not a square and y is not a square, then $x + y$ is not a square[14].

---

[14] This was pointed out in [8], which seems to be the first place where fields of this type were explicitly discussed.

(i) follows from 4.12, since $H(0 \;\; \infty, 1 \; -1)$. To prove (ii), consider first the case $x = y$, so that $x + y = 2x$. From the definition of addition, we have $H(x \; \infty, 0 \; 2x)$, so by 4.12 $x \infty \mid 0 \; 2x$, which implies (4.8) $x \; 2x \nmid 0 \; \infty$. By hypothesis, $x \; 1 \mid 0 \; \infty$, so that by 4.10, $1 \; 2x \mid 0 \; \infty$, which was to be proven.

The other case is $x \neq y$. By hypothesis, $x \; 1 \mid 0 \; \infty$, $y \; 1 \mid 0 \; \infty$,

so by 4.11 we have   (a) $x \; y \nmid 0 \; \infty$.

Since $H(\; \infty \; (x+y)/2, x \; y)$, we have   (b) $x \; y \mid (x+y)/2 \; \infty$.

By 4.10 and 4.8, (a) and (b) imply   (c) $x(x+y)/2 \nmid 0 \; y$.

By 4.8 alone, (b) implies   (d) $x \; (x+y)/2 \nmid y \; \infty$.

By 4.9, (c) and (d) yield   (e) $x \; (x+y)/2 \nmid 0 \; \infty$.

Combined with the hypothesis, (e) gives   (f) $1 \; (x+y)/2 \mid 0 \; \infty$.

Hence, $(x+y)/2$ is not a square, so, by the previous case, $x+y$ is not a square.

4.14 We now head toward a proof of Miquel's theorem (4.17). We assume that the reader is familiar with the linear fractional transformation from projective geometry, and can prove, using the same ideas as in projective geometry, that if $a = [0 \; 1 \; \infty]$ is a circle on a sphere k, if p, q, r, s are points of a other than ∞, and $ps - qr \neq 0$, then $x' = (px+q)/(rx+s)$ is the restriction to a of an automorphism ϕ such that $\phi \; k = k$, $\phi \; a = a$.

4.15 LEMMA. *Let k be an ordinary sphere, a an ordinary circle contained in k. Let ϕ be an automorphism of Λ such that $\phi \; k = k$, and let $b = \phi \; a$. Then, restricted to k, $\tau_a = \phi^{-1} \; \tau_b \; \phi$.*

The proof follows readily from 3.10, and is left to the reader.

4.16 LEMMA. *Let τ be an inversion under which $a = [0 \; 1 \; \infty]$ is anallagmatic, and let r and s be the double points of τ on a. Assume r, $s \neq \infty$. Then τ, restricted to a, is given by: $x' = ((r+s)x - 2rs)/(2x - (r+s))$.*

Proof. Let k be any sphere containing a. Let b be the circle of inversion of τ. By 4.14, there is an automorphism ϕ such that $\phi \; k = k$, $\phi \; a = a$, and ϕ is given on a by: $x' = (x-r)/(x-s)$. If $c = \phi \; b$, then c is the circle containing 0 and ∞ orthogonal to a, so that τ_c is given on a by: $x' = -x$. The rest of the proof consists of an application of 4.15.

4.17 MIQUEL'S THEOREM[15]. *If ϕ_1, ϕ_2, ϕ_3 are fundamental involutions such that there exist distinct circles a and b each anallagmatic under ϕ_i ($i = 1, 2, 3$), then $\phi_3 \; \phi_2 \; \phi_1$ is a fundamental involution.*

Proof. If the three fundamental involutions are the same, the theorem is immediate, so we assume the contrary. This implies $\{a \; b\}$. Hence, either

---

[15] In [10], Miquel's Theorem is taken as an axiom, and is the only assumption other than the planar analogue of (i) of our introduction. It is given in the following form: "let p, q, r, s, t, u, v, w be distinct points such that $[pqrs]$, $[tuvw]$, $[ptuq]$, $[qurv]$, $[rvws]$; then $[swtp]$." It is then shown that this implies that the affine plane "over" a point satisfies Pappus's Theorem, and that the circles are a family of conics in that affine plane.

$a \cap b$ is a pair (which case was treated in 3.11), or $a \cap b = 0$. We now consider the latter case. Let $k = \{a\ b\}$. The ϕ_i are clearly inversions, and by 2.9.3, 2.9.2, 2.9.5, and 2.14, k contains two distinct points p and q such that $k \cap A_p$ and $k \cap A_q$ are anallagmatic under each ϕ_i. Let c be any ordinary circle contained in k anallagmatic under each ϕ_i. Then p, $q \not< c$, and c contains two double points of each of the given inversions.

In what follows, we assume ϕ_1, ϕ_2, ϕ_3 are distinct; the cases in which they are not are easily handled. Let 0, ∞ be the double points of ϕ_1 on c; 1, s be the double points of ϕ_2 on c. Note that s must be negative, since the fundamental involution τ given by $\tau\colon 0 \to \infty$, $1 \to s$ has $p \cup q$ as an anallagmatic pair, and $H(0\ \infty,\ p\ q)$. If r is one of the double points of ϕ_3, then the other double point is clearly s/r. Let $\phi = \phi_3\phi_2\phi_1$. Applying 4.16, and examining the linear fractional transformation which describes the effect of ϕ on c, we see that there is an inversion ρ such that $\rho\phi$ is the identity on c and the circle of inversion of ρ contains p. Since $p < c$, it follows from 3.10 that $\rho\phi$ is the identity on k.

Let t be any point of Λ not contained in k. Then, letting d be the circle $\{t\ a\} \cap \{t\ b\}$, d is also anallagmatic with respect to the three given inversions. We have $\{c\ d\}$ because of the definition of coaxal circles. The previous reasoning now shows that $\rho\ \phi$ is the identity on $\{c\ d\}$ (note that the definition of ρ does not depend on the sphere containing c). This completes the proof.

5. Coordinates on a sphere and throughout Λ. In this section, we complete the proof that our axioms are sufficient to establish (ii) and (iii) of the introduction.

5.1 We first define the field V. We have previously defined the field of points contained in an ordinary circle, which consisted of a set of points and the defined laws of composition, addition and multiplication. We shall define V so that given any of the fields of points previously described, we have a natural isomorphism of V onto this field.

Let S be a set of fields F of points contained in circles and isomorphisms f of these fields, $S = \{F, f\}$, satisfying:

(i) if $F \in S$ and $G \in S$, then there is an isomorphism $f \in S$ such that $f\colon F \approx G$;

(ii) if $f \in S$ and $F \in S$, and if $f\colon F \approx F$, then f is the identity map of F onto itself;

(iii) if $f \in S$ and $g \in S$, and if the range of f is the domain of g, then $gf \in S$;

(iv) (a geometric condition) we first define: if F is the field of $[0\ 1\ \infty]$ and G is the field of $[0\ 1'\ \infty]$ (same 0 and ∞), and if there is an inversion ϕ such that $\phi\colon 0 \to 0$, $1 \to 1'$, $\infty \to \infty$, then F is said to be ϕ-related to G. We now require of our set S that if $F \in S$ and $G \in S$, and if F is ϕ-related to G, then the restriction of ϕ to F is an isomorphism of the set S[16].

There exists at least one set S, for example, a single F and the identity

---

[16] The usefulness of this condition will be seen in 5.4 (iii).

map. It follows by Zorn's lemma that there exists a maximal set S (which we denote by $\overline{S}$), and it will be shown in 5.10 that every $F \in \overline{S}$. We proceed with the construction of V.

Consider the F's in $\overline{S}$ as disjunct sets of elements; that is, points with labels attached to indicate the field. Let U be the set-theoretic union of all elements of all F in S. If $x, y \in U$, we define $x \sim y$ if there is an isomorphism $f \in \overline{S}$ such that $fx = y$. $\sim$ is an equivalence relation in U, and each equivalence class $\{x\}$ and each F have exactly one element in common. The equivalence classes form a field V in an obvious way, and if we define $\pi_F\{x\} = \{x\} \cap F$, then $\pi_F \colon V \approx F$([17]).

5.2 Let ∞ be any point of Λ, and let k be any ordinary sphere containing ∞. In the affine geometry "over" ∞ of 1.4, whose terminology we now adopt, k is a plane. An ordinary circle containing ∞ is called a line, and tangent lines are said to be parallel. We proceed to "coordinatize" k in the usual manner. Let 0, 1 be two other points contained in the plane k, and call the line $[0\ 1\ \infty]$ the x-axis. The line on k containing 0 orthogonal to the x-axis is called the y-axis. Of the two inversions that map the x-axis onto the y-axis, we select one, say ϕ, and consider the field G of $[0\ \phi 1\ \infty]$ on the y-axis. If we let F be the field of $[0\ 1\ \infty]$, then by 5.1 we have isomorphisms $\pi_F \colon V \approx F$, $\pi_G \colon V \approx G$. The 1-1 correspondence between ordered pairs of elements of V and points of the plane k (other than ∞) is given as follows: if $a, b \in V$, then (a, b) corresponds to p, where $p \cup \infty$ is the intersection of the line containing $\pi_F a$ orthogonal to the x-axis and the line through $\pi_G b$ orthogonal to the y-axis.

5.3 We prove that the points (a, b) constitute a field under the definitions to be given (5.3.2 and 5.3.3) of addition and multiplication. For this work, the following trivial lemma is helpful.

5.3.1 LEMMA. *Let K be nonempty set, $\{f\}$ a set of* 1-1 *mappings of K onto itself, and let K and $\{f\}$ satisfy the following conditions:*

(i) *there is a fixed* 1-1 *correspondence between the sets K and $\{f\}$. We designate the mapping $f \in \{f\}$ which is associated, in this fixed correspondence, with the element $x \in K$ by f_x;*

(ii) $f_y f_x = f_x f_y$;

(iii) K *contains an element e such that $f_x e = x$, for all $x \in K$.*

Then, if we define $x \circ y = f_x y$, K is an abelian group under $\circ$.

Proof. That e is a right unit and that every x has a right inverse is obvious. The operation $\circ$ is commutative, since $x \circ y = f_x y = f_x f_y e = f_y f_x e = f_y x = y \circ x$. The proof is completed by showing that $\circ$ is associative:

([17]) Our definition of the underlying field contains a certain element of arbitrariness, namely in the choice of the maximal $\overline{S}$. An alternative, and possibly superior, approach is to let the field F on the x-axis (see 5.2) play a forward role, prove that the axioms are sufficient for inversion geometry, and then define $\overline{S}$ to consist of all fields F and all isomorphisms f arising from composition of inversions. This set will satisfy (i), (ii), (iii), and (iv) by virtue of 2.12 (ii).

$$x \circ (y \circ z) = x \circ (z \circ y) = f_x f_z y = f_z f_x y = z \circ (x \circ y) = (x \circ y) \circ z.$$

5.3.2 *Addition.* Let p be any point contained in k other than ∞. Express p in terms of its coordinates, say, $p = (a, b)$. Let

ϕ_1 be the inversion "on" $y = 0$ (that is, the x-axis is the circle of inversion of ϕ_1);

ϕ_2 be the inversion on $y = b/2$;

ϕ_3 be the inversion on $x = 0$;

ϕ_4 be the inversion on $x = a/2$.

Let $\alpha_p = \phi_4 \phi_3 \phi_2 \phi_1$. Then the points of k other than ∞ and the transformations α (regarded as transformations of the set of points of k other than ∞) fulfill the conditions of 5.3.1.

Proof. $(0, 0)$, or briefly, 0, clearly plays the role of e in 5.3.1. All we need prove, then, is that if p, q are points of k other than ∞, then $\alpha_p \alpha_q = \alpha_q \alpha_p$. Let ϕ_1, ϕ_2, ϕ_3, ϕ_4 be the four inversions given above in the definition of α_p. Let τ_1, τ_2, τ_3, τ_4 be the four corresponding inversions in the definition of α_q. Note that $\tau_1 = \phi_1$, $\tau_3 = \phi_3$; that by 3.3 the inversions with subscripts 3 and 4 commute with the inversions with subscripts 1 and 2; that ϕ_2, ϕ_1, τ_2 satisfy the hypothesis of 4.17, and hence $\phi_2 \phi_1 \tau_2 = \tau_2 \phi_1 \phi_2$; and that, similarly, $\phi_4 \phi_3 \tau_4 = \tau_4 \phi_3 \phi_4$. Then

$$\begin{aligned}
\alpha_p \alpha_q &= \phi_4 \phi_3 \phi_2 \phi_1 \tau_4 \tau_3 \tau_2 \tau_1 \\
&= (\phi_4 \phi_3 \tau_4) \tau_3 (\phi_2 \phi_1 \tau_2) \tau_1 \\
&= \tau_4 \phi_3 \phi_4 \tau_3 \tau_2 \phi_1 \phi_2 \tau_1 \\
&= \tau_4 \phi_3 \tau_2 \phi_1 \phi_4 \tau_3 \phi_2 \tau_1 \\
&= \tau_4 \tau_3 \tau_2 \tau_1 \phi_4 \phi_3 \phi_2 \phi_1 = \alpha_q \alpha_p.
\end{aligned}$$

We now define $p + q = \alpha_p q$, and by 5.3.1, the points p of k constitute an abelian group under $+$. Further, it is easy to see from the proof of 4.5 that $(a, b) + (c, d) = (a+c, b+d)$. Finally, we remark that by 3.7, α_p takes any line of k into itself or into a line parallel to itself.

5.3.3 *Multiplication.* Let p be any point of k other than 0 or ∞. We shall define a transformation μ_p. First observe that the line containing p and 0 intersects the unit circle (the circle containing $(1, 0)$ orthogonal to both axes) in two points q and r, where $H(q\,r, 0\,\infty)$. Then by 4.13, exactly one of the points q or r, say q, has the property $q = p$ or $p\ q \nmid 0\ \infty$. Let

ϕ_1 be inversion on $y = 0$;

ϕ_2 be inversion given by $\phi_2 \colon 0 \rightarrow 0$, $\infty \rightarrow \infty$, $(1, 0) \rightarrow q$;

ϕ_3 be inversion on unit circle;

ϕ_4 be inversion given by $\phi_4 \colon 0 \rightarrow \infty$, $q \rightarrow p$.

Let $\mu_p = \phi_4 \phi_3 \phi_2 \phi_1$.

Then, defining pq (p multiplied by q) to be $\mu_p q$, the points contained in k other than 0 and ∞ constitute an abelian group under multiplication, with (1, 0) serving as unit. The proof follows the same lines as 5.3.2.

It is noted that as a transformation of the points of k, μ_p leaves 0 and ∞ fixed.

5.3.4 DEFINITION. If $p \neq \infty$, we say $0p = p0 = 0$; $\infty + p = p + \infty = \infty$. If $p \neq 0$, we say $\infty \, p = p \, \infty = \infty$.

5.3.5 THEOREM. *Under the given definition of addition and multiplication, the points of k other than ∞ constitute a field C, whose zero is 0 and whose unit is (1, 0).*

Proof. In view of what has gone before, all that remains to be shown is that if p, q, r are points of k other than ∞, then $p(q+r) = pq + pr$. This is immediate if at least one of p, q, r is 0, so we assume the contrary. Let us further assume, temporarily, that q, r, and 0 are not contained in a line. By the concluding remark after 5.3.2, the line containing $(q+r)$ and q is parallel to the line containing 0 and r, and the line containing $(q+r)$ and r is parallel to the line containing 0 and q. But μ_p: $0 \rightarrow 0$, $\infty \rightarrow \infty$; hence $p(q+r)$ is a point on the line containing pq parallel to the line containing 0 and pr, and also $p(q+r)$ is a point on the line containing pr parallel to the line containing 0 and pq. Hence, by the preceding sentence, $p(q+r) = pq + pr$, which was to be proven.

It remains to consider the case in which q, r, and 0 are contained in a line. Let s be any point of k not on the line containing q, r, and 0. Then

$(q+r)$ and s are not collinear with 0;

$(q+s)$ and r are not collinear with 0;

q and s are not collinear with 0.

Hence, by 5.3.2 and the preceding case, $p(q+r) + ps = p((q+r)+s)$ $= p((q+s)+r) = p(q+s) + pr = pq + ps + pr$. Subtracting ps from the first and last members, we have the theorem.

5.4 THEOREM. *In C, multiplication obeys the rule $(a, b)(c, d) = (ac - bd, ad + bc)$.*

Proof. Note that we have already remarked in 5.3.2 that addition in C obeys the rule $(a, b) + (c, d) = (a+c, b+d)$. We leave it to the reader to verify that

(i) $(a, 0)(b, 0) = (ab, 0)$, and

(ii) $(0, 1)(0, 1) = (-1, 0)$.

Further, it follows from condition (iv) of 5.1 (indeed, it is precisely for this reason that condition (iv) was introduced) that

(iii) $(0, 1)(a, 0) = (0, a)$.

We now proceed to prove the theorem.

$$(a, b)(c, d) = (a, 0)(c, 0) + (a, 0)(0, d) + (0, b)(c, 0) + (0, b)(0, d),$$

by the distributivity of multiplication with respect to addition.

$(a, 0)(c, 0) = (ac, 0)$ by (i) above.

$(a, 0)(0, d) = (a, 0)(d, 0)(0, 1) = (ad, 0)(0, 1) = (0, ad)$, by (iii) and (i). In like manner, $(0, b)(c, 0) = (0, bc)$.

$(0, b)(0, d) = (0, 1)(b, 0)(0, 1)(d, 0) = (-bd, 0)$, by (iii), (ii), and (i). The theorem is obtained by combining these equations.

5.5 DEFINITION. If $z = (x, y)$, we define $\bar{z} = (x, -y)$. If $z = \infty$, we define $\bar{z} = \infty$. It is obvious that the mapping $z \to \bar{z}$ is the restriction to k of inversion on the x-axis.

5.6 THEOREM. *The mapping $z \to (1, 0)/z$ (extended to all points of k by defining the image of 0 to be ∞ and the image of ∞ to be 0) is the restriction to k of a composition of inversions.*

Proof. Let ϕ_1, ϕ_2, ϕ_3, ϕ_4 be defined as in 5.3.3 for multiplication by z. Let $z' = \phi_1 \phi_3 z$. We show that $zz' = (1, 0)$:

$$zz' = \phi_4\phi_3\phi_2\phi_1\phi_1\phi_3 z = \phi_4\phi_2 z = \phi_2\phi_4 z = (1, 0).$$

(Note that the definitions of ϕ_1 and ϕ_3 are independent of z.) Further, it is clear that $\phi_1 \phi_3$ interchange 0 and ∞. This completes the proof.

5.7 *Equations of ordinary circles other than lines.* We assume that the reader is familiar with the transformations

$$(1) \qquad z' = (pz + q)/(rz + s), \qquad\qquad p, q, r, s \in C,$$

$$(2) \qquad z' = (p\bar{z} + q)/(r\bar{z} + s), \qquad\qquad ps - qr \neq 0,$$

and can prove that (1) and (2) are restrictions to k of automorphisms of Λ which take k into itself.

Let p, q, r be any three distinct points of C. Inversion on $[p\,q\,r]$ can be represented by

$$(3) \qquad z' = ((b\bar{c} - d\bar{a})\bar{z} + (b\bar{d} - d\bar{b}))/((c\bar{a} - a\bar{c})\bar{z} + (c\bar{b} - a\bar{d}))$$

where $a = q - r$, $b = p(r - q)$, $c = q - p$, $d = r(p - q)$. For this transformation is the composition of $\phi: z' = (az+b)/(cz+d)$, $\tau: z' = \bar{z}$ and ϕ^{-1}, and we apply Lemma 4.15.

Assume that p, q, r are not on a line. Then by solving (3) for double points, one sees that there exists $h, k, r \in V, r \neq 0$, such that $(x, y) \in [p\,q\,r]$ if and only if

$$(4) \qquad (x - h)^2 + (y - k)^2 = r^2.$$

(h, k) is the center of $[p\,q\,r]$, that is, the image of ∞ under (3).

Conversely, if $r \neq 0$, $h, k \in V$ are given arbitrarily, one can reverse this process to show that the set of points (x, y) satisfying (4) is the set of points on an ordinary circle not containing ∞, whose center is (h, k).

5.8 *Equations of lines.* Let p and q be distinct points of C. Then inversion on the line containing p and q can be represented by

(5) $$z' = ((q - p)\bar{z} - \bar{p}(q - p) + p(\bar{q} - \bar{p}))/(\bar{q} - \bar{p}),$$

since this transformation is the composition of $\phi\colon z' = (x - p)/(q - p)$, $\tau\colon z' = \bar{z}$, and ϕ^{-1}. By solving (5) for double points, one sees that there exist a, b, $c \in V$, a and b not both 0, such that $(x, y) < [p\ q\ \infty]$ if and only if

(6) $$ax + by + c = 0.$$

Conversely, if a, b, $c \in V$ are given arbitrarily, with $a^2 + b^2 \neq 0$, one can show that the set of points satisfying (6) is the set of points on a line.

5.9 Before we can discuss the coordinate system for 3-space, a few extensions of previous ideas are needed. We first remark that if ϕ is any inversion, then there exists a unique ordinary sphere k which is the locus of all fixed points of ϕ. The "sphere of inversion" has properties analogous to the circle of inversion. Since it is unique, we shall speak of ϕ_k. We shall show later that given any ordinary sphere k, ϕ_k exists.

5.9.1 DEFINITION. If a is an ordinary circle, k is an ordinary sphere, we say $a \perp k$ or $k \perp a$ if a is anallagmatic under ϕ_k.

It is obvious that $a \cap k$ is an ordinary pair, say $p \cup q$. If b is any circle such that $p \cup q < b < k$, then $b \perp a$, for on $\{a\ b\}$, ϕ_k is ϕ_b. Further, if p and q are arbitrary distinct points contained in a circle a, then there is one and only one sphere $k \perp a$ such that $p \cup q < k$. For ϕ_k is determined by the requirement that $A_p \cap a$ and $A_q \cap a$ must be anallagmatic pairs.

5.9.2 THEOREM. *If a, b, c are three distinct circles, and if there exists an ordinary pair $P < a$, b, c, then $b \perp a$ and $c \perp a$ imply $\{b\ c\} \perp a$.*

Proof. On $\{b\ a\}$, a is anallagmatic under ϕ_b. Hence $\{a\ c\}$ is anallagmatic under ϕ_b, so that $\{a\ c\}$ contains a circle of inversion of ϕ_b. This circle of inversion must be c, since $c \perp a$, $P < c$. Hence $\{b\ c\}$ is the sphere of inversion of ϕ_b.

5.9.3 THEOREM. *Given an ordinary sphere k, ϕ_k exists.*

Proof. Let p, q be distinct points contained in k, and let a, b be distinct circles such that $p \cup q < a$, $b < k$. Let the circle c be the intersection of the sphere containing p and q orthogonal to a and the sphere containing p and q orthogonal to b. As remarked in 5.9.1, $c \perp a$, $c \perp b$. It follows from 5.9.2 that $k \perp c$. Hence, by 5.9.1, ϕ_k is the inversion that has $A_p \cap c$ and $A_q \cap c$ as anallagmatic pairs.

5.9.4 DEFINITION. If k and j are ordinary spheres such that j is anallagmatic under ϕ_k, we say $k \perp j$.

We leave it to the reader to prove that $k \perp j$ implies $j \perp k$.

5.9.5 THEOREM. *Given an ordinary pair P, there exist at most 3 mutually orthogonal circles, each containing P. The sphere containing any two of the circles is orthogonal to the third circle; also, the sphere containing any two of the* *circles is orthogonal to any other sphere containing two of the circles.*

The proof is left to the reader.

5.9.6 LEMMA. *Let p and q be distinct points, ϕ_1 and ϕ_2 be inversions, each of which admits p and q as fixed points; let a be any circle containing p and q. Then*

(i) *If r is any point contained in a other than p or q, there is an inversion ϕ_3: $p\to p$, $q\to q$, $\phi_1 r\to\phi_2 r$;*

(ii) *further, if x is any point contained in a, then ϕ_3: $\phi_1 x\to\phi_2 x$.*

Proof of (i). If $\phi_1 r =\phi_2 r$, the result is immediate, so assume the contrary. Let k be the sphere of inversion of ϕ_k: $p\to q$, $r\to r$. Let b be any circle containing p and q anallagmatic under ϕ_1. Since $p\cup q<b$, b is anallagmatic under ϕ_k, so by 5.9.1, $b\cap k$ is an ordinary pair, say $s\cup t$. Let j be any ordinary sphere containing b, and let $c=k\cap j$, so that $c\perp b$ (5.9.1). Hence $H(s\,t,\,p\,q)$ so $H(p\,q,\,s\,t)$, which implies that $s\cup t$ is anallagmatic under ϕ_1. Since $s\cup t<k$ and $r<k$, it follows that $\phi_1 r<k$. Similarly, $\phi_2 r<k$. Let m be a sphere containing p, q, $\phi_1 r$, $\phi_2 r$; then by 2.16, $A_p\cap m$, $m\cap k$, $A_q\cap m$ are coaxal circles. Now $\phi_1 r\cup\phi_2 r<m\cap k$. It follows from 2.9.2 that if ϕ_3 is given by ϕ_3: $p\to p$, $\phi_1 r\to\phi_2 r$, then ϕ_3: $q\to q$.

Proof of (ii). We may assume $x\neq p$, q, r. Let h be the sphere of inversion of ϕ_h: $p\to q$, $x\to x$. Then $h\cap a$ contains two points, say x and y, and $H(x\,y,\,p\,q)$, so that $x\,y\,|\,p\,q$. Further, one can show that h is anallagmatic under ϕ_1, ϕ_2, ϕ_3, so that $\phi_3(\phi_1 x\cup\phi_1 y) =\phi_3(h\cap\phi_1 a) = h\cap\phi_3\phi_1 a = h\cap\phi_2 a = \phi_2 x\cup\phi_2 y$. By 2.12.4, we can prove (ii) by showing that the assumption ϕ_3: $\phi_1 x\to\phi_2 y$ leads to a contradiction. Setting $\rho =\phi_2\,\phi_3\,\phi_1$, we have ρ: $p\to p$, $q\to q$, $r\to r$, $x\to y$. Assume now that $p\,q\,\nmid\,r\,x$. Then, since nonseparation of pairs of points is preserved by ρ, we have $p\,q\,\nmid\,r\,y$. On the other hand, $p\,q\,|\,x\,y$ and $p\,q\,\nmid\,r\,x$ implies $p\,q\,|\,r\,y$, which is a contradiction. The other case, namely $p\,q\,|\,r\,x$, also leads to a contradiction, in a similar way.

5.10 *Digression.* Returning to 5.1, we are now in a position to prove that every $F\in\overline{S}$. Assume that there is a field of points $F\notin\overline{S}$. We shall show that this contradicts the maximality of $\overline{S}$.

Consider first the case in which F is not ϕ-related to any field in $\overline{S}$. Let G be any field of $\overline{S}$, then there is clearly a composition of inversions which maps G isomorphically onto F. Let us call this isomorphism f. Then the set S' consisting of: $\overline{S}$, F, the identity mapping of F on F, all isomorphisms of the form fg (where g is any isomorphism of $\overline{S}$ whose range is G), and $(fg)^{-1}$ satisfies conditions (i)–(iv) of 5.1, which violates the maximality of $\overline{S}$.

The other case is: F is ϕ-related to some field G of S. Let f be the mapping ϕ with domain and range cut down to G and F respectively, and form the set S' as in the previous case. That S' satisfies (i)–(iii) of 5.1 follows as in the previous case, and that S' satisfies (iv) is a consequence of 6.9.6. For any field E of $\overline{S}$ is, by 5.9.6 (i), ϕ-related to F if and only if it is ϕ-related to G.

242 A. J. HOFFMAN

And by 5.9.6 (ii), for these fields E the composed mappings, fg and $(fg)^{-1}$, fulfill the requirements of condition (iv).

5.11 *Sufficiency of the axioms.* Let ∞ be any point of Λ, and let 0 be any other point of Λ. Ordinary circles containing ∞ are called lines, ordinary spheres containing ∞ are called planes. Let the x-axis, the y-axis, and the z-axis be three mutually orthogonal lines containing 0 (see 5.9.5). Let 1 be any other point contained in the x-axis, determining a field E on the x-axis, and let F and G be fields on the y-axis and z-axis respectively, each of which is ϕ-related to E; by 5.9.6, F and G are also ϕ-related to each other. By 5.1, we have isomorphisms $\pi_E\colon V \approx E$, $\pi_F\colon V \approx F$, $\pi_G\colon V \approx G$. We now erect a 3-dimensional cartesian system of coordinates in the usual manner, noting that every point of Λ other than ∞ corresponds to an ordered triple of elements of V. The details are left to the reader. Note that the coordinatization of each of the coordinate planes is in accordance with 5.2, so that the equations of lines, and of circles not containing ∞, in each of the coordinate planes is in agreement with 5.8 and 5.7 respectively. Using simple analytic geometry, one can then prove that every plane is the locus of a linear equation in x, y, z with coefficients in V, and conversely. This proves (ii) of the introduction. Using the material of 5.9, one shows that every ordinary sphere not containing ∞ is the usual Euclidean sphere, and conversely. This proves (iii) of the introduction.

BIBLIOGRAPHY

1. G. Birkhoff, *Lattice theory*, enlarged and completely rev. ed., Amer. Math. Soc. Colloquium Publications, vol. 25, New York, 1948.

2. W. Blaschke, *Vorlesungen über Differentialgeometrie*, 3d ed., vol. 1, Berlin, 1930.

3. S. Gorn, *On incidence geometry*, Bull. Amer. Math. Soc. vol. 46 (1940) pp. 158–167.

4. B. Hesselbach, *Über zwei Vierecksätze der Kreisgeometrie*, Abh. Math. Sem. Hamburgischen Univ. vol. 9 (1933) pp. 265–271.

5. S. Izumi, *Lattice theoretic foundation of circle geometry*, Proc. Imp. Acad. Tokyo vol. 16 (1940) pp. 515–517.

6. B. Petkantschin, *Axiomatischer Aufbau der zweidimensionalen Möbiusschen Geometrie*, Annuaire de Université Sofia I. Faculté Physico-Mathématique. Livre 1 (Mathématique et Physique) vol. 36 (1940) pp. 219–325.

7. M. Pieri, *Nuovi principii di geometria delle inversioni*, Giornale di Matematiche di Battaglini vol. 49 (1911) pp. 49–96, vol. 50 (1912) pp. 106–140.

8. O. Veblen, *The square root and the relations of order*, Trans. Amer. Math. Soc. vol. 7 (1906) pp. 197–199.

9. O. Veblen and J. W. Young, *Projective geometry*, Boston, 1910, 2 vols.

10. B. L. van der Waerden and L. J. Smid, *Eine Axiomatik der Kreisgeometrie*, Math. Ann. vol. 110 (1935) pp. 753–776.

BARNARD COLLEGE, COLUMBIA UNIVERSITY,
    NEW YORK, N. Y.

Reprinted from
The Canadian Journal of Mathematics
Vol. IV, No. 3 (1952), pp. 295–301

CYCLIC AFFINE PLANES

A. J. HOFFMAN

1. Introduction. Let Π be an affine plane which admits a collineation τ such that the cyclic group generated by τ leaves one point (say X) fixed, and is transitive on the set of all other points of Π. Such "cyclic affine planes" have been previously studied, especially in India, and the principal result relevant to the present discussion is the following theorem of Bose [2]: every finite Desarguesian affine plane is cyclic. The converse seems quite likely true, but no proof exists. In what follows, we shall prove several properties of cyclic affine planes which will imply that for an infinite number of values of n there is no such plane with n points on a line. Our results are approximately parallel to those obtained by Hall [6] in his investigation of cyclic projective planes (projective planes admitting a collineation τ such that the group generated by τ is transitive on *all* the points), and our methods are derived from his stimulating and penetrating work. One contrast with the projective case, however, is that every cyclic plane is necessarily finite; for if P and Q are points collinear with X, and if $Q = \tau^d P$, then τ^d leaves the line $P \, Q \, X$ fixed, which implies that the orbit of P under τ belongs to a finite set of lines, and hence cannot contain all points of an infinite plane.

2. Affine difference sets. The integer n will always be the number of points on a line of the cyclic affine plane Π and $N = n^2 - 1$. We now show that the study of cyclic affine planes is equivalent to the study of "affine difference sets" of order n. (This was done by Bose [2] under the additional hypothesis that Π was Desarguesian.) The connection between cyclic projective planes and difference sets was first pointed out by Singer [7] in a paper which essentially inaugurated the subject.

The order of the cyclic collineation τ is N. Let P be any point of Π other than X. Then each point of Π other than X can be expressed uniquely as $\tau^r P$, where r runs over all residue classes mod N. If we write r in place of $\tau^r P$, the lines of the plane may be tabulated as follows:

(2.1)   *The ith line containing X $(i = 0, \ldots, n)$ consists of X and all $r \equiv i$ (mod $n + 1$)*;

(2.2)   *The ith line not containing X $(i = 0, \ldots, N - 1)$ consists of $d_1 + i, \ldots, d_n + i$ (mod N), where the members of the "affine difference set" $\{d_\nu\}$ are residue classes mod N such that the $n^2 - n$ differences $d_\alpha - d_\beta$ ($\alpha \neq \beta$) are precisely the $n^2 - n$ residues mod N which are not 0 (mod $n + 1$). Further, the $\{d_\nu\}$ are in "standard form"*:

$$d_\nu \not\equiv 0 \quad (\text{mod } n + 1), \qquad\qquad \nu = 1, \ldots, n.$$

Received January 29, 1951. This work was done under a grant from ONR.

Proof. Let m be the smallest positive power of τ leaving PX fixed. It is clear that τ^r leaves PX fixed if and only if r is a multiple of m. In particular, $m|N$. Since PX contains $n-1$ points other than X, there are exactly $n-1$ distinct residue classes mod N congruent to 0 (mod m). Hence $m = n+1$, and PX contains X and all residues mod N congruent to 0 (mod $n+1$); $\tau^i PX$ contains X and all residues mod N congruent to i (mod $n+1$). This proves (2.1).

Let l be any line of Π parallel to PX, and let the points of l be $d_1, \ldots, d_n$. Let $d \not\equiv 0$ (mod $n+1$). Then $\tau^d l \neq l$, nor is $\tau^d l$ parallel to l. For in either case we would have $\tau^d PX = PX$, contrary to 2.1. Hence $\tau^d l$ meets l in a single point $d_\alpha \equiv d_\beta + d$ (mod N); thus $d_\alpha - d_\beta \equiv d$ (mod N). It is also easy to see that $\tau^i l = l$ if and only if $i \equiv 0$ (mod N). This proves (2.2).

Conversely, if an affine difference set is given, it can be put in standard form [2], and (2.1) and (2.2) describe an affine plane with the cyclic collineation $X \to X$, $i \to i+1$ (mod N). Some trivial remarks follow at once:

(2.3)   *The projective extension of Π admits a polarity ρ.*

Proof. Define ρ to be the following correspondence: $X \leftrightarrow$ the line at infinity, the point $i \leftrightarrow$ the $(-i)$th line of (2.2), the intersection of the line at infinity with the ith line of (2.1) $\leftrightarrow$ the $(-i)$th line of (2.1).

(2.4)   *A necessary and sufficient condition that Π be Desarguesian is that it admit a collineation that moves X.*

Proof. The necessity is clear. For the sufficiency, it is easy to verify that if the vertices of two triangles are perspective from X, and if two of the pairs of corresponding sides are parallel, then the third pair of corresponding sides is parallel. It is then obvious that the given condition is sufficient for the validity of the affine Desargues' theorem.

(2.5)   *Let s be a number and σ the mapping $\sigma : X \to X$, $i \to si$ (mod N). Then a necessary and sufficient condition that σ be a collineation is that there exist a number k such that $sd_1, \ldots, sd_n$ (mod N) is a rearrangement of $d_1 + k, \ldots, d_n + k$ (mod N).*

Proof. The necessity is immediate. For the sufficiency, it is clear that all we need show is that $(s, N) = 1$. But the given condition implies that the set $\{sr\} = \{s(d_\alpha - d_\beta)\}$, where r runs over all residues mod N except multiples of $n+1$, is again the set $\{r\} = \{d_\alpha - d_\beta\}$ (mod N). If $(s, N) = t > 1$, then $s1 \equiv s(1 + N/t)$ (mod N), violating the condition.

A number s with the property described in (2.5) is called a *multiplier* of the difference set (or a multiplier of the plane).

3. Multipliers. We first prove a theorem conjectured by Chowla [4] in 1945. We wish to thank Dr. Gerald Estrin for a valuable suggestion contributed to the proof.

     3.1. THEOREM. *$p|n$ implies p is a multiplier.*

Proof. For convenient reference, we list several ideas:

(3.1.1)   If a and b are non-negative integers and m is a positive integer, then $a \equiv b \pmod{m}$ if and only if $x^a \equiv x^b \pmod{x^m - 1}$.

(3.1.2)   If $f(x)$ is a polynomial all of whose coefficients are non-negative, and if $g(x)$ is a polynomial of degree less than m, then $f(x) \equiv g(x) \pmod{x^m - 1}$ implies that the coefficients of $g(x)$ are non-negative.

(3.1.3) $f(x) \equiv g(x) \pmod{x^m - 1}$ implies that the sum of the coefficients of $f(x)$ equals the sum of the coefficients of $g(x)$.

(3.1.4)   If $d \mid m$, $f(x) = 1 + x^d + \ldots + x^{(m/d-1)d}$, and $g(x)$ is any polynomial, then there is a polynomial $g_1(x)$ of degree less than d such that $f(x)\, g(x) \equiv f(x)\, g_1(x) \pmod{x^m - 1}$.

(3.1.5)   In the special case of (3.1.4) in which all the non-zero terms of $g(x)$ are of the form cx^{kd}, we have $g(x)\, f(x) \equiv gf(x) \pmod{x^m - 1}$, where g is the sum of the coefficients of $g(x)$.

Now for the proof proper. We may assume that $0 < d_i < N$ (recall that $\{d_\nu\}$ is a standard difference set), and that p is a prime. Let $\theta(x) = x^{d_1} + \ldots x^{d_n}$. Then

$$(3.1.6) \qquad \theta(x)\theta(x^{N-1}) \equiv n + P(x)(R(x) - 1) \qquad \mathrm{mod}\; x^N - 1,$$

where $P(x) = 1 + x^{n+1} + \ldots + x^{(n-2)(n+1)}$, and

$$R(x) = 1 + x + \ldots + x^n.$$

Since p is prime to $n^2 - 1$, the numbers $pd_1, \ldots, pd_n$ form an affine difference set; hence by (3.1.1)

$$(3.1.7) \qquad \theta(x^p)\theta(x^{(N-1)p}) \equiv n + P(x)(R(x) - 1) \qquad \mathrm{mod}\; x^N - 1.$$

$p \mid n$ and $P(x) \mid x^N - 1$, so we may change the modulus of (3.1.6) to the double modulus $p, P(x)$, obtaining

$$(3.1.8) \qquad \theta(x)\theta(x^{N-1}) \equiv 0 \qquad \mathrm{modd}\; p, P(x).$$

Hence, $\theta(x^p)\theta(x^{N-1}) \equiv \theta(x)^p\theta(x^{N-1}) \equiv \theta(x)^{p-1}\theta(x)\theta(x^{N-1}) \equiv 0 \pmod{\mathrm{d}\; p, P(x)}$, which can be expressed as

$$(3.1.9) \qquad \theta(x^p)\theta(x^{N-1}) \equiv pf(x) + P(x)g(x) \qquad \mathrm{mod}\; x^N - 1.$$

By (3.1.4), we may assume $g(x) = g_0 + g_1 x + \ldots + g_n x^n$, and because of the presence of the term $pf(x)$, we may take $0 \leqslant g_1 \leqslant p - 1$. Further, writing $f(x) = C_0 + C_1 x + \ldots + C_{N-1}x^{N-1}$, we have $C_i \geqslant 0$, by (3.1.2). Since $R(x) \mid x^{n+1} - 1$, and $P(x) \equiv n - 1 \pmod{x^{n+1} - 1}$, (3.1.9) yields

$$(3.1.10) \qquad \theta(x^p)\theta(x^{N-1}) \equiv -g(x) \qquad \mathrm{modd}\; p, R(x).$$

On the other hand, since $d_1, \ldots, d_n$ is a standard difference set,

$$\theta(x) \equiv R(x) - 1 \qquad \mathrm{mod}\; x^{n+1} - 1;$$

a fortiori,

$$(3.1.11) \qquad \theta(x) \equiv -1 \qquad \mathrm{modd}\; p, R(x).$$

                                           A. J. HOFFMAN

From (3.1.6), we obtain

$$\theta(x)\theta(x^{N-1}) \equiv 1 \qquad \text{modd } p, R(x),$$

which combines with (3.1.11) to give

$$(3.1.12) \qquad \theta(x^p)\theta(x^{N-1}) \equiv (-1)^{p-1} \equiv 1 \qquad \text{modd } p, R(x).$$

Hence the right side of (3.1.10) is congruent to the right side of (3.1.12) (modd p, $R(x)$), so using (3.1.5) with $d = 1$, we have

$$(3.1.13) \qquad g(x) + 1 \equiv ph(x) + kR(x) \qquad \text{mod } x^{n+1} - 1,$$

where k is an integer (which we may take $0 \leqslant k \leqslant p - 1$) and $h(x)$ is some polynomial. Hence, $g_1 = \ldots = g_n = k \equiv g_0 + 1 \pmod{p}$. We cannot have $g_0 + 1 = p$, for from (3.1.9) and (3.1.3), this would imply

$$n^2 = pf + (n-1)(p-1),$$

where f is the sum of the coefficients of $f(x)$; that is $p \mid (n-1)(p-1)$, which is impossible. Therefore, $g_0 + 1 = k$, and

$$n^2 = pf + (n-1)(nk + k - 1);$$

that is, $p \mid k - 1$, so $k = 1$, $f = n/p$, $g(x) = R(x) - 1$. Therefore (3.1.9) can be rewritten

$$(3.1.14) \qquad \theta(x^p)\theta(x^{N-1}) \equiv pf(x) + P(x)(R(x) - 1) \qquad \text{mod } x^N - 1.$$

Replace x in (3.1.14) by x^{N-1}, and from (3.1.1) obtain

$$(3.1.15) \qquad \theta(x)\theta(x^{(N-1)p}) = pf(x^{N-1}) + P(x)(R(x^{N-1}) - 1) \quad \text{mod } x^N - 1.$$

The product of the left-hand sides of (3.1.14) and (3.1.15) equals the product of the left-hand sides of (3.1.6) and (3.1.7). Hence, the respective right-hand products are congruent.

$$(3.1.16) \qquad n^2 + 2nP(x)(R(x) - 1) + P^2(x)(R(x) - 1)^2 \equiv p^2 f(x) f(x^{N-1})$$
$$+ pP(x)[f(x)(R(x^{N-1}) - 1) + f(x^{N-1})(R(x) - 1)]$$
$$+ P^2(x)(R(x) - 1)(R(x^{N-1}) - 1) \qquad \text{mod } x^N - 1.$$

But $P(x)R(x) \equiv P(x)R(x^{N-1}) \equiv 1 + x + \ldots + x^{N-1} \pmod{x^N - 1}$. Using this and (3.1.5), (3.1.16) becomes

$$(3.1.17) \quad n^2 - 2nP(x) \equiv p^2 f(x)f(x^{N-1}) - pP(x)[f(x) + f(x^{N-1})] \; \text{mod } x^N - 1.$$

Change the modulus to $x^{n+1} - 1$, and (3.1.17) reads

$$(3.1.18) \quad n^2 - 2n(n-1) \equiv p^2 \bar{f}(x)\bar{f}(x^{N-1}) - p(n-1)[\bar{f}(x) + \bar{f}(x^{N-1})]$$

$$\text{mod } x^{n+1} - 1,$$

where

$$\bar{f}(x) = \sum_{i=0}^{n} e_i x^i, \quad e_i = \sum_{j=0}^{n-2} C_{j(n+1)+i}.$$

The term on the left of (3.1.18) must be the same as the constant term on the right of (3.1.18) after reduction mod $x^{n+1} - 1$. Therefore,

$$n^2 - 2n(n - 1) = p^2 \sum_{i=0}^{n} e_i^2 - 2p(n - 1)e_0.$$

Add $(n - 1)^2$ to both sides. Then

$$1 = p^2 \sum_{i=1}^{n} e_i^2 + [pe_0 - (n - 1)]^2.$$

Hence $e_1 = \ldots = e_n = 0$. Therefore,

$$f(x) = C_0 + C_{n+1}x^{n+1} + \ldots + C_{(n-2)(n+1)}x^{(n-2)(n+1)}.$$

By (3.1.5), this implies

$$P(x)f(x) \equiv P(x)f(x^{N-1}) \equiv \frac{n}{p}P(x) \qquad \mathrm{mod}\ x^N - 1.$$

Therefore, (3.1.17) becomes

$$n^2 \equiv p^2 f(x)f(x^{N-1}) \qquad \mathrm{mod}\ x^N - 1,$$

and recalling that the coefficients of $f(x)$ are non-negative, this obviously implies that $f(x)$ consists of a single term, say

$$f(x) = \frac{n}{p}x^{t(n+1)}.$$

Substituting in (3.1.14),

$$(3.1.19) \qquad \theta(x^p)\theta(x^{N-1}) \equiv nx^{t(n+1)} + P(x)(R(x) - 1) \qquad \mathrm{mod}\ x^N - 1.$$

But by (3.1.1), this means that n of the differences $pd_\alpha - d_\beta$ have the same value mod N, namely, $t(n + 1)$. Further, if $pd_\alpha - d_\beta \equiv pd_\mu - d_\nu \pmod{N}$, then $\alpha = \mu$ if and only if $\beta = \nu$. Hence the set of numbers $pd_1, \ldots, pd_n$ is a rearrangement of $d_1 + t(n + 1), \ldots, d_n + t(n + 1) \pmod{N}$. By (2.5), p is a multiplier.

3.2. **THEOREM.** *Let σ be the collineation of Π corresponding to the multiplier s. The fixed elements of σ form a sub-plane Π_1 if and only if $(s - 1, n + 1) \geqslant 3$. In this case, $(s - 1, N) = ((s - 1, n + 1) - 1)^2 - 1$.*

Proof. Since σ leaves X fixed, the set of fixed elements forms a sub-plane Π_1, if and only if σ fixes at least three lines of (2.1), that is, if and only if $sx \equiv x \pmod{n + 1}$ has at least three solutions, which is equivalent to $(s - 1, n + 1) \geqslant 3$. The second part of the theorem follows from a simple counting.

It is also possible to show that Π_1 is cyclic, and every multiplier of Π is also a multiplier of Π_1, but we omit the details.

3.3. **THEOREM.** *There is always at least one line of (2.2) left fixed by all multipliers.*

Proof. Let n be even. Then 2 is a multiplier, and the corresponding collineation fixes $X, 0$, the intersection of $X0$ and the line at infinity, and no other points

                          A. J. HOFFMAN

of the projective extension of Π. It fixes the line $X0$ and the line at infinity, so by a theorem of Baer [1, p. 155] exactly one other line must be left fixed. This line must be parallel to $X0$ (and hence belong to (2.2)), or another point would be fixed. By the reasoning of Hall [6, p. 1089], this line is fixed by all multipliers.

Let n be odd. Then S is a multiplier only if S is odd, so the $\frac{1}{2}(n+1)$th line of (2.1) is fixed by all multipliers. Therefore the line of (2.2) containing 0, parallel to this line, is fixed by all multipliers.

4. Non-existence theorems. It is known from the projective case that the preceding results imply that, for various values of n, there is no cyclic affine plane with n points on a line.

4.1. COROLLARY. *There is no cyclic affine plane with n points on a line if n is divisible by both of the primes in any one of the following pairs:*

(2,3), (2,5), (2,7), (2,11), (2,13), (2,17), (2,19), (2,23), (2,29), (2,31), (2,47), (2,61), (2,67), (2,71), (2,79), (3,5), (3,7), (3,11), (3,13), (3,17), (3,19), (3,29), (5,7), (5,11), (5,13), (5,29).
Proof as in [6, p. 1089].

4.2. COROLLARY. *There is no cyclic affine plane with n points on a line if there exist primes p and q such that* $n \equiv 1 \pmod{p}$, $p \equiv 3 \pmod 4$, *q divides the square-free part of n and* $(-p|q) = 1$.

4.3. COROLLARY. *There is no cyclic affine plane with n points on a line if n is not a square and there is an odd prime p such that* (i) $n \equiv 1 \pmod{p}$, *and* (ii) *some product of divisors of n is a primitive root of p.*

Proof. Let $\epsilon = e^{2\pi i/p}$. Substituting in (3.6), we have $\theta(\epsilon)\,\overline{\theta(\epsilon)} = \theta(\epsilon)\,\theta(\epsilon^{-1}) = n$; 4.2 then follows from the method of Chowla and Ryser [4, p. 95]; 4.3 follows from the method of Hall [6, p. 1089].
The theorems of this section, and the celebrated theorem of Bruck and Ryser [3], which established the non-existence of affine planes (whether cyclic or not) for various values of the number of points on a line, along with 3.2, are sufficient to decide the question of existence for all $n < 212$.

5. Remark. It is natural in the context of investigations on cyclic planes to inquire what can be said about an affine plane that admits a collineation cyclic on *all* its points. The answer is easily given: such a plane can only have two points on a line. We leave to the reader the proof that no infinite plane has this property. Here is a sketch of the proof for finite planes, with n points on a line:
Designating the points of the plane by residue classes mod n^2 in the familiar manner, it turns out, using the theorem of Baer quoted in 3.3 and the methods of §2, that the plane contains one pencil of n parallel lines such that the ith line consists of all residues congruent to $i \pmod{n}$. Further, if l is any other line, with points $d_0, \ldots, d_{n-1}$, then the differences $d_\alpha - d_\beta$ $(\alpha \neq \beta)$ yield all residues mod n^2 exactly once, except multiples of n. Accordingly, we choose our notation

so that $d_i \equiv i \pmod{n}$. Now precisely n of these differences will yield residues mod n^2 which are congruent to 1 (mod n), namely,

$$(5.1) \qquad \begin{aligned} d_1 - d_0 &\equiv a_0 n + 1, \\ d_2 - d_1 &\equiv a_1 n + 1, \\ &\cdots \qquad \cdots \\ d_0 - d_{n-1} &\equiv a_{n-1} n + 1 \end{aligned} \qquad \mathrm{mod}\ n^2.$$

The a^v form a complete system of residues mod n. Adding equations (5.1), we obtain

$$0 \equiv \tfrac{1}{2} n^2(n - 1) + n \qquad \mathrm{mod}\ n^2,$$

which is impossible if $n > 2$. For $n = 2$, such a collineation clearly exists.

References

1. R. Baer, *Projectivities of finite projective planes*, Amer. J. Math., vol. 69 (1947), 653-684.
2. R. C. Bose, *An affine analogue of Singer's Theorem*, J. Indian Math. Soc., vol. 6 (1942), 1-15.
3. R. H. Bruck and H. J. Ryser, *The nonexistence of certain finite projective planes*, Can. J. Math., vol. 1 (1949), 88-93.
4. S. Chowla, *On difference-sets*, J. Indian Math. Soc., vol. 9 (1945), 28-31.
5. S. Chowla and H. J. Ryser, *Combinatorial problems*, Can. J. Math., vol. 2 (1950), 93-99.
6. M. Hall, *Cyclic projective planes*, Duke Math. J., vol. 14 (1947), 1079-1090.
7. J. Singer, *A theorem in finite projective geometry and some applications to number theory*, Trans. Amer. Math. Soc., vol. 43 (1938), 377-385.

Institute for Advanced Study
Princeton, N.J.

Reprinted from
Pacific J. Math., Vol. 6, No. 1 (1956) pp. 83–96

ON THE NUMBER OF ABSOLUTE POINTS OF
A CORRELATION

A. J. Hoffman, M. Newman, E. G. Straus
and O. Taussky

1. Introduction. In 1948, R. W. Ball [2] presented methods for obtaining information about the number of absolute points of a correlation of a finite projective plane in which neither the theorem of Desargues nor any other special property (except, of course, the existence of the correlation) is assumed. This work was, in a sense, a continuation of an earlier investigation by R. Baer [1] of the case that the correlation is a polarity.

We shall show how, using an incidence-matrix approach[1], one may obtain the principal results of [2] somewhat more directly. Some of the results are strengthened. In addition, our method is sufficiently general to apply at once to the so-called symmetric group divisible designs, a class of combinatorial configurations including the finite projective planes. For simplicity, we shall present our main discussion in the language of planes, reserving to the end indications of the generalization.

As pointed out in §§ 3 and 4 the geometric problem with which we are concerned leads naturally to the question: What are the irreducible polynomials whose roots are roots of natural numbers? This question is treated in the following section.

2. Polynomials whose roots are roots of natural numbers. Let $f(x)$ be an irreducible polynomial with integral coefficients and let one of its roots be $z = n^{1/k}\zeta$, (n, k natural numbers, ζ a root of unity). Clearly z satisfies the equation

$$(1) \qquad z^k/n = \zeta^k = \zeta_h$$

for some h, where from now on we use ζ_h to denote a primitive hth root of unity. From (1) we see that $\Phi_h(z^k/n) = 0$, where Φ_h is the cyclotomic polynomial of order h. Hence

$$(2) \qquad f(x) \,|\, n^{\varphi(h)}\Phi_h(x^k/n) \,.$$

. The problem is therefore reduced to that of finding the irreducible factors of $\Phi_h(x^k/n)$ for arbitrary positive integers h, k, n. It will suffice

Received August 16, 1954. The work of the first two authors was supported (in part) by the Office of Naval Research.

[1] Arithmetic properties of the incidence matrix have been exploited with conspicuous success ([4], [5]). In this paper we study its characteristic polynomial.

for our purpose here to consider only the reducibility of $\Phi_h(x^2/n)$ (that is the case $k=2$). The general case is settled in the note following this paper [9].

If $n^{\varphi(h)}\Phi_h(x^2/n)$ is divisible by an irreducible polynomial $g(x)$, then $g(x)$ is not a polynomial in x^2. Hence $g(-x)$, which also divides $n^{\varphi(h)}\Phi_h(x^2/n)$, is different from $g(x)$ and is irreducible. Therefore,

$$(3) \qquad n^{\varphi(h)}\Phi_h(x^2/n) = \pm g(x)g(-x) ,$$

for $g(x)g(-x)$ is a polynomial in x^2, and $n^{\varphi(h)}\Phi_h(x^2/n)$ is irreducible in x^2. Then by (3), $\sqrt{n\zeta_h}$ or $-\sqrt{n\zeta_h}$ is a root of $g(x)$; thus $\zeta_h = (\pm\sqrt{n\zeta_h})^2/n$ is in the splitting field for $g(x)$. Thus the splitting field for $g(x)$ contains the hth roots of unity; but by (3), the degree of this splitting field is $\varphi(h)$. Therefore the splitting field for $g(x)$ is the same as $R(\zeta_h)$. Conversely since $\sqrt{n\zeta_h}$ is a root of $\Phi_h(x^2/n)$, $\sqrt{n\zeta_h} \in R(\zeta_h)$ implies that $\Phi_h(x^2/n)$ is reducible. We are thus led to the following lemma:

LEMMA 1. *The polynomial $\Phi_h(x^2/n)$ is reducible if and only if $\sqrt{n\zeta_h}$ is contained in $R(\zeta_h)$.*

LEMMA 2. *The polynomial $\Phi_h(x^2/n)$ where $n=n^{*2}n'$, n' squarefree, is reducible if and only if $n'\,|\,h$ and one of the following conditions holds:*
      (a) $h \equiv 1 \pmod 2$ *and* $n' \equiv 1 \pmod 4$;
      (b) $h \equiv 2 \pmod 4$ *and* $n' \equiv 3 \pmod 4$;
      (c) $h \equiv 4 \pmod 8$ *and* $n' \equiv 0 \pmod 2$.

Proof. We first list for convenience several facts to which we shall make reference in the course of this proof and subsequently.

(i) The discriminant of a subfield of an algebraic number field divides the discriminant of the whole field [7, p. 95, Satz 39].

(ii) The discriminant of $R(\sqrt{m})$, m a squarefree integer, is $4m$ if $m \equiv 2, 3 \pmod 4$, and m if $m \equiv 1 \pmod 4$ [7, p. 157, Satz 95].

(iii) The discriminant of the field of the mth roots of unity is divisible only by primes which divide m [7, p. 146, Satz 88].

$$(iv) \qquad \sum_{j=0}^{m-1} \zeta_m^{j^2} = \begin{cases} (1+i)\sqrt{m} & \text{if} \quad m \equiv 0 \pmod 4 \\ \sqrt{m} & \text{if} \quad m \equiv 1 \pmod 4 \\ i\sqrt{m} & \text{if} \quad m \equiv 3 \pmod 4 \end{cases}$$

[8, p. 177, Theorem 99].

(v) If $(r, s)=1$, then $\zeta_r\zeta_s$ is a primitive rsth root of unity.

(vi) If m is odd and squarefree, $m\,|\,r$ then $\{(-1)^{(m-1)/2}m\}^{1/2} \in R(\zeta_r)$ (This can be shown in a variety of ways: for example, from (iv) or from (i), (ii), (iii)).

We now turn to the proof proper. We first prove the necessity. Assume $\Phi_h(x^2/n)$ is reducible; that is, by Lemma 1,

$$(4) \qquad \sqrt{n'\zeta_h} \in R(\zeta_h).$$

Therefore $\sqrt{n'} \in R(\sqrt{\zeta_h})$, so by (i), (ii), (iii), n' is the product of primes each of which divides $2h$. If h is even, then $n'|h$. If h is odd, then since $\varphi(2h)=\varphi(h)$, we have $R(\sqrt{\zeta_h})=R(\zeta_h)$, so that by (i), (ii) and (iii) we have again $n'|h$. Next,

(a) Assume h odd. Then $\sqrt{\zeta_h} \in R(\zeta_h)$, so that (4) implies $n' \in R(\zeta_h)$. Further n' is odd, since $n'|h$, so either $n' \equiv 1 \pmod 4$ or $n' \equiv 3 \pmod 4$. But we cannot have $n' \equiv 3 \pmod 4$, for, by (ii), (i), and (iii), this would imply $2|h$.

(b) Assume $h \equiv 2 \pmod 4$. Then, since $\varphi(2h) > \varphi(h)$, it follows that $\sqrt{\zeta_h} \notin R(\zeta_h)$, so $\sqrt{n'\zeta_h} \in R(\zeta_h)$ implies $\sqrt{n'} \notin R(\zeta_h)$. If n' is odd, this implies $n' \equiv 3 \pmod 4$, by the fact that $n'|h$ and (vi). Further n' cannot be even. If n' were even, write $n'=2n''$. There are two cases: $n'' \equiv 1 \pmod 4$, $n'' \equiv 3 \pmod 4$. If $n'' \equiv 1 \pmod 4$, then $\sqrt{n'\zeta_h}= \sqrt{2}\sqrt{n''}\sqrt{\zeta_h} \in R(\zeta_h)$ implies $\sqrt{2}\sqrt{\zeta_h} \in R(\zeta_h)$, since $\sqrt{n''} \in R(\zeta_h)$ by the fact that $n''|h$ and (vi). But this means $\sqrt{2} \in R(\zeta_h)$, which is impossible. For $R(\sqrt{\zeta_h})$ contains i; and if it also contains $\sqrt{2}$, it would contain $\zeta_8=(1+i)/\sqrt{2}$. By (v), it follows that $R(\sqrt{\zeta_h})$ would then contain a primitive $8(h/2)=4h$th root of unity; therefore the degree of $R(\zeta_h)$ would be at least $\varphi(4h) > \varphi(2h)$; the actual degree of $R(\sqrt{\zeta_h})$.

If $n'' \equiv 3 \pmod 4$, then $\sqrt{n''\zeta_h} \in R(\zeta_h)$, for $i\sqrt{n''} \in R(\zeta_h)$ by the fact that $n''|h$ and (vi), and it is easy to see (for example by (v)) that $i\sqrt{\zeta_h} \in R(\zeta_h)$. Therefere $\sqrt{n''\zeta_h}=(i\sqrt{n''})(-i\sqrt{\zeta_h}) \in R(\zeta_h)$. Hence, $\sqrt{2} \in R(\zeta_h)$, and a fortiori $\sqrt{2} \in R(\sqrt{\zeta_h})$, and the preceding argument applies.

(c) Assume finally $h \equiv 4 \pmod 8$. Then n' cannot be odd. For since $R(\zeta_h)$ contains i, and $n'|h$, we learn from (vi) that $\sqrt{n'} \in R(\zeta_h)$. Therefore, $\sqrt{n'\zeta_h} \in R(\zeta_h)$ implies $\sqrt{\zeta_h} \in R(\zeta_h)$, which is impossible, since $\varphi(2h) > \varphi(h)$.

It remains to show that if $h \equiv 0 \pmod 8$, then $\sqrt{n'\zeta_h} \notin R(\zeta_h)$ for any n'. The argument used in (c) shows that n' cannot be odd. If n' were even, $n'=2n''$, then since $\zeta_8=(1+i)/\sqrt{2}$ and $i \in R(\zeta_h)$, we have $\sqrt{2} \in R(\zeta_h)$. Hence $\sqrt{n'\zeta_h}=\sqrt{2n''\zeta_h} \in R(\zeta_h)$ implies $\sqrt{n''\zeta_h} \in R(\zeta_h)$. Then we may use the argument just given to cover the case in which n' is odd. Hence we cannot have $h \equiv 0 \pmod 8$.

The sufficiency is established simply by constructing $g(x)$ of (3). We first prove that in cases (a), (b)

$$(5) \qquad z=n^*\zeta_h \sum_{j=0}^{n'-1} \zeta_h^{hj^2/n'}=\frac{n^*n'}{h}\zeta_h \sum_{j=0}^{h-1} \zeta_h^{hj^2/n'}$$

is a zero of $\Phi_h(x^2/n)$. Since $\zeta_h^{h/n'}$ is a primitive n'th root of unity we obtain from (iv):

$$\text{in case (a)} \quad z=n^*\sqrt{n'}\zeta_h=\sqrt{n}\,\zeta_h=\text{a zero of } \Phi_h(x^2/n);$$

$$\text{in case (b)} \quad z=n^*\sqrt{n'}i\zeta_h=\sqrt{n}\,\zeta_4\zeta_h=\text{a zero of } \Phi_h(x^2/n).$$

In case (c) we have

$$(6) \qquad z=\frac{1}{2}n^*\zeta_h\sum_{j=0}^{2n'-1}\zeta_h^{hj^2/2n'}=\frac{1}{2}n^*\sqrt{2n'}(1+i)\zeta_h=n^*\sqrt{n'}\zeta_8\zeta_h,$$

a zero of $\Phi_h(x^2/n)$. The conjugates $z^{(l)}$ of z in $R(\zeta_h)$ are now obtained simply by substituting ζ_h^l for ζ_h in (5) or (6) where $(l,h)=1$. Thus we obtain

$$g(x)=\prod_{\substack{l=1\\(l,h)=1}}^{h-1}(x-z^{(l)}).$$

Later on we shall need the sum of the $z^{(l)}$. We therefore establish the following lemma:

LEMMA 3. *If* (3) *holds, then the sum of the roots of* $g(x)$ *is*
  (a) $\pm n^*n'$ *if* $h\not\equiv 0$ (mod 4) *and squarefree,*
  (b) 0 *if* $h\not\equiv 0$ (mod 4) *and* h *is not squarefree,*
  (c) $\pm n^*n'$ *if* $h\equiv 0$ (mod 4) *and* $h/4$ *is odd and squarefree,*
  (d) 0 *if* $h\equiv 0$ (mod 4) *and* $h/4$ *is odd and not squarefree.*

Proof. Let us first note that, by Lemma 2, the foregoing enumeration accounts for all cases in which $n^{\varphi(h)}\Phi_h(x^2/n)$ may be reducible. Also, the $\pm$ in (a) and (c) is to be expected, since we are clearly unable to distinguish between $g(x)$ and $g(-x)$.

We now set $h=2^e p_1^{e_1}\cdots p_k^{e_k}$, $n'=2^\varepsilon p_1^{\varepsilon_1}\cdots p_k^{\varepsilon_k}$ where $\varepsilon, \varepsilon_i=0,1$; and write $h_0=2^e$, $h_i=p_i^{e_i}$; $n_0'=2^\varepsilon$, $n_i'=p_i^{\varepsilon_i}$; $\zeta_{(0)}=\zeta_{h_0}$, $\zeta_{(j)}=\zeta_{h_j}$.

Then $\zeta_h=\zeta_{(0)}\zeta_{(1)}\cdots\zeta_{(k)}$ and its conjugates ζ_h^l are the products of the conjugates $\zeta_{(0)}^{l_0}, \zeta_{(1)}^{l_1}, \cdots, \zeta_{(k)}^{l_k}$ where $l\equiv l_i$ (mod h_i). Cases (a), (b) of this lemma correspond to cases (a), (b) of Lemma 2. Here $\zeta_0=\pm 1$ so that we obtain from (5)

$$(7) \qquad \sum z^{(l)}=\pm n^*\sum_{j=0}^{n'-1}\prod_{i=1}^{k}\sum_{\substack{l_i=1\\(l_i,p_i)=1}}^{h_i-1}\zeta_{(i)}^{l_i[hj^2/n'+1]}.$$

As j runs from 0 to $n'-1$ its residues (mod n_i') run independently from 0 to $n_i'-1$; hence we can write

$$(8) \qquad a = \sum z^{(l)} = \pm n^* \prod_{i=1}^{k} \sum_{j_i=0}^{n_i'-1} \sum_{\substack{l_i=1 \\ (l_i,\, p_i)=1}}^{h_i-1} \zeta_{(i)}^{l_i [h j_i^2/n'+1]} = \pm n^* a_1 \cdots a_k \ .$$

In order to evaluate the a_i we first observe that the sum of the primitive mth roots of unity

$$(9) \qquad \sum_{\substack{l=1 \\ (l,m)=1}}^{m-1} \zeta_m^l = \mu(m) \ .$$

This is seen most simply by observing that

$$\Phi_m(x) = \prod_{d \mid m} (x^d - 1)^{\mu(m/d)} = x^{\varphi(m)} - \mu(m) x^{\varphi(m)-1} + \cdots \ .$$

Now for $h_i > n_i'$ we have $\zeta_{(i)}^{[h j^2/n'+1]}$ a primitive h_ith root of unity and therefore

$$(10) \qquad a_i = \sum_{j_i=0}^{n_i'-1} \sum_{\substack{l_i=1 \\ (l_i,\, p_i)=1}}^{h_i-1} \zeta_{(i)}^{l_i [h j_i^2/n'+1]} = \sum_{j_i=0}^{n_i'-1} \mu(h_i) = n_i' \mu(h_i) \ .$$

For $h_i = n_i'$ we have h/n' relatively prime to p_i so that

$$(11) \qquad \sum_{j_i=0}^{n_i'-1} \zeta_{(i)}^{h j_i^2/n'} = \pm \sum_{j_i=0}^{n_i'-1} \zeta_{(i)}^{j_i^2} = \pm \begin{cases} \sqrt{p_i} & \text{if } p_i \equiv 1 \ (\text{mod } 4) \\ i\sqrt{p_i} & \text{if } p_i \equiv 3 \ (\text{mod } 4) . \end{cases}$$

Where the sign depends on whether h/n' is or is not a quadratic residue (mod p_i). Similarly

$$(12) \qquad \sum_{j_i=0}^{n_i'-1} \zeta_{(i)}^{l_i j_i^2} = \left(\frac{l_i}{p_i}\right) \sum_{j_i=0}^{n_i'-1} \zeta_{(i)}^{j_i^2} \ , \qquad \left(\frac{l_i}{p_i}\right) = \text{Legendre symbol.}$$

From (11) and (12) we obtain

$$(13) \qquad a_i = \pm \sum_{\substack{l_i=1 \\ (l_i,\, p_i)=1}}^{n_i'-1} \left(\frac{l_i}{p_i}\right) \zeta_{(i)}^{l_i} \sum_{j_i=0}^{n_i'-1} \zeta_{(i)}^{j_i^2} \ .$$

Now

$$(14) \qquad \sum \left(\frac{l_i}{p_i}\right) \zeta_{(i)}^{l_i} = \sum_1 \zeta_{(i)}^s - \sum_2 \zeta_{(i)}^t \ ,$$

where $\sum_1$ ranges over those s in $1, \cdots, p_i-1$ which are quadratic residues (mod p_i) and $\sum_2$ ranges over those t in $1, \cdots, p_i-1$, which are quadratic nonresidues (mod p_i). According to (9)

$$(15) \qquad \sum_1 \zeta_{(i)}^s + \sum_2 \zeta_{(i)}^t = \mu(p_i) = -1$$

and obviously

$$(16) \qquad \sum_{j_i=0}^{n_i'-1} \zeta_{(i)}^{j_i^2} = 1 + 2 \sum_1 \zeta_{(i)}^s \ .$$

Combining (15) and (16) we have

$$(17) \qquad \sum_1 \zeta_{(i)}^s - \sum_2 \zeta_{(i)}^t = \sum_{j_i=0}^{n_i'-1} \zeta_{(i)}^{j_i^2} \ .$$

Substitution in (13) now yields

$$(18) \qquad a_i = \pm \left(\sum_{j_i=0}^{n_i'-1} \zeta_{(i)}^{j_i^2} \right)^2 = \pm p_i = \mp n_i' \mu(h_i) \ .$$

From (8), (10) and (18) we now obtain

$$(19) \qquad a = \pm n^* n' \mu(h) \ ,$$

which proves cases (a), (b). In cases (c), (d) we have case (c) of Lemma (2) and therefore equation (6) obtains. We now have $a = \pm n^* a_0 a_1 \cdots a_k$ where $a_1, \cdots, a_k$ are the same as in (10) and (18). The only new factor is according to (6)

$$(20) \qquad a_0 = \frac{1}{2} \sum_{\substack{l_0=1 \\ l_0 \text{ odd}}}^{3} \sum_{j_0=0}^{3} \zeta_{(0)}^{l_0[h_0 j_0^2/4 + 1]} \ .$$

If $h_0 > 4$ then, as in (10), we obtain

$$(21) \qquad a_0 = 2\mu(h_0) = n_0 \mu(h_0) = n_0 \mu(h_0/2) = 0 \ .$$

If $h_0 = 4$ then $\zeta_{(0)} = i$ and

$$(22) \qquad a_0 = \frac{1}{2} [\zeta_{(0)} + \zeta_{(0)}^2 + \zeta_{(0)}^5 + \zeta_{(0)}^{10} + \zeta_{(0)}^3 + \zeta_{(0)}^6 + \zeta_{(0)}^{15} + \zeta_{(0)}^{30}] = -2 = n_0 \mu(h_0/2) \ .$$

Thus, finally, in cases (c), (d)

$$(23) \qquad a = \pm n^* n' \mu(h/2)$$

which proves these cases.

3. **The incidence matrix.** We assume that we have a finite projective plane $\mathit{\Pi}$ with $n+1$ points on a line, $n > 1$, and consequently $N = n^2 + n + 1$ points in the plane. We further assume that the plane admits a correlation ρ, that is a one-to-one mapping of the set of points of $\mathit{\Pi}$ onto the set of lines of $\mathit{\Pi}$, together with a one-to-one mapping of the set of lines of $\mathit{\Pi}$ onto the set of points of $\mathit{\Pi}$ such that a point is on a line if and only if the image of the point is on the image of the line.

Our attack on the study of the number of absolute points of a correlation, that is, the set of points each of which lies on its image, is based on the following:

LEMMA 4. *Let ρ be a correlation of a finite projective plane Π, and let the points $P_1, \cdots, P_N$ and lines $l_1, \cdots, l_N$ of Π be so numbered that $\rho P_i = l_i$ ($i = 1, \cdots, N$). Let $A = (a_{ij})$ be a square matrix of order N defined by the rule $a_{ij} = 1$ if P_i is on l_j, and 0 otherwise, and let $P = (p_{ij})$ be a permutation matrix defined by $p_{ij} = 1$ if $\rho^2 P_i = P_j$, and 0 otherwise. Then if A^T denotes the transpose of A, we have* (i) $A^T = PA$, *and* (ii) *the number of absolute points of ρ is* tr A *(the trace of A).*

Proof. The second part of the lemma is immediate. To prove (i), observe that the (i, j)th element of A^T is $1 \Leftrightarrow a_{ji} = 1 \Leftrightarrow P_j$ is on $l_i \Leftrightarrow l_j = \rho P_j$ is on $\rho l_i = \rho^2 P_i$. But from the definition of P, the (i, j)th element of PA, is $1 \Leftrightarrow \rho^2 P_i$ is on l_j. Hence $A^T = PA$.

Of course, it is also true that if A is an incidence matrix of a finite projective plane, and there exists a permutation matrix $P = (p_{ij})$ such that $A^T = PA$, then the mappings $P_i \rightarrow l_i$; $l_i \rightarrow P_j$, where $p_{ij} = 1$, define a correlation.

Because of (ii), it is clear that knowledge of the eigenvalues of A will contribute to the solution of our problem. Now, $A^T = PA$ implies A is normal. For if $A^T = PA$, then $A = A^T P^T$. Hence $AA^T = A^T P^T P A = A^T A$. Thus the eigenvalues of AA^T are the squares of the moduli of the eigenvalues of A. But the eigenvalues of AA^T can easily be computed from the fact that the incidence properties of a plane imply

$$(24) \qquad\qquad AA^T = nI + J$$

where I is the identity matrix and J is the matrix every element of which is unity [4]. The eigenvalues of AA^T are

$$(25) \qquad\qquad (n+1)^2, n, n, \cdots, n \;.$$

But by (24), $n+1$ is an eigenvalue of A with $(1, 1, \cdots, 1)$ as corresponding eigenvector; hence the eigenvalues of A are

$$(26) \qquad\qquad n+1, \sqrt{n}\, e^{i\alpha_1}, \sqrt{n}\, e^{i\alpha_2}, \cdots, \sqrt{n}\, e^{i\alpha_{N-1}}$$

Let the permutation P split up into cycles of length $d_1, d_2, \cdots, d_r$; $d_1 + d_2 + \cdots + d_r = N$. Then the eigenvalues of P are the d_1th roots of unity, the d_2th roots of unity, $\cdots$, and the d_rth roots of unity. If we write out these eigenvalues of P as

$$(27) \qquad\qquad 1, e^{i\theta_1}, e^{i\theta_2}, \cdots, e^{i\theta_{N-1}}$$

then it follows from $A^T A^{-1} = P$, the normality of A, (26), and (27) that

          A. J. HOFFMAN, M. NEWMAN, E. G. STRAUS AND O. TAUSSKY

$$(28) \qquad\qquad e^{-i\theta_j}=e^{2i\alpha_j} \qquad\qquad j=1, 2, \cdots, N-1.$$

These elementary consideraticns alone suffice to prove the following:

THEOREM 1 (see [2, Theorem 2.1] and [1, Theorem 4]). *If $n=n^{*2}n'$, where n' is squarefree, and M is the number of absolute points of ρ, then $M \equiv 1 \pmod{n^*n'}$.*

Proof. By (26) and Lemma 4, we have

$$(29) \qquad\qquad M=n+1+\sqrt{n}\,t\,,$$

where $t=\sum_{j=1}^{N-1}e^{i\alpha_j}$ is an algebraic integer, by (27) and (28). Therefore, $(M-(n+1))^2 \equiv 0 \pmod{n}$, which implies the theorem.

4. The characteristic polynomial. By virtue of (26), the characteristic polynomial of A may be written

$$(30) \qquad\qquad (x-(n+1))Q(x)\,,$$

where $Q(x)=(x-\sqrt{n}\,e^{i\alpha_1})(x-\sqrt{n}\,e^{i\alpha_2})\cdots(x-\sqrt{n}\,e^{i\alpha_{N-1}})$. Then since $N-1=n^2+n$ is even, we have

$$(31) \qquad Q(x)Q(-x)=(x^2-ne^{2i\alpha_1})(x^2-ne^{2i\alpha_2})\cdots(x^2-ne^{2i\alpha_{N-1}})\,.$$

From (27), the fact that the complex conjugate of a dth root of unity is a dth root of unity, and the definition of $d_1, d_2, \cdots, d_r$, we may write the characteristic polynomial of P as

$$(32) \qquad \prod_{i=1}^{r}(x^{d_i}-1)=(x-1)(x-e^{-i\theta_1})(x-e^{-i\theta_2})\cdots(x-e^{-i\theta_{N-1}})\,.$$

In (22), replace x by x^2/n and multiply both sides by n^N. There results

$$(33) \qquad \prod_{i=1}^{r}(x^{2d_i}-n^{d_i})=(x^2-n)(x^2-ne^{-i\theta_1})\cdots(x^2-ne^{-i\theta_{N-1}})\,.$$

Comparing (33) and (31) we deduce

$$(34) \qquad \frac{1}{x^2-n}\prod_{i=1}^{r}(x^{2d_i}-n^{d_i})=Q(x)Q(-x)\,,$$

so that the irreducible factors of $Q(x)$ are of the type discussed in §2.

5. The number of absolute points of ρ. In this section we apply the results of §2 to present criteria sufficient to insure that $M=n+1$. If we write

$$Q(x) = x^{N-1} + ax^{N-2} + bx^{N-3} + \cdots,$$

Then by (30), $M = n+1-a$.

We wish to prove that, under certain circumstances, $a=0$, and this will certainly hold if every irreducible factor of the left side of (34) is a polynomial in x^2. These factors are the irreducible factors of $\Phi_h(x^2/n)$, $h|d_i$, which were investigated in §2.

On the basis of Lemma 2, we can assert the following.

THEOREM 2. *If, for each divisor of the orders $d_1, d_2, \cdots, d_r$ of the cycles of P, none of the conditions of Lemma 2 holds, then $M = n+1$. In particular* (see [2]), *$M = n+1$ if n' and $d = $ l.c.m. $\{d_i\}$ satisfy one of the following*:

(a) $n' \nmid d$;

(b) $2n' \nmid d$ and $n' \not\equiv 1 \pmod 4$;

(c) *there exist odd primes p and q such that $p \equiv q \pmod{2d}$ and $(n'/p)(n'/q) = -1$, where (a/b) is the generalized Legendre-Jacobi symbol*;

(d) $d = 1, 2,$ *or p^k, where p is a prime $\equiv 3 \pmod 4$, k a positive integer, $n' > 1$.*

Proof. The principal statement is an immediate consequence of Lemma 2.

Proof of (a): Since $n' \nmid d$ implies $n' \nmid h$ for any $h|d_i$, the irreducibility of each $\Phi_h(x^2/n)$ follows from Lemma 2.

Proof of (b): Assume (b) false. Then by virtue of (a), we may assume there exists a positive integer h such that for some d_i we have $n'|h|d_i$, and $\Phi_h(x^2/n)$ reducible. If h is odd, then we obtain the contradiction $n' \equiv 1 \pmod 4$ by Lemma 2. If h is even, then n' must be even, otherwise $2n'|h$. But by Lemma 2 (c), n' even implies $h \equiv 0 \pmod 8$, hence we are forced to the contradiction $2n'|h$.

Proof of (c): We have $(n'/p)(n'/q) = -1$. Assume $\Phi_h(x^2/n)$ reducible for some $h|d$. Then if h is odd, $n' \equiv 1 \pmod 4$, thus $(n'/p) = (p/n')$, $(n'/q) = (q/n')$, by the quadratic reciprocity law. Hence $-1 = (n'/p)(n'/q) = (p/n')(q/n')$. But $p \equiv q \pmod{2d}$ implies $p \equiv q \pmod{n'}$, since $n'|h|d$. Therefore $(p/n') = (q/n')$. Combined with $-1 = (p/n')(q/n')$, this yields a contradiction.

Now let h be even, $h \equiv 2 \pmod 4$. Then by Lemma 2 (b), $n' \equiv 3 \pmod 4$. By the quadratic reciprocity law

$$-1 = (n'/p)(n'/q) = (-1)^{(p+q-2)/2}$$

implies $p + q \equiv 0 \pmod 4$.

But $p \equiv q \pmod{2d}$ implies $p - q \equiv 0 \pmod 4$, since $h|d$. Therefore, $2p \equiv 0 \pmod 4$, contrary to the fact that p is an odd prime.

92   A. J. HOFFMAN, M. NEWMAN, E. G. STRAUS AND O. TAUSSKY

Finally, let $h \equiv 0$ (mod 4). Then by Lemma 2 (c), n' is even. Write $n' = 2n''$. Then

$$-1 = (n'/p)(n'/q) = (2/p)(2/q)(n''/p)(n''/q) = (n''/p)(n''/q) \, ,$$

since $p \equiv q$ (mod 8). If $n'' \equiv 1$ (mod 4), we obtain a contradiction as in the first case considered above. If $n'' \equiv 3$ (mod 4), we obtain a contradiction as in the second case. Note that the hypothesis $p \equiv q$ (mod d) (instead of $p \equiv q$ (mod $2d$)) is sufficient in all cases except when simultaneously $n' \equiv 3$ (mod 4) and $d \not\equiv 0$ (mod 4).

Proof of (d): If $d = 1$ (see [1, Theorem 6]) or $d = 2$, then the only $h \mid d$ are $h = 1$ or $h = 2$. If $h = 1$ we cannot have $n' \mid h$. If $h = 2$, then $n' \mid h$ implies $n' = 2$, contrary to Lemma 2 (b). If $d = p^k$, p a prime $\equiv 3$ (mod 4), then $h \mid d$ implies h is also of this form. Assume now $\Phi_h(x^2/n)$ reducible. Since $n' \mid h$, $n' = p$. By Lemma 2 (b), this implies h is even, a contradiction.

Even in case one or more of the polynomials $n^{\varphi(h)} \Phi_h(x^2/n)$ where h divides some d_i is reducible, we may still obtain information about M. We can use the results of Lemma 3 as follows. Let $d_1, \cdots, d_r$ be the lengths of the disjoint cycles of P. For each $i = 1, \cdots, r$ let k_i be defined as follows:

(i)   if $n' \equiv 1$ (mod 4), let k_i be the number of divisors of d_i each of which is odd, squarefree and a multiple of n';

(ii)   if $n' \equiv 3$ (mod 4), let k_i be the number of divisors of d_i each of which is even, squarefree and a multiple of n';

(iii)   if $n' \equiv 2$ (mod 4), let k_i be the number of divisors of d_i each of which is a multiple of n', and of the form $4t$, t odd and squarefree. Then we have the following theorem.

THEOREM 3.   *If k_i is defined as above, then $M = n + 1 + sn^*n'$, where*

$$-\sum_{i=1}^{r} k_i \leq s \leq \sum_{i=1}^{r} k_i \, . \quad \textit{Further,} \quad s \equiv \sum_{i=1}^{r} k_i \ (\text{mod } 2).$$

Proof. All that remains to be verified is the second sentence, which follows immediately from the fact that the sum of the roots of $Q(x)$ in (34) is the sum of $\sum k_i$ numbers $\pm n^*n'$.

6.   In this section, we compare the number of absolute points of ρ^j, where j is any number prime to twice the order of ρ^2, with the number of absolute points of ρ. The results obtained coincide with those of [2], so we shall merely sketch the present approach.

The index j in what follows is an integer prime to twice the order of $\rho^2 = 2d$. Let M_j be the number of absolute points of ρ^j, so that $M_1 = M$ in our previous notation. If we let $j = 2c + 1$, then $P^{-c}A$ is an incidence matrix for II that bears the same relation to ρ^j that A does to ρ. In particular, $M_j = \operatorname{tr} P^{-c} A$. Referring back to (26), (27), and (28), we see that

$$M_1 = n + 1 + \sqrt{n}\,(e^{i\alpha_1} + \cdots + e^{i\alpha_{N-1}})\,,$$

$$M_j = n + 1 + \sqrt{n}\,(e^{ij\alpha_1} + \cdots + e^{ij\alpha_{N-1}})\,.$$

But from Theorem 1, $n^{-1/2}(M_1 - (n+1))$ is of the form $u\sqrt{n'}$, where u is a rational integer. Further, if m is the least common multiple of the orders of the α's, then $n^{-1/2}(M_j - (n+1))$ is the image of $u\sqrt{n'}$ under the automorphism of $R(\zeta_m)$ which sends $\zeta_m \to \zeta_m^j$.

Now $m = d$ if d is odd, $m = 2d$ if d is even. In either case, however, the indices j considered correspond biuniquely to all automorphisms of $R(\zeta_m)$. Thus, if $M_1 \neq n + 1$ (so that we know $\sqrt{n'} \in R(\zeta_m)$), we have

$M_j = M_1$ if the automorphism $\zeta_m \to \zeta_m^j$ fixes $\sqrt{n'}$,

$M_j = 2(n+1) - M_1$ the automorphism $\zeta_m \to \zeta_m^j$ sends $\sqrt{n'}$ into $-\sqrt{n'}$.

One may use the Gauss sums of Lemma 2 (iv) to show explicitly that in general

$$M_j = (n'/j)(M_1 - (n+1)) + (n+1)\,,$$

where (n'/j) is defined to be 1 if $(j, n') > 1$. Among other things, this formula includes the equation $M_j = M_1$ if n is a square.

7. **We now show** how the preceding results may be extended to symmetric group divisible designs. (See [3] and [6] for a definition and discussion of the interesting properties of these designs.) For our purpose, it is appropriate to employ the following:

DEFINITION. A *symmetric group divisible design* Δ is a combinatorial configuration consisting of a set with v elements and v distinguished subsets such that

(i)   each subset is incident with exactly k elements, and

(ii)   the subsets can be partitioned into g groups, each group containing s subsets ($gs = v$), such that two distinct subsets in the same group have exactly λ_1 elements in common, two subsets in different groups have exactly λ_2 elements in common.

We assume that the design Δ admits a correlation ρ; that is, a one-to-one mapping of the elements of Δ onto the distinguished subsets of Δ, together with a one-to-one mapping of the subsets onto the elements such that an element is in a subset if and only if the image of the element contains the image of the subset. Now the existence of ρ implies that in the definition given above, we may interchange, in (i) and (ii) the words subset and element. Number the elements $E_1, E_2, \cdots, E_v$ such that $E_1, E_2, \cdots, E_s$ are the elements of the first group, E_{s+1}, E_{s+2}, $\cdots, E_{2s}$ are the elements of the second group, and so on. Number the subsets $S_1, S_2, \cdots, S_v$ so that $\rho E_i = S_i$. Define the incidence matrix $A = (a_{ij})$ of order v, by the stipulation $a_{ij} = 1$ if E_i is in S_j, 0 otherwise, and the permutation matrix $P = (p_{ij})$ such that $p_{ij} = 1$ if and only if $\rho^2 E_i = E_j$. Then as in the case of planes, we have

$$(35) \qquad A^T = PA, \text{ so } A \text{ is normal. Further}$$

$$(36) \qquad AA^T = (k - \lambda_1)I + (\lambda_1 - \lambda_2)K + \lambda_2 J ,$$

where I and J are as before, and K is the direct sum of g matrices of order s each of which consists entirely of 1's.

Our object, as before, is to obtain a count on the number of absolute points of $\rho = \operatorname{tr} A = M$.

Since the vector $(1, 1, \cdots, 1)$ is an eigenvector of A and A^T corresponding to the eigenvalue k, and is also an eigenvector of K with eigenvalue s, we have from (27) that $k^2 - \lambda_2 v = k - \lambda_1 + s(\lambda_1 - \lambda_2)$. Hence, we may compute [1] that

$$(37) \qquad |AA^T - xI| = (k^2 - x)[k + \lambda_1 + s(\lambda_1 - \lambda_2) - x]^{g-1}(k - \lambda_1 - x)^{v-g} .$$

Henceforth, let us assume $v > g > 1$. This is no restriction for the combinatorial configurations apparently so excluded are realized by allowing $\lambda_1 = \lambda_2$. (Indeed, the case $\lambda_1 = \lambda_2$ with the further trivial restrictions $v > k > \lambda_1 = \lambda_2 > 0$ is an important class of designs known as balanced symmetric incomplete block designs. Further, $\lambda_1 = \lambda_2 = 1$ characterizes finite projective planes.)

Because A is normal, the eigenvalues of AA^T are the squares of the moduli of the eigenvalues of A. Hence, by (37), the eigenvalues of A are

$$k, \sqrt{n_1}e^{i\alpha_1}, \sqrt{n_1}e^{i\alpha_2}, \cdots, \sqrt{n_1}e^{i\alpha_{g-1}}, \sqrt{n_2}e^{i\alpha_g}, \cdots, \sqrt{n_2}e^{i\alpha_{v-1}} ,$$

where $n_1 = k - \lambda_1 + s(\lambda_1 - \lambda_2)$, $n_2 = k - \lambda_1$.

On the other hand, if P is a product of disjoint cycles of lengths $d_1, d_2, \cdots, d_r$, $d_1 + \cdots + d_r = v$, then the eigenvalues of P are the d_1th roots of unity, the d_2th roots of unity, $\cdots$, the d_rth roots of unity, namely

$$(38) \qquad 1, e^{i\theta_1}, e^{i\theta_2}, \cdots, e^{i\theta_{v-1}} .$$

Now by (35) and (36) we have

$$(39) \qquad A^2 = (k - \lambda_1)P^T + (\lambda_1 - \lambda_2)P^T K + (\lambda_1 - \lambda_2)J .$$

Further, each of A, P^T, K, J commutes with the three others (for example, to check that P^T commutes with K multiply (39) on the left and right by P and apply (35)).   Hence all four of these normal matrices can be simultaneously diagonalized.   Let us imagine then that (39) is in diagonal form, and examine the diagonal elements.   Note that one eigenvalue of J is v, the rest are 0, and that g eigenvalues of K are s, the rest are 0.   Clearly, then, we have

$$(40) \qquad \begin{aligned} &k^2 = (k-\lambda_1)\cdot 1 + (\lambda_1-\lambda_2)\cdot s + \lambda_2\cdot v\ , \\ &n_1 e^{2i\alpha}{}_t = (k-\lambda_1)e^{-i\theta j}{}_t + (\lambda_1-\lambda_2)se^{-i\theta j}{}_t \end{aligned}$$

for $t=1, 2, \cdots, g-1$ and some $g-1$ indices j_t in the set $1, \cdots, v-1$, and also

$$(41) \qquad n_2 e^{2i\alpha}{}_\mu = (k-\lambda_1)e^{-i\theta j}{}_\mu$$

for $u=g, g+1, \cdots, v-1$, and $\{j_\mu\}$ the indices in $1, 2, \cdots, v-1$ not in $\{j_t\}$.

We contend that the $e^{-i\theta j}$ appearing in (40) can be partitioned into classes, each class consisting of a conjugate set of roots of unity.   For the characteristic polynomial of $P^T K$ is $(x-\zeta)x^{v-g}f(x)$, where

$$(42) \qquad f(x) = \prod_{t-1}^{g-1} (x - se^{-i\theta j}{}_t)\ .$$

But since $P^T K$ has rational coefficients, its characteristic polynomial is rational, hence $f(x)$ has rational coefficients.   Let $h(x)=s^{\varphi(h)}\varPhi_h(x/s)$ be the irreducible polynomial satisfied by $se^{-i\theta j_1}$;. that is, $e^{-i\theta j_1}$ is a primitive hth root of unity.   Then $h(x)$ and $f(x)$ have a root in common, so, by the irreducibility of $h(x)$, the set of roots of $f(x)$ contains all roots of $h(x)$, namely all numbers $s\zeta_h$.   Divide $f(x)$ by $h(x)$, apply the same argument to the quotient, and continue.   This verifies our statement.

We may now imitate our previous polynomial construction in § 4 for the case of planes as follows: If the characteristic polynomial of A is written as $(x-k)Q(x)$ and the characteristic polynomial of P as $\prod_{i=1}^{r}\varPhi_{a_i}(x)$, then from the foregoing we have

$$(43) \qquad \pm\, Q(x)Q(-x) = n_1^{g-1}n_2^{v-g} \prod_i \varPhi_{h_i}(x^2/n_1) \prod_j \varPhi_{h_j}(x^2/n_2)$$

where the h_i and h_j are divisors of the cycle lengths $d_1, d_2, \cdots, d_r$, $\sum_i \varphi(h_i)=g-1$, $\sum_j \varphi(h_j)=v-g$.   One can then proceed from (43) by the techniques previously used in studying the consequences of (34).

REFERENCES

1. R. Baer, *Polarities in finite projective planes*, Bull. Amer. Math. Soc., **52** (1946), 77–93.

2. R. W. Ball, *Dualities of finite projective planes*, Duke. Math. J., **15** (1948), 929.

3.   R. C. Bose and W. S. Connor, *Combinatorial properties of group divisible incomplete block designs*, Ann. Math. Stat., **23** (1952), 367.

4.   R. H. Bruck and H. J. Ryser, *The nonexistence of certain finite projective planes*, Canad. J. Math., **1** (1949), 88.

5.   S. Chowla and H. J. Ryser, *Combinatorial problems*, Canad. J. Math., **2** (1950), 93.

6.   W. S. Connor, *Some relations among the blocks of symmetrical group divisible designs*, Ann. Math. Stat., **23** (1952), 602.

7.   O. Hilbert, *Gesammelte Abhandlungen*, Vol. 1, Berlin, 1932.

8.   T. Nagell, *Introduction to number theory*, Uppsala, 1951.

9.   E. G. Straus and O. Taussky, *Remark on the preceding paper. Algebraic equations satisfied by roots of natural numbers*, Pacific J. Math., **6** (1956), 000.

NATIONAL BUREAU OF STANDARDS
UNIVERSITY OF CALIFORNIA, LOS ANGELES

ON UNIONS AND INTERSECTIONS OF CONES*

A. J. Hoffman

IBM WATSON RESEARCH CENTER
YORKTOWN HEIGHTS, NEW YORK

1. INTRODUCTION

Let $m \leq n$ be given positive integers, and let $C_1, \ldots, C_n$ be closed, convex pointed cones in R^m (a cone is pointed if it contains no line). We tacitly assume that each C_i contains at least one nonzero vector. The problem considered is that of finding conditions on the pattern of intersections of the cones $\{C_i\}$ and $\{-C_i\}$ which will ensure that $\bigcup C_i \cup \bigcup -C_i = R^m$. We solve this problem only for certain values of m and n, as stated in Theorem 1.

A special case of a theorem of Ky Fan [1] on closed antipodal sets of the $(n-1)$ sphere is the inspiration for the present investigation.

FAN'S THEOREM. *If $\bigcup C_i \cup \bigcup -C_i = R^m$, then*

$$\text{for every choice of } \varepsilon_n = \pm 1 (j = 1, \ldots, n) \text{ and}$$

$$\text{of a permutation } \sigma \text{ of } \{1, \ldots, n\},$$

$$\text{there exist indices } 1 \leq i_1 < \cdots < i_m \leq n \text{ such that} \qquad (1.1)$$

$$\bigcap_{h=1}^{m} (-1)^k \varepsilon_{\sigma ik} C_{\sigma ik} \neq 0.$$

The main result of the present paper is a partial converse.

THEOREM 1. *If $n = m$ or $n = m + 1$, or if $m = 2$ or $m = 3$, then* (1.1) *implies* $\bigcup C_i \cup \bigcup -C_i = R^m$.

* This paper contains a section of the talk "Bounds on the Rank and Eigenvalues of Matrices, and the Theory of Linear Inequalities" given at the Third Waterloo Symposium on Combinatorial Mathematics, May, 1968. The research reported was sponsored in part by the Office of Naval Research under contract Nonr-3775(00).

A. J. HOFFMAN

We do not know if the theorem holds for other values of m. A corollary of the proof of Theorem 1 is the following curiosity.

COROLLARY. Let $P_1, \ldots, P_{t+3}$ be points in R^t in general position (i.e., no hyperplane contains $t + 1$ of the points). Let G be the graph whose vertices are the points, with two vertices P_i and P_j adjacent if and only if the $t + 1$ hyperplanes determined by the remaining points have the property that all separate the vertices P_i and P_j or none separate P_i and P_j. Then G is a simple polygon.

2. PRELIMINARIES

We first prove a theorem on the rank of real matrices, generalizing results given in [2] and [3], that may be of some independent interest.

THEOREM 2. *Let $\mathcal{M}$ be a set of matrices of order n, closed, convex, and not containing 0, and let $1 \leq m \leq n$. Then the following are equivalent:*

$$
\begin{aligned}
&\textit{For every real matrix } A, \text{ Tr } AM^T > 0 \textit{ for all} \\
&M \in \mathcal{M} \textit{ implies rank } A > m - 1.
\end{aligned}
\tag{2.1}
$$

$$
\begin{aligned}
&\textit{For every } n - m + 1 \textit{ dimensional subspace} \\
&L_{n-m+1} \subseteq R^n, \textit{ there exists a matrix } M \in \mathcal{M} \\
&\textit{all of whose rows are in } L_{n-m+1}.
\end{aligned}
\tag{2.2}
$$

PROOF: Suppose (2.2) holds and Tr $AM^T > 0$. If rank $A \leq m - 1$, then there exists a subspace L_{n-m+1} such that $Ax = 0$ for all $x \in L_{n-m+1}$. Let M be the matrix whose existence is assured by (2.2), then $AM^T = 0$, contradicting Tr $AM^T > 0$.

Conversely, assume (2.1). Let $\mathcal{L} = \mathcal{L}(L_{n-m+1})$ be the set of all matrices each of whose rows belong to a given L_{n-m+1}. Then $\mathcal{L}$ is obviously a linear subspace of the n^2-dimensional space of all matrices. If (2.2) is false, there exists an L_{n-m+1} such that $\mathcal{L} \cap \mathcal{M} = \emptyset$. By a hyperplane separation theorem there exists a linear function which is zero on $\mathcal{L}$ and positive on $\mathcal{M}$; i.e., there exists a matrix A such that

$$
\begin{aligned}
\text{Tr } AX^T &= 0 \quad \text{for all} \quad X \in \mathcal{L}, \\
\text{Tr } AM^T &> 0 \quad \text{for all} \quad M \in \mathcal{M}.
\end{aligned}
$$

The second statement is the hypothesis of (2.1); the first statement says that A annihilates L_{n-m+1}, which contradicts the conclusion of (2.1). Hence, if (2.2) is false, so is (2.1).

LEMMA 1. *Let D_i be the convex hull of $C_i \cap \{x \mid \|x\| = 1\}$ (note that D_i is nonvoid and does not contain zero). Let $\mathcal{M}_i$ be the set of all matrices of order n in which each entry is zero except for row i; in row i, the first m coordinates are given by an arbitrary vector in D_i, the remaining coordinates are zero. Let $\mathcal{M}$ be the convex hull of $\{\mathcal{M}_i\}$, $i = 1, \ldots, n$. Then $\mathcal{M}$ satisfies (2.2) if and only if $\bigcup C_i \cup \bigcup -C_i = R^m$.*

PROOF: Suppose $\mathcal{M}$ satisfies (2.2). Let x be any vector in R^m (where we identify R^m with the space spanned by the first m unit coordinate vectors of R^n), and $L^{(x)}_{n-m+1}$ be the space generated by x and the last $n - m$ coordinate vectors of R^n. From (2.2) it follows that some positive or negative multiple of x is in D_i; i.e., $\bigcup C_i \cup \bigcup -C_i$ contains x.

Conversely, assume $\bigcup C_i \cup \bigcup -C_i = R^n$. Let L_{n-m+1} be any $(n - m + 1)$-dimensional subspace of R^n; $L_{n-m+1} \cap R^m$ has dimension at least 1. Let $x \in L_{n-m+1} \cap R^m$, $x \neq 0$. Then some positive or negative multiple of x is contained in some D_i; hence, $\mathscr{L}(L_{n-m+1}) \cap \mathcal{M} \neq \varnothing$.

LEMMA 2. *Let $\mathcal{M}$ be a closed convex set of matrices not containing zero, and let $\mathcal{M}^T = \{M \mid M^T \in \mathcal{M}\}$. Then $\mathcal{M}^T$ satisfies (2.2) if and only if $\mathcal{M}$ satisfies (2.2).*

PROOF: Invoke from Theorem 2 the equivalence of (2.1) and (2.2), and recall that any matrix A and its transpose A^T have the same rank.

We define, for any real number x,

$$\operatorname{sgn} x \begin{cases} +1 & \text{if} & x > 0 \\ 0 & \text{if} & x = 0 \\ -1 & \text{if} & x < 0. \end{cases}$$

LEMMA 3. *If each L_{n-m+1} contains a vector $x = (x_1, \ldots, x_n)$, $x \neq 0$, such that*

$$\bigcap_{\operatorname{sgn} x_i \neq 0} \operatorname{sgn} x_i \, C_i \neq 0, \tag{2.3}$$

then $\bigcup C_i \cup \bigcup -C_i = R^m$.

PROOF: Let $\mathcal{M}$ be as in Lemma 1. Consider $\mathcal{M}^T$. Let

$$y \in \bigcap_{\operatorname{sgn} x_i \neq 0} \operatorname{sgn} x_i \, C_i, \qquad \|y\| = 1.$$

Then the matrix which consists entirely of zeros in the last $n - m$ rows, and in the first m rows has (i, j)th entry $y_i x_j / \sum_j |x_j|$ is clearly in $\mathcal{M}^T$. Hence, $\mathcal{M}^T$

satisfies (2.2). By Lemma 2, $\mathcal{M}$ satisfies (2.2), so the conclusion follows from Lemma 1.

3. Proof of Theorem 1

CASE 1 ($n = m$). In this case, (1.1) says $\bigcap_i \varepsilon_i C_i \neq 0$ for all choices of $\varepsilon_i = \pm 1$. Further, in this case $n - m + 1 = 1$, so (2.3) is also the statement $\bigcap_i \varepsilon_i C_i \neq 0$ for arbitrary choice of $\varepsilon_i = \pm 1$. Hence, the theorem follows from Lemma 3.

COROLLARY. If $\bigcup_{i=1}^m C_i \cup \bigcup_{i=1}^m -C_i = R^m$, then there exist polyhedral subcones each with at most 2^{m-1} generators $E_i \subset C_i$ such that $\bigcup E_i \cup \bigcup -E_i = R^m$. [More generally $\bigcup_{i=1}^n C_i \cup \bigcup_{i=1}^n -C_i = R^m$ implies that there exist polyhedral subcones $E_i \subset C_i$ such that $\bigcup E_i \cup \bigcup -E_i = R^m$. But this we shall prove elsewhere. The case $n = m$ is of special interest, however, since it shows that the use of "round" cones to prove the nonsingularity of real square matrices (using Theorem 2 and Lemma 1) is superfluous.]

PROOF: By Fan's theorem, for any subset $S \subset \{1, \ldots, n\}$ and its complement $\bar{S}$, there exists a vector $y_{(s;\,\bar{s})} \in \bigcap_{i \in s} C_i \bigcap_{i \in \bar{s}} -C_i$, with $y_{(\bar{s};\,s)} = -y_{(s;\,\bar{s})}$. The generators of E_i are all $y_{(s;\,\bar{s})}$, with $i \in S$.

CASE 2 ($n = m + 1$). Then an arbitrary L_{n-m+1} becomes an arbitrary L_2. Assume that L_2 can be conceived as generated by two independent vectors $a = (a_1, \ldots, a_n)$ and $b = (b_1, \ldots, b_n)$, in which no $b_i = 0$. We choose a permutation σ so that

$$\frac{a_{\sigma 1}}{b_{\sigma 1}} \geq \cdots \geq \frac{a_{\sigma n}}{b_{\sigma n}}. \tag{3.1}$$

Define $\varepsilon_{\sigma t} = (-1)^t \operatorname{sgn} b_{\sigma t}$. We shall show that with this choice of σ and ε_i, (1.1) implies (2.3) for some x. Namely, we consider the n vectors x that are obtained from linear combinations of a and b of the form

$$x^{\sigma k} = b_{\sigma k} a - a_{\sigma k} b, \qquad k = 1, \ldots, n. \tag{3.2}$$

Note that $x_{\sigma k}^{\sigma k} = 0$ for all k, but $x^{\sigma k} = 0$ for no k. To verify (2.3) it is sufficient to show that, for at least one k,

$$\bigcap_{l \neq k} \operatorname{sgn} x_{\sigma l}^{\sigma k} C_{\sigma l} \neq 0. \tag{3.3}$$

But with our choice of σ and ε_i, (1.1) becomes the statement that, for at least one k,

$$0 \neq \bigcap_{r<k} (-1)^r \varepsilon_{\sigma r} \, C_{\sigma r} \cap \bigcap_{r>k} (-1)^{r+1} \varepsilon_{\sigma r} \, C_{\sigma r}$$

$$= \bigcap_{r<k} (-1)^{2r} \operatorname{sgn} b_{\sigma r} \, C_{\sigma r} \cap \bigcap_{r>k} (-1)^{2r+1} \operatorname{sgn} b_{\sigma r} \, C_{\sigma r}$$

$$= \bigcap_{r<k} \operatorname{sgn} b_{\sigma r} \, C_{\sigma r} \cap \bigcap_{r>k} (-\operatorname{sgn} b_{\sigma r}) C_r .$$

To show this implies (3.3), it is sufficient to prove that

$$(\operatorname{sgn} x^{\sigma k}_{\sigma r})(\operatorname{sgn} x^{\sigma k}_{\sigma s}) = \operatorname{sgn}(r - k) \operatorname{sgn} b_{\sigma r} \operatorname{sgn}(s - k) \operatorname{sgn} b_{\sigma s} \qquad \text{or} \qquad 0$$

$$\text{for all} \quad r \neq s. \qquad (3.4)$$

But the left side of (3.4) is

$$\operatorname{sgn} (a_{\sigma r} b_{\sigma k} - b_{\sigma r} a_{\sigma k})(a_{\sigma s} b_{\sigma k} - b_{\sigma s} a_{\sigma k})$$

$$= \operatorname{sgn} \frac{b_{\sigma r} b_{\sigma s}(a_{\sigma r} b_{\sigma k} - b_{\sigma r} a_{\sigma k})(a_{\sigma s} b_{\sigma k} - b_{\sigma s} a_{\sigma k})}{b^2_{\sigma k} b_{\sigma r} b_{\sigma s}}$$

$$= \operatorname{sgn} b_{\sigma r} b_{\sigma s} \left(\frac{a_{\sigma r}}{b_{\sigma r}} - \frac{a_{\sigma k}}{b_{\sigma k}} \right) \left(\frac{a_{\sigma s}}{b_{\sigma s}} - \frac{a_{\sigma k}}{b_{\sigma k}} \right)$$

$$= \operatorname{sgn}(b_{\sigma r} b_{\sigma s}(r - k)(s - k)) \qquad \text{or} \qquad 0,$$

from (3.1).

There remains to consider the case in which there is some index i such that $x_i = 0$ for all $x \in L_2$. But such an L_2 can be approximated by a sequence of two-dimensional subspaces for which this does not occur. Since each such subspace will satisfy (2.3), so will L_2.

CASE 3 ($m = 2$). For the study of this case, and also the case $m = 3$, it is convenient to prove first the following lemma.

LEMMA 4. *Let L_{n-m+1} be spanned by the row vectors of a matrix L with $n - m + 1$ rows and n columns, and assume that every $n - m + 1$ columns of L are linearly independent. Let P be a matrix with $m - 1$ rows and n columns, whose rows span the orthogonal complement of L_{n-m-1}, and whose columns are denoted by P_j.*
For each $S \subset \{1, \ldots, n\}$, with $|S| = m$, let $x \neq 0$, $x \in L_{n-m+1}$ satisfy $x_j = 0$ for all $j \notin S$. Then $\sum_{j \in S} x_j P_j = 0$.

PROOF: Let $x' = z'L$. Then, since $LP^T = 0$, it follows that $x'P^T = z'LP^T = 0$. But this is precisely $\sum_{j \in S} x_j P_j = 0$.

A. J. HOFFMAN

Note that our hypotheses on L ensure that an x satisfying the hypothesis exists for each S when $|S| = m$, is unique (up to multiplication by a constant), and $x_j \neq 0$ for any $j \in S$.

In applying Lemma 4, it may happen that L does not satisfy the hypothesis of Lemma 4 that every $n - m + 1$ columns are independent. Then replace L by a nearby matrix that does satisfy this hypothesis. It is clear that if we prove (2.3) for a sequence of linear spaces of dimension $n - m + 1$ approaching L_{n-m+1}, then (2.3) holds for L_{n-m+1}. So in both the cases $m = 2$ and $m = 3$, we make the assumption that the hypothesis of Lemma 4 holds.

In case $m = 2$, $n - m + 1$, and P consists of a single row $p = (p_1, \ldots, p_n)$ in which all p_i are different and no p_i is zero. Choose σ so that

$$p_{\sigma 1} > p_{\sigma 2} > \cdots > p_{\sigma n}. \tag{3.5}$$

Choose $\varepsilon_t = \operatorname{sgn} p_t$. With this choice of σ and $\{\varepsilon_i\}$, (1.1) becomes the statement that there exist indices $k \neq l$ such that

$$\varepsilon_{\sigma k} C_{\sigma k} \cap -\varepsilon_{\sigma l} C_{\sigma l} \neq 0 \tag{3.6}$$

$$x_{\sigma k} p_{\sigma k} + x_{\sigma l} p_{\sigma l} = 0. \tag{3.7}$$

Let $x = (x_1, \ldots, x_n)$ arise from Lemma 4 with $S = \{\sigma k, \sigma l\}$. To verify (2.3) we must show, in view of (3.6),

$$\operatorname{sgn} x_{\sigma k} = \varepsilon_{\sigma k}, \qquad \operatorname{sgn} x_{\sigma l} = -\varepsilon_{\sigma l} \tag{3.8}$$

or

$$\operatorname{sgn} x_{\sigma k} = -\varepsilon_{\sigma k}, \qquad \operatorname{sgn} x_{\sigma l} = \varepsilon_{\sigma l}$$

From (3.7), $\operatorname{sgn} x_{\sigma k} x_{\sigma l} = -\operatorname{sgn} p_{\sigma k} p_{\sigma l} = -\varepsilon_{\sigma k} \varepsilon_{\sigma l}$ [from (3.5)], which proves (3.8).

CASE 4 ($m = 3$). In this case P is a matrix of two rows $a = (a_1, \ldots, a_n)$ and $b = (b_1, \ldots, b_n)$, and we may assume all b_i different from zero. Choose σ so that

$$\frac{a_{\sigma 1}}{b_{\sigma 1}} > \cdots > \frac{a_{\sigma n}}{b_{\sigma n}}. \tag{3.9}$$

Note that $b_i \neq 0$ and the strict inequality in (3.9) are consequences of the stipulations on L.

Let $\varepsilon_i = \operatorname{sgn} b_i$. From (1.1) we know there exist indices $j < k < l$ such that

$$\varepsilon_{\sigma j} C_{\sigma j} \cap -\varepsilon_{\sigma k} C_{\sigma k} \cap \varepsilon_{\sigma l} C_{\sigma l} \neq 0. \tag{3.10}$$

Let $x = (x_1, \ldots, x_n)$ arise from Lemma 4 with $S = \{\sigma j, \sigma k, \sigma l\}$. To verify (2.3) we must show, in view of (3.6), that if A_t denotes the tth column of P, then

$$x_{\sigma j} A_{\sigma j} + x_{\sigma k} A_{\sigma k} + x_{\sigma l} A_{\sigma l} = 0 \tag{3.11}$$

implies

$$\operatorname{sgn} x_{\sigma j} x_{\sigma k} = -\operatorname{sgn} b_{\sigma j} b_{\sigma k}, \qquad \operatorname{sgn} x_{\sigma k} x_{\sigma l} = -\operatorname{sgn} b_{\sigma k} b_{\sigma l}. \tag{3.12}$$

But (3.11) may be rewritten as

$$\frac{x_{\sigma j}}{x_{\sigma k}} A_{\sigma j} + \frac{x_{\sigma l}}{x_{\sigma k}} A_{\sigma l} = -A_{\sigma k}. \tag{3.13}$$

Using Cramer's rule on (3.13), and invoking (3.9), one demonstrates (3.12).

In case $m = 4$, $n = 6$, it is easy to make up examples which show that there exists L_{n-m+1} with the property that the intersection patterns guaranteed by (1.1) are by themselves insufficient to ensure (2.3). This remark, of course, is not yet a disproof of the converse of Fan's theorem.

PROOF OF COROLLARY: Let $n = t + 3$, $m = 3$, and L be the matrix with $n - m + 1 = t + 1$ rows and n columns, one of whose rows consists entirely of ones, the remainder of the matrix given by the column vectors P_j. Let σ be the permutation given in Case 4 of the proof of Theorem 1. Then it is easy to see from the proof of Case 4 that $P_{\sigma i}$ is adjacent to $P_{\sigma j}$ if and only if $|i - j| = 1$ or $n - 1$.

ACKNOWLEDGMENT

We are very grateful to Benjamin Weiss for stimulating conversations about this material. An alternative proof of the corollary has been kindly communicated to us by Professor H. D. Ursell.

REFERENCES

1. KY FAN, A Generalization of Tucker's Combinatorial Lemma with Topological Applications, *Ann. Math.* **56** (1952), 431–437.
2. A. J. HOFFMAN, On the Nonsingularity of Real Matrices, *Math. Comp.* **19** (1965), 56–61.
3. A. J. HOFFMAN, On the Nonsingularity of Real Partitioned Matrices, *ICC Bull.* **4** (1965), 7–17.

BINDING CONSTRAINTS AND HELLY NUMBERS

A. J. Hoffman

IBM T. J. Watson Research Center
Yorktown Heights, New York 10598

INTRODUCTION

Bell [1, 2] has shown that if a collection of half spaces in $\mathbb{R}^n$ contains no point of $\mathbb{Z}^n$ in their intersection, the same is true for a subset of at most 2^n of the given half spaces. Herbert Scarf [3] has recently shown that if the integer linear program

$$\text{Maximize } (c, x)\,|\,Ax \geq b, \, x \in \mathbb{Z}^n \tag{1}$$

is feasible and has an optimum value of c_0, then there exists a subset S, of cardinality at most $2^n - 1$, of the inequalities $Ax \geq b$, say $A^s x \geq b^s$, such that the integer linear program

$$\text{Minimize } (c, x)\,|\,A^s x \geq b^s, \, x \in \mathbb{Z}^n \tag{2}$$

has the same optimum value c_0.

The theorems of Bell and Scarf are intimately related, and it is easy to deduce one from the other (see next section). Our purpose is to give a new proof of these theorems which highlights their axiomatic foundation. Scarf's proof uses an artful labeling algorithm. Todd [4] has given an ingenious alternate proof based on programming concepts without using labeling. Bell has given several elegant proofs, and ours is closest in spirit to that of Bell [1]. We start with an abstract intersectional system [5] and formulate the question of the *number* of binding constraints in terms of that system; that number is one less than the *Helly number* [5] of that system. If the system satisfies some additional hypotheses, Bell's and Scarf's theorems follow. This abstract viewpoint enables us to progress on concrete problems as well. For instance, we can find the number of binding constraints in linear programming problems with m real variables and n integer variables: $(m + 1)2^n - 1$. We can also prove "skeleton theorems" in linear approximation theory [6] for the case where the coefficients are restricted to integers, and in other cases. Another application is to centrally symmetric convex bodies in $\mathbb{R}^n$: we can show that, if $K_1, \ldots, K_t, t > 2^{n-1}$ are such bodies with the property that every $2^n - 1$ of them contains a nonzero integral point, then the intersection of all of them contains a nonzero integral point (proofs and discussions of these and related theorems will be published elsewhere.)

HELLY SYSTEMS

Let U be a set, $\mathcal{K}$ a family of subsets (called *pseudoconvex*) of U containing ϕ and U, and closed with respect to arbitrary intersection. The pair $[U, \mathcal{K}]$ will be called a *Helly system*. The Helly system $[U', \mathcal{K}']$ is said to be embedded in $[U, \mathcal{K}]$ if $U' \subset U$ and the sets in $\mathcal{K}'$ are the distinct sets in $\{U' \cap \mathcal{K}\}_{K \in \mathcal{K}}$.

We now formulate the binding constraint question for a general Helly system $[U, \mathcal{K}]$. Suppose there exists an integer b such that: if $t > b$ and $K_1, \ldots, K_t \in \mathcal{K}$ satisfy,

$$\bigcap_1^t K_j \neq \phi \tag{3}$$

284

and $\{K_\alpha\}$ is a collection of pseudoconvex sets simply ordered by inclusion, then there exist integers $1 \le j_i < j_2 < \cdots < j_b \le t$ such that, for each α,

$$\bigcap_1^b K_{j_i} \cap K_\alpha = \phi \text{ if and only if } \bigcap_1^t K_j \cap K_\alpha = \phi \tag{4}$$

If there exists no such integer, we say that $b[U, \mathcal{K}] = \infty$; otherwise $b[U, \mathcal{K}]$ will be the smallest integer b satisfying the stipulations. Note that if $U = \mathbb{Z}^n \subset \mathbb{R}^n$, and our Helly system is embedded in $\mathbb{R}^n$, Scarf's theorem states $b[\mathbb{Z}^n, \mathcal{K}] = 2^n - 1$. In general, $b[U, \mathcal{K}]$ is called the *binding constraint number* of the Helly system $[U, \mathcal{K}]$, and seems to capture what that number should mean for programming problems in Helly systems embedded in Euclidean space.

Suppose there exists an integer h such that, if $t > h$ and $K_1, \ldots, K_t \in \mathcal{K}$ satisfy

$$\bigcap_1^t K_j = \phi \tag{5}$$

then there exist $1 \le j_1 < j_2 < \cdots < j_h \le t$ such that,

$$\bigcap_1^h K_{j_i} = \phi \tag{6}$$

If no such h exists, we say $h[U, \mathcal{K}] = \infty$; otherwise, $h[U, \mathcal{K}]$ is the smallest integer satisfying the stipulations. The term *Helly number* has been used for $h[U, \mathcal{K}]$ (Danzer [5]).

PROPOSITION 1. If $[U, \mathcal{K}]$ is a Helly system, $b[U, \mathcal{K}] = h[U, \mathcal{K}] - 1$.

Proof: Suppose $h = h[U, \mathcal{K}]$ finite. Let $t > h - 1$, and assume (3) and a family $\{K_\alpha\}$ of pseudoconvex sets simply ordered by inclusion. (For ease of notation, assume the subscripts α form a simply ordered set, and $\alpha < \alpha'$ implies $K_\alpha \subset K_{\alpha'}$). If we prove (4) with $b = h - 1$, we shall have established $b[U, \mathcal{K}] \le h[U, \mathcal{K}] - 1$. Let $A = \{\alpha \mid \bigcap_1^t K_j \cap K_\alpha = \phi\}$. If (4) is false, then, $A \ne \phi$ and for each $S \subset \{1, \ldots, t\}$, with $|S| = h - 1$, there is at least one $\alpha \in A$ such that

$$\bigcap_{j \in S} K_j \cap K_\alpha \ne \phi \tag{7}$$

Pick one such α satisfying (7) and call it $\alpha(S)$. Let $\alpha' = \max_S \{\alpha(S)\}$. Then $\alpha' \in A$, and (7) implies

$$\bigcap_{j \in S} K_j \cap K_{\alpha'} \ne \phi, \text{ for all } S \subset \{1, \ldots, t\}, \text{ with } |S| = h - 1 \tag{8}$$

But $\bigcap_1^t K_j \cap K_{\alpha'} = \phi$ implies, by (6), that some h of the $t + 1$ psuedoconvex sets $K_1, \ldots, K_t, K_{\alpha'}$ have an empty intersection. One of those h pseudoconvex sets must be $K_{\alpha'}$, otherwise (3) is false. But this violates (8).

To show that $h[U, \mathcal{K}] \le h[U, \mathcal{K}] + 1$, assume $b = b[U, \mathcal{K}]$ finite and the inequality is false. That means that there is an integer $t > b + 1$ and pseudoconvex sets $K_1, \ldots, K_t$ such that (5) holds but (6) is false for $h = b + 1$. Let $t > b + 1$ be the smallest integer for which such a collection of t pseudoconvex sets $K_1, \ldots, K_t$ exists. The minimality of t implies that $\bigcap_1^{t-1} K_j \ne \phi$. In applying (2.2), let the family $\{K_\alpha\}$ be the single set K_t, and note $t - 1 > b$. It follows from (4) that there is a subset $S \subset \{1, \ldots, t - 1\}$, with $|S| = b$ such that $\bigcap_{j \in S} K_j \cap K_t = \phi$. But this proves (6) with $h = b + 1$. $\square$

REMARK. It is worth noting that, for Helly systems embedded in $\mathbb{R}^n$, if b is the binding constraint number for *linear programming problems*, then $h \leq b + 1$. This can be proved as follows. As before let $t > b + 1$ and $K_1, \ldots, K_t$ be a minimal counterexample. Let $p_i \in \mathbb{R}^n$, $i = 1, \ldots, t$ satisfy $p_i \in \bigcap_{j=1, j \neq i}^t K_i$ (the existence of p_i follows from the minimality of the counterexample). Since $\bigcap_1^t K_j = \phi$, $p_i \notin K_i$. Let $\hat{K}_i = U \cap$ convex hull of $\{p_j\}_{j \neq i}$. Then $\hat{K}_i \subset K_i$ for all i, so that $\bigcap_1^t \hat{K}_i = \phi$.

Let $A_i x \geq b_i$ be a finite system of linear inequalities in $\mathbb{R}^n$ such that $x \in \hat{K}_i$ if and only if $x \in U$ and $A_i x \geq b_i$. The existence of such a finite system follows from the fact that, in $\mathbb{R}^n$, every convex polytope is the intersection of a finite number of closed half spaces. It follows that the finite system of linear inequalities $A_1 x \geq b_1, \ldots,$ $A_t x \geq b_t$ is not satisfied by any $x \in U$. Let $Lx \geq l$ be a subset of the foregoing inequalities which is not satisfied by any $x \in U$, and which is minimal with respect to this property. If L has $b + 1$ or fewer rows we are done, this would imply (6) with $h \leq b + 1$. So assume L has $u > b + 1$ rows, and consider the nonempty pseudo-convex set by:

$$K = \{x \mid x \in U, (L_1, x) \geq l_1, \ldots, (L_{u-1}, x) \geq l_{u-1}\}$$

and the simply ordered family of pseudoconvex sets $K_\alpha = \{x \mid x \in U, (L_u, x) \geq -\alpha\}$, indexed by real α. We know that there is no $x \in U$ which is in K and K_{-l_u}. By the definition of b, it follows that a subset of b of the inequalities defining K and K_{-l_u} are inconsistent. But this implies $h \leq b + 1$.

We now return to the general $[U, \mathcal{K}]$, without assuming embedding in $\mathbb{R}^n$. If $S \subset U$ is finite, we define the (pseudoconvex) polytope generated by S to be $\bigcap_{S \subset K} K$, and write this as $K(S)$. Define $h'[U, \mathcal{K}]$ to be the largest cardinality of a finite subset $S \subset U$ such that,

$$\bigcap_{s \in S} K(S - s) = \phi \tag{9}$$

If there exist finite sets of arbitrarily large cardinality satisfying (9), then $h'[U, \mathcal{K}] = \infty$.

PROPOSITION 2. If $[U, \mathcal{K}]$ is a Helly system, $h'[U, \mathcal{K}] = h[U, \mathcal{K}]$.

Proof: Let S satisfy (9). If $s' \in S$, then $s' \in \bigcap_{s \in S, s \neq s'} K(S - s)$. Consequently $h[U, \mathcal{K}] \geq h'[U, \mathcal{K}]$.

To prove the reverse inequality, assume it false, and let $t > h'$ and $K_1, \ldots, K_t$ be pseudoconvex sets of a counterexample with the smallest number of pseudo-convex sets. Let $p_i \in \bigcap_{j \neq i} K_j$, $P = \{p_1, \ldots, p_t\}$, $\hat{K}_i = K(P - p_i)$. Then $\hat{K}_1, \ldots, \hat{K}_t$ are polytopes also providing a minimal counterexample. But $t > h'$ asserts that $\bigcap_1^t \hat{K}_i \neq \phi$, a contradiction. $\square$

BELL'S AND SCARF'S THEOREMS

Before looking at $\mathbb{Z}^n$, let us add two additional hypotheses to a general Helly system $[U, \mathcal{K}]$:

(i) If P and Q are finite subsets of U, and $K(P) = K(Q)$, then $K(P) = K(Q) = K(P \cap Q)$.

It follows that, if K is a polytope, there is a unique minimal set $V \subset U$ such that $K = K(V)$. We call that set the vertices of K, and write $V = V(K)$.

(ii) If K is a polytope, K is finite.

The flats of a real projective space satisfy neither (i) nor (ii). The flats of a finite geometry form a Helly system satisfying (ii) but not (i). The convex sets of Euclidean

Hoffman: Binding Constraints   287

space satisfy (i) but not (ii). If $[U, \mathscr{K}]$ is imbedded in $\mathbb{R}^n$, and U is nowhere dense, then (i) and (ii) both hold. Hence these hypotheses will be useful in Scarf's theorem, since $\mathbb{Z}^n$ is nowhere dense.

We now define the Scarf number $s[U, \mathscr{K}]$ to be the largest cardinality of a finite subset $S \subset U$ such that,

$$s \in \mathscr{K}, \; S - p \in \mathscr{K}, \text{ for each } p \in S \tag{10}$$

If these exist subsets S of arbitrarily large cardinality satisfying **(10)**, then $s[U, S] = \infty$.

PROPOSITION 3. If $[U, \mathscr{K}]$ satisfies (i) and (ii), then $s[U, \mathscr{K}] = h'[U, \mathscr{K}]$.

Proof: Let S satisfy **(10)**. If $p \in S$, and $p \in K(S - p)$, then $K(S) = K(S - p)$, contradicting $S - p \in \mathscr{K}$. Next, **(9)** must hold, since $p \in \bigcap_{q \in S} K(S - q)$ implies that $p \notin S$, $p \in K(S)$, contradicting $S \in \mathscr{K}$. So we have proved $s[U, \mathscr{K}] \le h'[U, \mathscr{K}]$.

To prove the reverse inequality, let $t > s$. We shall prove below that, if K is a polytope with $|V(K)| = t$, then,

$$\bigcap_{q \in V(K)} K(V(K) - q) \ne \phi \tag{11}$$

This implies $h'[U, \mathscr{K}] \le s[U, \mathscr{K}]$.

Let P_t be the set of all polytopes K with t vertices, partially ordered by inclusion. We prove **(11)** by induction, assuming that K is minimal in P_t or that **(11)** has been proved for all predecessors in P_t. Note that hypothesis (ii) assures us that any K in P_t has only a finite number of predecessors.

Let $V(K) = \{p_1, \ldots, p_t\}$. Since $t > s$ and $V\{K[V(K)]\} = V(K)$, it follows from **(10)** that K contains a point p', $p' \notin V(K)$. For all $p' \in K - V(K)$, let $m(p')$ be the number of indices i such that $p' \notin K(V(K) - p_i)$, and let $\bar{p} \in K - V(K)$ be such that

$$m(\bar{p}) \le m(p') \text{ for all } p' \in K - V(K). \tag{12}$$

Our object is to show $m(\bar{p}) = 0$. Assume $m(\bar{p}) > 0$, and that,

$$\bar{p} \notin K(V(K) - p_i), \; i = 1, 2, \ldots, m \tag{13}$$

and $$\qquad \bar{p} \in K(V(K) - p_i), \; i = m + 1, \ldots, t \tag{14}$$

Let $\bar{K} = K(\{\bar{p}, p_2, \ldots, p_t\})$. By **(13)**, $\bar{p} \notin K(\{p_2, \ldots, p_t\})$. Further, if $k > 1$, $p_k \notin K(\{\bar{p} \cup \bigcup_{j=2, j \ne k}^{t} p_j\})$. Otherwise, $K = K(\{p_1, \ldots, p_t\}) \subset K(\{\bar{p} \cup \bigcup_{j=1, j \ne k}^{t} p_j\})$. Therefore, $K = K(\{\bar{p} \cup \bigcup_{j=1, j \ne k}^{t} p_j\})$, implying, from (i), that $K = K(\bigcup_{j=1, j \ne k}^{t} p_j)$, a contradiction. Therefore, $V(\bar{K}) = \{\bar{p}, p_2, \ldots, p_t\}$, and $\bar{K} \in P_t$, $\bar{K} < K$ in the partial ordering of P_t. By the induction hypothesis, **(11)** holds for $\bar{K}$ (not that K could not be minimal in P_t). It follows that there exists a point p'' such that

$$p'' \in \bigcap_{q \in V(\bar{K})} K(V(\bar{K}) - q). \tag{15}$$

From **(15)**, setting $q = \bar{p}$, we infer,

$$p'' \in K[V(K) - p_1] \tag{16}$$

Setting $q = p_j$, $j \ge m + 1$ and using **(14)**,

$$p'' \in K[V(\bar{K}) - p_j] \subset K[V(K) - p_j]. \tag{17}$$

Comparing **(16)** and **(17)** with **(13)** and **(14)**, we see that $m(p'') < (\bar{p})$. This contradicts **(12)**. $\quad\square$

288   Annals New York Academy of Sciences

PROPOSITION 4. If $U = \mathbb{Z}^n$ imbedded in $\mathbb{R}^n$, then $b[U, \mathcal{K}] = 2^n - 1$ and $h[U, \mathcal{K}] = 2^n$.

Proof: By PROPOSITIONS 1, 2 and 3, all we need prove is that $s[U, \mathcal{K}] = 2^n$. The vertices of the unit cube satisfy (10) and show that $s[U, \mathcal{K}] \geq 2^n$. Further, any $2^n + 1$ points S in $\mathbb{Z}^n$ contain a pair which are congruent mod 2, hence their midpoint is in $K(S)$, so $s[U, \mathcal{K}] \leq 2^n$ (see Bell [1, 2] and Scarf [3]). $\square$

SUMMARY

Starting with axioms for an abstract intersectional system, we define the *Helly number*, *Scarf number*, and *binding constraint number* of such a system. The last concept is based on a definition of a mathematical programming problem in the system. From these definitions, we deduce (1) *Bell's theorem* that a collection of half spaces contains a point of $\mathbb{Z}^n$ if the intersection of every subset of 2^n of the half spaces does, and (2) *Scarf's theorem* that an integer programming problem on $\mathbb{Z}^n$ has at most $2^n - 1$ binding constraints. Our arguments use coordinates only at the last moment.

REFERENCES

1. BELL, D. E. 1974. Intersections of Corner Polyhedra. Int. Inst. Appl. Syst. Anal. Laxenburg, Austria, Res. Memo. RM-74-14.
2. BELL, D. E. 1977. A theorem concerning the integer lattice. Stud. Appl. Math. **56:** 187–188.
3. SCARF, H. 1977. An observation on the structure of production sets with indivisibilities. Proc. Math. Acad. Sci. USA **74:** 3637–3641.
4. TODD, M. J. 1977. The number of necessary constraints in an integer program: a new proof of Scarf's theorem. Tech. Rep. 35. School of Operations Research and Industrial Engineering, Cornell University.
5. DANZER, L. W., B. GRUNBAUM & V. KLEE. 1963. Helly's theorem and its relatives. Proc. Symp. Pure Math. Am. Math. Soc. Providence, R. I. **VII:** 101–180.
6. RIVLIN, T. J. 1974. The Chebyshev Polynomials. Wiley, New York.

Combinatorics

1. A characterization of comparability graphs and of interval graphs

In 1956 I attended the first meeting of the Austrian Mathematical Society after the end of the Soviet occupation. I danced a wildly rapid Viennese waltz with Mary Cartwright (on the plane trip back to London we asked each other such questions as: if you were on a desert island and knew for certain that rescue was impossible, would you continue to do mathematics?) and heard a charming lecture by G. Hajós raising the question of characterizing interval graphs. I knew I could do this if I could characterize comparability graphs; indeed I had a conjecture for this characterization, which (about seven or eight years later) Paul Gilmore and I proved. Meanwhile, A. Ghouila-Houri had also proved the conjecture; hence the characterization has been called the GH theorem for his initials and ours.

2. Some properties of graphs with multiple edges

I recall (maybe incorrectly) that the motivation for this work was to see if, for some graphs, it would be possible to derive theorems on general matching without introducing the additional inequalities (cuts) of Edmonds' celebrated work. We succeeded, gathering along the way the theorem of Erdös and Gallai characterizing the possible sets of degrees of the nodes of a graph. What we could not realize at the time (and no one has yet explored) was that our proofs relate intimately to the important concept of Hilbert basis as used by Giles and Pulleyblank.

3. Self-orthogonal latin squares

The introduction to the paper explains why we called these combinatorial objects "spouse avoiding mixed doubles round robins", a term I have since seen in some tennis magazines. The problem we solved was to construct for as many values of n as possible, a SAMDRR for n couples. The rules for a SAMDRR require that n be at least 4, and we began by constructing round robins for the case $n = 4$ and the case $n = 5$, but failed for $n = 6$. We showed that $n = 6$ was not possible, but all larger n were, and the key to our success was to represent the pairings in the matches in a nice way (by a matrix, of course!). Then it became apparent that a SAMDRR for n couples corresponded to a latin square of order n orthogonal to its tranpose. Since it was long known that there do not exist any pair of orthogonal latin squares of order 6, we had the perfect explanation for why we failed in the case $n = 6$. And for the other values of n, we had available all the methods used by Bose, Parker and Shrikhande in their famous disproof of Euler's conjecture about latin squares.

Working on this paper was the only time in my life when I have been able to tell family and friends what I was doing.

4. On partitions of a partially ordered set

In the early fifties, a group of mathematicians (including Ray Fulkerson, George Dantzig, David Gale, Harold Kuhn and I), all involved in aspects of linear programming, were thrilled to discover that we could prove some combinatorial theorems as corollaries of the duality theorem of linear programming. For me personally, the epiphany occurred when I realized that the König–Egerváry theorem was a special case of linear programming duality (George Dantzig's explanation of duality by referring to diet pills and shadow prices had not moved me; Shizuo Kakutani had told me of König–Egerváry on an automobile trip from Princeton to Gainesville in 1950; fortunately I remembered what Kakutani had told me).

George and Ray in the first volume of the Naval Research Logistics Quarterly ingeniously formulated a tanker scheduling problem in linear programming terms in such a way that it was easy to adapt their formulation to give a linear programming proof (using duality, of course) of Dilworth's theorem that the smallest number of chains covering a partially ordered set is the largest cardinality of an anti-chain in that set. Over the years I have found various situations in which that approach is useful, and the work in this paper is one of my favorites. It shows that the little machine called linear programming duality not only can prove the results that Greene and Kleitman derived with great ingenuity, but also do it without any sweat at all. Another favorite is "Path partitions and packs of acyclic digraphs", with Ron Aharoni and Irith Hartman: in the coloring problem encountered there, the names of the colors assigned to the nodes are, in fact, numbers derived from the actual numerical values of dual variables. I thought this was lovely.

5. Variations on a theorem of Ryser

Since my year at the Institute for Advanced Study, I had been fascinated by Ryser's theorem that the design dual to a symmetric balanced incomplete block design was also such a design. For more than 45 years I had challenged myself and my students to find a proof which did not use matrices, even in the most trivial way, such as using the concept of inverse. Finally, with help from my friends, we did it. Success was eventually achieved by the following incredible route: we wrote out various identities using incidence matrices and their transposes and multiplying them in diverse ways, almost like the legendary monkeys typing in the British Museum. And finally one sequence of identities worked! And the only way matrices seemed to be used was as a notation for counting incidences in two ways several times. And when we realized this, we could write so that the forbidden word "matrix" was never mentioned in the proof.

Reprinted from
The Canadian Journal of Mathematics
Vol. 16 (1964), pp. 539–548

A CHARACTERIZATION OF COMPARABILITY GRAPHS AND OF INTERVAL GRAPHS

P. C. GILMORE AND A. J. HOFFMAN

1. Introduction. Let $<$ be a non-reflexive partial ordering defined on a set P. Let $G(P, <)$ be the undirected graph whose vertices are the elements of P, and whose edges (a, b) connect vertices for which either $a < b$ or $b < a$. A graph G with vertices P for which there exists a partial ordering $<$ such that $G = G(P, <)$ is called a comparability graph.

In §2 we state and prove a characterization of those graphs, finite or infinite, which are comparability graphs. Another proof of the same characterization has been given in **(2)**, and a related question examined in **(6)**. Our proof of the sufficiency of the characterization yields a very simple algorithm for directing all the edges of a comparability graph in such a way that the resulting graph partially orders its vertices.

Let O be any linearly ordered set. By an interval α of O is meant any subset of O with the same ordering as O and such that, for all a, b, and c, if b is between a and c and a and c are in α, then b is in α. Two intervals of O are said to intersect if and only if they have an element in common.

Let I be any set of intervals on a linearly ordered set O and let $G(O, I)$ be the undirected graph whose vertices are the intervals in I and whose edges (α, β) connect intersecting intervals α and β. A graph G is an interval graph if there exists such an O and I for which $G = G(O, I)$.

In §3 we state and prove a characterization of those graphs, finite or infinite, which are interval graphs. This solves a problem closely related to one first proposed in **(4)**, and independently in **(1)**. A different characterization was given in **(5)**. As a corollary of our result, we are able to determine for any interval graph G the minimum cardinality of a linearly ordered set O for which there is a set of intervals I such that $G = G(O, I)$.

All graphs considered in this paper have no edge joining a vertex to itself.

2. Comparability graphs. By a *cycle* of a graph G is meant here any finite sequence of vertices $a_1, a_2, \ldots, a_k$ of G such that all of the edges (a_i, a_{i+1}), $1 \leqslant i \leqslant k - 1$, and the edge (a_k, a_1) are in G, and for no vertices a and b and integers $i, j < k$, $i \neq j$, is $a = a_i = a_j$, $b = a_{i+1} = a_{j+1}$ or $a = a_i = a_k$, $b = a_{i+1} = a_1$. A cycle is odd or even depending on whether k is odd or even.

Received May 13, 1963. This research was supported in part by the Office of Naval Research under Contract No. Nonr 3775(00), NR 047040. The results were announced, without proofs, in **(3)**.

                    P. C. GILMORE AND A. J. HOFFMAN

Note that there can exist cycles in which a vertex appears more than once. For example, in Figure 1, $d, a, b, e, b, c, f, c, a$ is a cycle with nine vertices.

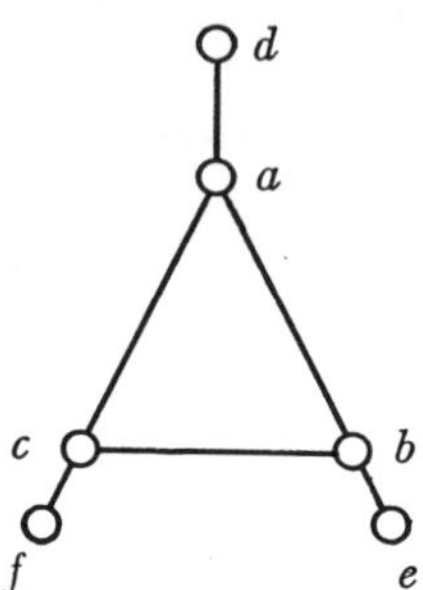

FIGURE 1

By a *triangular chord* of a cycle $a_1, a_2, \ldots, a_k$ of G is meant any one of the edges (a_i, a_{i+2}), $1 \leqslant i \leqslant k - 2$, or (a_{k-1}, a_1) or (a_k, a_2). For example, the cycle of nine vertices in Figure 1 has no triangular chords.

THEOREM 1. *A graph G is a comparability graph if and only if each odd cycle has at least one triangular chord.*

Proof. The necessity of the condition is not difficult to establish. For if an odd cycle $a_1, \ldots, a_k$ without a triangular chord occurs in a graph G, then any orientation of the edges of G which is to partially order the vertices of G must give any successive pair of edges of the cycle opposite orientations in the sense that both are directed towards or away from the common vertex of the pair. For if (a, b) and (b, c) are edges of G while (a, c) is not, then if $a \rightarrow b$ is the direction given to (a, b), $c \rightarrow b$ must be the direction given to (b, c). For the direction $b \rightarrow c$ would require, by the transitivity of partial ordering, that (a, c) also be an edge of G. Similarly also, if $b \rightarrow a$ is the direction given to (a, b). But only in an even cycle can all successive pairs of edges be given opposite orientations.

Several definitions and lemmas are useful for the argument that the condition of Theorem 1 is also sufficient for G to be a comparability graph.

Two edges (a, b) and (b, c) of a graph G are said to be strongly joined if and only if $(a, c) \notin G$. A path $a_1, \ldots, a_k$ in G is a strong path if and only if for all i, $1 \leqslant i \leqslant k - 2$, $(a_i, a_{i+2}) \notin G$. Two edges (a, b) and (c, d) are strongly connected with ends a and d if and only if there exists a strong path $a_1, a_2, \ldots, a_k$, where k is odd and where $a_1 = a$, $a_2 = b$, $a_{k-1} = c$, and $a_k = d$. Two edges (a, b) and (c, d) are said to be strongly connected if and only if they are strongly connected with ends a and d or strongly connected with ends a and c.

The justification for the apparently restricted definition of "strongly connected with ends" can be seen in the following simple consequences of the definitions. An edge (a, b) is strongly connected to itself with ends a and a

since a, b, a is a strong path. If $a_1, \ldots, a_k$ is a strong path, then so is $a_2, a_1,$ $\ldots, a_k$, or $a_1, \ldots, a_k, a_{k-1}$, or $a_2, a_1, \ldots, a_k, a_{k-1}$. If (a, b) and (c, d) are strongly connected with ends a and d, then they are also strongly connected with ends b and c.

An immediate property of strong connectedness we state as a lemma.

LEMMA 1. *If (a, b) and (e, f) are strongly connected with ends a and f and if (c, d) and (e, f) are strongly connected with ends d and f, then (a, b) and (c, d) are strongly connected with ends a and d.*

Under the assumption that every odd cycle in G has a triangular chord, the following lemmas can be established.

LEMMA 2. *No edges (a, b) and (c, d) of G are both strongly connected with ends a and d and strongly connected with ends a and c.*

Proof. If $a, b_1 (= b), b_2, \ldots, b_k (= c), d$ and $a, b_1' (= b), b_2', \ldots, b_m' (= d), c$ were strong paths with k and m odd, then $a, b_1, b_2, \ldots, b_k, b_m', \ldots, b_1'$ would be an odd cycle without any triangular chords.

LEMMA 3. *Let a, b, c be any triangle in G and let (d, e) be any edge strongly connected to (a, b) with ends b and e. Then one of the following three possibilities must occur:*

(1) (a, b) *is strongly connected to* (a, c) *with ends b and c;*

(2) (a, b) *is strongly connected to* (b, c) *with ends a and c;*

(3) (c, d) *and* (c, e) *are both edges of G and (c, d) is strongly connected to (a, c) with ends a and d and (c, e) is strongly connected to (c, b) with ends b and e.*

Proof. Let $a_1 = b$, $a_2 = a$, $a_3, \ldots, a_{k-1} = d$, $a_k = e$ be a strong path with k odd.

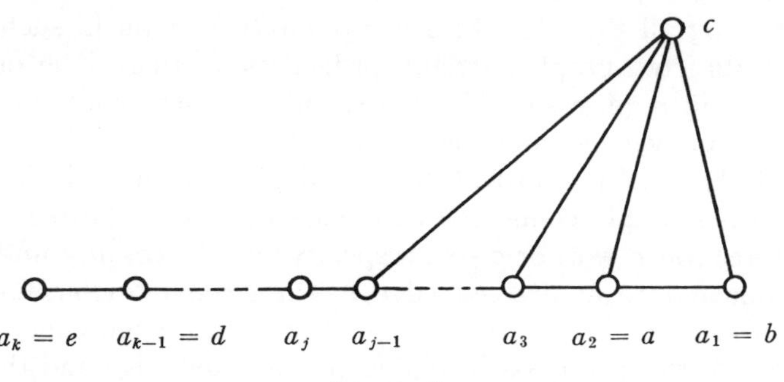

FIGURE 2

Let $j, j \leqslant k$, be such that $(a_i, c) \in G$ for $1 \leqslant i \leqslant j - 1$ and $(a_j, c) \notin G$. If j were odd, $a_1, a_2, a_3, \ldots, a_{j-1}, a_j, a_{j-1}, c, a_{j-3}, c, \ldots, c, a_4, c, a_2, c$ would be a strong path with an odd number of vertices and, therefore, (a, b) and (a, c)

would be strongly connected with ends b and c. If j were even, $a, a_1, a_2, a_3,$ $\ldots, a_{j-1}, a_j, a_{j-1}, c, a_{j-3}, c, \ldots, c, a_3, c, a_1, c$ would be a strong path with an odd number of vertices and, therefore, (a, b) and (b, c) would be strongly connected with ends a and c. Thus, if neither (1) nor (2) is to be the case, there can exist no such j. In particular, therefore, no a_i can be identical with c, since that would require that (a_{i-2}, c) not be an edge of G. Hence, we can assume that (a_{k-1}, c) and (a_k, c) are edges of G. But then $a_1, c, a_3, c, \ldots, c, a_k$ and $a_2, c, a_4, c, \ldots, c, a_{k-1}$ are both strong paths with an odd number of vertices. Therefore, (3) must be the case.

Two corollaries follow immediately from the lemma.

COROLLARY 1. *Let a, b, c be any triangle of G and let d be any vertex for which (c, d) is an edge strongly connected to (a, b). Then one of the possibilities (1) or (2) of Lemma 3 must occur.*

Proof. Since (c, d) is strongly connected to (a, b), it is strongly connected either with ends b and c or with ends b and d. But, in either case, possibility (3) would require that there be an edge joining c to c.

COROLLARY 2. *In a triangle a, b, c of G, if (a, b) and (a, c) are strongly connected with ends b and a, then (a, b) and (b, c) are strongly connected with ends a and c.*

Proof. Let d in Corollary 1 be taken to be a. By hypothesis (a, b) and (a, c) are strongly connected and hence, by Corollary 1, either (a, b) and (b, c) are strongly connected with ends a and c or (a, b) and (a, c) are strongly connected with ends b and c. But, the latter alternative is not possible by Lemma 2 and the hypothesis of the corollary, so that the former alternative is necessarily true.

The proof of the sufficiency for Theorem 1 will provide an algorithm for actually directing all the edges of a comparability graph in such a way that the resulting directed graph partially orders its vertices. The description of the algorithm will require some further definitions involving graphs G' which consist of the same vertices and edges of G but with some of the edges directed.

An edge (a, b) of G' is said to have a strongly determined direction $b \to a$, or $a \to b$, if it is strongly connected with ends a and d to a directed edge (c, d) of G' with direction $c \to d$, or $d \to c$ respectively. Hence, any undirected edge strongly connected to a directed edge has a strongly determined direction which depends upon the direction assigned to the directed edge, and depends upon the ends of the strong path joining the directed edge and the undirected edge.

An edge (a, b) of G' is said to have a transitively determined direction $a \to b$ if there are directed edges (a, c) and (c, b) in G' with directions $a \to c$ and $c \to b$.

G' is consistent if and only if there is no directed cycle; that is, there is no cycle $a_1, \ldots, a_k$ such that $a_1 \to a_2, a_2 \to a_3, \ldots, a_{k-1} \to a_k, a_k \to a_1$ are the directions assigned to its edges. Note that if an edge has two directions in G', then there is a directed cycle in G'.

G' is complete with respect to strong connection if no undirected edge of G' has a direction strongly determined from G'. G' is complete if it is complete with respect to strong connection, and further no undirected edge of G' has a direction transitively determined from G'.

For any edge (a, b) of G let $G(a \to b)$ be the graph obtained from G by giving the edge (a, b) the direction $a \to b$ and by then giving any edge with a determined direction, whether it is already directed or not, that determined direction. By Lemma 2 it follows that no edge of $G(a \to b)$ has two directions assigned to it. For any G' let $G' \cup G(a \to b)$ be the graph obtained from G by giving any of its edges that are directed in either G' or $G(a \to b)$ the directions it has in G' and $G(a \to b)$. For some G' and $G(a \to b)$ it is, therefore, possible for some edges of $G' \cup G(a \to b)$ to receive two directions.

LEMMA 4. *For any edge* (e, f) *of* G, $G(e \to f)$ *is consistent and complete.*

Proof. Let $F = G(e \to f)$. We shall show first that F is complete. By definition, F is complete with respect to strong connection. To show that it is complete with respect to transitive connections, let $c, a,$ and b be any three vertices of G for which (c, a) and (a, b) are edges of G which have been assigned the directions $c \to a$ and $a \to b$ in F. Necessarily, (c, b) is also an edge of F, for otherwise (a, c) and (a, b) would be strongly joined and therefore each would have been assigned two directions, which, as we noted above, is not possible. Further, (a, c) and (e, f) are strongly connected with ends a and f, and (a, b) and (e, f) are strongly connected with ends b and f, so that from Lemma 1 it follows that (a, c) and (a, b) are strongly connected with ends a and b. By Corollary 2 to Lemma 3, therefore, (a, b) and (b, c) are strongly connected with ends a and c. Again, by Lemma 1, then (b, c) and (e, f) are strongly connected with ends c and e. Hence, the edge (b, c) must have received the direction $c \to b$ in F.

The consistency of F is then immediate. For, if $c, a,$ and b are consecutive vertices of a directed cycle in F, then from $c \to a$ and $a \to b$ will follow that (c, b) is in G and is directed $c \to b$. Hence, for any directed cycle in F there is a smaller one, and since there cannot be one with two vertices, there can be none at all.

LEMMA 5. *If* G' *is complete and consistent and* (e, f) *is any undirected edge in* G', *then* $G' \cup G(e \to f)$ *is consistent.*

Proof. Let $F = G(e \to f)$. There are certainly no directed cycles of two vertices in $G' \cup F$ since that would require that a directed edge of F be strongly connected to a directed edge of G' and, therefore, that (e, f) be directed in G'.

Let there be a directed cycle of more than two vertices in $G' \cup F$. Since both G' and F are consistent, the cycle must have edges both (directed) in

G' and in F. If any two consecutive edges of a directed cycle are in G', then since G' is complete, necessarily the chord joining their ends is in G' and so directed that a smaller cycle can be found. We can, therefore, assume that there are consecutive vertices a, b, c, and d in a directed cycle such that $a \rightarrow b$ and $c \rightarrow d$ are directions assigned in F and $b \rightarrow c$ is the direction assigned in G'. Then (a, b) and (c, d) are strongly connected, while (a, b) and (b, c) are not. Further, (a, c) must exist; otherwise, (a, b) and (b, c) would be strongly joined, contradicting $a \rightarrow b$ in F, $b \rightarrow c$ in G'. From Corollary 1 of Lemma 3 it follows that (a, b) and (a, c) are strongly connected with ends b and c. Hence (a, c) is assigned the direction $a \rightarrow c$ in F. But this argument permits one to obtain from any directed cycle in $G' \cup F$ a directed cycle in F, which is not possible.

LEMMA 6. *If G' is consistent and complete with respect to strong connections and the undirected edge (a, b) has $a \rightarrow b$ as a transitively determined direction, then $G' \cup G(a \rightarrow b)$ is consistent.*

Proof. Let $T = G(a \rightarrow b)$. We shall show first that every directed edge in T has a transitively determined direction in G' which is the same as the direction given to it in T. For, let (d, e) be any directed edge in T. We can assume without loss in generality that (a, b) and (d, e) are strongly connected with ends b and e. Since (a, b) is undirected in G', it is necessarily not strongly connected to the directed edges (a, c) and (b, c) in G', which gave (a, b) its transitively determined direction. Possibility (3) of Lemma 3 must, therefore, occur. But, since (a, c) and (c, b) have the directions $a \rightarrow c$ and $c \rightarrow b$ in G', necessarily (c, d) and (c, e) have the directions $d \rightarrow c$ and $c \rightarrow e$, while (d, e) has the direction $d \rightarrow e$ in T.

But, it is therefore possible to replace any directed cycle in $G' \cup T$ by a directed cycle in G' since each edge in the cycle which is in T can be replaced by the two directed edges of G' which transitively determine its direction. This completes the proof of Lemma 6.

Consider now the following algorithm for assigning directions to all the edges of G. Initially in the algorithm G' is G.

(1) Choose any undirected edge (a, b) of G' and a direction $a \rightarrow b$ for it; let $G' = G' \cup G(a \rightarrow b)$ and go to (2). If there is no undirected edge in G', then stop.

(2) If there is an edge (a, b) of G' with a transitively determined direction $a \rightarrow b$, then let $G' = G' \cup G(a \rightarrow b)$ and go to (2). If there is no such edge, then go to (1).

It is evident that $G' \cup G(e \rightarrow f)$ in Lemma 4 and $G' \cup G(a \rightarrow b)$ in Lemma 5 are complete with respect to strong connections. Hence, from Lemmas 4, 5, and 6, one sees that in the finite case the algorithm will produce a partial ordering of the vertices of G consonant with the edges of G. In the infinite case (and the argument embraces the finite case as well), we could partially order all consistent G' with $G' < G''$ if $a \rightarrow b$ in G' implies $a \rightarrow b$ in G''. This partially ordered set has a maximal simply ordered set, by Zorn's lemma, and it is easy to see that the union of the G' in this simply ordered set is a G_0', which is also consistent. If not, every edge in G has been assigned a direction in G_0'; then, using either Lemma 5 or Lemma 6, we would have a contradiction of the maximality of G_0'.

COROLLARY. *Let G' be G with some of its edges directed, where G satisfies the hypothesis of Theorem 1. A necessary and sufficient condition that it be possible to give all edges of G' a direction which partially orders its vertices is that the completion of G' with respect to all strongly determined directions has no directed cycle.*

For, the algorithm given above (or the use of Zorn's lemma) could have begun with any consistent G'.

3. Interval graphs.

3. Interval graphs. If G is any graph, then G^c is the complementary graph; that is, G^c has the same vertices as G but has an edge connecting two vertices if and only if that edge does not occur in G.

THEOREM 2. *A graph G is an interval graph if and only if every quadrilateral in G has a diagonal and every odd cycle in G^c has a triangular chord.*

Proof. The necessity of the conditions is readily seen. For, let α, β, and γ be three intervals such that both α and β and β and γ overlap while α and γ do not overlap. Then, any interval overlapping both α and γ must of necessity overlap β. Also, if α and β are any two intervals that do not overlap, i.e. in G^c an edge joins the vertices corresponding to α and β, then we say $\alpha < \beta$ if every element of α precedes (in O) every element of β. This is clearly a partial ordering; hence G^c is a comparability graph.

To prove the sufficiency of the conditions, we shall show how to construct for any G satisfying the conditions a linearly ordered set O and a set of intervals I from O such that $G = G(O, I)$.

Since G^c is a comparability graph, we can by Theorem 1 assume that all of its edges have been directed in such a way as to partially order its vertices. Because G satisfies the characterizing conditions, the directing of the edges of G^c will also be such as to satisfy the following lemma.

LEMMA. *Let a, b, c, and d be any vertices of G for which (a, b) is an edge of G, (c, d) is an edge of G if $c \neq d$, and for which (a, c) and (b, d) are edges of G^c. Then (a, c) and (b, d) are both directed towards or away from (a, b).*

Proof. If $c \rightarrow a$ and $b \rightarrow d$ are the directions assigned to the edges, then necessarily $c \neq d$, since otherwise transitivity would require that (a, b) be an edge of G^c rather than of G. Also, necessarily, either (a, d) or (b, c) is an edge of G^c, since otherwise a, d, c, b would be a quadrilateral of G without a diagonal. But neither (a, d) nor (b, c) can be an edge of G^c, since neither could be assigned a direction which would not require by transitivity that either (a, b) or (c, d) be an edge of G.

Define a complete subgraph of a graph to be a set of vertices each pair of which is joined by an edge, and a maximally complete subgraph to be one properly contained in no other complete subgraph.

Consider now any set of maximally complete subgraphs of G such that every vertex and edge of G is in at least one of them. Form a graph $\bar{G}$ with vertices such a set of maximally complete subgraphs, and with an edge joining each pair of vertices of G. Every pair of maximally complete subgraphs of G necessarily has at least one edge of G^c connecting a vertex of one to a vertex of the other. Hence, each edge of $\bar{G}$ can be given one or more directions depending upon the directions of edges of G^c joining the maximally complete subgraphs of G corresponding to the ends of the edge. But, from the lemma, it is immediate that each edge in $\bar{G}$ receives a unique direction so that $\bar{G}$ can be regarded as a complete graph with every edge directed.

The directed graph $\bar{G}$ is transitive. For, if not, there would exist three maximally complete subgraphs G_1, G_2, and G_3 of G and six vertices (possibly not all distinct) a, b in G_1, c, d in G_2, and e, f in G_3 such that (b, c), (d, e), and (f, a) are all edges in G^c and have the directions $b \to c$, $d \to e$, and $f \to a$ as in Figure 3, where, if $a \neq b$ (a, b) is an edge of G, if $c \neq d$ (c, d) is an edge of G, and if $e \neq f$ (e, f) is an edge of G. But $a = b$, $c = d$, and $e = f$ is not

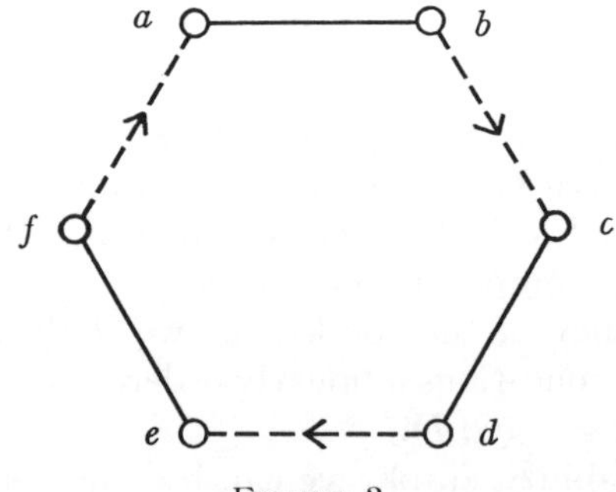

Figure 3

possible since transitivity would be violated in G^c; assume, therefore, that $a \neq b$. We may assume that $a \neq d$ and (a, d) is an edge of G^c, since otherwise the vertices $a, d, e,$ and f would contradict the lemma. Again from the lemma it follows that $a \to d$ is the direction assigned to (a, d) in G^c. From transitivity in G^c, therefore, it follows that (a, e) is in G^c and is directed $a \to e$. But then the vertices a, e, f contradict the lemma. Hence, $\bar{G}$ is transitive.

Since $\bar{G}$ is directed and transitive and since every pair of vertices in $\bar{G}$ has an edge joining them, it linearly orders its vertices. Let O be the vertices of $\bar{G}$ linearly ordered by $\bar{G}$.

We shall say that a vertex of G is a member of an element of O if and only if it is a vertex of the maximally complete subgraph of G corresponding to the element of O. If a vertex a of G is a member of two elements G_1 and G_3 of O, then it is a member of every element G_2 of O lying between G_1 and G_3. For, if not, there would be a vertex b of G_2 not connected to a in G and the edge (a, b) of G^c would have to receive two different directions since G_2 lies between G_1 and G_3. Hence, for any vertex a of G, the set $\alpha(a)$ of all elements of O of which a is a member is an interval of O. Let I be the set of all such intervals of O.

It is immediate that $G = G(O, I)$, for the elements of O correspond to a set of maximally complete subgraphs of G which cover all the vertices and edges of G. Hence, two intervals $\alpha(a)$ and $\alpha(b)$ of I overlap if and only if (a, b) is an edge of G.

COROLLARY. *There is a set O of cardinality equal to the least cardinality of a set of maximally complete subgraphs that contain all the vertices and edges of G. This is the set O of least cardinality.*

Proof. If $G = G(O, I)$, then the set of intervals in I containing a given element of O is a maximally complete subgraph of G.

When G is finite and an interval graph, the only set of maximally complete subgraphs containing all the vertices and edges of G is the set of all maximally complete subgraphs. For, let O and I be as constructed in the proof of the theorem and let G_1 be a maximally complete subgraph which does not correspond to an element of O. The directed edges of G^c, as above, will linearly order $O \cup \{G_1\}$ in such a way that if a vertex of G is a member of any two elements of $O \cup \{G_1\}$, then it is a member of all elements lying between the two. Hence, G_1 cannot be an end-point of $O \cup \{G_1\}$ since it would be necessary that its immediate neighbour in $O \cup \{G_1\}$ contain all of its vertices. But also G_1 cannot be between two other elements G_2 and G_3 since there must be a vertex a of G_1 which is not in G_3 and a vertex b of G_1 which is not in G_2. Since the vertices of G_1 must be contained in its immediate neighbours if they are to be contained in any elements of $O \cup \{G_1\}$, it follows that a is in G_2 and b is in G_3. But the edge (a, b) is in G_1 and, hence, must be in some member G_4 of $O \cup \{G_1\}$, which, therefore, necessarily contains both a and b. G_4 cannot be between G_2 and G_3, since we assumed G_2 and G_3 to be immediate neighbours of G_1. Yet, neither can G_2 lie between G_4 and G_3, nor can G_3 lie between G_2 and G_4, since the first case would imply that b is in G_2, while the second case would imply that a is in G_3.

When G is infinite, however, and an interval graph, then a proper subset of the set of all maximally complete subgraphs may cover all edges and vertices of G. For example, consider the interval graph R arising from the set of all open intervals on the real line. Let S be the set of all maximally complete subgraphs of R, each of which is generated by the intervals containing a rational point. Then S covers all the vertices and edges of R even though the cardinality of A is strictly less than the cardinality of R.

548   P. C. GILMORE AND A. J. HOFFMAN

References

1. S. Benzer, *On the topology of the genetic fine structure*, Proc. Natl. Acad. Sci. U.S., *45* (1959), 1607–1620.
2. A. Ghouilà-Houri, *Caractérisation des graphes nonorientés dont on peut orienter les arêtes de manière à obtenir le graphe d'une relation d'ordre*, C. R. Acad. Sci. Paris, *254* (1962), 1370–1371.
3. P. C. Gilmore and A. J. Hoffman, *Characterizations of Comparability and Interval Graphs*, Abstract, Internat. Congress Mathematicians (Stockholm, 1962), p. 29.
4. G. Hajos, *Über eine Art von Graphen*, Intern. Math. Nachr., *11* (1957), Sondernummer 65.
5. C. G. Lekkerkerker and J. Ch. Bohland, *Representation of a finite graph by a set of intervals in the real line*, Fund. Math., *51* (1962), 45–64.
6. E. S. Wolk, *The comparability graph of a tree*, Proc. Am. Math. Soc., 13 (1962), 789–795.

IBM Research Center

Reprinted from
The Canadian Journal of Mathematics
Vol. 17 (1965), pp. 166–177

SOME PROPERTIES OF GRAPHS WITH MULTIPLE EDGES

D. R. FULKERSON,* A. J. HOFFMAN,† AND M. H. McANDREW

1. Introduction. In this paper we consider undirected graphs, with no edges joining a vertex to itself, but with possibly several edges joining pairs of vertices. The first part of the paper deals with the question of characterizing those sets of non-negative integers $d_1, d_2, \ldots, d_n$ and $\{c_{ij}\}$, $1 \leqslant i < j \leqslant n$, such that there exists a graph G with n vertices whose valences (degrees) are the numbers d_i, and with the additional property that the number of edges joining i and j is at most c_{ij}. This problem has been studied extensively, in the general case **(1, 2, 9, 11)**, in the case where the graph is bipartite **(3, 5, 7, 10)**, and in the case where the c_{ij} are all 1 **(6)**. A complete answer to this question has been given by Tutte in **(11)**. The existence conditions we obtain (Theorem 2.1) are simplifications of Tutte's conditions but are less general, being applicable only in case the graph G_c corresponding to positive c_{ij} satisfies a certain distance requirement on its odd cycles. Our primary interest in Theorem 2.1, however, attaches to the method of proof. For our proof depends on studying properties of certain systems of linear equations and inequalities, in a context which previously has been exploited only in the case when the matrix of the system is totally unimodular, i.e. when every square submatrix has determinant 0, 1, or -1 **(8)**. That similar results can be achieved when this is not so seems to us the principal point of interest of Theorem 2.1 and its proof.

In the second part of the paper we consider the question of performing certain simple transformations on a graph, called "interchanges," so that, by a sequence of interchanges one can pass from any graph in the class $\mathfrak{G}$ of all graphs with prescribed valences $d_1, d_2, \ldots, d_n$ and at most c_{ij} edges joining i and j, to any other graph in $\mathfrak{G}$. It is shown (Theorem 4.1) that if the graph G_c satisfies a certain cycle condition, this is always possible. The cycle condition required here is sufficiently general to include the case of the complete bipartite graph and hence Theorem 4.1 generalizes the interchange theorem of Ryser for (0, 1)-matrices having prescribed row and column sums **(10)**. The cycle condition also includes the case of an ordinary complete graph ($c_{ij} = 1$ for $1 \leqslant i < j \leqslant n$). Thus, following Ryser, one can deduce from Theorem 4.1 that, for any of the well-known integral-valued functions of a graph (such

Received October 15, 1963.

*The work of this author was supported in part by The United States Air Force Project RAND under contract AF 49(638)-700 at The RAND Corporation, Santa Monica, California.

†The work of this author was supported in part by the Office of Naval Research under Contract No. Nonr 3775(00), NR 047040.

as the colouring number), the set of values attained by all graphs having prescribed valences is a consecutive set of integers.

The last part of the paper discusses other applications to the case in which all $c_{ij} = 1$. The existence conditions of Theorem 2.1 simplify considerably in this special case. They are stated explicitly in Theorem 5.1. It is also shown that one can transform an ordinary graph into a certain canonical form by interchanges. This result, suggested by a theorem of Hakimi **(6)** completes a lacuna in Hakimi's proof.

2. Graphs with prescribed valences. Let

$$(2.1) \qquad\qquad d = (d_1, d_2, \ldots, d_n),$$

$$(2.2) \qquad c = (c_{12}, c_{13}, \ldots, c_{1n}, c_{23}, c_{24}, \ldots, c_{2n}, \ldots, c_{n-1, n})$$

be two vectors of non-negative integers, the vector c having $n(n-1)/2$ components. Denote by

$$(2.3) \qquad\qquad \mathfrak{G} = \mathfrak{G}(d, c)$$

the class of all graphs on n vertices having the properties:

(a) the valence (degree) of vertex i is d_i, $1 \leqslant i \leqslant n$;

(b) the number of edges joining vertices i and j is at most c_{ij}, $1 \leqslant i < j \leqslant n$.
We call d the *valence vector* and c the *capacity vector*.

Throughout this paper we adopt the convention that $c_{ji} = c_{ij}$, $1 \leqslant i < j \leqslant n$, and $c_{ii} = 0$. This will simplify matters in writing sums. We also use this convention for other vectors whose components correspond to pairs (i, j), $1 \leqslant i < j \leqslant n$.

Let G_c denote the graph on n vertices in which there is an edge joining vertex i and vertex j if and only if $c_{ij} > 0$. We shall say that the capacity vector c satisfies the *odd-cycle condition* if the graph G_c has the property that any two of its odd (simple) cycles either have a common vertex, or there exists a pair of vertices, one from each cycle, which are joined by an edge. In other words, the distance between any two odd cycles of G_c is at most 1. In particular, if G_c is bipartite (has no odd cycles) or if G_c is complete (all $c_{ij} = 1$), then c obviously satisfies the odd-cycle condition.

THEOREM 2.1. *Assume that c satisfies the odd-cycle condition. Then* $\mathfrak{G}(d, c)$ *is non-empty if and only if*

(i) $\sum^n_{i=1} d_i$ *is even, and*

(ii) *for any three subsets S, T, U which partition $N = \{1, 2, \ldots, n\}$, we have*

$$(2.4) \qquad\qquad \sum_{i \in S} d_i \leqslant \sum_{i \in T} d_i + \sum_{\substack{i \in S \\ j \in S \cup U}} c_{ij}.$$

(Empty sets are not excluded.)

Proof. The cases $n = 1$ and $n = 2$ can easily be handled separately, so in the course of this proof we shall assume that $n \geqslant 3$. Let A be the n by

        D. R. FULKERSON, A. J. HOFFMAN, AND M. H. MCANDREW

$n(n-1)/2$ incidence matrix of all pairs selected from $N = \{1, 2, \ldots, n\}$, Let

$$B = \begin{bmatrix} A & O \\ I & I \end{bmatrix},$$

where I is the identity matrix of order $n(n-1)/2$, and define the vector

$$b = \begin{pmatrix} d \\ c \end{pmatrix}.$$

Then $\mathfrak{G}$ is non-empty if and only if there is a non-negative integral vector z satisfying

(2.5) $$Bz = b.$$

We now break the proof into a series of three lemmas.

LEMMA 2.2. *The equations* (2.5) *have an integral solution if and only if* (i) *holds.*

Lemma 2.2 does not require the non-negativity of b.

Assume first that the equations (2.5) have an integral solution z, and let x be the vector of the first $n(n-1)/2$ components of z. Let u be a vector with n components, each of which is 1. Then

$$u'Ax = 2 \sum_{i<j} x_{ij} = \sum_{i=1}^{n} d_i.$$

Since each x_{ij} is an integer, (i) follows.

To prove Lemma 2.2 in the reverse direction, we exhibit a specific integral solution of $Ax = d$. Clearly such a vector x can be extended to an integral vector z which is a solution of (2.5).

Let $s = \sum_i d_i$. Let

$$\begin{aligned}
x_{12} &= d_1 + d_2 - \tfrac{1}{2}s, \\
x_{13} &= d_1 + d_3 - \tfrac{1}{2}s, \\
x_{23} &= \tfrac{1}{2}s - d_1, \\
x_{1j} &= d_j \qquad \text{for } 3 < j \leqslant n, \\
x_{ij} &= 0 \qquad \text{otherwise.}
\end{aligned}$$

Then this integral vector x clearly satisfies $Ax = d$.

LEMMA 2.3. *The equations* (2.5) *have a non-negative solution if and only if* (ii) *holds.*

It is a consequence of the duality theorem for linear equations and inequalities that (2.5) has a non-negative solution if and only if every vector y satisfying

(2.6) $$y'B \geqslant 0$$

also satisfies

(2.7) $$(y, b) \geqslant 0.$$

Let C be the cone of all vectors y satisfying (2.6). In order to check (2.7), it suffices to look at the extreme rays of C. Let w be a vector on an extreme ray of C, so chosen that all its components are integers and have 1 as their greatest common divisor. Then it can be shown (we omit the details of the proof, since we shall give in §3 another proof of Lemma 2.3) that either every component of y is non-negative (in which case (2.7) is automatic), or else w has the following appearance. Denote the first n components of w by w_i and the last $n(n-1)/2$ components by w_{ij}, $1 \leqslant i < j \leqslant n$. Then there is a partition S, T, U of $N = \{1, 2, \ldots, n\}$ such that

$$w_i = \begin{cases} -1 & \text{for } i \in S, \\ 1 & \text{for } i \in T, \\ 0 & \text{for } i \in U, \end{cases}$$

$$w_{ij} = \begin{cases} 2 & \text{for } i \in S, j \in S, \\ 1 & \text{for } i \in S, j \in U, \\ 0 & \text{otherwise.} \end{cases}$$

If we take the inner product of w with b, then (2.7) is the same as (2.4).

Lemmas 2.2 and 2.3 make no use of the odd-cycle condition imposed on c. But this assumption is essential in Lemma 2.4.

LEMMA 2.4. *Let c satisfy the odd-cycle condition. If the equations (2.5) have both a non-negative solution and an integral solution, then they have a non-negative integral solution.*

Let $Ax = d$, $0 \leqslant x \leqslant c$. The proof proceeds constructively by reducing the number of non-integral components of x. Let G be the graph on n vertices in which an edge joins i and j if and only if x_{ij} is non-integral. Since each d_i is an integer, it follows that if G has edges, then it must contain a cycle, i.e. there is a sequence of distinct integers $i_1, i_2, \ldots, i_k$ such that $x_{i_1 i_2}, x_{i_2 i_3}, \ldots, x_{i_k i_1}$ are non-integral. We now consider cases.

Case 1. G contains an even cycle. Then we alter x by alternately adding and subtracting a real number ϵ around this cycle. This preserves the valence at each vertex, and ϵ can be selected so that (a) the bounds on components of x are not violated, and (b) at least one component of x corresponding to the even cycle has been made integral.

Case 2. G has only odd cycles. Let $1, 2, \ldots, k, 1$ represent an odd cycle of G. Suppose first that two components of x which are adjacent in this cycle have a non-integral sum, say x_{12}, x_{1k}. Then there is a j, distinct from 2 and k, such that x_{1j} is non-integral. It follows from this and the case assumption that G contains a subgraph which consists of two odd cycles joined by exactly one path (which may be of length 0). Let us denote the two odd cycles by $1, 2, \ldots, k, 1$ and $1', 2', \ldots, l', 1'$, and the path joining them by $1, j_1, j_2, \ldots, j_r, 1'$. (Thus $1 = 1'$ if the path has zero length.) Now consider the sequence

170   D. R. FULKERSON, A. J. HOFFMAN, AND M. H. MCANDREW

$$(2.9) \qquad 1, 2, \ldots, k, 1, j_1, j_2, \ldots, j_r, 1', 2', \ldots, l', 1', j_r, j_{r-1}, \ldots, j_1, 1$$

and the components of x corresponding to adjacent pairs of this sequence. Again we alter components of x corresponding to adjacent pairs of (2.9) by alternately adding and subtracting ϵ. This time components of x corresponding to the path joining the two odd cycles are alternately decreased and increased by 2ϵ, whereas components corresponding to the odd cycles are changed by ϵ. The valence at each vertex is preserved, and ϵ may be selected to decrease the number of non-integral components of x without violating $0 \leqslant x \leqslant c$.

It remains to consider the case in which each pair of components of x which are adjacent in the odd cycle $1, 2, \ldots, k, 1$ sum to an integer. Thus we have

$$(2.10) \qquad \begin{aligned} x_{12} + x_{23} \qquad\qquad &= d_2', \\ x_{23} + x_{34} \qquad\qquad &= d_3', \\ \cdots \qquad\qquad& \\ x_{k-1,\,k} + x_{1k} \;\;\; &= d_k', \\ x_{12} \qquad\qquad + x_{1k} &= d_1', \end{aligned}$$

for integers $d_1', d_2', \ldots, d_k'$. The system of equations (2.10) has a unique solution in which, for example,

$$x_{12} = \tfrac{1}{2}(d_2' - d_3' + d_4' - \ldots + d_1').$$

Thus, x_{12} is half of an odd integer, and similarly for other components of x corresponding to the odd cycle. Now, since $\sum^k_{i=1} d_1'$ is odd and $\sum^n_{i=1} d_i$ is even (by Lemma 2.2), the integer $\sum^n_{i=1} d_i - \sum^k_{i=1} d_i'$ is odd. Hence, there must be another component of x not yet accounted for which is also non-integral, and which is consequently contained in another cycle of G, having vertices $1', 2', \ldots, l'$, say. We may assume that this new cycle is odd, disjoint from the first, and that each component of x corresponding to the new cycle is half an odd integer, since otherwise we would be in a situation previously examined. Now, by the odd-cycle assumption on c, we may also assume that $c_{11'} > 0$. If $x_{11'}$ is non-integral, again we have a sequence of form (2.9). If $x_{11'} = 0$, change x as follows: add 1 to $x_{11'}$, subtract 1/2 from x_{12}, add 1/2 to $x_{23}, \ldots$, subtract 1/2 from $x_{1k'}$ subtract 1/2 from $x_{1'2'}$, add 1/2 to $x_{2'3'}, \ldots$, subtract 1/2 from $x_{1'l'}$. If $x_{11'}$ is a positive integer, reverse the alteration just described.

Hence, in all cases the number of non-integral components of x can be decreased. This proves Lemma 2.4 and hence Theorem 2.1.

It can be seen from examples that the odd-cycle assumption on c is essential for the sufficiency part of Theorem 2.1. For let $i_1, i_2, \ldots, i_k$ and $j_1, j_2, \ldots, j_l$ be two odd cycles of G_c violating the odd-cycle condition. Let

$$d_{i_1} = d_{i_2} = \ldots = d_{i_k} = d_{j_1} = d_{j_2} = \ldots = d_{j_l} = 1,$$

all other $d_i = 0$. Thus (i) holds. Moreover, taking components of x corresponding to the two cycles equal to 1/2 and all other components equal to 0 gives a solution of $Ax = d$, $0 \leqslant x \leqslant c$. Hence (ii) holds. But there is no integral solution to $Ax = d$, $0 \leqslant x \leqslant c$.

If each component of the valence vector d is 1, then an integral solution of $Ax = d$, $0 \leqslant x \leqslant c$, corresponds to a perfect matching (1-factor) of the graph G in which c_{ij} edges join i and j. Suppose that G is regular, having valence k at each vertex. Then taking $x_{ij} = c_{ij}/k$ yields a non-negative solution of equations (2.5). Hence Lemma 2.4 implies

THEOREM 2.5. *A regular graph on an even number of vertices which satisfies the odd-cycle condition contains a perfect matching.*

Theorem 2.5 is a generalization of a well-known theorem for bipartite graphs which, rephrased in terms of incidence matrices, asserts that an n by n $(0, 1)$-matrix having k 1's per row and column contains a permutation matrix.

3. Remarks on the connection with bipartite graphs. Let $d_1, d_2, \ldots, d_m$ and $d_{m+1}, d_{m+2}, \ldots, d_n$ be given non-negative integers such that

$$(3.1) \qquad \sum_{i=1}^{m} d_i = \sum_{i=m+1}^{n} d_i,$$

and let s denote this common sum. Let

$$c_{ij} \geqslant 0, 1 \leqslant i \leqslant m, m + 1 \leqslant j \leqslant n,$$

be given non-negative integers. Does there exist a bipartite graph such that the number of edges joining vertex i of $A = \{1, 2, \ldots, m\}$ and vertex j of $B = \{m + 1, m + 2, \ldots, n\}$ is at most c_{ij}, and such that the valence of vertex i is d_i, $1 \leqslant i \leqslant n$? It is well-known **(7)** that such a graph exists if and only if, for every $I \subseteq A$ and $J \subseteq B$ we have

$$(3.2) \qquad \sum_{\substack{i \in I \\ j \in J}} c_{ij} \geqslant \sum_{i \in I} d_i + \sum_{j \in J} d_j - s.$$

Let us illustrate how this result is a consequence of Theorem 2.1. We only treat the sufficiency, since the necessity is, as usual, trivial. The cycle condition on c is, of course, satisfied, and (i) holds, since the sum of the valences is $2s$. We need only show that (3.2) implies (ii). Let S, T, U partition $\{1, 2, \ldots, n\}$. Let $S_1 = S \cap A$, $S_2 = S \cap B$, and similarly define T_1, T_2, U_1, U_2. Take $I = S_1$ and $J = S_2 \cup U_2$. Then, by (3.2), we have

$$(3.3) \qquad \sum_{\substack{i \in S_1 \\ j \in S_2 \cup U_2}} c_{ij} \geqslant \sum_{i \in S_1} d_i + \sum_{j \in S_2 \cup U_2} d_j - s.$$

Now take $I = S_1 \cup U_1$, $J = S_2$. Then, by (3.2), we have

$$(3.4) \qquad \sum_{\substack{i \in S_1 \cup U_1 \\ j \in S_2}} c_{ij} \geqslant \sum_{i \in S_1 \cup U_1} d_i + \sum_{j \in S_2} d_j - s.$$

         D. R. FULKERSON, A. J. HOFFMAN, AND M. H. MCANDREW

Adding (3.3) and (3.4), we obtain

$$\sum_{\substack{i \in S \\ j \in S \cup U}} c_{ij} \geqslant 2 \sum_{i \in S_1} d_i + \sum_{i \in U_1} d_i + 2 \sum_{j \in S_2} d_j + \sum_{j \in U_2} d_j - 2s,$$

or

$$\sum_{\substack{i \in S \\ j \in S \cup U}} c_{ij} \geqslant \sum_{i \in S_1} d_i - \sum_{i \in T_1} d_i + \sum_{j \in S_2} d_j - \sum_{j \in T_2} d_j = \sum_{i \in S} d_i - \sum_{i \in T} d_i,$$

which is inequality (2.4).

On the other hand, we can show that (ii) is sufficient for the existence of a non-negative solution to (2.5), by using the sufficiency of (3.2) for bipartite graphs. Thus, let d and c be the given valence and capacity vector, respectively, for a graph on n vertices. Now consider the bipartite graph on $2n$ vertices, so paired that the ith vertex of part A and the ith vertex of part B are both required to have valence d_i, $1 \leqslant i \leqslant n$. For this bipartite graph, let y_{ij}, $1 \leqslant i, j \leqslant n$, be the number of edges joining vertex i of A and vertex j of B, and suppose that $y_{ij} \leqslant c_{ij}$. Then setting

$$(3.5) \qquad x_{ij} = \tfrac{1}{2}(y_{ij} + y_{ji}), \qquad 1 \leqslant i < j \leqslant n,$$

yields a non-negative solution to (2.5). Hence, it suffices to show that (ii) implies (3.2). Let $I \subseteq \{1, 2, \ldots, n\}$, $J \subseteq \{1, 2, \ldots, n\}$ be given. Let $S = I \cap J$ and let $U = (I - S) \cup (J - S)$, $T = \overline{S \cup U}$. By (ii) we have

$$(3.6) \qquad \sum_{\substack{i \in S \\ j \in S \cup U}} c_{ij} \geqslant \sum_{i \in S} d_i - \sum_{i \in T} d_i.$$

But

$$\sum_{\substack{i \in S \\ j \in S \cup U}} c_{ij} \leqslant \sum_{\substack{i \in I \\ j \in J}} c_{ij}$$

and

$$\sum_{i \in S} d_i - \sum_{i \in T} d_i = 2 \sum_{i \in S} d_i + \sum_{i \in U} d_i - \sum_{i=1}^{n} d_i = \sum_{i \in I} d_i + \sum_{j \in J} d_j - \sum_{i=1}^{n} d_i.$$

Thus, (3.6) implies (3.2).

This connection between Theorem 2.1 and bipartite subgraph theory shows, among other things, that an efficient construction is available for subgraphs, having prescribed valences, of a graph satisfying the odd-cycle condition. For, one can first construct the appropriate bipartite graph by methods known to be efficient **(3)**, and then apply the procedure outlined in the proof of Lemma 2.4 to remove any fractions resulting from (3.5). See also **(1, 2)**.

4. An interchange theorem. Our object in this section is to prove that if the capacity vector c satisfies a certain cycle condition, then for any two graphs $G_1, G_2 \in \mathfrak{G} = \mathfrak{G}(d, c)$, one can pass from G_1 to G_2 by a sequence of simple transformations, each of which produces a graph in $\mathfrak{G}$. These transformations we call "interchanges," following **(10)**, and they are defined as follows. For $G \in \mathfrak{G}$, let y_{ij} denote the number of edges joining i and j. If i, j, k, l are distinct vertices of G with $y_{ij} < c_{ij}$, $y_{jk} > 0$, $y_{kl} < c_{kl}$, and $y_{li} > 0$, an *interchange* adds 1 to y_{ij} and y_{kl}, and subtracts 1 from y_{jk} and y_{li}. Thus, an interchange is the simplest kind of transformation that can produce a new graph in $\mathfrak{G}$.

We now describe the condition to be imposed on the capacity vector c. Let us call a subgraph of G_c which is either an even cycle, or two odd cycles joined by exactly one path P (which may be of length zero), an *even set* of G_c. Observe that the latter kind of even set can be represented as a generalized even cycle, in which the vertices of P are repeated, as was done in the proof of Lemma 2.4. If the two odd cycles consist of vertices $1, 2, \ldots, k$ and $1', 2', \ldots, l'$ respectively, and the path, joining 1 and $1'$, has vertices $1, a_1, a_2, \ldots, a_m, 1'$, then a representation is

$$(4.1) \qquad 1, 2, \ldots, k, 1, a_1, a_2, \ldots, a_m, 1', 2', \ldots, l', 1', a_m, a_{m-1}, \ldots, a_1, 1.$$

We say that c satisfies the *even-set condition* if, for every even set E of G_c, there is a representation of the vertices of E as a generalized even cycle

$$(4.2) \qquad\qquad\qquad b_1, b_2, \ldots, b_{2p}, b_1$$

in which, for some i, b_i and b_{i+3} (the subscripts taken mod $2p$) are joined by an edge of G_c.

THEOREM 4.1. *Let c satisfy the even-set condition. If $G_1, G_2 \in \mathfrak{G}(d, c)$, then G_1 can be transformed into G_2 by a finite sequence of interchanges.*

Proof. We first introduce a distance between pairs of graphs in $\mathfrak{G}$. If x_{ij} is the number of edges joining i and j in one graph, y_{ij} the corresponding number in the other graph, then the *distance* between the graphs is

$$(4.3) \qquad\qquad\qquad \sum_{i<j} |x_{ij} - y_{ij}|.$$

Let $\mathfrak{G}_1$ be the set of all graphs into which G_1 is transformable by finite sequences of interchanges, and let $\mathfrak{G}_2$ be the corresponding set arising from G_2. Let $H_1 \in \mathfrak{G}_1$ and $H_2 \in \mathfrak{G}_2$ be such that the distance between them is the minimum distance between graphs in $\mathfrak{G}_1$ and $\mathfrak{G}_2$. If the distance between H_1 and H_2 is zero, we are finished. Assume, therefore, that it is positive.

We now introduce some notation. If the number of edges joining i and j is greater in H_1 than in H_2, we shall write $(i, j)_1$. If the number is greater in H_2 than in H_1, we shall write $(i, j)_2$. Since H_1 and H_2 are not the same, there must exist at least one pair of vertices i and j such that $(i, j)_1$. Since the valence of j is the same in both graphs, there must exist a vertex k such that $(j, k)_2$. Continuing this way, we must finally obtain a cycle of distinct vertices

$$(4.4) \qquad\qquad\qquad i_1, i_2, \ldots, i_k, i_1$$

174   D. R. FULKERSON, A. J. HOFFMAN, AND M. H. MCANDREW

such that

(4.5) $(i_1, i_2)_1,$ $(i_2, i_3)_2,$ $(i_3, i_4)_1,$ ....

We now consider cases.

Case 1. In (4.4), k is even. We first examine the case $k = 4$. We then have

$$(i_1, i_2)_1,   (i_2, i_3)_2,   (i_3, i_4)_1,   (i_4, i_1)_2.$$

Hence, an interchange on H_1 involving the vertices i_1, i_2, i_3, i_4 yields a graph H_1' in $\mathfrak{G}_1$ which is closer to H_2, violating our assumption on the minimality of the distance between H_1 and H_2. Thus $k > 4$. Suppose now that we have established the impossibility of a cycle (4.4) of length l for all even $l < k$. We shall prove the impossibility of such a cycle of length k. Since c satisfies the even-set condition, and our cycle is an even set in G_c, we may assume without loss of generality that $c_{i_1 i_4} > 0$. Let $x_{i_1 i_4}$ be the number of edges in H_1 joining i_1 and i_4. If $x_{i_1 i_4} < c_{i_1 i_4}$, then we may perform an interchange on H_1 involving i_1, i_2, i_3, i_4 to produce a graph H_1' in $\mathfrak{G}_1$ which is closer to H_2. Hence $x_{i_1 i_4} = c_{i_1 i_4}$. Let $y_{i_1 i_4}$ be the number of edges joining i_1 and i_4 in H_2. An analogous argument shows that $y_{i_1 i_4} = 0$. Since $c_{i_1 i_4} > 0$, we have $(i_1, i_4)_1$. Now consider the sequence $i_1, i_4, i_5, \ldots, i_k, i_1$. This is an even cycle of form (4.4) with length less than k, a contradiction.

Case 2. In (4.4), k is odd. Then we have

$$(i_1, i_2)_1,   (i_2, i_3)_2,   \ldots,   (i_{k-1}, i_k)_2,   (i_k, i_1)_1.$$

Since the valence of i_1 is the same in both graphs, there must be a vertex j_1 such that $(i_1, j_1)_2$. If j_1 is i_r for some $r \neq 1$, then either $i_1, i_2, \ldots, i_r, i_1$ or $i_1, i_k, i_{k-1}, \ldots, i_r, i_1$ is an even alternating cycle which we have shown to be impossible. Similarly, we must have $(j_1, j_2)_1, (j_2, j_3)_2, \ldots$ for new vertices $j_2, j_3, \ldots$ until our sequence terminates with a vertex j_r which is either i_1 or j_t for $t < r$. If $j_r = i_1$, r even, or if $j_r = j_t$, $t < r$, $r - t$ even, again we have an even cycle. In the remaining cases $j_r = i_1$, r odd, or $j_r = j_t$, $t < r$, $r - t$ odd, we have an even set of G_c consisting of two odd cycles joined by just one path. Without loss of generality let

(4.6) $i_1, \ldots, i_k, i_1, j_1, \ldots, j_t, j_{t+1}, \ldots, j_r = j_t, j_{t-1}, \ldots, j_1, i_1$

be that representation of the set which exhibits the even-set condition. Again we shall proceed inductively to show the impossibility of (4.6). The smallest case to consider consists of five vertices arising in the order 1, 2, 3, 1, 4, 5, 1. The even-set condition implies that either $c_{24} > 0$ or $c_{35} > 0$. Without loss of generality, assume $c_{24} > 0$. Since $(2, 3)_2$, $(3, 1)_1$, $(1, 4)_2$, we conclude (reasoning as in Case 1) that $(2, 4)_2$. But then $(1, 2)_1$, $(2, 4)_2$, $(4, 5)_1$, and $(5, 1)_2$ form an even cycle, which we know to be impossible.

Next consider (4.6), assuming inductively that we have established the impossibility of sequences of this type having a smaller number of vertices. Using the even-set condition, the basic line of reasoning we have been following shows that a new even set with a smaller number of vertices in which edges are alternately $(\ )_1$ and $(\ )_2$ (which may or may not be an ordinary even cycle) would also exist, so that either the Case 1 argument applies or the induction assumption is violated.

This completes the proof of Theorem 4.1. We remark that, when G_c is a bipartite graph, if c does not satisfy the even-set condition, then there is a choice of $\{d_i\}$ so that interchanges are not possible. For the only even sets possible in the bipartite case are simple even cycles, and one can easily show by induction that if there is such a cycle $b_1, \ldots, b_{2k}, b_1$, with no edge in G_c joining b_i and b_{i+3} for any i, then there is an even cycle of length >4 for which G_c contains no edges joining vertices of the cycle except vertices adjacent in the cycle. Set $d_i = 1$ for all i in the latter cycle, 0 otherwise. The two graphs are possible, but one cannot reach either from the other by interchanges.

5. Applications to ordinary graphs. In this section we confine attention to the case in which all components of the capacity vector c are 1. Thus, G_c is the complete graph on n vertices. Since the odd-cycle condition and the even-set condition are both satisfied by c, Theorems 2.1 and 4.1 are applicable.

The existence conditions (ii) of Theorem 2.1 simplify enormously in this special case. For, arranging the components of the valence vector in monotonically decreasing order,

$$(5.1) \qquad d_1 \geqslant d_2 \geqslant \ldots \geqslant d_n,$$

it follows at once that all the inequalities (2.4) are equivalent to the $n(n+1)/2$ inequalities

$$(5.2) \qquad \sum_{i=1}^{k} d_i \leqslant \sum_{i=l+1}^{n} d_i + k(l-1), \qquad 1 \leqslant k \leqslant l \leqslant n.$$

If we use the term "ordinary graph" to mean a graph in which at most one edge joins a pair of vertices, we then have

THEOREM 5.1. *There is an ordinary graph on n vertices having valences (5.1) if and only if $\sum_{i=1}^{n} d_i$ is even and the inequalities (5.2) hold.*

The inequalities (5.2) can be further simplified to a system of n inequalities, as follows. Represent the valences (5.1) by an n by n $(0,1)$-matrix whose ith row contains d_i 1's, these being filled in consecutively from the left, except that a 0 is placed in the main diagonal position. Let $\bar{d}_i$, $1 \leqslant i \leqslant n$, be the column sums of this matrix. One can then show that

$$(5.3) \qquad \sum_{i=1}^{k} \bar{d}_i = \operatorname*{Min}_{0 \leqslant l \leqslant n} \left\{ \sum_{i=l+1}^{n} d_i + kl - \operatorname{Min}(k, l) \right\}.$$

176   D. R. FULKERSON, A. J. HOFFMAN, AND M. H. MCANDREW

On the other hand, (5.2) holds for all k, l in $1 \leqslant k \leqslant l \leqslant n$ if and only if the left side of (5.2) is at most the right side of (5.3) for all k in $1 \leqslant k \leqslant n$. Hence, inequalities (5.2) are equivalent to

$$(5.4) \qquad \sum_{i=1}^{k} d_i \leqslant \sum_{i=1}^{k} \overline{d_i}, \qquad 1 \leqslant k \leqslant n.$$

We turn now to the notion of an interchange as applied to ordinary graphs. Here an interchange replaces edges (i, j) and (k, l) with (i, k) and (j, l), the latter pairs being non-edges originally. From Theorem 4.1 we have

THEOREM 5.2. *Let G_1 and G_2 be two ordinary graphs having the same valences. Then one can pass from G_1 to G_2 by a finite sequence of interchanges.*

In connection with Theorem 5.2, we note that an ordinary graph can be transformed by interchanges into a simple canonical form suggested by Hakimi **(6)**. This canonical form, which is the analogue of a similar one for the case of $(0, 1)$-matrices having prescribed row and column sums **(4, 5)**, can be described informally as follows. Assume (5.1). Then there will be edges from vertex 1 to vertices $2, 3, \ldots, d_1 + 1$. Reduce valences appropriately, arrange the new valences in decreasing order, and repeat the process. To prove that this canonical form can be realized, it is sufficient to carry out the first step of distributing the edges at vertex 1 to vertices $2, 3, \ldots, d_1 + 1$. Assume that, by interchanges, we have gone as far as possible in this direction, so there are edges from 1 to $2, \ldots, k$, $k < d_1 + 1$, and no edge from 1 to $k + 1$. Let t be any vertex other than $2, \ldots, k$ which is joined to 1 by an edge. Let u be any vertex joined to $k + 1$ by an edge. If t and u are not joined by an edge, an interchange involving $1, k + 1, u, t$, contradicts our assumption on k. Hence, t and u are joined by an edge. But since u was an arbitrary vertex joined to $k + 1$ by an edge and since t is joined to 1, it follows that the valence of t exceeds that of $k + 1$. This contradicts our scheme for numbering vertices, and hence proves the validity of the canonical form.

This argument provides another proof of Theorem 5.2, since any two ordinary graphs G_1 and G_2 having the same valences can be transformed into the canonical form by interchanges, and hence G_1 can be transformed into G_2.

We also observe that any vertex could play the role of vertex 1 in the construction of the canonical form outlined above, and hence there are a variety of "canonical forms," obtainable by selecting an arbitrary vertex, distributing its edges among other vertices having greatest valences, and repeating the procedure in the reduced problem.

A consequence of Theorem 5.2 is that, for any integer-valued function of a graph which changes by at most 1 under an interchange (e.g., the colouring number), the values attained within the class of all ordinary graphs having prescribed valences form a consecutive set of integers.

REFERENCES

1. J. R. Edmonds, *Paths, trees, and flowers*, presented at the Graphs and Combinatorics Conference (Princeton, 1963).
2. —— *Maximum matchings and a polyhedron with* (0, 1)*-vertices*, presented at the Graphs and Combinatorics Conference (Princeton, 1963).
3. L. R. Ford, Jr. and D. R. Fulkerson, *Flows in networks* (Princeton, 1962).
4. D. R. Fulkerson and H. J. Ryser, *Multiplicities and minimal widths for* (0, 1)*-matrices*, Can. J. Math., *14* (1962), 498–508.
5. D. Gale, *A theorem on flows in networks*, Pacific J. Math. *7* (1957), 1073–1082.
6. S. L. Hakimi, *On realizability of a set of integers as the degrees of the vertices of a linear graph—* I, J. Soc. Ind. and Appl. Math., *10* (1962), 496–507.
7. A. J. Hoffman, *Some recent applications of the theory of linear inequalities to extremal combinatorial analysis*, Proc. Symp. Appl. Math., *10* (1960), 317–327.
8. A. J. Hoffman and J. B. Kruskal, Jr., *Integral boundary points of convex polyhedra; Linear inequalities and related systems*, Annals of Math. Study *38* (Princeton, 1956).
9. O. Ore, *Graphs and subgraphs*, Trans. Amer. Math. Soc., *84* (1957), 109–137.
10. H. J. Ryser, *Combinatorial properties of matrices of zeros and ones*, Can. J. Math., *9* (1957), 371–377.
11. W. T. Tutte, *The factors of graphs*, Can. J. Math., *4* (1952), 314–329.

RAND Corporation
and IBM Research Center

Reprinted from
Colloquio Internazionale Sulle
Teorie Combinatorie (1976), pp. 509–517

R. K. Brayton, Don Coppersmith and A. J. Hoffman [*]

SELF-ORTHOGONAL LATIN SQUARES

RIASSUNTO. — Un quadrato latino auto–ortogonale di ordine n è un quadrato latino di ordine n ortogonale al suo trasposto. Si può vedere che l'esistenza di tale quadrato implica $n \doteq 2, 3, 6$. Noi dimostriamo, viceversa, che se $n \neq 2, 3, 6$, tale quadrato esiste.

Questo lavoro è stato ispirato dal seguente problema (isomorfo) riguardante la programmazione degli sport: organizzare un torneo di tennis all'italiana di doppi misti per n coppie, che eviti i coniugi. In tale torneo, marito e moglie non giocano mai insieme né come compagni né come avversari, ogni coppia di giuocatori dello stesso sesso si incontra (come avversari) esattamente in un match ed ogni coppia di giuocatori di sesso opposto (non sposati) giuoca esattamente un match insieme come compagni e esattamente un match come avversari.

I. INTRODUCTION

E. Nemeth [15] has used the term "self-orthogonal" latin square to denote a latin square orthogonal to ist transpose. The problem of constructing self-orthogonal latin squares is a natural question to consider, was first posed (we believe) by S. K. Stein in [20], and has been treated in [3], [7]–[11], [13]–[16], [18]–[21].

Without being aware of this literature, we were led to examine this question by John Melian [11], director of the Briarcliff Racquet Club, Briarcliff, N. Y., who asked if it were possible to design what might be termed a spouse-avoiding mixed doubles round robin for n couples (SAMDRR(n)) playing tennis. In such a round robin, there are n couples, and each match consists of a pair of players of opposite sex playing a pair of players of opposite sex, with the surnames of all four players different. (Such matches enhance sociability, avoid family tensions and ameliorate the baby-sitter problem). Every two players of the same sex oppose each other exactly once. Every two players of opposite sex (if they are not husband and wife) play together exactly once as partners and exactly once as opponents. Let A $= (a_{ij})$ be a matrix of order n, in which $a_{ii} = i$ and a_{ij} is the surname of the woman who plays with Mr. i in his match with Mr. j. Then it is easy to see that A is a latin square orthogonal to its transpose, and that, conversely, given such a latin square of order n (where we may assume without loss of generality that $a_{ii} = i$), we may construct by the above association a SAMDRR(n).

(*) The work of this author was supported (in part) by the U. S. Army under contract DAHC 04–72–C–0023.

— 510 —

As we will see, the techniques of the celebrated disproof by Bose, Parker and Shrikhande of the Euler conjecture on orthogonal latin squares, combined with the methods of Hanani and Wilson in their remarkable work on block design construction, combined with earlier work on self-orthogonal latin squares can be adapted to solve the problem completely.

THEOREM. *There exists a self–orthogonal latin square of order n if and only if $n \neq 2, 3, 6$.*

So far as we are aware, most previous results on this problem have disposed of various infinite classes of n, or some isolated values of n. An exception is [3], in which the first manuscript outlines a method, based on [18], for treating all sufficiently large n; and the second manuscript reports that calculations based on the method prove the existence of a self-orthogonal latin square of order n for all but 217 values of n. The remarkable work of Wilson [22] also readily implies that a self-orthogonal latin square of order n exists for all n sufficiently large.

2. NOTATIONS AND LEMMAS

We shall adhere to the notation of [5], and make frequent reference to it as well.

(2.1) DEFINITION. A special orthogonal array is an OA (n, s) in which n columns consist of $(i, i, \cdots, i)$, $i = 1, \cdots, n$. We shall delete such columns in the special orthogonal array, so only $n^2 - n$ columns remain. A spouse-avoiding special orthogonal array of order n SOA (n) is a special orthogonal array OA $(n, 4)$ in which, whenever (a, b, c, d) is a column of the array, then so is (c, d, a, b). Note that the interpretation (i, a_{ij}, j, a_{ji}) for the columns of a SOA (n) shows that the set of SOA (n) is isomorphic with the set of SAMDRR (n).

(2.2) DEFINITION. $B = \{n \mid \text{there exists a SAMDRR } (n)\}$.

LEMMA 2.3. *If $n_1, n_2 \in B$, then $n_1 n_2 \in B$. (The usual proof of MacNeish's theorem ([5], p. 191) applies to SOA's).*

LEMMA 2.4. *If $m \in B$, then $3m + 1 \in B$. (The proof in [5], pp. 195, 196 applies to SOA's).*

LEMMA 2.5. *If n is a prime power, $n \neq 2, 3, n \in B$ ([13], [14]).*

(2.5) DEFINITION. A pseudo-geometry $\Pi(v)$ of order v is a collection of v points, together with some distinguished subsets, called lines, such that two distinct points are contained in exactly one line.

— 511 —

This concept goes back at least to Parker [17], and has been fundamental in subsequent work.

LEMMA 2.7. [22] *If the cardinality of each line of* $\Pi(v)$ *is in* B, *the* $v \in$ B.

Proof. Construct a SAMDRR (n) on the points of each line. Then the matches so arranged yield a SAMDRR (v). To find the other players in the match in which Mr. i opposes Mr. j, or Mrs. i opposes Mrs. j, or Mr. i opposes Mrs. j, or Mr. i partners Mrs. j, consult the SAMDRR on the unique line containing i and j.

LEMMA 2.8. [22] *If* $n \in$ B, OA $(n, 5)$ *exists,* $0 \leqq m \leqq n$, $m \in$ B, *then* $4n + m \in$ B.

Proof. We first construct a $\Pi(5n)$. Our points will be all ordered pairs of integers (r, s) where $1 \leqq r \leqq n$, $1 \leqq s \leqq 5$. We will have $5 + n^2$ lines. Line $l_1, \cdots, l_5$ are defined by $l_s = \{(r, s) \mid r = 1, \cdots, n\}$, $s = 1, \cdots, 5$. The other n^2 lines $k_1, \cdots, k_{n^2}$ are determined by the columns of OA $(n, 5)$ in the following way. If the jth column of OA $(n, 5)$ is $(a_{1j}, \cdots, a_{5j})$, then k_j consists of the five points $(a_{1j}, 1)$, $(a_{2j}, 2)$, $\cdots, (a_{5j}, 5)$. Next delete $n - m$ points from l_1, yielding a line l'_1, and let $k'_1, \cdots, k'_{n^2}$ be what is left of $k_1, \cdots, k_{n^2}$. The geometry $\Pi(4n + m)$ has each line cardinality m, n, 4 or 5. By lemmas 2.5 and 2.7, $4n + m \in$ B.

LEMMA 2.9. *For all* $k \geqq 1$, $4k \in$ B *and* OA $(4k, 5)$ *exists.*

Proof. By [5], p. 192, OA $(4k, 5)$ exists except possibly if k is divisible by 3, but not by 9. We shall show OA $(12,5)$ and OA$(24,5)$ exist, which implies OA $(4k, 5)$ exists for all k. But OA $(12,5)$ exists [4], and OA$(24,5)$ exists by deleting a point from EG$(25,2)$ to define a $\Pi(24)$ and apply [5], p. 196, with the clear set consisting of the lines with 4 points.

By lemmas 2.3 and 2.5, $4k \in$ B for all k if $12 \in$ B and $24 \in$ B. To prove $12 \in$ B a special construction is given in the next section. To prove $24 \in$ B, we use $\Pi(24)$ described above.

3. SOME SPECIAL CONSTRUCTIONS

We exhibit in this section examples of self-orthogonal latin squares of orders 10, 12, 14, 15, 18. This example for case 10 is due to Hedayat [7], (an earlier example was constructed by Weisner [21]) and 14 and 18 were constructed by exploiting Hedayat's idea.

$$- \ 512 \ -$$

(3.10)

0	3	9	7	8	1	2	5	6		4
1	6	0	9	2	3	5	4	7		8
2	5	1	6	9	4	0	7	8		3
3	4	7	5	1	9	8	6	2		0
4	0	8	2	7	5	9	3	1		6
5	8	6	3	4	2	7	9	0		1
6	7	3	1	0	8	4	2	9		5
9	1	2	0	5	6	3	8	4		7
8	9	5	4	6	7	1	0	3		2
7	2	4	8	3	0	6	1	5		9

(3.12)

0	8	3	6	2	9	11	1	10	5	7	4
10	1	9	4	7	3	5	5	2	11	0	8
4	11	2	10	5	8	9	0	7	3	6	1
9	5	6	3	11	0	2	10	1	8	4	7
1	10	0	7	4	6	8	3	11	2	9	5
7	2	11	1	8	5	0	9	11	6	3	10
5	7	4	11	1	10	6	2	9	0	8	3
11	0	8	5	6	2	4	7	3	10	1	9
3	6	1	9	0	7	11	5	8	4	11	7
8	4	7	2	10	1	3	11	0	9	5	6
2	9	5	8	3	11	7	4	6	1	10	0
6	3	10	0	9	4	1	8	5	7	2	11

(3.14)

1	9	4	13	10	3	6	11	7	12	2	5	14	8
14	2	10	5	1	11	4	7	12	8	13	3	6	9
7	14	3	11	6	2	12	5	8	13	9	1	4	10
5	8	14	4	12	7	3	13	6	9	1	10	2	11
3	6	9	14	5	13	8	4	1	7	10	2	11	12
12	4	7	10	14	6	1	9	5	2	8	11	3	13
4	13	5	8	11	14	7	2	10	6	3	9	12	1
13	5	1	6	9	12	14	8	3	11	7	4	10	2
11	1	6	2	7	10	13	14	9	4	12	8	5	3
6	12	2	7	3	8	11	1	14	10	5	13	9	4
10	7	13	3	8	4	9	12	2	14	11	6	1	5
2	11	8	1	4	9	5	10	13	3	14	12	7	6
8	3	12	9	2	5	10	6	11	1	4	14	13	7
9	10	11	12	13	1	2	3	4	5	6	7	8	14

(3.15)

13	1	8	5	11	10	3	0	7	4	14	12	6	9	2
3	14	2	9	6	12	11	4	1	8	5	0	13	7	10
11	4	0	3	10	7	13	12	5	2	9	6	1	14	8
9	12	5	1	4	11	8	14	13	6	3	10	7	2	0
1	10	13	6	2	5	12	9	0	14	7	4	11	8	3
4	2	11	14	7	3	6	13	10	1	0	8	5	12	9
10	5	3	12	0	8	4	7	14	11	2	1	9	6	13
14	11	6	4	13	1	9	5	8	0	12	3	2	10	7
8	0	12	7	5	14	2	10	6	9	1	13	4	3	11
12	9	1	13	8	6	0	3	11	7	10	2	14	5	4
5	13	10	2	14	9	7	1	4	12	8	11	3	0	6
7	6	14	11	3	0	10	8	2	5	13	9	12	4	1
2	8	7	0	12	4	1	11	9	3	6	14	10	13	5
6	3	9	8	1	13	5	2	12	10	4	7	0	11	14
0	7	4	10	9	2	14	6	3	13	11	5	8	1	12

(3.18)

1	3	12	7	15	14	2	10	5	8	16	11	17	4	9	13	18	6
18	2	4	13	8	16	15	3	11	6	9	17	12	1	5	10	14	7
15	18	3	5	14	9	17	16	4	12	7	10	1	13	2	6	11	8
12	16	18	4	6	15	10	1	17	5	13	8	11	2	14	3	7	9
8	13	17	18	5	7	16	11	2	1	6	14	9	12	3	15	4	10
5	9	14	1	18	6	8	17	12	3	2	7	15	10	13	4	16	11
17	6	10	15	2	18	7	9	1	13	4	3	8	16	11	14	5	12
6	1	7	11	16	3	18	8	10	2	14	4	5	9	17	12	15	13
16	7	2	8	12	17	4	18	9	11	3	15	6	5	10	1	13	14
14	17	8	3	9	13	1	5	18	10	12	4	16	7	6	11	2	15
3	15	1	9	4	10	14	2	6	18	11	13	5	17	8	7	12	16
13	4	16	2	10	5	11	15	3	7	18	12	14	6	1	9	8	17
9	14	5	17	3	11	6	12	16	4	8	18	13	15	7	2	10	1
11	10	15	6	1	4	12	7	13	17	5	9	18	14	16	8	3	2
4	12	11	16	7	2	5	13	8	14	1	5	10	18	15	17	9	3
10	5	13	12	17	8	3	6	14	9	15	2	7	11	18	16	1	4
2	11	6	14	13	1	9	4	7	15	10	16	3	8	12	18	17	5
7	8	9	10	11	12	13	14	15	16	17	1	2	3	4	5	6	18

4. SOME MORE SPECIAL CONSTRUCTIONS

We use here the method of differences, as explained in [5], p. 201 for the case 26, 30, 38, 42. Consider, for instance, (4.26) describing matrix P_0 whose numbers are taken modulo 19, with indeterminate $x_1, \cdots, x_7$. Let P_1, P_2, P_3 be obtained from P_0 by cyclic permutations of the rows, let $A_0 = (P_0, P_1, P_2, P_3)$, A_i be obtained by adding i to each number in A_0 modulo 19, E be the SOA on $x_1, \cdots, x_7$. The $[A_0, A_1, \cdots, A_{18}, E]$ is the desired SOA (26).

$$- 514 -$$

(4.26)

0	x_1	x_2	x_3	x_4	x_5	x_6	x_7
1	0	0	0	0	0	0	0
3	15	10	7	8	12	9	6
6	1	2	4	6	7	8	10

(4.30)

0	0	x_1	x_2	x_3	x_4	x_5	x_6	x_7
1	10	0	0	0	0	0	0	0
3	7	9	5	7	11	22	19	8
6	2	22	21	19	17	14	10	12

(4.38)

0	0	0	0	x_1	x_2	x_3	x_4	x_5	x_6	x_7
3	1	12	4	0	0	0	0	0	0	0
8	3	6	1	11	27	9	19	6	18	23
15	11	16	18	2	26	7	9	20	13	16

(4.42)

0	0	0	0	0	x_1	x_2	x_3	x_4	x_5	x_6	x_7
3	2	1	14	17	0	0	0	0	0	0	0
8	10	5	3	4	6	13	19	21	27	16	7
17	3	15	26	33	2	28	9	11	22	15	18

5. PROOF OF THEOREM

The proof will rest on Lemmas 2.8, 2.9, the constructions given in § 3 and § 4, and some other constructions based on lemma 2.7. We first remark that the impossibility of $n = 2,6$ follows from the fact that there is no pair of orthogonal latin squares of order 2 or of order 6. The impossibility of 3 (also 2) comes from the fact that each match in a SAMDRR (n) consists of four players with different names.

Now, let $n \neq 2 , 3 , 6$. Write $n = 16k + c$. If $c = 0$, we know already that $n \in B$, since $16k = 4(4k)$, $4k \in B$ by Lemma 2.9, $4 \in B$ by Lemma 2.5, and Lemma 2.3 applies. If $c = 1$, then since $4k \in B$, $1 \in B$, it follows (lemma 2.8) that $n \in B$.

Suppose $c = 2$. If $k = 1$, $n \in B$ by (3.18). If $4k \geq 18$, $n \in B$, by (3.18) and lemma 2.8 (with $m = 18$). So we need only check cases $k = 2 , 3 , 4$, namely $n = 34$, 50, 66. The case $n = 34$ is covered by lemma 2.4, since $11 \in B$ by lemma 2.5. The case $n = 50$ is covered by adding one point at infinity to EG $(2 , 7)$, yielding a pseudo geometry $\Pi (50)$ in which every line has cor-

— 515 —

dinality 7 or 8, with 7, 8 ∈ B by lemma 2.5. To do $n = 66$, consider all 66 points on 5 concurrent lines of PG (2 , 13), together with the intersections of the lines of PG (2, 13) with these points. The resulting Π (66) has every line cardinality 5 or 14. Since 5 ∈ B, and 14 ∈ B by (3.14), it follows from lemma 2.7 that 66 ∈ B.

Suppose $c = 3$. Then 19 ∈ B by lemma 2.5, and $n = 16k + 3 ∈ B$ for all k such that $4k \geq 19$, by reasoning similar to that in case $c = 2$. Therefore, we need only check that 35, 51, 67 ∈ B. But $35 = 5 \times 7$, and 67 is prime, so 35, 67 ∈ B. To prove 51 ∈ B take the points on 5 parallel lines in EG (2, 11) and delete 4 points from this set which are collinear (but the line they are on is not one of the 5 parallel lines). Call " lines " the intersection of lines of EG (2, 11) with this point set. The resulting Π (51) has line cardinalities 11, 5, 4, so 51 ∈ B.

Suppose $c = 4$. Then $n = 16k + 4 ∈ B$ provided $4k \geq 4$, so we need only check that 20 ∈ B, which is true since $20 = 5 \times 4$.

Suppose $c = 5$. Then $n = 16k + 5 ∈ B$ if $4k \geq 5$, so we need only check that 21 ∈ B. This has been shown elsewhere [10], [11], [17]. But it can also be shown by Π (21) = PG (2, 4).

Suppose $c = 6$. Now 22 ∈B by lemmas 2.5 and 2.4. So $n = 16k + 6 ∈ B$ for $k \geq 1$ whenever $4k \geq 22$, so we need only check $n = 38, 54, 70, 86$. But 38 ∈ B by (4.38), 54 ∈ B by deleting one point from a set of 5 parallel lines in EG (2, 11), 70 ∈ B by deleting two points from one of eight parallel lines in EG (2, 9) 86 ∈ B by taking all points on 5 concurrent lines of PG (2, 17).

Suppose $c = 7$. Since 7 ∈ B and $n = 16k + 7 ∈ B$ if $4k \geq 7$, so we need only check $n = 23 ∈ B$ by lemma 2.5.

Suppose $c = 8$. Since 8 ∈ B we need only verify $n = 24 ∈ B$, which was already done in proving lemma 2.9.

Suppose $c = 9$, we need only check $n = 9, 25, 41 ∈ B$, by lemma 2.5.

Suppose $c = 10$, we need only check $n = 10, 26, 42 ∈ B$, which we learn from (3.10), (4.26) and (4.42).

Suppose $c = 11$, we need only check $n = 11, 27, 43 ∈ B$, by lemma 2.5.

Suppose $c = 12$, we need only check 12, 28, 44 ∈ B. But 12 ∈ B follows from (3.12), and 28, 44 ∈ B follow from lemmas 2.3 and 2.5.

Suppose $c = 13$, we need only check 13, 29, 45, 61 ∈ B which follow from lemmas 2.3 and 2.5.

Suppose $c = 14$, we need only check 14, 30, 46, 62, Now (3.14) yields 14 ∈ B, (4.30) yields 30 ∈ B. Take all points on 5 concurrent lines of PF(2,9)

— 516 —

to yield $46 \in B$ with the help of lemma 2.7 and (3.10). Finally delete 3 points from one line of a set of 5 parallel lines of EG (2, 13), to obtain $62 \in B$.

Finally, suppose $c = 15$. We need only check 15, 31, 47, 63, But $15 \in B$ by (3.15), and 31, 47, 63 $\in B$ by lemmas 2.3 and 2.5.

We are very grateful for help received from A. J. W. Hilton, D. Knuth, N. S. Mendolsohn, R. C. Mullin and especially C. C. Lindner.

Added in proofs:

An error in our manuscript occurs in the proof of the theorem in Cases $c = 3$, 3, 6, where we did not cover respectively the cases $n = 82, 83, 102$. Gut these yield respectively to lemmas 2.4, 2.5 and 2.8. We thank Philip Benjamin for calling these lacunae to our attention.

REFERENCES

[1] R. C. BOSE, E. T. PARKER and S. SHRIKHANDE (1960) – *Further results on the construction of mutually orthogonal Latin squares and the falsity of Euler's conjecture*, «Can. J. Math. », *12*, 189–203.

[2] R. C. BOSE and S. SHRIKHANDE (1960) – *On the construction of sets of mutually orthogonal Latin squares and the falsity of a conjecture of Euler*. «Trans. Amer. Math. Soc. », *95*, 191–209.

[3] D. J. CRAMPIN and A. J. W. HILTON – *The spectrum of latin squares orthogonal to their transposes*, manuscript; *Remarks on Sade's disproof of the Euler conjecture with an application to latin squares orthogonal to their transpose*, manuscript.

[4] A. L. DULMAGE, D. M. JOHNSON and N. S. MENDELSOHN (1961) – *Orthomorphisms of groups and orthogonal Latin squares*, I, «Can. J. Math. », *13*, 356–372.

[5] M. HALL JR. (1967) – *Combinatorial Theory*, Blaisdell Publishing Co., Waltham.

[6] H. HANANI (1961) – *The existence and construction of balanced incomplete block designs*, «Ann. Math. Stat. », *32*, 361–386.

[7] A. HEDAYAT (1973) – *An application of sum composition: a self orthogonal latin square of order ten*, « J. Combinatorial Theory », Series A, *14*, 256–260.

[8] J. D. HORTON (1970) – *Variations on a theme by Moore*, Proceedings of the Louisiana Conference on Graph Theory, Combinatorics and Computing, Louisiana State University, Baton Rouge, March 1-5.

[9] C. C. LINDNER (1971) – *The generalized singular direct product for quasigroups*, «Canad. Math. Bull. », *14*, 61–63.

[10] C. C. LINDNER (1971) – *Construction of quasigroups satisfying the identity $x(xy) = yx$*, «Canad. Math. Bull. », *14*, 57–59.

[11] C. C. LINDNER (1972) – *Application of the singular direct product to constructing various types of orthogonal latin squares*, Memphis State University Combinatorial Conference.

[12] JOHN MELIAN, – Oral communications.

[13] N. S. MENDELSOHN (1969) – *Combinatorial designs as models of universal algebras*, Recent progress in combinatorics, Academic Press, Inc., New York.

[14] N. S. MENDELSOHN (1971) – *Latin squares orthogonal to their transposes*, « J. Comb. Theory», Ser. A, *11*, 187–189.

[15] R. C. MULLIN and E. NEMETH (1970) – *A construction for self orthogonal latin squares from certain Room squares*, Proceedings of the Louisiana Conference on Graph Theory,

— 517 —

Combinatorics and computing, Louisiana State University, Baton Route, March 1–5, 213–225.

[16] E. NEMETH – *Study of Room Squares*, Ph. D. Thesis, University of Waterloo.

[17] E. PARKER (1951) – *Construction of some sets of mutually orthogonal Latin squares*, « Proc. Amer. Math. Soc. », *10*, 946–949.

[18] A. SADE (1960) – *Produit direct–singulier de quasigroupes orthogonaux et anti–abéliens*, « Ann. Soc. Sci. Bruxelles », Ser. 1, *74*, 91–99.

[19] A. SADE (1972) – *Une nouvelle construction des quasigroupes orthogonaux à leur conjoint*, « Notices, American Mathamatical Society », *19*, 72T–A105.

[20] S. K. STEIN (1957) – *On the foundations of quasigroups*, « Trans. Amer. Math. Soc. », *85*, 228–256.

[21] L. WEISNER (1963) – *Special orthogonal latin squares of order 10*, « Can. Math. Bull. », *6*, 61–63.

[22] R. M. WILSON (1972) – *An existence theory for pairwise balanced designs*, Part I e II, « J. Comb. Theory », Ser. A, *13*, 220–273.

Reprinted from JOURNAL OF COMBINATORIAL THEORY, Series B
All Rights Reserved by Academic Press, New York and London

Vol. 23, No. 1, August 1977

On Partitions of a Partially Ordered Set

A. J. HOFFMAN*

IBM T. J. Watson Research Center, Yorktown Heights, New York 10598

AND

D. E. SCHWARTZ

City University of New York

Received November 9, 1976

Using linear programming, we prove a generalization of Greene and Kleitman's generalization of Dilworth's theorem on the decomposition of a partially ordered set into chains.

1. INTRODUCTION

In [7], Greene and Kleitman prove an interesting extension of Dilworth's theorem on decompositions of partially ordered sets. Let P be a finite partially ordered set (where the notation $a < b$ will imply $a \neq b$). Let t be a non-negative integer, and let $f(t)$ be the largest cardinality of a subset S of P satisfying the condition that no more than t elements of S are contained in a chain of P. For any collection $\mathscr{C}$ of disjoint chains $C_1 ,..., C_s$ of P such that $P = \cup C_i$, let $g(\mathscr{C}, t) = \sum_{i=1}^{s} \min(t, | C_i |)$. (Here and throughout $| S |$ denotes the cardinality of the set S.) Denote by $g(t)$ the minimum of $g(C, t)$ over all collections $\mathscr{C}$ of disjoint chains whose union is P. It is obvious that $g(t) \geqslant f(t)$.

THEOREM 1.1 [7]. *In the above notation, $g(t) = f(t)$ for all integers $t \geqslant 0$.*

Note that Dilworth's theorem is the case $t = 1$. In proving Theorem 1.1, Greene and Kleitman establish another result interesting in its own right.

THEOREM 1.2. [7]. *For every integer $t \geqslant 0$, there exists a collection $\mathscr{C}$ of disjoint chains whose union is P such that $g(t) = g(C, t)$ and $g(t+1) = g(C, t+1)$*

* This work was supported (in part) by the Army Research Office under Contract DAAG 29-74-C-0007.

3

4   HOFFMAN AND SCHWARTZ

The purposes of this note are twofold. In the first place, the form of Theorem 1.1 suggests that it is a special case of the duality theorem of linear programming; likewise, Theorem 1.2 is redolent of concepts from parametric linear programming. We shall show this is indeed the case, so that ideas from linear programming may be substituted for Greene and Kleitman's ingenious combinatorial arguments.

In the second place, the use of linear programming makes it possible to generalize the Greene–Kleitman theorems in a way which we will explain below.

Before doing so, we first remark that Greene in [6] proves analogs of Theorems 1.1 and 1.2, in which the word chain is replaced by antichain (a subset of P no two elements of which are comparable). A generalization of these analogs, based on switch functions, will be given elsewhere. We also note that [10] contains generalizations of Theorems 1.1 and its analog in a different direction.

First, we introduce a nonnegative integral function defined on all chains $\mathscr{C}$ of P. If C and D are chains with at least one element x, we define (see [8] for a similar idea)

$$(C, x, D) = \{y \mid y \in C, y < x\} \cup \{x\} \cup \{y \mid y \in D, x < y\}.$$

Clearly, (C, x, D) is also a chain of P.

DEFINITION 1.1. A nonnegative integer function $r(C)$ defined on all chains of P is said to be a switch function on P if the following hold:

$$\text{if } C \text{ is a subchain of } D, r(C) \leqslant r(D), \tag{1.1}$$

and

$$\text{if } x \in \mathscr{C} \cap D, r(C) + r(D) = r(C, x, D) + r(D, x, C). \tag{1.2}$$

Note that if r is a switch function, $r + 1$ is also a switch function. Also, the constant function t is a switch function. Let $f(r)$ be the largest cardinality of a subset Q of P such that, for all chains C,

$$|Q \cap C| \leqslant r(C). \tag{1.3}$$

For any collection $\mathscr{C} = \{C_1 ,..., C_s\}$ of disjoint chains whose union is P, define

$$g(C, r) = \sum_{i=1}^{s} \min(r(C_i), |C_i|). \tag{1.4}$$

and

$$g(r) = \min_{\mathscr{C}} g(C, r). \tag{1.5}$$

The results we shall prove, in view of the remarks following Definition 1.1, contain Theorems 1.1 and 1.2.

THEOREM 1.3.   *For every switch function r on P,*

$$g(r) = f(r).$$

THEOREM 1.4.   *For every switch function r on P, there exists a collection $\mathscr{C}$ of disjoint chains whose union is P such that*

$$g(r) = g(C, r) \qquad and \qquad g(r + 1) = g(C, r + 1).$$

The idea behind the generalization is based on [8], which was an exploitation of the original Ford and Fulkerson concepts in the max flowmincut theorem [3]. And the idea behind the proof goes back to the paper by Dantzig and Fulkerson [1], which provided a framework [2] for a (cumbersome) proof of Dilworth's theorem. It is tempting to try to use Fulkerson's elegant proof [4] of Dilworth's theorem to derive Theorems 1.1 and 1.2, but we have not succeeded.

2. PRELIMINARY LEMMAS

We first derive a canonical form for a switch function r.

LEMMA 2.1.   *Let $r(a)$ be a nonnegative integral function defined on the elements of P, and $r(a, b)$ a nonnegative integral function defined on all pairs (a, b) where $a < b$. For any chain $C = \{a_0 < a_1 < \cdots < a_s\}$ in P, define*

$$r(C) = r(a_0) + r(a_1, a_2) + \cdots + r(a_{s-1}, a_s). \tag{2.1}$$

Then (2.1) defines a switch function P. Conversely, every switch function on P arises in this way.

Proof.   Let C and D be chains, with $x \in C \cap D$. This means

$$C = \{a_0 < a < \cdots < a_{u-1} < x < a_{u+1} < \cdots < a_s\},$$
$$D = \{b_0 < b_1 < \cdots < b_{r-1} < x < b_{r+1} < \cdots < b_t\}.$$

Then

$$(C, x, D) = \{a_0 < \cdots < a_{u-1} < x < b_{r+1} < \cdots < b_t\}$$

and

$$(D, x, C) = \{b_0 < \cdots < b_{r-1} < x < a_{u+1} < \cdots < a_s\}.$$

6   HOFFMAN AND SCHWARTZ

Consequently, by (2.1)

$$r(C, x, D) + r(D, x, C)$$
$$r(a_0) + r(a_0, a_1) + \cdots + r(a_{u-1}, x)$$
$$+ r(x, b_{r+1}) + \cdots + r(b_{t-1}, b_t)$$
$$+ r(b_0) + r(b_0, b_1) + \cdots + r(b_{r-1}, x)$$
$$+ r(x, a_{u+1}) + \cdots + r(a_{s-1}, a_s)$$
$$= r(a_0) + \cdots + r(a_{u-1}, x) + r(x, a_{u+1}) + \cdots + r(a_{s-1}, a_s)$$
$$+ r(b_0) + \cdots + r(b_{r-1}, x) + r(x, b_{r+1}) + \cdots + r(b_{t-1}, b_t)$$
$$= r(C) + r(D),$$

verifying (1.2). Of course (1.1) is obvious.

Conversely, assume r is a switch function, and $r(a)$ is the value of r on the one-element chain $\{a\}$. Let $r(\{a < b\})$ be the value of r on the two-element chain $\{a < b\}$, and define $r(a, b) = r(\{a < b\}) - r(a)$. By (1.1), $r(a, b)$ is nonnegative, so all we need show is that (2.1) holds for all C, which we shall establish by induction on the number of elements in C. We know it holds if $|C| = 1$ or 2. Suppose we know it true if $|C| = s$. Now consider a chain D such that $|D| = s + 1$. Then

$$D = \{a_0 < a_1 < \cdots < a_{s-1} < a_s\}.$$

Let $C = \{a_0 < a_1 < \cdots < a_{s-1}\}$, $E = \{a_{s-1} < a_s\}$.
By (1.2)

$$r(C, a_{s-1}, E) + r(E, a_{s-1}, C) = r(C) + r(E). \tag{2.2}$$

But $(C, a_{s-1}, E) = D$, and $(E, a_{s-1}, C) = \{a_{s-1}\}$, so (2.2) becomes

$$r(D) + r(a_{s-1}) = r(C) + r(E). \tag{2.3}$$

Therefore,

$$r(D) = r(C) + r(E) - r(a_{s-1}).$$

By the induction hypotheses,

$$r(D) = r(a_0) + (a_0, a_1) + \cdots + r(a_{s-2}, a_{s-1}) + r(\{a_{s-1} < a_s\}) - r(a_{s-1})$$
$$= r(a_0) + r(a_0, a_1) + \cdots + r(a_{s-1}, a_s),$$

which verifies (2.1).

The foregoing lemma will be used in proving Theorem 1.3, the next in proving Theorem 1.4. The lemma will give certain sufficient conditions for the "t-phenomenon"—i.e., Theorems 1.2 and 1.4 or similar results—to hold.

LEMMA 2.2.   *Let A be an $m \times n$ matrix of rank m, b an m-vector, c and d n-vectors. Assume $P(A, b) \equiv \{x \mid Ax = b, x \geqslant 0\}$ is not empty and that, for each real t, $0 \leqslant t \leqslant 1$, $\min(c + td, x)$, $x \in P(A, b)$ exists. Further, assume that b, d, c are integral, and that the $(m + 1) \times n$ matrix*

$$\begin{bmatrix} A \\ d \end{bmatrix}$$

is totally unimodular.

   Then there exists an integral vector x^0 such that

$$(c, x^0) = \min_{x \in P(A,b)} (c, x)$$

and

$$(c + d, x^0) = \min_{x \in P(A,b)} (c + d, x).$$

   Proof.   What we must show is that, in considering the parametric objective function $(c + td, x)$ on $P(A, b)$ (see [5]) there is a vertex optimal for both $t = 0$ and $t = 1$. This vertex will be integral because A is totally unimodular. Choose a value of t (say $\frac{1}{2}$) between 0 and 1, and let x^0 be the vertex which optimizes $(c + \frac{1}{2}d, x^0)$ on $P(A, b)$. To determine all values of t for which this vertex is optimal, one proceeds as follows. Let B be a basis corresponding to x^0. Find vectors u and v such that

$$u'B = \tilde{c}', \quad v'B = \tilde{d}', \tag{2.4}$$

where $\tilde{c}$ and $\tilde{d}$ are the respective restrictions of c and d to the columns of the basis. If the columns of A are denoted by $A_1, A_2, ..., A_n$, then the set of all t for which x^0 minimizes $(c + td, x)$ on $P(A, b)$ is

$$\{t \mid \forall j, c_j + td_j - (u' + tv') A_j \geqslant 0\}. \tag{2.5}$$

We know (2.5) is nonempty, since $t = \frac{1}{2}$ satisfies all the inequalities in (2.5). We will be done if we can show that, for each j,

$$c_j + td_j - (u' + tv')A_j = e_j + tf_j,$$

where $f_j = 0, \pm 1$ and e_j is an integer. Now,

$$e_j = c_j - u'A_j = c_j - \tilde{c}'B^{-1}A_j. \tag{2.6}$$

Since A is totally unimodular, $B^{-1}A_j$ is an integral vector; further, c is an integral vector. Hence from (2.6), e_j is integral. We also have

$$f_j = d_j - v'A_j = d_j - \tilde{d}'B^{-1}A_j. \tag{2.7}$$

8   HOFFMAN AND SCHWARTZ

If A_j is a column of B, $f_j = 0$. So assume A_j not a column of B. Consider the following matrix with $m + 1$ rows and $m + 2$ columns, and denote its

$$M = \begin{bmatrix} 0 \\ \vdots \\ 0 \\ \hline 1 \end{bmatrix} \begin{bmatrix} B & A \\ \hline \tilde{d}' & d_j \end{bmatrix}_j$$

columns by M_0, M_1,..., M_m, M_{m+1}. The first $m + 1$ columns are linearly independent, so we may write

$$M_{m+1} = a_0 M_0 + a_1 M_1 + \cdots + a_m M_m . \tag{2.8}$$

Clearly, $a_0 = f_j$ from (2.7). Further, f_j is an integer, since B is unimodular, and d is integral. If $f_j = 0$, we are done, so assume otherwise. From (2.8), we may write.

$$M_0 = (a_1/f_j) M_1 - \cdots - (a_m/f_j) M_m + (1/f_j)M_{m+1} . \tag{2.9}$$

But the matrix formed by columns M_1,..., M_{m+1} is unimodular, by hypothesis. Hence, from (2.9), $1/f_j$ is also an integer. Hence, $f_j = \pm 1$.

We remark that, just as the work pioneered by Fulkerson and Edmonds showed that the uses of linear prrogamming in polyhedral combinatorics need not be confined to cases where the matrix of inequalities was totally unimodular, it seems reasonable to believe that interesting instances of the t-phenomenon can arise in cases where the hypotheses of this lemma are not satisfied.

3. PROOF OF THEOREM 1.3

Our approach is to apply the duality theorem to a suitably chosen transportation problem with $n + 1$ rows and columns, indexed 0, 1,..., n, and where 1,..., n refer to the n elements of P. The 0th row and columns have sum n, all other rows and columns have sum 1. The costs c_{ij} are given as follows:

$$c_{00} = c_{1,0} = \cdots = c_{n0} = 0,$$
$$c_{0j} = r(j), \qquad j = 1,..., n,$$

where $r(j)$ comes from Lemma 2.1;

$$c_{ii} = 1, \qquad i = 1,\dots, n,$$

$$c_{ij}\,(i \neq j) = \begin{cases} \infty & \text{if } i \not< j, \\ r(i,j) & \text{if } i < j \text{ (Lemma 2.1)}. \end{cases}$$

In the usual fashion of exhibiting transportation problems in a table listing costs and sums, we have

0	r(1)	. . .	r(n)	n
0	1	r(1,2) . . .	r(1,n)	1
.	r(2,1)	.	.	.
.	.	.	.	.
.	.	.	r(n-1,n)	
0	r(n,1)	. . .	1	1
n	1	. . .	1	

In this table, if $i \not< j$, $r(i,j)$ should be replaced by ∞.

We now minimize $\sum_{i=0}^{n} \sum_{j=0}^{n} c_{ij}x_{ij}$, subject to

$$x_{ij} \geq 0, \qquad \text{all } i \text{ and } j,$$

$$\sum_{j} x_{0j} = \sum_{j} x_{i0} = n,$$

$$\sum_{j} x_{ij} = 1 \qquad \text{for all } i = 1,\dots, n,$$

$$\sum_{i} x_{ij} = 1 \qquad \text{for all } j = 1,\dots, n.$$

At a minimizing vertex, all x_{ij} are integers. Clearly all x_{ij} other than x_{00} will be 0 or 1, and $x_{ij} = 0$ if $i \not< j$. Let $x_{0j_1} > 0$ for some $j_1 > 0$. Then $x_{j_1 j_2} = 1$ for some $j_2 \neq j_1$. If $j_2 = 0$, stop. Otherwise, $x_{j_2 j_3} = 1$ for some $j_3 \neq j_1, j_2$. If $j_3 = 0$, stop. Otherwise, continue in this fashion. Eventually, we must stop. Then we have a chain C of P, $C\{ j_1 < j_2 < \cdots < j_r\}$ (when $x_{j_r 0} = 1$), where $x_{0j_1} = x_{j_1 j_2} = \cdots = x_{j_{r-1} j_r} = x_{j_r 0} = 1$, and the contribution of these positive x_{ij} to the objective function is $r(C)$ by Lemma 2.1. (Note that all other entries in rows and columns $j_1,\dots, j_r$ are 0.)

In this manner, from each nonzero x_{0j}, $j > 0$, we construct a corresponding

10   HOFFMAN AND SCHWARTZ

chain. This gives us a collection of disjoint chains $C_1,...,C_r$ of P. Now consider those (i,j) such that $x_{ij} = 1$, $i > 0$, $j > 0$, but neither i nor j is contained in $C_1 \cup C_2 \cup \cdots \cup C_r$.

Suppose there is such an $x_{i_1 i_2} = 1$, $i_1 \neq i_2$. Then we must have a cycle $i_1 < i_2 < i_3 < \cdots < i_1$, where all $i_k > 0$. But this is impossible, since P is partially ordered. Hence the only nonzero remaining elements x_{ij}, $i > 0$, $j > 0$ are all x_{ii}, $i \notin \bigcup_{i=1}^{r} C_i$. Let us think of these as one-element chains $C_{r+1},...,C_{r+s}$. As for x_{00}, whatever its value, it contributes nothing to the objective function since $c_{00} = 0$. Thus the value of the objective function is

$$\sum_{i=1}^{r} r(C_i) + \sum_{i=r+1}^{r+s} |C_i|. \tag{3.1}$$

If we let $\mathscr{C} = \{C_1,...,C_{r+s}\}$, $\mathscr{C}$ is a collection of disjoint chains including all elements of P. We claim (3.1) is $g(\mathscr{C}, r)$.

Suppose, for $1 \leqslant i \leqslant r$, $|C_i| < r(C_i)$. Let $C_i = \{a_1 < a_2 < \cdots < a_q\}$, which means

$$x_{0a_1} = x_{a_1 a_2} = \cdots = x_{a_q 0} = 1.$$

Set these x's to 0, replace the 0 value of $x_{a_1 a_1},..., x_{a_q a_q}$ by 1; change x_{00} to $x_{00} + 1$, and leave all other x's unchanged. The row and column sum conditions will still be satisfied, and the value of the objective function decreased.

Similarly, suppose for $r + 1 \leqslant i \leqslant r + s$, $r(C_i) = r(i) < |C_i| = 1$; i.e., $r(i) = 0$. Then change x_{ii} from 1 to 0, change x_{0i} and x_{i0} from 0 to 1, change x_{00} to $x_{00} - 1$, and the objective function is decreased. (Note that as long as $x_{ii} = 1$ for some i, $x_{00} > 0$.)

Therefore, the value of the objective function is $g(\mathscr{C}, r)$ for some $\mathscr{C}$.

We shall now prove that the value of the objective function in the dual problem is $|Q|$ for some $Q \subset P$ satisfying (1.3). Since Q satisfying (1.3) and $\mathscr{C}$ a collection of disjoint chains covering P implies $|Q| \leqslant f(r)$, the duality theorem will show $|Q| = f(r)$, which will prove Theorem 1.3.

The dual problem is

$$\text{maximize } n(\xi_0 + \eta_0) + \sum_{i>0} \xi_i + \sum_{j>0} \eta_j,$$

where $\xi_i + \eta_j \leqslant c_{ij}$.

Clearly, we may set ξ_0 to 0 without disturbing either the objective function or the inequalities. Since $c_{00} = 0$, $\eta_0 \leqslant 0$. If $\eta_0 < 0$, replace it by 0, and lower all ξ_i, $i > 0$, by $-\eta_0$. The inequalities are still satisfied and the objective function is unchanged. Hence, we may assume $\xi_0 = \eta_0 = 0$.

We claim that, for each $i = 1,...,n$,

$$\xi_i + \eta_i = 0 \text{ or } 1. \tag{3.2}$$

Suppose (3.2) false, i.e., for some i,

$$\xi_i + \eta_i < 0. \tag{3.3}$$

Since our problem is to maximize, (3.3) would permit us to raise ξ_i unless

$$\xi_i + \eta_j = c_{ij} \qquad \text{for some} \quad j \neq i. \tag{3.4}$$

Similarly, we may assume

$$\xi_k + \eta_i = c_{ki} \qquad \text{for some} \quad k \neq i. \tag{3.5}$$

Suppose $j > k > 0$. Then (3.4) and (3.5) become

$$\xi_i + \eta_j = r(i, j), \qquad \xi_k + \eta_i = r(k, i). \tag{3.6}$$

But $k < i < j$ implies $k < j$, so

$$\xi_k + \eta_j \leqslant r(k, j). \tag{3.7}$$

Now (3.3), (3.6), and (3.7) imply $r(i, j) + r(k, i) = (\xi_i + \eta_i) + (\xi_k + \eta_j <
r(k, j)$.

Therefore, $r(\{k < i < j\}) = r(k) + r(k, i) + r(i, j) < r(\{k < j\}) =
r(k) + r(k, j)$, which violates (1.1).

Next, assume $j = 0$, $k > 0$. Then (3.4) and (3.5) become

$$\xi_i = r(i), \qquad \xi_k + \eta_i = r(k, i).$$

Together with (3.3) and $\xi_k + \eta_0 = \xi_k \leqslant 0$, this means $r(i) + r(k, i) =
\xi_i + \eta_i + \xi_k < r(k, i)$, which implies $r(i) < 0$, impossible.

Next, assume $j > 0$, $k = 0$. From (3.4) and (3.5), $\xi_i + \eta_j = r(i, j)$,
$\eta_i = r(i)$. Therefore, $r(i) + r(i, j) < r(j)$, violating (1.1).

Finally, if $j = 0$ and $k = 0$, we have from (3.4) and (3.5) $\xi_i = 0$, $\eta_i = r(i)$,
so $\xi_i + \eta_i = r(i)$, which cannot be negative. So (3.2) is true.

Let $Q = \{i \mid i > 0, \xi_i + \eta_i = 1\}$. We will be done if we prove (1.3) holds
for all chains C. Let $C = (\{a_1 < a_2 < \cdots < a_s\}$. Recall $\xi_0 = 0 = \eta_0$. Then

$$
\begin{aligned}
|Q \cap C| &\leqslant \xi_0 + \eta_0 + \sum_1^s \xi_{a_i} + \sum_1^s \eta_{a_i} \\
&= (\xi_0 + \eta_{a_1}) + (\xi_{a_1} + \eta_{a_2}) + \cdots + (\xi_{a_{s-1}} + \eta_{a_s}) + (\xi_{a_s} + \eta_0) \\
&\leqslant r(a_1) + r(a_1, a_2) + \cdots + r(a_{s-1}, a_s) + 0 \\
&= r(C),
\end{aligned}
$$

which is (1.3).

12   HOFFMAN AND SCHWARTZ

4. Proof of Theorem 1.4

Retain the same transportation problem as in the preceding section, except that the entries $r(1),..., r(n)$ in the 0th cost row are replaced by $r(1) + t,..., r(n) + t$. We must show that there is a vertex which minimizes the objective function when $t = 0$ and when $t = 1$. Let $C = \{c_{ij}\}$ be the cost vector of the original problem, $d = \{d_{ij}\}$ be defined by

$$d_{01} = \cdots = d_{0n} = 1, \qquad \text{all other } d_{ij} \text{ are } 0.$$

We are minimizing $\sum_{i=0}^{n} \sum_{j=0}^{n} (c_{ij} + t d_{ij}) x_{ij}$, where

$$x_{ij} \geqslant 0,$$

$$\sum_{j} x_{ij} = 1, \qquad i = 1,..., n,$$

$$\sum_{i} x_{ij} = 1, \qquad j = 1,..., n, \tag{4.1}$$

$$\sum_{i} x_{i0} = n.$$

Note that we have not included in (4.1) the equation $\sum_{j} x_{0j} = n$, since it is implied by the others. Thus the matrix of the Eqs. (4.1) has $2n + 1$ rows and is of rank $2n + 1$. All entries in that matrix A are $(0, 1)$, all data are integral, and the matrix

$$M = \begin{bmatrix} A \\ d \end{bmatrix}$$

is totally unimodular, since the rows of the $(0, 1)$ M can be partitioned into two parts, such that every column has at most two nonzeroes, and if two occur they are in different parts [9]. Hence, Lemma 2.2 applies and we are done.

References

1. G. B. Dantzig and D. R. Fulkerson, Minimizing the number of tankers to meet a fixed schedule, *Naval Res. Logist. Quart.* **1** (1954), 217–222.
2. G. B. Dantzig, D. R. Fulkerson, and A. J. Hoffman, Dilworth's theorem on partially ordered sets, *in* "Linear Inequalities and Related Systems" (H. W. Kuhn and A. W. Tucker, Eds.), pp. 207–214, Princeton Univ. Press, Princeton, N. J., 1956.
3. L. R. Ford, Jr. and D. R. Fulkerson, Maximal flow through a network, *Canad. J. Math.* **8** (1956), 399–404.
4. D. R. Fulkerson, Note on Dilworth's decomposition theorem for partially ordered sets, *Proc. Amer. Math. Soc.* **7** (1956), 701–702.
5. S. I. Gass, "Linear Programming: Methods and Applications," McGraw-Hill, New York, 1958.

6. C. Greene, Some partitions associated with a partially ordered set, *J. Combinatorial Theory Ser. A* **20** (1976), 69–79.

7. C. Greene and D. J. Kleitman, Strong versions of Sperner's theorem, *J. Combinatorial Theory Ser. A* **20** (1976), 80–88.

8. A. J. Hoffman, A generalization of max flow–min cut, *Math. Programming* **6** (1974), 352–359.

9. A. J. Hoffman and J. B. Kruskal, Integral boundary points of convex polyhedra, *in* "Linear Inequalities and Related Systems" (M. W. Kuhn and A. W. Tucker, Eds.), pp. 223–246, Princeton Univ. Press, Princeton, N.J., 1956.

10. A. J. Hoffman, J. B. Kruskal, and D. E. Schwartz, On lattice polyhedra, *in* "Proceedings 5th Hungarian Colloquium on Combinatorics, 1976," to appear.

NORTH-HOLLAND

Variations on a Theorem of Ryser

Dasong Cao
Algorithms, Combinatorics and Optimizations
School of Industrial and System Engineering
Georgia Institute of Technology
Atlanta, Georgia 30332

V. Chvátal
Department of Computer Science
Rutgers University
New Brunswick, New Jersey 08903

A. J. Hoffman
IBM Thomas J. Watson Research Center
Yorktown Heights, New York 10598

and

A. Vince
Department of Mathematics
University of Florida
Gainesville, Florida 32611

Submitted by Richard A. Brualdi

ABSTRACT

A famous theorem of Ryser asserts that a $v \times v$ zero-one matrix A satisfying $AA^T = (k - \lambda)I + \lambda J$ with $k \neq \lambda$ must satisfy $k + (v - 1)\lambda = k^2$ and $A^TA = (k - \lambda)I + \lambda J$; such a matrix A is called the incidence matrix of a symmetric block design. We present a new, *elementary* proof of Ryser's theorem and give a characterization of the incidence matrices of symmetric block designs that involves eigenvalues of AA^T. © Elsevier Science Inc., 1997

LINEAR ALGEBRA AND ITS APPLICATIONS 260:215–222 (1997)

                                                    DASONG CAO ET AL.

1. INTRODUCTION

In the first volume of the *Proceedings of the American Mathematical Society*, Ryser [3] proved the following theorem.

RYSER'S THEOREM (Version 1). *Let V be a set of size v, and let S_1, $S_2, \ldots, S_v$ be subsets of V. If there are distinct k and λ such that $|S_i| = k$ for all i and $|S_i \cap S_j| = \lambda$ whenever $i \neq j$, then $k + (v - 1)\lambda = k^2$, each point of V is included in precisely k of the sets S_i, and each pair of distinct points of V is contained in precisely λ of the sets S_i.* ∎

This paper explores variations on Ryser's theorem, in two different spirits. Ryser's original proof, and all other proofs that we have seen or concocted, resort to notions such as determinants, matrix inverses, linear independence, or eigenvalues and rely on results of linear algebra such as

if C is a square matrix such that the equation $Cx = 0$ has a nonzero solution, then $\det CC^T = 0$

or

if A is a square matrix such that equation $AA^Tx = 0$ has no nonzero solution, then there is a matrix B such that $BA = I$.

While use of algebraic techniques to prove a combinatorial theorem is surely not reprehensible, it is natural to wonder if such techniques are necessary. In this particular case, the answer is negative: in Section 2, we shall present an elementary proof of Ryser's theorem.

A *symmetric block design* is any pair $(V, \{S_1, S_2, \ldots, S_v\})$ that satisfies the hypothesis (and the conclusion) of Ryser's theorem. The *incidence matrix A* of this design is the $v \times v$ matrix, with rows indexed by $i = 1, 2, \ldots, v$ and columns indexed by the elements of V, such that the ith row of A is the incidence vector of S_i; equivalently, $A = (a_{ix})$ with

$$a_{ix} = \begin{cases} 1 & \text{if} \quad x \in S_i, \\ 0 & \text{if} \quad x \notin S_i. \end{cases}$$

Note that A is the incidence matrix of a symmetric block design if and only if A is a square zero-one matrix and there are distinct integers k, λ such that

$$AA^T = (k - \lambda)I + \lambda J,$$

where I and J denote as usual the identity and all ones matrix, respectively. In Section 3, we shall prove that these conditions can be weakened: A is the incidence matrix of a symmetric block design if and only if A is a zero-one matrix, A is nonsingular, A has constant row sums, AA^T has precisely two distinct eigenvalues, and AA^T is *irreducible*, meaning that it cannot be permuted to assume the form

$$\begin{pmatrix} B & C \\ 0 & D \end{pmatrix}$$

where B, D are square matrices of positive order. The "only if" part is trivial [in particular, $k + (v - 1)\lambda$ and $k - \lambda$ are the only eigenvalues of a $v \times v$ matrix $(k - \lambda)I + \lambda J$]; our proof of the "if" part relies heavily on the Perron-Frobenius theorem.

2. FIRST VARIATION

Here, we offer an elementary proof of the following generalization of Ryser's theorem:

RYSER'S THEOREM (Version 2).   *Let A be a real $v \times v$ matrix. If there are distinct k and λ such that $AJ = kJ$ and $AA^T = (k - \lambda)I + \lambda J$, then $k + (v - 1)\lambda = k^2$, $JA = kJ$, and $A^TA = (k - \lambda)I + \lambda J$.*

Proof.   Writing $A = (a_{ix})$ and

$$d_x = \sum_i a_{ix}, \qquad d_{xy} = \sum_i a_{ix}a_{iy}, \qquad t = k + (v - 1)\lambda,$$

note that, as $\sum_x(\sum_i a_{ix}) = \sum_i(\sum_x a_{ix})$,

$$\sum_x d_x = vk, \tag{1}$$

and that, as $\sum_x(\sum_i a_{ix})(\sum_j a_{jx}) = \sum_i(\sum_j \sum_x a_{ix}a_{jx})$,

$$\sum_x d_x^2 = vt, \tag{2}$$

and that, as $\Sigma_x(\Sigma_i a_{ix}^2) = \Sigma_i(\Sigma_x a_{ix}^2)$,

$$\sum_x d_{xx} = vk, \tag{3}$$

and that, as $\Sigma_x\Sigma_y(\Sigma_i a_{ix}a_{iy}) = \Sigma_i(\Sigma_x a_{ix})(\Sigma_y a_{iy})$,

$$\sum_x \sum_y d_{xy} = vk^2, \tag{4}$$

and that, as $\Sigma_x\Sigma_y(\Sigma_i a_{ix}a_{iy})(\Sigma_j a_{jx}a_{jy}) = \Sigma_i[\Sigma_j(\Sigma_x a_{ix}a_{jx})(\Sigma_y a_{iy}a_{jy})]$,

$$\sum_x \sum_y d_{xy}^2 = v\left[k^2 + (v-1)\lambda^2\right], \tag{5}$$

and that, as $\Sigma_x\Sigma_y(\Sigma_i a_{ix}a_{iy})(\Sigma_r a_{rx})(\Sigma_s a_{sy}) = \Sigma_i(\Sigma_r\Sigma_x a_{ix}a_{rx})(\Sigma_s\Sigma_y a_{iy}a_{sy})$,

$$\sum_x \sum_y d_{xy}d_x d_y = vt^2. \tag{6}$$

We propose to show that the identities (1)–(6) imply the desired conclusions:

$$t = k^2, \tag{7}$$

$$d_x = k \qquad \text{for all } x, \tag{8}$$

$$d_{xy} = \begin{cases} k & \text{if} \quad x = y, \\ \lambda & \text{if} \quad x \neq y. \end{cases} \tag{9}$$

For this purpose, let us set

$$c_{xy} = \begin{cases} t(k - \lambda) & \text{if} \quad x = y, \\ 0 & \text{if} \quad x \neq y. \end{cases}$$

From (5), (6), (3), and (2), we have

$$\sum_x \sum_y \left(td_{xy} - \lambda d_x d_y - c_{xy}\right)^2 = 0,$$

which, since each $td_{xy} - \lambda d_x d_y - c_{xy}$ is a real number, implies

$$td_{xy} - \lambda d_x d_y - c_{xy} = 0 \qquad \text{for all } x \text{ and } y. \tag{10}$$

From (4) and (1), we have

$$\sum_x \sum_y \left(td_{xy} - \lambda d_r d_y - c_{xy} \right) = v(k - \lambda)(k^2 - t),$$

which, along with (10) and the assumption that $k \neq \lambda$, implies (7). Next, from (2) and (1), we have

$$\sum_x \left(d_x - k \right)^2 = v(t - k^2),$$

which, along with (7) and the assumption that each $d_x - k$ is a real number, implies (8). Finally, writing

$$b_{xy} = \begin{cases} k & \text{if} \quad x = y, \\ \lambda & \text{if} \quad x \neq y, \end{cases}$$

we obtain from (3), (4), (5)

$$\sum_x \sum_y \left(d_{xy} - b_{xy} \right)^2 = 2v\lambda(t - k^2),$$

which, along with (7) and the assumption that each $d_{xy} - b_{xy}$ is a real number, implies (9). ∎

In Version 2, the assumption that A is a *real* matrix can be dropped; see Ryser's second proof of his theorem [4, theorem 2.1, p. 103] or Marshall Hall's extension of the theorem [1, Theorem 10.2.3, p. 104]. However, this assumption is indispensable in our proof; we know of no elementary proof of the generalization of Version 2 where A can be a *complex* matrix.

3. SECOND VARIATION

LEMMA. *If M is a nonnegative irreducible symmetric matrix with exactly two distinct eigenvalues, then $M = uu^T + sI$ for some positive u and some s.*

Proof. Let n denote the order of M, If $n = 2$, then the conclusion follows by setting

$$u = \left[\left(\frac{d + m_{11} - m_{22}}{2} \right)^{1/2}, \quad \left(\frac{d - m_{11} + m_{22}}{2} \right)^{1/2} \right]^{T},$$

$$s = \frac{m_{11} + m_{22} - d}{2}$$

with $M = (m_{ij})$ and $d = [(m_{11} - m_{22})^2 + 4m_{12}]^{1/2}$. Hence we may assume that $n \geqslant 3$.

By the Perron-Frobenius theorem [2, Theorem 9.2.1, p. 285], the characteristic equation of any nonnegative irreducible matrix has a simple root; in particular, the characteristic equation of M has a simple root, r. Every real symmetric matrix of order k has k linearly independent eigenvectors [2, Theorem 29.4, p. 76]; in particular, M has n linearly independent eigenvectors. Since only one of these n eigenvectors corresponds to r, the remaining $n - 1$ eigenvectors must correspond to the other root, s. In other words, the rank of $M - sI$ is 1. Hence $M - sI = ab^T$ for some real vectors a and b. Since M is symmetric, a and b are multiples of each other, and so $M - sI = \pm uu^T$ for some real vector u. Since M is ireducible, no component of u is zero. For any choice of three components u_i, u_j, u_k of u, the three products $u_i u_j$, $u_i u_k$, $u_j u_k$ are off-diagonal entries of M; since M is nonnegative, the three products are nonnegative, and so u_i, u_j, u_k must have the same sign. Hence all components of u have the same sign; replacing u by $-u$ if necessary, we conclude that u is a positive vector and, since M is nonnegative, $M - sI = uu^T$. ∎

THEOREM. *A is the incidence matrix of a symmetric block design if and only if A is a zero-one matrix, A is nonsingular, A has constant row sums, AA^T is irreducible, and AA^T has precisely two distinct eigenvalues.*

Proof. As noted in the introduction, the "only if" part is trivial. To prove the "if" part, we use the Lemma with AA^T in place of M to find that $AA^T = uu^T + sI$ for some positive vector u and some s. Since A is zero-one, the diagonal elements of AA^T equal the row sums of A; since A has constant row sums, it follows that all diagonal elements of AA^T are the same. In turn, since u is a positive vector, it follows that all components of u are the same. Hence $AA^T = sI + tJ$ for some t; since A is nonsingular, $s \neq 0$. We conclude that A is the incidence matrix of a symmetric block design with $k = s + t, \lambda = t$. ∎

This theorem is best possible in the sense that none of its five conditions,

(a) A is a zero-one matrix,
(b) A is nonsingular,
(c) A has constant row sums,
(d) AA^T is irreducible,
(e) AA^T has precisely two distinct eigenvalues,

is implied by the four others:

To see that (a) cannot be dropped, consider

$$
A = \begin{pmatrix}
a & b & b & \cdots & b \\
1 & 2 & 1 & \cdots & 1 \\
1 & 1 & 2 & \cdots & 1 \\
\multicolumn{5}{c}{\cdots\cdots\cdots\cdots\cdots\cdots} \\
1 & 1 & 1 & \cdots & 2
\end{pmatrix}
$$

with $a = 2 - (v - 1)c$, $b = 1 + c$, $c = 2v(v + 2)/(v^3 + v^2 - 2v - 1)$. Since

$$
\frac{a^2 + (v - 1)b^2 - 1}{a + vb} = \frac{a + vb}{v + 2},
$$

the rank of $AA^T - I$ is 1; hence 1 is an eigenvalue of AA^T, and its multiplicity is $v - 1$. The other eigenvalue of AA^T, corresponding to the eigenvector $[a + vb, v + 2, v + 2, \ldots, v + 2]^T$, is $a^2 + (v - 1)b^2 + (v - 1)(v + 2)$; hence A is nonsingular.

To see that (b) cannot be dropped, consider any zero-one matrix A, other than the all ones or the all zeros matrix, such that all the rows of A are the same.

To see that (c) cannot be dropped, take the incidence matrix B of a symmetric block design with $k = \lambda^2 + 3\lambda + 1$ and $v = \lambda^3 + 6\lambda^2 + 10\lambda + 4$. (If $\lambda = 0$ then $B = I$; if $\lambda = 1$, then the design is the projective plane of order four. We do not know for what other values of λ such designs exist.) Then let e denote the all ones vector, and consider

$$
A = \begin{pmatrix} 1 & e^T \\ e & B \end{pmatrix}.
$$

Since

$$
\frac{v + 1 - k + \lambda}{k + 1} = \frac{k + 1}{\lambda + 1},
$$

                                          DASONG CAO ET AL.

the rank of $AA^T - (k - \lambda)I$ is 1; hence $k - \lambda$ is an eigenvalue of AA^T, and its multiplicity is $v - 1$. The other eigenvalue of AA^T, corresponding to eigenvector $[k + 1, \lambda + 1, \lambda + 1, \ldots, \lambda + 1]^T$, is $v + 1 + v(\lambda + 1)$; hence A is nonsingular.

To see that (d) cannot be dropped, consider

$$A = \begin{pmatrix} B & 0 \\ 0 & B \end{pmatrix}$$

such that B is the incidence matrix of a symmetric block design.

To see that (e) cannot be dropped, consider

$$A = \begin{pmatrix} 0 & 1 & 1 & 0 & 0 & \cdots & 0 \\ 1 & 0 & 1 & 0 & 0 & \cdots & 0 \\ 1 & 1 & 0 & 0 & 0 & \cdots & 0 \\ 1 & 0 & 0 & 1 & 0 & \cdots & 0 \\ 1 & 0 & 0 & 0 & 1 & \cdots & 0 \\ \cdots & \cdots & \cdots & \cdots & \cdots & \cdots & \cdots \\ 1 & 0 & 0 & 0 & 0 & \cdots & 1 \end{pmatrix}.$$

We thank D. Coppersmith and A. Krishna for valuable conversations.

REFERENCES

1   M. Hall, Jr., *Combinatorial Theory*. Blaisdell, Waltham, Mass., 1967.
2   P. Lancaster, *Theory of Matrices*, Academic, New York, 1969.
3   H. J. Ryser, A note on a combinatorial problem, *Proc. Amer. Math. Soc.* 1:422–424 (1950).
4   H. J. Ryser, *Combinatorial Mathematics*, Math. Assoc. Amer., 1963.

Received 8 February 1996; final manuscript accepted 14 May 1996

Matrix Inequalities and Eigenvalues

1. The variation of the spectrum of a normal matrix

This paper became popular because Jim Wilkinson had kind words for it in his book on the algebraic eigenvalue problem. What pleased me most about the paper was that it showed that the methods of linear programming could be used to study a problem on estimation of eigenvalues, a fact not widely recognized in 1953. In fact, alternate proofs for the symmetric case were offered by others because the concepts of linear programming were at that time considered exotic by experts in numerical linear algebra! For the symmetric case, our theorem is a consequence of Lidskiï's famous result restricting the spectrum of the sum of two symmetric matric with prescribed spectra, which neither Wielandt nor I realized at that time. I don't think many people realize that now.

It was also my hope that the paper would stimulate professional interaction between two people I admired passionately, George Dantzig and Olga Taussky, but it didn't.

2. Some metric inequalities in the space of matrices

There is a well-known connection between numbers and matrices, in which the polar decomposition of a matrix is analogized to expressing a complex number in polar form, and the "real part" of a complex matrix is analogized to the real part of a complex number. I wish I could recall why Ky Fan and I decided to examine "nearest matrix (of a certain class)" questions in view of that analogy, but I don't. Of course, the relevance of singular values came as no surprise. I first learned about singular values from the work of Robert Schatten, who was a young instructor or assistant professor at Columbia when I was a student.

3. On the nonsingularity of complex matrices

There is a very famous result in matrix theory sometimes called the Lévy–Desplanques theorem, sometimes Gerschgorin's theorem. It asserts that a complex matrix in which, for each row, the modulus of the diagonal entry is larger than the sum of the moduli of the off-diagonal entries, is nonsingular. Equivalently (and this is Gerschgorin's formulation) it asserts that each eigenvalue of a matrix is contained in the union of the disks with centers the diagonal entries and radii the sum of the moduli of the off-diagonal entries. Olga Taussky publicized this theorem in a beautiful 1948 note in the American Mathematical Monthly, and Gerschgorin's theorem is beloved by a generation of matrix theorists. To some extent, this is because it is occasionally useful (indeed, Olga learned of Gerschgorin's theorem and used it in research on flutter problems during World War II). But to a greater extent, it is because of admiration and affection for Olga.

This paper and the two that follow are examples of what I like to call Gerschgorin Variations, as in "variations on a theme of ...". We prove here (using tools from the theory of linear inequalities) that, if you are just paying attention to the moduli of the entries, diagonal dominance as defined above is essentially the only reason all matrices with given moduli can be nonsingular. When we first conjectured this, we were fairly sure the conjecture would be incorrect, arguing ad hominem that, if true, such a fundamental result would have been discovered long ago. But we were wrong, the conjecture wasn't.

Let me use this occasion to state my metaprinciple that conditions for nonsingularity are likely to be convex conditions. Most regions in n-space that we are likely to imagine are convex; hence, to specify a region in matrix space (e.g., a region consisting solely of nonsingular matrices) we are likely to think of convex conditions, are we not?

4. Combinatorial aspects of Gerschgorin's theorem

Nowosad had introduced (and I named) the concept of "Gerschgorin family" or "G-function" to generalize the concept of diagonal dominance further than had been achieved in some famous papers of Ostrowski. (I believe it was Ostrowski who began the practice of generalizing Gerschgorin. I also think that none of the generalizations, including mine, have had much practical impact, but I would be delighted to learn otherwise). The precise definitions of G-family are given in the paper. Here, for various specifications about the G-functions (were they continuous, or homogeneous, etc.?), the necessary patterns of dependence of the functions on the off-diagonal positions were derived. This work simply investigated the questions: we had no premonition of the answers, and simply worked them out.

5. Linear G-functions

Most Gerschgorin Variations have very short proofs. This one is on the long side, but the reward is that all of Ostrowski's Gerschgorin Variations, and many generalizations thereof, are special cases. I first did this work for a Gatlinburg conference, later wrote it up for UC Santa Barbara lectures in 1969. I remember that I thought the question to be attacked was reasonable and doable, and formulated it precisely as the main theorem states, except that I confined myself to patterns of dependence where the kth function depended only on entries in row k and column k. Then the difficulty of handling a huge system of inequalities was solved, after many sleepless nights, by invoking Helly's theorem. The extension to dependence on all off-diagonal entries required some research (in the lawyer's sense of research: note the results quoted in the definition of equillibrant), but no night work.

6. On the relationship between the Hausdorff distance and matrix distance of ellipsoids

There is a mystery about this result. Does the constant have to depend on n? We tried hard to settle this question and failed. I hope someone succeeds.

7. Bounds for the spectrum of normal matrices

I had forgotten the genesis of this paper and checked with Earl Barnes. He told me that (1) he had noticed that a paper we published in 1981 on bounds for eigenvalues of real symmetric matrices could be seen to follow from a paper by de Bruijn published in 1980; (2) I had then commented "if we had to be anticipated, I am glad it was by de Bruijn". Also, these results suggested that the theorem of Mirsky mentioned in this paper could be strengthened. Which we did, in various directions. I was very pleased (but not surprised) that we made strong use of a theorem of Alfred Horn, one of the most creative scholars in matrix theory. I still have copies of some of Horn's letters to me around the time he was formulating his famous conjectures (now proved by Klyachko and Knutson and Tao) about all relations among the spectra of hermitian matrices A, B, C, where $A = B + C$.

Reprinted from
Duke Mathematical Journal
Vol. 20, No. 1 (1953), pp. 37–40

THE VARIATION OF THE SPECTRUM OF A NORMAL MATRIX

By A. J. Hoffman and H. W. Wielandt

If A and B are two normal matrices, what can be said about the "distance" between their respective eigenvalues if the "distance" between the matrices is known? An answer is given in the following theorem (in what follows, all matrices considered are $n \times n$; the Frobenius norm $\| K \|$ of a matrix K is $(\sum_{ij} | k_{ij} |^2)^{1/2}$).

THEOREM 1. *If A and B are normal matrices with eigenvalues $\alpha_1 , \cdots , \alpha_n$ and $\beta_1 , \cdots , \beta_n$ respectively, then there exists a suitable numbering of the eigenvalues such that* $\sum_i | \alpha_i - \beta_i |^2 \leq \| A - B \|^2$.

Proof. Let A_0 and B_0 denote the diagonal matrices with diagonal elements $\alpha_1 , \cdots , \alpha_n$ and $\beta_1 , \cdots , \beta_n$ in arbitrarily fixed order. Since A and B are normal, there are unitary matrices U and V such that $A = UA_0U^*$ and $B = UVB_0V^*U^*$. Then we have $\| A - B \| = \| A_0 - VB_0V^* \|$; hence, Theorem 1 is equivalent to

(1) *The minimum of* $\| A_0 - VB_0V^* \|^2$, *where V ranges over the set of all unitary matrices, is attained for $V = P$, where P is an appropriate permutation matrix.*

To prove (1), observe that

$$\| A_0 - VB_0V^* \|^2 = \text{Trace} \, (A_0 - VB_0V^*)(A_0^* - VB_0V^*)$$

$$= \text{Trace} \, (A_0A_0^* + B_0B_0^*) + r(V),$$

where $r(V) = \sum_{ij} d_{ij}w_{ij} \, ; \, d_{ij} = -(\alpha_i\bar{\beta}_j + \bar{\alpha}_i\beta_j), \, w_{ij} = v_{ij}\bar{v}_{ij} , \, V = (v_{ij})$. Hence, min $\| A_0 - VB_0V^* \|^2$ is attained at a V for which $r(V)$ is a minimum.

Let $\mathfrak{X}_n$ be the set of all matrices $X = (x_{ij})$ such that

(2) $$\sum_i x_{ij} = 1, \qquad \sum_j x_{ij} = 1, \qquad x_{ij} \geq 0 \qquad (i, j = 1, \cdots , n).$$

Let $\mathfrak{W}_n$ be the set of all matrices $W = (w_{ij}) = (v_{ij}\bar{v}_{ij})$, with $V = (v_{ij})$ a unitary matrix. Then $\mathfrak{W}_n$ is a subset of $\mathfrak{X}_n$ (indeed, $\mathfrak{W}_n$ is a proper subset, if $n \geq 3$, in view of

Received April 2, 1952; the work of A. J. Hoffman was supported (in part) by the Office of Scientific Research, USAF.

A. J. HOFFMAN AND H. W. WIELANDT

$$\begin{bmatrix} \frac{1}{2} & \frac{1}{2} & 0 & & & & \\ \frac{1}{2} & 0 & \frac{1}{2} & & & & \\ 0 & \frac{1}{2} & \frac{1}{2} & & & & \\ & & & 1 & & & \\ & & & & \ddots & & \\ & & & & & & 1 \end{bmatrix}$$

which is in $\mathfrak{X}_n$, but not in $\mathfrak{W}_n$).

Consider each X as a point in n^2-dimensional affine space whose co-ordinates are its coefficients. Then (2) implies that $\mathfrak{X}_n$ is a closed, bounded, convex polyhedron, and we shall show that (1) is implied by the following lemma.

LEMMA. *The vertices of $\mathfrak{X}_n$ are the permutation matrices.*

Proof. Other proofs of this lemma or generalizations of it are in the literature (see, for example, [1], [2]), but for the reader's convenience we give a simple *ad hoc* demonstration.

The polyhedron $\mathfrak{X}_n$ is the intersection of the $2n - 1$ hyperplanes and n^2 half-spaces given in (2) (the $2n$ equations given by the first two relations in (2) are clearly dependent). Hence, every vertex of $\mathfrak{X}_n$ must lie on the bounding hyperplane of at least $n^2 - (2n - 1)$ of the half-spaces; that is, $x_{ij} = 0$ for at least $(n - 1)^2$ pairs i, j. This shows that at least one column of any vertex consists entirely of 0 except for one entry, which must be 1; and the same must be true for the row containing that 1. If we delete this row and column, we obtain a matrix of order $n - 1$ that also satisfies conditions (2) if n is replaced by $n - 1$, and must also be a vertex of $\mathfrak{X}_{n-1}$. Hence, by induction, every vertex of $\mathfrak{X}_n$ has the property that each column (row) consists entirely of 0 except for one entry which is 1, *i.e.* every vertex is a permutation matrix. Since it is trivial that every permutation matrix is a vertex, the lemma is proven.

The set of points at which a linear form defined on a convex body attains its minimum always includes a vertex. Hence $\sum_{ij} d_{ij} x_{ij}$ attains its minimum at some permutation matrix). But since $\mathfrak{W}_n$ is a subset of $\mathfrak{X}_n$

$$\min_{\mathfrak{W}_n} \sum_{ij} d_{ij} w_{ij} \geq \min_{\mathfrak{X}_n} \sum_{ij} d_{ij} x_{ij} \, .$$

Since P is in $\mathfrak{W}_n$ as well as $\mathfrak{X}_n$, $\min_{\mathfrak{W}_n} \sum_{ij} d_{ij} w_{ij}$ is attained for $W = P$, thus $r(V)$ reaches its minimum for $V = P$. This completes the proof of (1) and hence of Theorem 1.

Remarks. 1. It is clear that essentially the same proof, with obvious changes, will also show that it is possible to renumber the eigenvalues so that

$$\sum_i |\,\alpha_i - \beta_i\,|^2 \geq \|\,A - B\,\|^2.$$

2.   Although the arrangement of the eigenvalues mentioned in Theorem 1 is difficult, in general, to describe more explicitly, it is easy in the special case that A is Hermitian.   Then a "best" arrangement is

$$(3) \qquad\qquad \alpha_1 \geq \cdots \geq \alpha_n \ ; \qquad \operatorname{Re} \beta_1 \geq \cdots \geq \operatorname{Re} \beta_n$$

Proof.   Assume the α_i are in the order given in (3) and the β_i are not; say $\operatorname{Re} \beta_1 \not\geq \operatorname{Re} \beta_2$.   Because

$$|\,\alpha_1 - \beta_1\,|^2 + |\,\alpha_2 - \beta_2\,|^2 \geq |\,\alpha_1 - \beta_2\,|^2 + |\,\alpha_2 - \beta_1\,|^2,$$

$\sum_i |\,\alpha_i - \beta_i\,|^2$ is not increased by interchanging β_1 and β_2 .   Hence, by successive steps, each consisting of an interchange of two β_i , we can bring the β_i to the order in (3) without increasing $\sum_i |\,\alpha_i - \beta_i\,|^2$.   If the original arrangement is "best", then so is (3).

3.   Theorem 1 is false if we do not require both A and B to be normal.   Let $A = \left(\begin{smallmatrix} 0 & 0 \\ 0 & 4 \end{smallmatrix}\right)$, $B = \left(\begin{smallmatrix} -1 & -1 \\ +1 & +1 \end{smallmatrix}\right)$.   Then A is normal but B is not, and $\|\,A - B\,\|^2 = 12$; $\sum_i |\,\alpha_i - \beta_i\,|^2 = 16$ for any ordering of the eigenvalues.

4.   Let us make precise the notion of "distance" between spectra.   If $\alpha = \{\alpha_1 , \cdots , \alpha_n\}, \beta = \{\beta_1 , \cdots , \beta_n\}$ are each a set of n complex numbers, we define

$$d(\alpha, \beta) = \min \left(\sum_i |\,\alpha_i - \beta_{\sigma i}\,|^2 \right)^{1/2}$$

where $(\sigma 1, \cdots , \sigma n)$ runs through all permutations of $(1, \cdots , n)$.   Using this concept, Theorem 1 essentially gives a complete solution to the question: If A is a normal matrix with spectrum α and k is a positive number, what spectrum β can occur for a normal matrix B such that $\|\,A - B\,\| \leq k$?

THEOREM 2.   *If A is a normal matrix with spectrum α and k is a positive number, then β is the spectrum of a normal matrix B with $\|\,A - B\,\| \leq k$ if and only if $d(\alpha, \beta) \leq k$.*

Proof.   The necessity is given by Theorem 1.   The sufficiency is easily demonstrated by letting A_0 be the diagonal matrix with entries $\alpha_1 , \cdots , \alpha_n$, B_0 be the diagonal matrix with entries $\beta_1 , \cdots , \beta_n$, numbered so that $d(\alpha, \beta) = \left(\sum_i |\,\alpha_i - \beta_i\,|^2\right)^{1/2}$.   We know there is a unitary U such that $A = U A_0 U^*$.   Then $B = U B_0 U^*$ has the required property.

REFERENCES

1. GARRETT BIRKHOFF, *Three observations on linear algebra*, Universidad Nacional de Tucumán, Revista.   Serie A. Matemáticas y Física Teórica, vol. 5(1946), pp. 147–151.
2. G. B. DANTZIG, *Application of the Simplex Method to a Transportation Problem*, Chapter XXIII of Activity Analysis of Production and Allocation, edited by T. C. Koopmans, New York, 1951.

NATIONAL BUREAU OF STANDARDS,
AMERICAN UNIVERSITY AND UNIVERSITY OF TÜBINGEN.

Reprinted from the
PROCEEDINGS OF THE AMERICAN MATHEMATICAL SOCIETY
Vol. 6, No. 1, pp. 111-116
February, 1955

SOME METRIC INEQUALITIES IN THE SPACE OF MATRICES[1]

KY FAN AND A. J. HOFFMAN

1. In this note we shall prove three inequalities suggested by the well-known analogy between matrices and complex numbers. These are the matricial analogues of the following simple numerical inequalities:

(a) If $z = |z| \cdot e^{i\theta}$, θ real, and if α is any real number, then

$$\left| z - e^{i\theta} \right| \leq \left| z - e^{i\alpha} \right| \leq \left| z + e^{i\theta} \right|.$$

(b) If z is complex, x real, then

$$\left| z - \mathrm{Re}\, z \right| \leq \left| z - x \right|.$$

(c) If x and y are real, then

$$\frac{1}{2} \left| \frac{x - i}{x + i} - \frac{v - i}{y + i} \right| \leq \left| x - y \right|.$$

In developing the matricial statements corresponding to (a), (b), (c), we must replace the modulus of a complex number by a suitably chosen norm for matrices. Let M_n denote the linear space of all square matrices of order n with complex coefficients. A norm on M_n is a real-valued function $\| \cdot \|$ defined on M_n such that: (i) $\|A\| \geq 0$; (ii) $\|A\| = 0$ if and only if $A = 0$; (iii) $\|cA\| = |c| \cdot \|A\|$ for any complex number c; (iv) $\|A + B\| \leq \|A\| + \|B\|$. A norm $\| \cdot \|$ is *unitarily invariant* if it satisfies the additional condition: (v) $\|A\| = \|UA\| = \|AU\|$ for every unitary matrix U of order n. It is rather noteworthy that *the matricial analogues of* (a), (b), (c) *hold for any norm that is unitarily invariant*.

For any matrix $A \in M_n$, the non-negative square roots of the eigenvalues of A^*A will be called *singular values* of A. The following result of J. von Neumann [3] characterizes all unitarily invariant norms on M_n. For any symmetric gauge function[2] Φ of n real variables, the function $\| \cdot \|$ defined on M_n by

---

Received by the editors, January 25, 1954.

[1] The preparation of this paper was sponsored in part by the Office of Scientific Research, USAF; and in part by the Office of Naval Research, USN.

[2] Following von Neumann, a gauge function Φ (in the sense of Minkowski) is called symmetric if $\Phi(t_1, t_2, \cdots, t_n) = \Phi(\epsilon_1 t_{j_1}, \epsilon_2 t_{j_2}, \cdots, \epsilon_n t_{j_n})$ for any combination of signs $\epsilon_i = \pm 1$ and for any permutation $(j_1, j_2, \cdots, j_n)$ of $(1, 2, \cdots, n)$. For general properties of symmetric gauge functions, see [4, pp. 84–92].

 KY FAN AND A. J. HOFFMAN [February

$$(1) \qquad \|A\| = \Phi(\alpha_1, \alpha_2, \cdots, \alpha_n) \qquad (A \in M_n),$$

where $\alpha_1, \alpha_2, \cdots, \alpha_n$ are the singular values of A, is a unitarily invariant norm. Conversely, every unitarily invariant norm on M_n can be obtained in this way; let $\Phi(\alpha_1, \alpha_2, \cdots, \alpha_n) = \|A\|$, where A is a diagonal matrix with diagonal entries $\alpha_1, \alpha_2, \cdots, \alpha_n$.

Let $A, B \in M_n$. Let $\alpha_1 \geq \alpha_2 \geq \cdots \geq \alpha_n$ and $\beta_1 \geq \beta_2 \geq \cdots \geq \beta_n$ be the singular values of A and B respectively. Then it is known [1, Theorem 4] that $\|A\| \leq \|B\|$ for every unitarily invariant norm $\|\cdot\|$ if and only if

$$(2) \qquad \sum_{i=1}^{k} \alpha_i \leq \sum_{i=1}^{k} \beta_i \qquad (1 \leq k \leq n).$$

According to these known results, the proof of the matricial analogues of (a), (b), (c) amounts to showing certain inequalities involving singular values.

In our proof of the matricial analogue of (a), we shall need the following theorem:[3] If X, Y, Z are Hermitian matrices of order n, with eigenvalues

$$x_1 \geq x_2 \geq \cdots \geq x_n, \quad y_1 \geq y_2 \geq \cdots \geq y_n, \quad z_1 \geq z_2 \geq \cdots \geq z_n$$

respectively, and if $X - Y = Z$, then

$$(3) \qquad \mathrm{Max}_{j_1 < j_2 < \cdots < j_k} \sum_{i=1}^{k} (x_{j_i} - y_{j_i}) \leq \sum_{i=1}^{k} z_i \qquad (1 \leq k \leq n).$$

2. **Theorem 1.** *Let $A \in M_n$ and $A = UH$, where U is unitary and H is Hermitian positive semi-definite. Then for any unitary matrix $W \in M_n$,*

$$(4) \qquad \|A - U\| \leq \|A - W\| \leq \|A + U\|$$

holds for every unitarily invariant norm.

Since the norm is unitarily invariant, we have

$$\|A \mp U\| = \|U(H \mp I)\| = \|H \mp I\|,$$
$$\|A - W\| = \|U(H - U^*W)\| = \|H - U^*W\|.$$

It follows that Theorem 1 is equivalent to the following apparently less general theorem:

Theorem 1′. *Let $H, V \in M_n$. If H is Hermitian positive semi-defi-*

---

[3] See [5, Theorem 2]. An equivalent geometric formulation of this result is stated in [2, Theorem 1].

nite, and V is unitary, then

$$(5) \qquad \|H - I\| \leq \|H - V\|,$$

$$(6) \qquad \|H - V\| \leq \|H + I\|$$

hold for every unitarily invariant norm.

PROOF OF (5). We first digress to define, for any matrix M of order n, the Hermitian matrix

$$\tilde{M} = \begin{pmatrix} 0 & M \\ M^* & 0 \end{pmatrix}$$

of order $2n$. Then it is easy to see that the eigenvalues of $\tilde{M}$ are precisely the singular values of M and their negatives.[4]

Let $A = H - I$, $B = H - V$. Let $\alpha_1 \geq \alpha_2 \geq \cdots \geq \alpha_n$ be the singular values of A; $\beta_1 \geq \beta_2 \geq \cdots \geq \beta_n$ be the singular values of B; $\eta_1 \geq \eta_2 \geq \cdots \geq \eta_n$ be the eigenvalues (also singular values) of H. Then

$$\sum_{i=1}^{k} \alpha_i = \operatorname*{Max}_{i_1 < i_2 < \cdots < i_k} \sum_{i=1}^{k} |\eta_{i_i} - 1| \qquad (1 \leq k \leq n).$$

To prove (2), which will imply (5), we must show

$$(7) \qquad \operatorname*{Max}_{i_1 < i_2 < \cdots < i_k} \sum_{i=1}^{k} |\eta_{i_i} - 1| \leq \sum_{i=1}^{k} \beta_i \qquad (1 \leq k \leq n).$$

This inequality (7) will be obtained, if we apply the theorem mentioned above to $\tilde{H} - \tilde{V} = \tilde{B}$. In fact, according to the remark made at the beginning of this proof, the eigenvalues of $\tilde{H}$, $\tilde{V}$, and $\tilde{B}$ are

$$\eta_1, \eta_2, \cdots, \eta_n, -\eta_n, -\eta_{n-1}, \cdots, -\eta_1,$$
$$1, 1, \cdots, 1, -1, -1, \cdots, -1,$$
$$\beta_1, \beta_2, \cdots, \beta_n, -\beta_n, -\beta_{n-1}, \cdots, -\beta_1$$

respectively. Thus (5) is proved.

PROOF OF (6). Let $\alpha_1 \geq \alpha_2 \geq \cdots \geq \alpha_n$ and $\beta_1 \geq \beta_2 \geq \cdots \geq \beta_n$ be the singular values of $H - V$ and $H + I$ respectively. Let $\eta_1 \geq \eta_2 \geq \cdots \geq \eta_n$ be the eigenvalues (also singular values) of H. We are to prove (2).

It is known [1, Theorem 2] that if X, Y, Z are any three matrices of order n, with singular values

$$x_1 \geq x_2 \geq \cdots \geq x_n, \quad y_1 \geq y_2 \geq \cdots \geq y_n, \quad z_1 \geq z_2 \geq \cdots \geq z_n$$

---

[4] The authors are grateful to H. Wielandt for calling this useful fact to their attention.

respectively, and if $X + Y = Z$, then

$$z_{i+j+1} \leqq x_{i+1} + y_{j+1},$$

and in particular:

$$z_i \leqq x_i + y_1 \qquad\qquad (1 \leqq i \leqq n).$$

If we apply this fact to $H - V = H + (-V)$, then

$$\alpha_i \leqq \eta_i + 1 \qquad\qquad (1 \leqq i \leqq n).$$

As $\eta_i + 1 = \beta_i$, we have not only (2), but actually

$$(8) \qquad\qquad \alpha_i \leqq \beta_i \qquad\qquad (1 \leqq i \leqq n).$$

THEOREM 2. *Let* $A, H \in M_n$. *If H is Hermitian, then*

$$(9) \qquad \left\| A - \frac{A + A^*}{2} \right\| \leqq \| A - H \|$$

holds for every unitarily invariant norm.

PROOF. Observe first that the singular values of a matrix X are the same as those of X^*. Combining this fact with von Neumann's characterization of all unitarily invariant norms on M_n, it follows that $\| X \| = \| X^* \|$ for every unitarily invariant norm.

We write

$$A - \frac{A + A^*}{2} = \frac{A - H}{2} + \frac{H - A^*}{2},$$

which implies $\| A - (A + A^*)/2 \| \leqq \| A - H \|/2 + \| H - A^* \|/2$. This is precisely (9), since $\| H - A^* \| = \| A - H \|$.

REMARK 1. Corresponding to the inequality $| \operatorname{Re} z | \leqq | z |$ for complex numbers z, we have the trivial inequality $\| (A + A^*)/2 \| \leqq \| A \|$ for matrices. In this connection, we mention the following less trivial proposition: *Let $A \in M_n$. If $\lambda_1 \geqq \lambda_2 \geqq \cdots \geqq \lambda_n$ are the eigenvalues of* $(A + A^*)/2$, *and if $\alpha_1 \geqq \alpha_2 \geqq \cdots \geqq \alpha_n$ are the singular values of A, then*

$$(10) \qquad\qquad \lambda_i \leqq \alpha_i \qquad\qquad (1 \leqq i \leqq n).$$

Observe that $\| (A + A^*)/2 \| \leqq \| A \|$ insures only that $\sum_{i=1}^{k} \lambda_i \leqq \sum_{i=1}^{k} \alpha_i$ $(1 \leqq k \leqq n)$. To prove (10), let $A = UH$, where U is unitary and H is Hermitian positive semi-definite. Let $x_1, x_2, \cdots, x_n$ be n orthonormal eigenvectors of A^*A such that $A^*Ax_i = \alpha_i^2 x_i$ $(1 \leqq i \leqq n)$. Let μ_i denote the maximum of the inner product $((A + A^*)y/2, y)$, when the vector y varies under the conditions

$$(11) \qquad\qquad \| y \| = 1; \qquad (x_j, y) = 0 \qquad \text{for } 1 \leqq j \leqq i - 1.$$

Then by the minimum-maximum principle:

$$(12) \qquad\qquad \lambda_i \leq \mu_i \qquad\qquad (1 \leq i \leq n).$$

On the other hand, since $A = UH$, we have

$$\left(\frac{A + A^*}{2}\, y,\, y\right) = \mathrm{Re}\,(Ay,\, y) = \mathrm{Re}\,(Hy,\, U^*y)$$

$$\leq \|Hy\| \cdot \|U^*y\| = \|Hy\| \cdot \|y\|.$$

If $\|y\| = 1$, then

$$\left(\frac{A + A^*}{2}\, y,\, y\right) \leq \|Hy\| = (A^*Ay,\, y)^{1/2}.$$

Hence μ_i is not greater than the maximum of $(A^*Ay,\, y)^{1/2}$, when y varies under conditions (11). But this maximum is precisely α_i, so we have $\mu_i \leq \alpha_i$, which together with (12) proves (10).

REMARK 2. If $\rho_1 \geq \rho_2 \geq \cdots \geq \rho_n$ and $\alpha_1 \geq \alpha_2 \geq \cdots \geq \alpha_n$ denote the singular values of $(A + A^*)/2$ and A respectively, then inequalities $\rho_i \leq \alpha_i$ $(1 \leq i \leq n)$ are generally false. This can be seen by taking

$$A = \begin{pmatrix} 1 & 0 \\ 1 & 0 \end{pmatrix}.$$

THEOREM 3. *Let H, $K \in M_n$ be both Hermitian, and let U, V be their Cayley transforms*:

$$U = (H - iI)(H + iI)^{-1}, \qquad V = (K - iI)(K + iI)^{-1}.$$

If $\alpha_1 \geq \alpha_2 \geq \cdots \geq \alpha_n$ and $\beta_1 \geq \beta_2 \geq \cdots \geq \beta_n$ are the singular values of $(U - V)/2$ and $H - K$ respectively, then

$$(13) \qquad\qquad \alpha_i \leq \beta_i \qquad\qquad (1 \leq i \leq n).$$

Consequently, we have

$$(14) \qquad\qquad \|U - V\| \leq 2\|H - K\|$$

for every unitarily invariant norm.

PROOF. We write

$$U = I - 2i(H + iI)^{-1}, \qquad V = I - 2i(K + iI)^{-1}$$

so that

$$(U - V)/2i = (K + iI)^{-1} - (H + iI)^{-1}$$

$$= (K + iI)^{-1}[(H + iI) - (K + iI)](H + iI)^{-1},$$

or

$$(15) \qquad (U - V)/2i = (K + iI)^{-1}(H - K)(H + iI)^{-1}.$$

It is known [1, Theorem 2] that if X, Y, Z are any three matrices of order n, with singular values

$$x_1 \geqq x_2 \geqq \cdots \geqq x_n, \quad y_1 \geqq y_2 \geqq \cdots \geqq y_n, \quad z_1 \geqq z_2 \geqq \cdots \geqq z_n$$

respectively, and if $XY = Z$, then

$$z_{i+j+1} \leqq x_{i+1} \cdot y_{j+1}.$$

The singular values of $(U-V)/2i$ are obviously also those of $(U-V)/2$. Let $\eta_1 \geqq \eta_2 \geqq \cdots \geqq \eta_n$ and $\kappa_1 \geqq \kappa_2 \geqq \cdots \geqq \kappa_n$ be the singular values of $(H+iI)^{-1}$ and $(K+iI)^{-1}$ respectively. Applying the inequality just mentioned to (15), we get

$$\alpha_{i+j+k+1} \leqq \kappa_{i+1}\beta_{j+1}\eta_{k+1}.$$

In particular:

$$(16) \qquad\qquad \alpha_i \leqq \kappa_1\beta_i\eta_1 \qquad\qquad (1 \leqq i \leqq n).$$

On the other hand, from

$$(H + iI)^{-1*}(H + iI)^{-1} = (H^2 + I)^{-1},$$

we infer that $\eta_1 \leqq 1$. Similarly, $\kappa_1 \leqq 1$. Hence (16) implies (13).

REFERENCES

1. K. Fan, *Maximum properties and inequalities for the eigenvalues of completely continuous operators*, Proc. Nat. Acad. Sci. U.S.A. vol. 37 (1951) pp. 760–766.

2. V. B. Lidskiĭ, *The proper values of the sum and product of symmetric matrices* (in Russian), Doklady Akad. Nauk SSSR vol. 75 (1950) pp. 769–772.

3. J. von Neumann, *Some matrix-inequalities and metrization of matric-space*, Tomsk Univ. Rev. vol. 1 (1937) pp. 286–300.

4. R. Schatten, *A theory of cross-spaces*, Princeton University Press, 1950.

5. H. Wielandt, *An extremum property of sums of eigenvalues*, Proc. Amer. Math. Soc. vol. 6 (1955) pp. 106–110.

UNIVERSITY OF NOTRE DAME,
AMERICAN UNIVERSITY, AND
NATIONAL BUREAU OF STANDARDS

PACIFIC JOURNAL OF MATHEMATICS
Vol. 17, No. 2, 1966

ON THE NONSINGULARITY OF COMPLEX MATRICES

PAUL CAMION* AND A. J. HOFFMAN**

Let $A = (a_{ij})$ be a real square matrix of order n with nonnegative entries, and let $M(A)$ be the class of all complex matrices $B = (b_{ij})$ of order n such that, for all i, j, $|b_{ij}| = a_{ij}$. If every matrix in $M(A)$ is nonsingular, we say $M(A)$ is regular, and it is the purpose of this note to investigate conditions under which $M(A)$ is regular.

Many sufficient conditions have been discovered (cf., for instance, [8] and [3], and their bibliographies), motivated by the fact that the negation of these conditions, applied to the matrix $B - \lambda I$, yields information about the location of the characteristic roots. We shall show that a mild generalization of the most famous conditions [2] is not only sufficient but also necessary. (The application of our result to characteristic roots will not be discussed here, but is contained in [5]. See also [7] and [9]).

If

$$(1.1) \qquad a_{ii} > \sum_{i \neq j} a_{ij}, \qquad\qquad i = 1, \cdots, n,$$

then ([2]) $M(A)$ is regular. Clearly if P is a permutation matrix, and D a diagonal matrix with positive diagonal entries, such that PAD satisfies (1.1), then $M(A)$ is regular. We shall show that, conversely, if $M(A)$ is regular, there exist such matrices P and D so that (1.1) holds.

2. **Notation and lemmas.** If $x = (x_1, \cdots, x_n)$ is a vector, x^D is the diagonal matrix whose ith diagonal entry is x_i. If $M = (m_{ij})$ is a matrix, M^v is the vector whose ith coordinate is m_{ii}. A vector $x = (x_1, \cdots, x_n)$ is positive if each $x_j > 0$; x is semi-positive if $x \neq 0$ and each $x_j \geq 0$. A diagonal matrix D is positive (semipositive) if D^v is positive (semi-positive). If $A = (a_{ij})$ is a matrix with nonnegative entries, a particular entry a_{ij} is said to be dominant in its column if

$$a_{ij} > \sum_{k \neq i} a_{kj}.$$

LEMMA 1. *If $e_1, \cdots, e_n$ are nonnegative numbers such that the largest does not exceed the sum of the others, then there exist complex numbers z_i such that*

Received June 29, 1964. *Euratom
** The work of this author was supported in part by the Office of Naval Research under Contract No. Nonr 3775(00), NR 047040.

$$(2.1) \qquad\qquad |z_i| = e_i \qquad\qquad i = 1, \cdots, n$$

and

$$(2.2) \qquad\qquad \Sigma z_i = 0 \; .$$

Proof. It is geometrically obvious (and can easily be proved by induction) that the conditions on $\{e_i\}$ imply there exists a (possibly degenerate) polygon in the complex plane whose successive sides have length $e_1, e_2, \cdots, e_n$. Let the vertices $x_1, \cdots, x_n$ be so numbered that $|x_i - x_{i+1}| = e_i, i = 1, \cdots, n-1, |x_n - x_1| = e_n$. Setting $z_i = x_i - x_{i+1}$ obviously satisfies (2.1) and (2.2).

LEMMA 2. *Let M be a real matrix with m rows and n columns. Then*

$$(2.3) \qquad\qquad Mx \leqq 0 \; , \qquad x \; semi\text{-}positive$$

is inconsistent if and only if

$$(2.4) \qquad\qquad w'M > 0 \; , \qquad w \geqq 0$$

is consistent. Further, if (2.4) holds, we may assume there exists a w satisfying (2.4) with at most n coordinates of w positive.

This lemma is well known in the theory of linear inequalities.

3. THEOREM. *Let $A = (a_{ij})$ be a matrix of order n with each entry nonnegative. The following statements are equivalent:*

(3.1) *$M(A)$ is regular:*

(3.2) *if D is any semi-positive diagonal matrix, then DA contains an entry dominant in its column;*

(3.3) *there exists a permutation matrix P and a positive diagonal matrix D such that PAD satisfies (1.1).*

Proof. (3.1) $\Rightarrow$ (3.2). Assume (3.2) false for some semipositive D with $D^v = (d_1, \cdots, d_n)$. Let $(a_{1j}d_1, \cdots, a_{nj}d_n)$ be any column vector of DA. The coordinates of this vector satisfy the hypotheses of Lemma 1, so there exist complex numbers $z_1, \cdots, z_n$ satisfying

$$(3.4) \qquad\qquad \Sigma z_i = 0 \; ,$$

and

$$(3.5) \qquad\qquad |z_i| = a_{ij}d_i \; , \qquad\qquad i = 1, \cdots, n \; .$$

Let

$$b_{ij} = a_{ij}z_i/|z_i| \, ,$$

with $z_i/|z_i| = 1$, if $z_i = 0$. Then (3.4) and (3.5) become

$$(3.6) \qquad\qquad \sum d_i b_{ij} = 0 \, ,$$

and

$$(3.7) \qquad\qquad |b_{ij}| = a_{ij} \, , \qquad\qquad i, j = 1, \cdots, n \, .$$

But (3.7) states $B \in M(A)$, and (3.6)—since not all d_i are 0—asserts a linear dependence among the rows of B. Thus $B \in M(A)$ would be singular, violating (3.1).

$(3.2) \Rightarrow (3.3)$. Let K be a matrix of order n with $k_{ii} = 1$, $k_{ij} = -1$ for $i \neq j$, and let A_j be the jth column of A. Consider the system of n^2 linear inequalities in the semi-positive vector x

$$(3.7) \qquad\qquad KA_j^D x \leqq 0 \, , \qquad\qquad j = 1, \cdots, n \, .$$

Notice that (3.2) is identical with the statement that (3.7) is inconsistent. By Lemma 2, there exist n nonnegative vectors $\mu^1, \cdots, \mu^n$ such

$$(3.8) \qquad\qquad \sum_j \mu^{j\prime} KA_j^D > 0 \, .$$

Let $\mu^j = (\mu_1^j, \cdots, \mu_n^j)$. By the last sentence of Lemma 2, we may assume at most n of the n^2 numbers $\{\mu_k^j\}$ are positive.

Since each row of each KA_j^D contains at most one positive entry, it follows from (3.8) that exactly n of the $\{\mu_k^j\}$ are positive. We now show that, for each j, there is exactly one k such that $\mu_k^j > 0$. Assume otherwise, then for (say) $j = j^*$, $\mu^{j^*} = 0$. Let $\tilde{A}$ be the matrix obtained from A by replacing A_{j*} by 0. Then (3.8) would still hold with A_j replaced by 0, so (from the "only if" part of Lemma 2), for any semi-positive diagonal matrix E, $E\tilde{A}$ contains an entry dominant in its column. Let y be a real nonzero vector orthogonal to the columns of $\tilde{A}$, let $N = \{i \mid y_i \geqq 0\}$, and N' the complementary set of indices. Then, for each j.

$$(3.9) \qquad\qquad \sum_{i \in N} y_i a_{ij} = \sum_{i \in N'} (-y_i)a_{ij} \, .$$

If E is the diagonal matrix with $E^r = (|y_1|, \cdots, |y_n|)$, then $E\tilde{A}$, from (3.9), would contain no entry dominant in its column, a contradiction.

Let σ be the mapping sending $j \to k$, where $\mu_k^j > 0$. By (3.8), σ is a permutation of $\{1, \cdots, n\}$, and

$$a_{i,\sigma^{-1}i}\mu_i^{\sigma^{-1}i} > \sum_{j \neq i} a_{i,\sigma^{-1}j}\mu_j^{\sigma^{-1}j} \, , \qquad\qquad i = 1, \cdots, n \, ,$$

which is (3.3).

$(3.3) \Rightarrow (3.1)$ was noted in the introduction.

4. Remarks. (i) It is perhaps worth pointing out that the permutation in (3.3) is unique. For, without loss of generality, assume P and D both the identity matrix, so that (1.1) holds. Assume Q and E given so that QAE satisfies (1.1). If Q is not the identity permutation, then there must exist some cycle such that (say)

$$a_{q_1 q_2} e_{q_2} > a_{q_1 q_1} e_{q_1}$$
$$\cdot \quad \cdot \quad \cdot$$

(4.1)
$$a_{q_{r-1} q_r} e_{q_r} > a_{q_{r-1} q_{r-1}} e_{q_{r-1}}$$
$$a_{q_r q_1} e_{q_1} > a_{q_r q_r} e_{q_r} .$$

Multiplying the inequalities (4.1) together, we obtain

$$a_{q_1 q_2} \cdots a_{q_r q_1} > a_{q_1 q_1} \cdots a_{q_r q_r} ,$$

which violates (1.1).

In fact, it is clear from the foregoing that the diagonal entries in the PAD of (3.3) will be that collection of n entries of A, one from each row and column, whose product is a maximum. Further, that collection is necessarily unique. Finding the collection amounts to solving the assignment problem of linear programming [1] where the "scores" are $\{\log a_{ij}\}$. In some cases this can be done easily ([4]), but not in general [6].

(ii) If we had confined our attention to real rather than complex matrices, our theorem does not apply, and the problem seems difficult. With somewhat stronger hypotheses than the real case of (3.1), the problem has been solved by Ky Fan [3].

REFERENCES

1. G. B. Dantzig, *Linear Programming and Extensions*, Princeton University Press, Princeton, 1963.

2. J. Desplanques, *Théoréme d'Algébre*, J. Math. Spec. (3) **1** (1887), 12–13.

3. Ky Fan, *Note on circular disks containing the eigenvalues of a matrix*, Duke Math J. **25** (1958), 441–445.

4. A. J. Hoffman, *On Simple Linear Programming Problems*, Proceeding, of Symposia in Pure Mathematics, Vol. III, Convexity, 1963, American Mathematical Society, pp. 317–327.

5. B. W. Levinger and R. S. Varga, *Minimal Gerschgorin sets* II, to be published.

6. T. S. Motzkin, *The Assignment Problem*, Proceedings of Symposia in Applied Mathematics, Vol. VI, Numerical Analysis, 1956, American Mathematical Society, pp. 109–125.

7. Hans Schneider, *Regions of exclusion for the latent roots of a matrix*, Proc. American Math. Soc. **5** (1954), 320–322.

8. Olga Taussky, *A recurring theorem on determinants*, Amer. Math. Monthly **56** (1949), 672–676.

9. R. S. Varga, *Minimal Gerschgorin sets*, to appear in Pacific J. Math.

IBM RESEARCH CENTER

Combinatorial Aspects of Gerschgorin's Theorem

A. J. Hoffman*

1. Introduction

A well-known theorem of Gerschgorin asserts that, if $A = (a_{ij})$ is a complex matrix of order n, every eigenvalue λ of A lies in the union of the n disks:

$$|a_{kk} - \lambda| \leq \sum_{j \neq k} |a_{k_j}|, \quad k = 1, \ldots, n. \tag{1.1}$$

There exist many generalizations of Gerschgorin's theorem (see [3]), and we shall be particularly concerned with generalizations in which the right-hand sides of (1.1) are replaced by more general functions of the moduli of the off-diagonal entries. Specifically, we shall speak of non-negative functions f of $n(n-1)$ non-negative arguments, and by $f(A)$ we shall mean the value of such a function when the arguments are the moduli of the off-diagonal entries of A. A set $\{f_1, \ldots, f_n\}$ of such functions will be called a G-generating family if, for every complex matrix A, every eigenvalue of A lies in the union of the n disks:

$$|a_{kk} - \lambda| \leq f_k(A), \quad k = 1, \ldots, n; \tag{1.2}$$

equivalently, $\{f_1, \ldots, f_n\}$ form a G-generating family if, for every complex matrix A,

$$|a_{kk}| > f_k(A), \quad k = 1, \ldots, n \quad \text{imples } A \text{ nonsingular.} \tag{1.3}$$

This concept appears to have been first introduced by Nowosad in [4], and a systematic investigation initiated in [1]. In [2], the following combinatorial problem was raised:

Let us say that f depends on (i, j) if there exist two complex matrices A and B of order n such that $|a_{k\ell}| = |b_{k\ell}|$ if $(k, \ell) \neq (i, j)$, but $f(A) \neq f(B)$. Then define

$$D(f) = \{(i, j) \mid f \text{ depends on } (i, j)\}.$$

What is the "pattern of dependencies" of a G-generating family $\{f_1, \ldots, f_n\}$? More formally, if each of $D_1, \ldots, D_n$ is a subset of $\{(i, j)\}_{i \neq j}$, what are the necessary

*This work was supported (in part) by the Office of Naval Research under Contract NONR 3775(00). It contains portions of material presented in a lecture under this title given at the *New York Graph Theory Conference*, sponsored by St. John's University, in June 1970.

and sufficient conditions that there exist a G-generating family $\{f_1, \ldots, f_n\}$ such that $D(f_k) = D_k$, $k = 1, \ldots, n$? In [2], the following result was established:

Theorem 1. *There exists a G-generating family $\{f_1, \ldots, f_n\}$ of functions each of which is homogeneous (of degree one) and bounded on bounded sets (e.g., continuous) such that $D(f_k) = D_k$, $k = 1, \ldots, n$, if and only if, for every subset $S \subset \{1, \ldots, n\}$, with $|S| \geq 2$, every cyclic permutation σ of S, and every subset $T \subset S$,*

$$\left| \{(i, \sigma i)\}_{i \in S} \cap \bigcup_{k \in T} D_k \right| \geq |T|. \tag{1.4}$$

(When (1.4) is satisfied, it is shown in [2] that the $\{f_k\}$ can be taken to be linear, so the requirement that the f's be linear, e.g., imposes no restriction on $\{D_k\}$ in addition to (1.4).)

The purposes of this note are: (i) to recast the general problem in terms of the language of directed graphs; (ii) in that language, to find a more perspicuous and more easily testable restatement of (1.4); (iii) to consider the problem with other (or no) restrictions on the G-generating family.

We shall denote by D the complete directed graph, without loops, on n vertices. Thus, the (directed) edges of D consist of all $n(n-1)$ ordered pairs (i, j), $i, j = 1, \ldots, n$, $i \neq j$. We shall say $E \subset D$ if E has the same vertex set as D, and every edge of E is an edge of D. If $E \subset D$, $\bar{E}$ is the graph, with the same vertex set, whose edges are precisely all ordered pairs (i, j), $i \neq j$, which are not edges of E. If E_1, $E_2 \subset D$, $E_1 \cap E_2$ is the graph with the same vertex set, whose edges are precisely all edges in both E_1 and E_2. A path in $E \subset D$ is a sequence $\{i_1, \ldots, i_k\}$, $k \geq 2$, of distinct vertices such that $(i_r, i_r + 1)$ is an edge of E, $r = 1, \ldots, k-1$, together with all these edges. The vertices i_1 and i_k are respectively initial and terminal vertices. A cycle in $E \subset D$ is a sequence $\{i_1, \ldots, i_k\}$, $k \geq 2$, of distinct vertices such that $(i_r, i_r + 1)$ is an edge of E, $r = 1, \ldots, k-1$, (i_k, i_1) is an edge of E, together with all these edges.

We can now restate Theorem 1.

Theorem 1′. *There exists a G-generating family $\{f_1, \ldots, f_n\}$ of functions each of which is homogeneous (of degree one) and bounded on bounded sets (e.g. continuous) such that $D(f_k) = D_k$, $k = 1, \ldots, n$ if and only if*
(1.4.1) for every $k = 1, \ldots, n$, $\bar{D}_k$ contains no cycle including k, and
(1.4.2) for every pair $i, j, i \neq j$, $\bar{D}_i \cap \bar{D}_j$ contains no path whose initial vertex is i and whose terminal vertex is j.

Our other results are stated in the following theorems.

Theorem 2. *There exists a G-generating family $\{f_1, \ldots, f_n\}$ with $D(f_k) = D_k$, $k = 1, \ldots, n$, if and only if for every $i \neq j$,*

$$(i, j) \text{ is not an edge of } \bar{D}_i \cap \bar{D}_j. \tag{1.5}$$

The statement of Theorem 3 is somewhat more complicated. Assume $D_i, \ldots, D_n$ given, let C be any cycle in D, and let V be the set of vertices contained in C. Let $G(C)$ be the following (undirected) graph: the vertex set of $G(C)$ is V, and two vertices $i \neq j$ of V are adjacent if $D_i \cap D_j$ contains an edge of C.

We shall say $G(C)$ is balanced if, for each connected component K of $G(C)$, we have

$$|V(K)| = |\text{ number of edges in } \bigcup_{i \in K} D_i \text{ which are also edges of } C|.$$

Theorem 3. *There exists a G-generating family $\{f_1, \ldots, f_k\}$ of functions each of which is homogeneous (of degree one) with $D(f_k) = D_k$, $k = 1, \ldots, n$, if and only if, for every cycle C in G*

$$G(G) \text{ is balanced.} \tag{1.6}$$

We shall prove these theorems in reverse order.

2. Proof of Theorem 3

We begin by first noting that it is not difficult to prove that (1.6) is equivalent to: for every subset $S \subset \{1, \ldots, k\}$, $|S| \geq 2$, every cyclic permutation σ of S, and every subset $T \subset S$, we have (1.4) or

$$\{(i, \sigma i)\}_{i \in S} \cap \bigcup_{k \in T} D_k \cap \bigcup_{k \in S-T} D_k \neq \emptyset. \tag{2.1}$$

We first prove necessity. Assume $\{f_1, \ldots, f_n\}$ G-generating, homogeneous, but there exists S, σ, T such that the $D(f_k)$ satisfy neither (1.4) nor (2.1). Let $|T| = t$, and let $i_1, \ldots, i_r \in S$, be such that $r < t$, $i \in S - \{i_1, \ldots, i_r\}$ implies $(i, \sigma i) \notin \bigcup_{k \in T} D(f_k)$. This follows from the negation of (1.4). Let $\varepsilon > 0$ be given, and define the off-diagonal entries of a matrix $A(\varepsilon)$ by

$$
\begin{aligned}
a_{i_j}, \sigma_{i_j} &= -\varepsilon & j &= 1, \ldots, r \\
a_{i, \sigma i} &= -1 & i &\in S - \{i_1, \ldots, i_n\} \\
a_{k\ell} &= 0 & &\text{otherwise, } k \neq \ell.
\end{aligned}
\tag{2.2}
$$

It follows from the homogeneity of f_k that

$$f_k(A(\varepsilon)) = \varepsilon f_k(A(1)), \quad \text{for all } k \in T. \tag{2.3}$$

From the negation of (1.6), we have

$$(A(\varepsilon)) = f_k(A(1)) \quad \text{for all } k \in S - T. \tag{2.4}$$

Hence, from (2.3) and (2.4)

$$\prod_{k \in S} f_k(A(\varepsilon)) = \varepsilon^t \prod_{k \in S} f_k(A(1)). \tag{2.5}$$

It follows from (2.5) that, since $r < t$, there exists a positive number ε such that

$$\varepsilon^r > \prod_{k \in S} f_k(A(\varepsilon)).$$

Hence, we can choose positive numbers $c_k = c_k(\varepsilon)$, $k \in S$ such that

$$c_k > f_k(A(\varepsilon)), \qquad k \in S, \tag{2.6}$$

$$\varepsilon^r = \prod_{k \in S} c_k. \tag{2.7}$$

Now define diagonal entries of $A(\varepsilon)$

$$a_{i,i}(\varepsilon) = c_i, \qquad i \in S \tag{2.8}$$

$$a_{i,i}(\varepsilon) > f_i(A(\varepsilon)), \; i \notin S. \tag{2.9}$$

From (2.6), (2.8), (2.9), $|a_{kk}| > f_k(A(\varepsilon))$ for all k. Since $\{f_1, \ldots, f_n\}$ is assumed a G-generating family, it follows from (1.3) that $A(\varepsilon)$ is nonsingular. But using (2.1), (2.7) and (2.8), $\det A = 0$. This contradiction establishes the necessity of at least one of (1.4) and (2.1).

To prove the sufficiency, we must exhibit a G-generating family $\{f_1, \ldots, f_n\}$ of homogeneous functions with each $D(f_k) = D_k$. To that end, we first define

$$D_k(A) = \{(i,j)|(i,j) \in D_k, \; a_{ij} \neq \emptyset\};$$

$$f_k(A) = \begin{cases} 0 & \text{if } D_k(A) = \emptyset \\[2ex] n! \, \dfrac{\max\limits_{(i,j)\in D_k(A)} |a_{ij}|^n}{\min\limits_{(i,j)\in D_k(A)} |a_{ij}|^{n-1}} & \text{if } D_k(A) \neq \emptyset \end{cases} \right\} \quad k = 1, \ldots, n.$$

Clearly, all we need to prove is that the f's so defined form a G-generating family. Recalling (1.3) and the definition of a determinant as the sum of products, all we need to show is

$$\text{for every } S \subset \{1, \ldots, k\}, \quad |S| \geq 2, \tag{2.10}$$

every cyclic permutation σ of S, and every matrix $A = (a_{ij})$ such that $\prod_{i \in S} |a_{i,\sigma i}| \neq 0$,

$$\prod_{i \in S} |a_{i,\sigma i}| \leq \prod_{k \in S} \left(\frac{\max\limits_{(i,\sigma i)\in D_k} |a_{i,\sigma i}|^n}{\min\limits_{(i,\sigma i)\in D_k} |a_{i,\sigma i}|^{n-1}} \right).$$

(Note that our assumptions that at least one of (1.4) and (2.1) holds guarantees that, for each $k \in S$, $\{i, \sigma i\}_{i \in S} \cap D_k \neq \emptyset$.)

Let

$$b_1 \geq b_2 \geq \cdots \geq b_{|S|} \tag{2.11}$$

be a rearrangement in descending order of the quantities $\{\log |a_{i,\sigma i}|_{i \in S}\}$.

For simplicity of notation, assume $S = \{1, \ldots, |S|\}$. Let $M = (m_{ij})$ be the $(0,1)$ matrix of order $|S|$ with $m_{ij} = 1$ $(i, j \in S)$ if an only if D_j contains $(\tau i, \sigma \tau i)$, where

τ is the permutation of S such that $b_i = \log |a_{\tau i, \sigma \tau i}|$. By taking the logarithm on both sides in (2.10), our problem reduces to proving that (2.11) implies

$$\sum_{i=1}^{|S|} b_i \leq n \sum_{k=1}^{|S|} b_{\alpha(k)} - (n-1) \sum_{k=1}^{|S|} b_{\beta(k)}, \tag{2.12}$$

where $\alpha(k)$ is the smallest i such that $m_{ik} = 1$,

$$\beta(k) \text{ is the largest } i \text{ such that } m_{ik} = 1.$$

We think of (2.12) as an inequality which we must show is satisfied by every vector b satisfying (2.11).

Let u^j be the vector with $|S|$ components, of which the first j are 1, the remainder 0. Any b satisfying (2.11) can be expressed as

$$b = \sum_{j=1}^{|S|} \lambda_j u^j, \quad \lambda \geq 0, \ j = 1, \ldots, |S| - 1. \tag{2.13}$$

Since (2.12) holds as an equality for $j = |S|$, it follows from (2.13) that all we need to prove is that (2.12) holds if $b = u^j$, $j < |S|$. We rewrite this case of (2.12) as

$$j \leq n \sum_{k=1}^{|S|} (u_{\alpha(k)}^J - u_{\beta(k)}^j) + \sum_{k=1}^{|S|} u_{\beta(k)}^j. \tag{2.14}$$

If, for some k_0, $\alpha(k_0) \leq j$, $\beta(k_0) > j$, then the right-hand side of (2.14) becomes

$$n + n \sum_{k \neq k_0} (u_{\alpha(k)}^j - u_{\beta(k)}^j) + \sum_{k} u_{\beta(k)}^j. \tag{2.15}$$

Since $u_{\alpha(k)}^j \geq u_{\beta(k)}^j$ for all k, and $u_{\beta(k)}^j \geq 0$ for all k, (2.19) is at least $n > j$, verifying (2.14).

Suppose such a k_0 does not exist. This means that, for each $k \in S$, either $\alpha(k) > j$ or $\beta(k) \leq j$. If we let $T = \{k \in S | \alpha(k) > j\}$, then (2.1) is violated.

It follows that (1.4) holds, which means

$$|S| - j \geq |T|. \tag{2.16}$$

But under these circumstances, the first term on the right-hand side of (2.14) vanishes and the second term is $|S| - |T|$. Thus, (2.16) proves (2.14).

3. Proof of Theorem 2

It is more convenient to restate (1.5) as

for every $S \subset \{1, \ldots, n\}, |S| \geq 2$, and every cyclic permutation σ of S,

$$\{(i, \sigma i)\}_{i \in S} \subset \bigcup_{k \in S} D_k. \tag{3.1}$$

We first show the necessity of (3.1). Assume that (3.1) is false, so that there exists $i_0 \in S$ with

$$(i_0, \sigma i_0) \notin \bigcup_{k \in S} D(f_k). \tag{3.2}$$

Let B be a matrix in which $b_{i,\sigma i} = -1$ for all $i \in S$, $i \neq i_0$, all other off-diagonal elements 0. Let $\{c_k\}_{k \in S}$ be positive numbers such that

$$c_k > f_k(B) \quad \text{for all } k \in S. \tag{3.3}$$

Let $D = (d_{ij})$ be a matrix in which all off-diagonal elements are 0 except

$$\begin{aligned} d_{i,\sigma i} &= -1 \quad \text{for } i \in S, i \neq i_0 \\ d_{i_0,\sigma i_0} &= - \prod_{k \in S} c_k . \end{aligned} \tag{3.4}$$

Further, let

$$d_{ii} = c_i \quad \text{for } i \in S \tag{3.5}$$

$$d_{ii} > f_i(D) \text{ for } i \notin S. \tag{3.6}$$

By comparing B with D, we see from (3.3), (3.5) and (3.6) that $|d_{kk}| > f_k(D)$ for all k. But (3.4) and (3.5) show that $\det D = 0$, a contradiction.

To prove sufficiency of (3.1), define

$$f_k(A) = n! \left(1 + \sum_{\emptyset \neq T \subset D_k} \prod_{(i,j) \in T} |a_{ij}| \right), \quad k = 1, \ldots, n.$$

It is clear that $\{f_k\}$ so defined form a G-generating family with each $D_k = D(f_k)$. (Indeed, we have other choices of f_k, and could have made them polynomials of degree at most two.)

Note that the hypothesis that the f's be continuous, or even polynomials, would not change the conditions (1.5) or (3.1).

4. Proof of Theorem 1

We must prove that (1.4) implies both (1.4.1) and (1.4.2) and conversely. Assume (1.4) holds. Suppose

(1.4.1) is false for some k, then this violates (1.4) with $T = \{k\}$. Suppose (1.4.2) is false for some $i \neq j$. Let

$$i = i_1, i_2, \ldots, i_k = j$$

be the sequence of vertices in a path contained in $\bar{D}_i \cap \bar{D}_j$. Then the cycle obtained by adjoining to the path the edge (j, i) violates (1.4) with $T = \{i, j\}$.

Conversely, assume (1.4.1) and (1.4.2). Suppose (1.4) is false for some cycle determined by an S and σ and some subset $T \subset S$. If $|T| = 1$, this violates (1.4.1). Assume $|T| = t > 1$, and let i_1, $i_2, \ldots, i_t$ be the vertices in T in the order in which they occur around the cycle. Let the number of edges in the subpath of the cycle from i_1 to i_2 be $a_1, \ldots$, from i_{t-1} to i_t be a_{t-1}, from i_t to i, be a_t. Since (1.4.2) holds, there are at most $a_k - 1$ edges of the cycle in $\bar{D}_{i_k} \cap \bar{D}_{i_{k+1}}$ (all k taken mod t) in the subpath from i_k to i_{k+1}. Consequently $\bigcap_{k=1}^{t} \bar{D}_{i_k}$ contains at most $\sum_{i=1}^{t}(a_i - 1) = |S| - t$ edges. There $\bigcup_{k=1}^{t} D_{i_k}$ contains at least t edges of the cycle, contradicting the presumed falsity of (1.4).

Acknowledgment

We are grateful to Ellis Johnson, Peter Lax, Michael Rabin and Richard Varga for useful conversation about this material.

References

[1] A. J. Hoffman, "Generalizations of Gerschgorin's Theorem: G-Generating Families," lecture notes, University of California at Santa Barbara, August, 1969, 46 pp.

[2] A. J. Hoffman and R. S. Varga, Patterns of Dependence in Generalizations of Gerschgorin's Theorem, to appear in *SIAM J. Numer. Anal.*

[3] M. Marcus and H. Minc, *A Survey of Matrix Theory and Matrix Inequalities* (Allyn and Bacon, Boston, 1964).

[4] P. Nowosad, "On the Functional (x^{-1}, Ax) and Some of Its Applications," *An. Acad. Brasil Ci.* **37** (1965) 163–165.

Linear and Multilinear Algebra, 1975, Vol. 3, pp. 45–52
© Gordon and Breach Science Publishers Ltd.

Linear G-Functions[†]

ALAN J. HOFFMAN [‡]

To Olga Taussky, who introduced me to this topic, in gratitude
for over 20 years of inspiration and friendship.

*Mathematical Sciences Department, IBM Watson Research Center,
Yorktown Heights, New York*

(*Received December 9, 1974*)

Let $\{d_{ij}^k\}_{i \neq j}$ be a given set of $n^2(n-1)$ nonnegative numbers. We characterize those sets $\{d_{ij}^k\}_{i \neq j}$ for which the following statement is true: for every complex matrix A, every eigenvalue of A lies in the union of the n disks

$$|c_{kk} - \lambda| \leq \sum_{j \neq i} d_{ij}^k |a_{ij}|, \quad k = 1, \ldots, n.$$

Many generalizations of Gerschgorin's theorem are shown to be consequences.

1. INTRODUCTION

We shall deal with nonnegative functions f of $n(n-1)$ nonnegative arguments. Let F_n be the set of such functions. If $f \in F_n$, and A is a complex matrix of order n, $f(A) \equiv f(\{|a_{ij}|\}_{i \neq j; \, i,j = 1, \ldots, n})$. A G-function (of order n) is a mapping of $C^{n,n}$ (the space of complex matrices of order n) into R_n^+, given by $f(A) = (f_1(A), \ldots, f_k(A))$, where each $f_i \in F_n$ and where the following proposition holds:

---

† The preparation of this manuscript was supported (in part) by U.S. Army contract ⧣DAHC04-72-C-0023.

‡ This paper is a revision of a portion of lectures given at the University of California, Santa Barbara, in August 1969 [4]; the lectures were an extension of material presented at a Gatlinburg Conference, sponsored by Oak Ridge National Laboratory, the preceding year. Revision of other parts of [4] have appeared in print in [1], [5], [6], and some of the material of Section 3 of this manuscript was announced in [3]. We believe that the concept of G-functions, introduced by Nowosad in [11], is the proper setting for studying many of the generalizations of Gergorin's theorem, and the purpose of the Santa Barbara notes was to lay the foundations of their theory. The appearance of [1], [2], and [10] have strengthened this belief.

46   A. J. HOFFMAN

For every complex matrix $A = (a_{ij})$ of order n, every eigenvalue λ of A lies in the union of the n disks

$$|a_{kk} - \lambda| \leqslant f_k(A), \quad k = 1, \ldots, n.$$

Equivalently, $f = (f_1, \ldots, f_n)$ is a G-function if, for every complex matrix A of order n,

$$|a_{kk}| > f_k(A), \quad k = 1, \ldots, n \tag{1.1}$$

implies A is nonsingular.

Of course, the best known G function comes from Gerschgorin's theorem: $f_k(A) = \sum_{j \neq k} |a_{kj}|$, $k = 1, \ldots, n$. In Section 2, we shall characterize all linear G-functions (i.e. $f_k(A) = \sum_{i \neq j} d_{ij}^k |a_{ij}|$, $k = 1, \ldots, n$, $\{d_{ij}^k\}_{i=j}$; $i, j, k = 1, \ldots, n$ a given set of $n^2(n-1)$ nonnegative numbers. This characterization depends on a numerical function of nonnegative matrices, which we call the equilibrant and which may be defined as follows: for any nonnegative matrix B, let $\rho(B)$ be its spectral radius; then the equilibrant of B is

$$\mathscr{E}(B) \equiv \inf_{\substack{F \text{ positive diagonal matrix} \\ \prod_i f_{ii} = 1}} \rho(FB). \tag{1.2}$$

The principal result

THEOREM 1.1 *Let $\{d_{ij}^k\}$, $i \neq j$; $i, j, k = 1, \ldots, n$ be a given set of $n^2(n-1)$ nonnegative numbers. Define*

$$f_k(A) = \sum_{i \neq j} d_{ij}^k |a_{ij}^d|, \quad k = 1, \ldots, n. \tag{1.3}$$

Then $f = (f_1, \ldots, f_n)$ is a G-function if and only if
for every subset $S \subset \{1, \ldots, n\}$, $|S| \geqslant 2$, and every cyclic permutation σ of S,
$$\tag{1.4}$$

$$\mathscr{E}((d_{i,\sigma,i}^k)_{\substack{i \in S \\ k \in S}}) \geqq 1.$$

Although we shall mention several equivalent definitions of the equilibrant, we have no easy formula for it except in special cases. Fortunately, one of the special cases applies to the important special choice of $\{d_{ij}^k\}$, where $d_{ij}^k = 0$ unless $i = k$ or $j = k$. This enables us to prove in section 3:

THEOREM 1.2 *Let (r_{ij}) and (c_{ij}) be nonnegative matrices. Define*

$$f_k(A) = \sum_{j \neq k} r_{kj} |a_{kj}| + \sum_{i \neq k} c_{ik} |a_{ik}|, k = 1, \ldots, n. \tag{1.5}$$

Then $f = (f_1, \ldots, f_n)$ is a G-function if and only if
for every subset $S \subset \{1, \ldots, n\}$, $|S| \geqslant 2$, and every cyclic permutation σ of S,
$$\tag{1.6}$$

$$(\prod_{i \in S} r_{i,\sigma i})^{1/|S|} + (\prod_{i \in S} c_{i,\sigma i})^{1/|S|} \geqq 1.$$

The rest of section 3 is devoted to showing that several well-known sufficient conditions for nonsingularity, even though they are not expressed as linear conditions, are nevertheless corollaries of Theorem 1.2.

2. PROOF OF THEOREM 1.2

We first note for the record equivalent definitions of the equilibrant. The equivalence of the last two has been previously noted in [9], and the last definition, which inspired the term "equilibrant", is made possible by the theorem ([8], [13]) which asserts that, if B is a positive matrix, there exist unique (up to scalar multiples) positive diagonal matrices D_1 and D_2 such that $D_1 B D_2$ is doubly stochastic.

Let B be a nonnegative matrix of order n. Then

$$\mathcal{E}(B) = \inf_{\substack{x>0 \\ \Pi x_j = 1}} \left(\prod_i (Bx)_i\right)^{1/n}$$

$$= \frac{1}{n} \inf_{\substack{x>0,\, \Pi x_i = 1 \\ y>0,\, \Pi y_i = 1}} x'By$$

$$= \inf_{\substack{\varepsilon \to 0+ \\ D_1(\varepsilon)(B+\varepsilon J)D_2(\varepsilon) \\ \text{doubly stochastic}}} (\det D_1(\varepsilon)D_2(\varepsilon))^{-1/n} \tag{2.1}$$

Now to prove the theorem. It is easy to see that, if (1.3) is a G-function, and $T \subset \{1, \ldots, n\}$, $T \neq \emptyset$, then defining

$$f_k^T(C) = \sum_{\substack{i \neq j \\ i,j \in T}} d_{ij}^k |c_{ij}|, \qquad k \in T \tag{2.2}$$

yields a G-function on matrices C whose rows and columns are indexed by T. Furthermore, we see that (1.4) holds for T. Also, the theorem is trivially true if $n = 1$. Consequently, we may assume by induction that (1.4) holds if $|s| < n$, and that (2.2) is a G-function provided $|T| < n$.

In addition, from the theory of M-matrices, we know that it is sufficient to prove the theorem under the additional hypothesis that A is a real matrix with positive diagonal and nonpositive off diagonal. Combined with the induction hypothesis, and the theory of M-matrices, we see that proving our theorem is equivalent to proving the equivalence of (1.4) with the following implication: for every real matrix, $A = (a_{ij})$ of order n, if

$$a_{kk} > 0, k = 1, \ldots, n \tag{2.3}$$

and
$$a_{ij} \leq 0, i \neq j, i,j = 1, \ldots, n \tag{2.4}$$

$$a_{kk} + \sum_{j \neq i} d_{ij}^k a_{ij} > 0, k = 1, \ldots, n, \tag{2.5}$$

then

$$x > 0 \text{ implies } Ax = 0 \text{ impossible.} \tag{2.6}$$

                            A. J. HOFFMAN

Let us define two sets of matrices

$$M^k = (m_{ij}^k) = \begin{cases} 1 & \text{if } i = j = k \\ 0 & \text{if } i = j \neq k \\ d_{ij}^k & \text{if } i \neq j \end{cases}, \qquad k = 1, \ldots, n$$

$$M^{ij} = (m_{rs}^{ij}) = \begin{cases} -1 & \text{if } i = r, j = s \\ 0 & \text{otherwise} \end{cases}, \qquad i \neq j, i, j = 1, \ldots, n.$$

We first prove that (2.3)–(2.5) imply (2.6) if and only if

for every $x > 0$, there exists a nonnegative, nonzero vector a and $\qquad$ (2.7)
nonnegative numbers $\{\lambda_k\}$ and $\{\lambda_{ij}\}$ such that

$$(a_i x_j) = \sum_k \lambda_k M^k + \sum_{i \neq j} \lambda_{ij} M^{ij}. \tag{2.7a}$$

Assume (2.7). Let A satisfy (2.3)–(2.5). By (2.5),

$$tr \, A(M^k)^T > 0 \quad \text{for} \quad k = 1, \ldots, n. \tag{2.8}$$

By (2.4) (note (2.3) follows from (2.4) and (2.5)),

$$tr \, A(M^{ij})^T \geqslant 0 \quad \text{for} \quad i \neq j, i, j = 1, \ldots, n. \tag{2.9}$$

Assume $x > 0$, $Ax = 0$. Then

$$tr \, A(a_i x_j)^T = 0,$$

so by (2.7a), (2.8), (2.9), all $\{\lambda_k\}$ and $\{\lambda_{ij}\}$ are 0. But since $a \geq 0$, $a \neq 0$, there exists at least one $\lambda_k > 0$, a contradiction. Thus, if (2.7) is true, it follows that (2.3)–(2.5) imply (2.6).

Next, assume (2.3)–(2.5) imply (2.6). Let $x > 0$ be given. Let $L(x)$ be the linear space (in matrix space) of all matrices of the form $(a_i x_j)$ for all possible choices of the real vector a. Let $\mathcal{M}$ be the polyhedron of all matrices of the form $\Sigma_k \lambda_k M^k + \sum_{i \neq j} \lambda_{ij} M^{ij}$, where each $\lambda_k \geqq 0$, $\lambda_{ij} \geqq 0$, and $\Sigma_k \lambda_k = 1$. If (2.7) were false, $L(x)$ is disjoint from $\mathcal{M}$. By a separation theorem for convex sets, there would exist a point (i.e. a matrix), which makes a positive inner product with each M^k, a nonnegative inner product with each M^{ij}, and is orthogonal to $L(x)$. Call this matrix A. We have $tr \, A(M^k)^T > 0$ for all k; $tr \, A(M^{ij})^T \geqslant 0$ for all $i \neq j$, $tr \, AB^T = 0$ for all $B \in L(x)$. This contradicts the statement that (2.3)–(2.5) imply (2.6). Hence, (2.7) is true.

Therefore, we have reduced our problem to proving that (2.7) holds if and only if (1.4) holds. If we examine the form of (2.7a), we see that $\lambda_k = a_k x_k$. Also, instead of requiring $\Sigma \lambda_k = 1$, we can change our normalization so that $\Sigma a_i = 1$. Thus, (2.7) holds if and only if (2.10) for every $x > 0$, there exists a nonnegative vector a satisfying $\Sigma a_i = 1$ and the $n(n - 1)$ inequalities

$$\sum_k (d_{ij}^k x_k) a_k \geqq a_i x_j, \qquad i \neq j, j = 1, \ldots, n. \tag{2.10a}$$

Assume (1.4) false. By the induction hypothesis, we know (1.4) holds if $|S| < n$, so assume $|S| = n$ and we have a cyclic permutation σ of $\{1, \ldots, n\}$

such that $\mathscr{E}(\{d^k_{i,\sigma i}\}) < 1$. By (1.2) and the Perron–Frobenius theory, there exist positive numbers $\{f_k\}$ and nonnegative numbers $\{b_k\}$ such that $\Pi_k f_k = 1$, and

$$\sum_i b_i f_i d^k_{i,\sigma i} < b_k, \qquad k = 1, \ldots, n. \tag{2.11}$$

Now, because of our stipulations on $\{f_k\}$, there is a vector $x > 0$ satisfying $x_i = f_i x_{\sigma i}$. Further, if we assume (2.10) holds, then (2.10a), for the cases $j = \sigma i$, can be rewritten.

$$f_i \sum_k d^k_{i,\sigma i} c_k \geq c_i, \qquad i = 1, \ldots, n, \tag{2.12}$$

where $c_i = a_i x_i \geq 0$, so the $\{c_i\}$ are nonnegative numbers, not all 0.

If we multiply the kth inequality in (2.11) by c_k and add, we get

$$\sum_{i,k} c_k b_i f_i d^k_{i,\sigma i} < \sum c_k b_k. \tag{2.13}$$

On the other hand, if we multiply the ith inequality in (2.12) by b_i and add, we get the negation of (2.13). Thus, if (2.10) is true, (1.4) is true.

Next, assume (2.10) is false. Then there exists a positive vector x such that the system of inequalities (2.10a), together with

$$a_k \geq 0, k = 1, \ldots, n \tag{2.14}$$

$$\sum_k a_k = 1 \tag{2.15}$$

is inconsistent. This is a system of inequalities on the variable vector $a = (a_1, \ldots, a_n)$ in R^{n-1} (since $\Sigma a_k = 1$). By Helly's theorem, there exist a set of n pairs $(i_j, j_1, \ldots, i_n, j_n)$, $i_t \neq j_t$, such that

$$\sum_k (d^k_{i_t j_t} x_k) a_k \geq a_{i_t} x_{j_t}, \qquad t = 1, \ldots, n, \tag{2.16}$$

together with (2.14) and (2.15) is inconsistent.

Suppose there exists some index k such that $i_t \neq k$, $t = 1, \ldots, n$. Then setting $a_k = 1$, all other $a_i = 0$ makes (2.14)–(2.16) consistent. So $\{i_1, \ldots, i_n\} = \{1, \ldots, n\}$, and $i_t \to j_t$, $t = 1, \ldots, n$ is a mapping of $\{1, \ldots, n\}$ into itself. Since $i_t \neq j_t$, this mapping contains a cycle of length at least 2, so there exists $S \subset \{1, \ldots, n\}$, $|s| \geq 2$, and a cyclic permutation σ of S such that $i_t \in S$ implies $\sigma i_t = j_t \in S$.

Next, consider the system of inequalities in the variables $\{b_k\}_{k \in S}$

$$\sum_{k \in S} (d^k_{i,\sigma i} x_k) b_k \geq b_i x_{\sigma i}, \qquad i \in S$$

$$b_i \geq 0, \qquad i \in S \tag{2.17}$$

$$\sum_{i \in S} b_i = 1.$$

If (2.17) were consistent, then setting $a_k = b_k$, $k \in s$, $a_k = 0$ for $k \notin S$ would make (2.14)–(2.16) consistent. So (2.17) is inconsistent. Define $\{f_i\}_{i \in S}$ by the

50   A. J. HOFFMAN

rule $f_i x_{\sigma i} = x_i$, $i \in S$, and substitute in (2.17), writing $c_k = b_k x_k$, $k \in S$. We conclude that

$$\sum_{k \in S} f_i d^k_{i, \sigma i} c_k \geqq c_i, \qquad i \in S \qquad (2.18)$$

cannot be satisfied in nonnegative variables $\{c_k\}_{k \in S}$, not all 0. But $f_i > 0$, $\prod_{i \in S} f_i = 1$. By (1.2), this means that (1.4) is false. And this contradiction completes the proof of Theorem 1.1.

3. PROOF OF THEOREM 1.2 AND SOME CONSEQUENCES

To prove Theorem 1.2, all that needs to be shown in view of Theorem 1.1, is that (3.1) implies (3.2)

If $B = (b_{ij})$ is a nonnegative matrix of order n such that (3.1) $b_{ij} = 0$ unless $i = j$ or $i = j - 1 (j = 2, \ldots, n)$ or $i = n$, $j = 1$, and $c \equiv (\prod_i b_{ii})^{1/n}$, $d = (b_{n1} \prod_i b_{i, i+1})^{1/n}$, then

$$\mathscr{E}(B) = c + d. \qquad (3.2)$$

It is clear from the definitions of equilibrant that it is sufficient to prove (3.2) in case c and d are both positive, so we assume this. Also, if X and Y are positive diagonal matrices so that det $XY = 1$, then $\mathscr{E}(XBY) = \mathscr{E}(B)$. Define recursively $2n$ positive numbers,

$$y_1 = 1$$
$$x_i = c/b_{ii} y_i \qquad i = 1, \ldots, n$$
$$y_i = d/b_{i-1 i} x_{i-1} \qquad i = 2, \ldots, n$$

Let $X = \text{diag}(x_1, \ldots, x_n)$, $Y = \text{diag}(y_1, \ldots, y_n)$. Then det $XY = 1$, and, if P is the cyclic permutation matrix $(p_{1,2} = p_{2,3} = \cdots = p_{n-1,n} = p_{n,1} = 1$, all other $p_{ij} = 0)$, then

$$XBY = cI + dP.$$

Set $u = (1, \ldots, 1)$. Then $(\prod_i ((cI + dP)u)_i)^{1/n} = c + d$. So all we need show, in view of (2.1), is that $x > 0$, $\prod x_i = 1$ implies

$$(\prod_i ((cI + dP)x)_i^{1/n} \geqq c + d. \qquad (3.3)$$

But

$$cx_i + dx_{i+1} = (c + d)\left(\frac{c}{c + d} x_i + \frac{d}{c + d} x_{i+1} \right)$$
$$\geqq (c + d) x_i^{c/(c+d)} x_{i+1}^{d/(c+d)}, \qquad (3.4)$$

by the inequality between arithmetic and geometric means. But since $\Pi x_i = 1$, multiplying (3.4) for all i yields (3.3).

It is amusing to consider the case when all c_{ij} are 0. In that case, Theorem 1.2 becomes:

$$|a_{kk}| > \sum_{j \neq k} r_{kj}|a_{kj}|$$

if and only if, for every subset $S \subset \{1, \ldots, n\}$, $|S| \geq 2$, and every cyclic permutation σ of S

$$\prod_{i \in S} r_{i,\sigma i} \geq 1. \tag{3.5}$$

It is interesting to note that (3.5) holds for all S and σ if and only if there exists a positive vector x such that

$$r_{kj} \geq \frac{x_j}{x_k}, \qquad j, k = 1, \ldots, n, \quad j \neq k. \tag{3.6}$$

We omit the proof because the result is known or can be easily derived from the duality theorem of linear programming.

COROLLARY 3.1 [12]   *Let $0 \leq \alpha \leq 1$. If $A = (a_{ij})$ satisfies*

$$|a_{ii}|a_{jj}| > (\sum_{k \neq i} |a_{ik}| \sum_{k \neq j} |a_{jk}|)^{\alpha}(\sum_{k \neq i} |a_{ki}| \sum_{k \neq j} |a_{kj}|)^{1-\alpha}$$

$$\text{for } i \neq j, \; i, j = 1, \ldots, n, \tag{3.7}$$

then A is nonsingular.

Proof   Define $P_i \equiv \sum_{k \neq i} |a_{ik}|$, $Q_i \equiv \sum_{k \neq i} |a_{ki}|$, $i = 1, \ldots, n$.

From (3.7), there is a positive number ε such that

$$|a_{ii}| \, |a_{jj}| > ((P_i + \varepsilon)(P_j + \varepsilon))^{\alpha}((Q_i + \varepsilon)(Q_j + \varepsilon))^{1-\alpha} \quad \text{for} \quad i \neq j, i, j = 1, \ldots, n \tag{3.8}$$

From (3.8), there is at most one i such that

$$|a_{ii}| \leq (P_i + \varepsilon)^{\alpha}(Q_i + \varepsilon)^{1-\alpha}.$$

This remark, together with (3.8), shows that, if $|S| \geq 2$, then

$$\prod_{i \in S} \frac{|a_{ii}|}{(P_i + \varepsilon)^{\alpha}(Q_i + \varepsilon)^{1-\alpha}} > 1. \tag{3.9}$$

Let

$$r_{ij} = \frac{\alpha|a_{ii}|}{P_i + \varepsilon} \qquad i \neq j, \; i, j = 1, \ldots, n.$$

$$c_{ji} = \frac{(1 - \alpha)|a_{ii}|}{Q_i + \varepsilon} \qquad i \neq j, \; i, j = 1, \ldots, n.$$

Then

$$|a_{ii}| > \sum_{k \neq i} r_{ik}|a_{ik}| + \sum_{k \neq i} c_{ki}|a_{ki}|, \qquad i = 1, \ldots, n.$$

52   A. J. HOFFMAN

By Theorem 1.2, we will be done if we show that, for any $S \subset \{1, \ldots, n\}$, $|S| \geq 2$ and any cyclic permutation σ of S

$$(\prod_{i \in S} r_{i,\sigma i})^{1/|S|} + (\prod_{i \in S} c_{i,\sigma i})^{1/|S|} \geq 1. \tag{3.10}$$

But

$$
\begin{aligned}
(\prod_{i \in S} r_{i,\sigma i})^{1/|S|} + (\prod_{i \in S} c_{i,\sigma i})^{1/|S|} &= \alpha \left(\prod_{i \in S} \frac{|a_{ii}|}{P_i + \varepsilon} \right)^{1/|S|} + (1 - \alpha) \left(\prod_{i \in S} \frac{|a_{ii}|}{Q_i + \varepsilon} \right)^{1/|S|} \\
&\geq \left(\left(\prod_{i \in S} \frac{|a_{ii}|}{P_i + \varepsilon} \right)^{\alpha} \left(\prod_{i \in S} \frac{|a_{ii}|}{Q_i + \varepsilon} \right)^{1-\alpha} \right)^{1/|S|} \\
&= \left(\prod_{i \in S} \frac{|a_{ii}|}{(P_i + \varepsilon)(Q_i + \varepsilon)} \right)^{1/|S|} \\
&> 1, \quad \text{by (3.9),}
\end{aligned}
$$

verifying (3.10).

Many other well-known criteria for nonsingularity, due mainly to Ostrowski (see [7] for the best known of these), are in principle also corollaries of Theorem 1.2, since in each case, as the original proofs show implicitly, it is possible to find a positive vector x (depending on the matrix), and let $r_{ij} = x_j/x_i$, $c_{ij} = 0$.

References

[1] D. H. Carlson and R. S. Varga, Minimal G-functions, *Lin. Alg. Appl.* **6** (1973), 97–117.

[2] D. H. Carlson and R. S. Varga, Minimal G-functions II, *Lin. Alg. Appl.* **7** (1973), 233–242.

[3] A. J. Hoffman, Bounds for the rank and eigenvalues of a matrix, *IFIP* **68**, North-Holland, Amsterdam (1969), 111–113.

[4] A. J. Hoffman, Generalizations of Gerschgorin's theorem: G-generating families, lecture notes, Universities of California at Santa Barbara, August, 1969, pp. 46.

[5] A. J. Hoffman, Combinatorial aspects of Gerschgorin's theorem, Recent trends in graph theory, *Lecture notes in Mathematics* **186**, Springer, New York (1971), 173–179.

[6] A. J. Hoffman and R. S. Varga, Patterns of dependence in generalizations of Gerschgorin's theorem, *SIAM J. Numer. Anal.* **7** (1970), 571–574.

[7] M. Marcus and H. Minc, *A Survey of Matrix Theory and Matrix Inequalities*, Boston, Allyn (1964).

[8] M. Marcus and M. Newman, Generalized functions of symmetric matrices, *Proc. Amer. Math. Soc.* **16** (1965), 826–830.

[9] A. W. Marshall and I. Olkin, Scaling of matrices to achieve specified row and column sums, *Num. Math.* **12** (1968), 83–90.

[10] H. I. Medley, A note on G-generating families and isolated Gersgorin disks, *Num. Math.* **21** (1973), 93–95.

[11] P. Nowosad, On the functional (x^{-1}, Ax) and some of its applications, *Ann. Acad. Brasil Ci.* **37** (1965), 163–165.

[12] A. Ostrowski, Ueber das Nichtverschwinden einer Klasse von determination und die Lokalisierung der charakteristichen Wurzeln von Matrizen, *Compositio Math.* **9** (1951), 209–226.

[13] R. Sinkhorn, A relationship between arbitrary positive matrices and doubly stochastic matrices, *Ann. Math. Statist.* **35** (1964), 876–879.

On the Relationship between the
Hausdorff Distance and Matrix Distances of Ellipsoids

Jean-Louis Goffin
McGill University
Montreal, Quebec, Canada

and

Alan J. Hoffman*
IBM T. J. Watson Research Center
Yorktown Heights, New York

Dedicated to Sasha Ostrowski on his 90th birthday.
His work and life have always been an inspiration.

Submitted by Walter Gautschi

ABSTRACT

The space of ellipsoids may be metrized by the Hausdorff distance or by the sum
of the distance between their centers and a distance between matrices. Various
inequalities between metrics are established. It follows that the square root of positive
semidefinite symmetric matrices satisfies a Lipschitz condition, with a constant which
depends only on the dimension of the space.

*This research was done while both authors were visiting the Department of Operations
Research, Stanford University.

Research and reproduction of this report were partially supported by the Department of
Energy Contract AM03-76SF00326, PA# DE-AT03-76ER72018; Office of Naval Research
Contract N00014-75-C-0267; National Science Foundation Grants MCS76-81259, MCS-7926009
and ECS-8012974 at Stanford University; and the D.G.E.S. (Quebec), the N.S.E.R.C. of Canada
under grant A 4152, and the S.S.H.R.C. of Canada.

Any opinions, findings, and conclusions or recommendations expressed in this publication are
those of the authors and do *not* necessarily reflect the views of the above sponsors.

Reproduction in whole or in part is permitted for any purposes of the United States
Government. This document has been approved for public release and sale; its distribution is
unlimited.

LINEAR ALGEBRA AND ITS APPLICATIONS 52/53:301–313 (1983)   301

JEAN-LOUIS GOFFIN AND ALAN J. HOFFMAN

1.   DISTANCE BETWEEN ELLIPSOIDS AS SETS

Ellipsoids in R^n may be viewed as elements of the set of subsets of R^n, subsets which could be restricted to be compact, convex, and centrally symmetric. The set of subsets of R^n is usually metrized by the Hausdorff metric [2]:

$$\delta(E, F) = \max\left\{ \sup_{y \in F} \inf_{x \in E} \|x - y\|, \sup_{x \in E} \inf_{y \in F} \|x - y\| \right\}$$

$$= \inf\{\delta \geqslant 0 : E + \delta S \supset F, F + \delta S \supset E\},$$

where E and F are subsets of R^n, $\|\ \|$ represents the Euclidean norm, and $S = \{x \in R^n : \|x\| \leqslant 1\}$ is the unit ball.

If E and F are convex, then

$$\delta(E, F) = \sup\{|h(x, E) - h(x, F)| : \|x\| = 1\},$$

where $h(x, E) = \sup\{(x, y) : y \in E\}$ is the support function of E (see Bonnensen and Fenchel [1]) and $(\cdot, \cdot)$ denotes the scalar product.

If E and F are convex and contain the origin in their interiors, then

$$\delta(E, F) = \sup\{|g(x, E^d) - g(x, F^d)| : \|x\| = 1\},$$

where $E^d = \{x \in R^n : (x, y) \leqslant 1 \ \forall y \in E\}$ is the dual of E, and

$$g(x, E^d) = \inf\{\mu \geqslant 0 : x \in \mu E^d\}$$

is the distance function, or gauge, of E^d; this follows because $h(x, E) = g(x, E^d)$.

If E and F are convex, full-dimensional, and centrally symmetric with respect to the origin, then E^d and F^d inherit the same properties, and $g(x, E^d)$ and $g(x, F^d)$ define norms on R^n. Thus $\delta(E, F)$ may be viewed as a distance between norms on R^n.

The Hausdorff distance is invariant under congruent, but not affine, transformations, and reduced by projection. It will be assumed throughout that the space of ellipsoids contains the degenerate ellipsoids. The space of ellipsoids is not closed under addition.

The following lemma indicates that it will be sufficient to study ellipsoids centered at the origin.

LEMMA 1. *Let E and F be two subsets of R^n, compact, convex, and symmetric with respect to the origin; let $E^1 = e + E$ and $F^1 = f + F$. Then*

$$\delta(E^1, F^1) \leqslant \delta(E, F) + \|e - f\| \leqslant 2\delta(E^1, F^1),$$

$$\delta(E, F) \leqslant \delta(E^1, F^1), \qquad \|e - f\| \leqslant \delta(E^1, F^1).$$

Proof.

$$\begin{aligned}
\delta(E^1, F^1) &= \sup\{|h(x, E^1) - h(x, F^1)| : \|x\| = 1\} \\
&= \sup\{|h(x, E) - h(x, F) + (e - f, x)| : \|x\| = 1\} \\
&\leqslant \sup\{|h(x, E) - h(x, F)| : \|x\| = 1\} + \sup\{|(e - f, x)| : \|x\| = 1\} \\
&= \delta(E, F) + \|e - f\|.
\end{aligned}$$

Conversely

$$-\delta(E^1, F^1) \leqslant h(x, E) - h(x, F) + (e - f, x) \leqslant \delta(E^1, F^1) \qquad \forall \|x\| = 1;$$

now $h(x, E) = h(-x, E)$ and $h(x, F) = h(-x, F)$, as E and F are symmetric with respect to the origin, and thus

$$-\delta(E^1, F^1) \leqslant h(x, E) - h(x, F) - (e - f, x) \leqslant \delta(E^1, F^1) \qquad \forall \|x\| = 1.$$

Adding and subtracting, one gets

$$-\delta(E^1, F^1) \leqslant h(x, E) - h(x, F) \leqslant \delta(E^1, F^1),$$

$$-\delta(E^1, F^1) \leqslant (e - f, x) \qquad\quad \leqslant \delta(E^1, F^1) \qquad \forall \|x\| = 1,$$

and hence $\delta(E, F) \leqslant \delta(E^1, F^1)$ and $\|e - f\| \leqslant \delta(E^1, F^1)$. $\blacksquare$

2. ELLIPSOIDS AS VECTORS AND MATRICES

An ellipsoid may also be represented by a vector (its center) and a matrix (its size, shape, and position):

$$E = e + \overline{A}S = \{x \in R^n : x = e + \overline{A}t, \|t\| \leqslant 1\};$$

note that $h(x, E) = (e, x) + \|\overline{A}^T x\|$. If $\overline{A}$ is nonsingular, then

$$E = \left\{ x \in R^n : (x - e)^T (\overline{A}\,\overline{A}^T)^{-1} (x - e) \leqslant 1 \right\}.$$

To any ellipsoid is associated an equivalence class of matrices; in fact $E = e + \overline{A}S = e + \overline{\overline{A}}S$ if and only if $\overline{A} = \overline{\overline{A}}O$ where O is an orthogonal matrix, or equivalently if $\overline{A}\,\overline{A}^T = \overline{\overline{A}}\,\overline{\overline{A}}^T$. Define now $H = \overline{A}\,\overline{A}^T$, and $A = H^{1/2}$; then in the remainder of this paper an ellipsoid will be defined by

$$E = e + AS = e + H^{1/2}S,$$

where A and H are positive semidefinite real symmetric matrices. Using any of these two definitions, there exists a one-to-one correspondence between ellipsoids and points (e, A) in $R^n \times p(R^n)$ [respectively $(e, H) \in R^n \times p(R^n)$], where $p(R^n)$ is the set of $n \times n$ positive semidefinite symmetric matrices.

One could have tried to associate to an ellipsoid a lower triangular matrix $L(H = LL^T)$; L is unique if H is nonsingular, but not necessarily so if H is singular. This is the key reason why the results of this paper will not extend to the case of Cholesky factors.

If A is nonsingular, then

$$E = \left\{ x \in R^n : (x - e)^T A^{-2} (x - e) \leqslant 1 \right\}$$

$$= \left\{ x \in R^n : (x - e)^T H^{-1} (x - e) \leqslant 1 \right\}.$$

We may now define two matric distances on the space of ellipsoids.

Let $E = e + AS = e + H^{1/2}S$ and $F = f + BS = f + K^{1/2}S$ be two ellipsoids in R^n, where A, B, H, and K are positive semidefinite symmetric matrices; then define

$$d(E, F) = \|e - f\| + \|A - B\|,$$

$$\Delta(E, F) = \|e - f\| + \|H - K\|^{1/2} = \|e - f\| + \|A^2 - B^2\|^{1/2},$$

where $\| \ \|$, for matrices, is the spectral norm.

It is clear that d and Δ satisfy the axioms for a metric (or distance).

Various inequalities between d, Δ, and δ will be proven in the next section; the relationship between d and δ is the closest one, as d and δ are related by inequalities involving constants depending only upon the dimension of the space.

DISTANCE OF ELLIPSOIDS   305

The inequalities imply that the three metrics define the same topology on the space of ellipsoids, but, more strongly, that rates of convergence can be related.

The inequalities between d and δ imply that the rates of convergence of a sequence of ellipsoids may be studied within a space of sets, or a space of matrices, and that the two rates are identical.

3. INEQUALITIES BETWEEN DISTANCES

If E and F are ellipsoids centered at the origin, and $E^1 = e + E$, $F^1 = f + F$, then

$$d(E^1, F^1) = \|e - f\| + d(E, F),$$

$$\Delta(E^1, F^1) = \|e - f\| + \Delta(E, F);$$

and Lemma 1 indicates that it is enough to study ellipsoids centered at the origin. In that case,

$$d(E, F) = \|A - B\|,$$

$$\Delta(E, F) = \|H - K\|^{1/2} = \|A^2 - B^2\|^{1/2},$$

$$\delta(E, F) = \sup\{\|\,\|Ax\| - \|Bx\|\,\| : \|x\| = 1\}.$$

THEOREM 2.   *Let $E = AS$ and $F = BS$ be two ellipsoids in R^n, centered at the origin, where A and B are $n \times n$ positive semidefinite symmetric matrices. Then*

$$k_n^{-1}\|A - B\| \leqslant \sup\{\|\,\|Ax\| - \|Bx\|\,\| : \|x\| = 1\} \leqslant \|A - B\|,$$

or

$$\delta(E, F) \leqslant d(E, F) \leqslant k_n \delta(E, F),$$

where $k_n = 2\sqrt{2}\, n(n + 2)$.

Proof.   For the first part, one has

$$\|\,\|Ax\| - \|Bx\|\,\| \leqslant \|Ax - Bx\| = \|(A - B)x\| \leqslant \|A - B\|\|x\|,$$

and $\sup\{\|\,\|Ax\| - \|Bx\|\,\| : \|x\| = 1\} \leqslant \|A - B\|$.

JEAN-LOUIS GOFFIN AND ALAN J. HOFFMAN

For the second part, let $\delta = \sup\{|\,\|Ax\| - \|Bx\|\,| : \|x\| = 1\}$, and $\alpha_n \leqslant \alpha_{n-1} \leqslant \cdots \leqslant \alpha_1, \beta_n \leqslant \beta_{n-1} \leqslant \cdots \leqslant \beta_1$ be the ordered eigenvalues of A and B (clearly all real and nonnegative numbers).

The maximum characterization for the eigenvalues of Hermitian matrices gives

$$\alpha_k^2 = \max_{S_k} \min_{x \in S_k} x^T A^2 x,$$

where S_k represents the intersection of any k-dimensional subspace with the unit spherical surface; assume that S_k^* gives the maximum. Now define x_k^* by

$$x_k^{*T} B^2 x_k^* = \min_{x \in S_k^*} x^T B^2 x.$$

Thus

$$\beta_k^2 = \max_{S_k} \min_{x \in S_k} x^T B^2 x \geqslant \min_{x \in S_k^*} x^T B^2 x = \|Bx_k^*\|^2$$

and

$$\alpha_k^2 = \min_{x \in S_k^*} x^T A^2 x \leqslant \|Ax_k^*\|^2;$$

it follows that

$$\alpha_k - \beta_k \leqslant \|Ax_k^*\| - \|Bx_k^*\| \leqslant \delta.$$

Reversing the argument, $\beta_k - \alpha_k \leqslant \delta$, and $|\alpha_k - \beta_k| \leqslant \delta \ \forall k = 1, \ldots, n$.

The content of the theorem is unchanged if A is replaced by $O^T A O$ and B by $O^T B O$, where O is any orthogonal matrix; hence we may assume that A is diagonal, and that

$$\alpha_i = a_{ii} \qquad \forall i = 1, \ldots, n.$$

Denote by $B_k = Be_k$ the kth column of B, where e_k is the kth column of the identity matrix; then

$$|\alpha_k - \|B_k\|\,| = |\,\|Ae_k\| - \|Be_k\|\,| \leqslant \delta \qquad \forall k = 1, \ldots, n.$$

Hence

$$|\beta_k - \|B_k\|| \leqslant |\beta_k - \alpha_k| + |\alpha_k - \|B_k\|| \leqslant 2\delta \qquad \forall k = 1,\ldots,n.$$

Now

$$\|B_k\|^2 = \left(\sum_{i=1}^{n} b_{ik}^2 \right) \geqslant b_{kk}^2;$$

thus

$$0 \leqslant \|B_k\| - b_{kk},$$

$$0 \leqslant \sum_{k=1}^{n} (\|B_k\| - b_{kk}) = \sum_{k=1}^{n} (\|B_k\| - \beta_k)$$

$$\leqslant \sum_{k=1}^{n} |\|B_k\| - \beta_k| \leqslant 2n\delta,$$

implying that

$$0 \leqslant \|B_k\| - b_{kk} \leqslant 2n\delta \qquad \forall k = 1,\ldots,n.$$

This leads to

$$|\alpha_k - b_{kk}| \leqslant |\alpha_k - \beta_k| + |\beta_k - \|B_k\|| + |\|B_k\| - b_{kk}| \leqslant (2n+3)\delta.$$

Let $D = \mathrm{diag}(b_{kk})$, and x be any vector of unit length; then

$$|\|Bx\| - \|Dx\|| \leqslant |\|Bx\| - \|Ax\|| + |\|Ax\| - \|Dx\||$$

$$\leqslant \delta + \|A - D\| \leqslant \delta + (2n+3)\delta = (2n+4)\delta,$$

as $\|A - D\| = \max_{k=1,\ldots,n} |\alpha_k - b_{kk}| \leqslant (2n+3)\delta$.

It remains to show that the off-diagonal elements of B are bounded by a multiple of δ. If $b_{ii} = 0$ or $b_{kk} = 0$, then $b_{ik} = 0$ $(i \neq k)$, as $b_{ik}^2 \leqslant b_{ii}b_{kk}$ by the positive semidefiniteness of B. So assume that $b_{ii} > 0$, $b_{kk} > 0$, and let $a = b_{ii}$, $b = b_{kk}$, $c = |b_{ik}|$, and $\sigma = +1 \, (-1)$ if b_{ik} is positive (negative).

JEAN-LOUIS GOFFIN AND ALAN J. HOFFMAN

Choose

$$z = \frac{1}{\sqrt{a^2 + b^2}} (be_i + \sigma a e_k);$$

then

$$\|Bz\| = \frac{1}{\sqrt{a^2 + b^2}} \|bB_i + \sigma a B_k\|$$

$$= \frac{1}{\sqrt{a^2 + b^2}} \left\| (ab + ac)e_i + \sigma(ab + bc)e_k + \sum_{j \neq i,k} (bb_{ji} + \sigma ab_{jk})e_j \right\|$$

$$\geq \frac{1}{\sqrt{a^2 + b^2}} \left\| (ab + ac)e_i + \sigma(ab + bc)e_k \right\|$$

$$= \frac{1}{\sqrt{a^2 + b^2}} \left[(ab + ac)^2 + (ab + bc)^2 \right]^{1/2}$$

$$= \left(c^2 + 2\frac{abc(a + b)}{a^2 + b^2} + 2\frac{a^2 b^2}{a^2 + b^2} \right)^{1/2}$$

$$= \left(\frac{\sqrt{2}\, ab}{\sqrt{a^2 + b^2}} + d \right),$$

where this last equation defines d $(d > 0)$.

Now, as $\|Dz\| = \sqrt{2}\, ab/\sqrt{a^2 + b^2}$, it follows that $d \leq \|Bz\| - \|Dz\| \leq (2n + 4)\delta$.

The value of d is given by the positive root of

$$d^2 + \frac{2\sqrt{2}\, ab}{\sqrt{a^2 + b^2}} d = c^2 + \frac{2abc(a + b)}{a^2 + b^2};$$

the left-hand side increases with d $(d \geq 0)$ and is less than the right-hand side for $d = 0$ and $d = c/\sqrt{2}$, implying that the value of d is greater than $c/\sqrt{2}$, and

$$c < d\sqrt{2} \leq 2\sqrt{2}\,(n + 2)\delta.$$

Thus $|b_{ik}| \leqslant 2\sqrt{2}\,(n+2)\delta \;\; \forall i, k,\; i \neq k$, and

$$
\begin{aligned}
\|A - B\|^2 &\leqslant \mathrm{Tr}\,(A - B)^2 \\[1mm]
&= \sum_{k} (a_{kk} - b_{kk})^2 + \sum\sum_{i \neq k} b_{ik}^2 \\[1mm]
&\leqslant n(2n+3)^2 \delta^2 + n(n-1)8(n+2)^2 \delta^2 \\[1mm]
&\leqslant 8n^2(n+2)^2 \delta^2;
\end{aligned}
$$

hence $\|A - B\| \leqslant 2\sqrt{2}\, n(n+2)\delta.$ ∎

The next result, which compares the distances δ and Δ, uses an operator-theory proof, and hence carries to infinite-dimensional Hilbert spaces.

Theorem 3. *Let $E = H^{1/2}S$ and $F = K^{1/2}S$ be two ellipsoids in R^n, centered at the origin, where H and K are positive semidefinite symmetric matrices. Then*

$$
\delta(E, F) \leqslant \|H - K\|^{1/2} \leqslant \left[\delta^2(E, F) + \delta(E, F)\max(D(E), D(F))\right]^{1/2},
$$

where $\delta(E, F) = \sup\{|(x^T H x)^{1/2} - (x^T K x)^{1/2}| : \|x\| = 1\}$, $\Delta(E, F) = \|H - K\|^{1/2}$, *and* $D(E) = 2\|H\|^{1/2}$ *is the diameter of E; this may also be written as*

$$
\frac{\|H - K\|}{\left[\|H - K\| + \max(\|H\|, \|K\|)\right]^{1/2} + \left[\max(\|H\|, \|K\|)\right]^{1/2}} \leqslant \delta(E, F) \leqslant \|H - K\|^{1/2}.
$$

Proof. Let $\delta = \delta(E, F)$; thus

$$
(x^T H x)^{1/2} - (x^T K x)^{1/2} \leqslant \delta \|x\| \qquad \forall x;
$$

hence

$$
\begin{aligned}
x^T H x &\leqslant \delta^2 \|x\|^2 + 2\delta \|x\|(x^T K x)^{1/2} + x^T K x \qquad \forall x, \\[1mm]
&\leqslant \delta^2 \|x\|^2 + x^T K x + \varepsilon^{-1}\delta^2 \|x\|^2 + \varepsilon(x^T K x) \qquad \forall x \;\; \forall \varepsilon > 0 \\[1mm]
&= \delta^2(1 + \varepsilon^{-1})\|x\|^2 + (1 + \varepsilon)(x^T K x) \qquad \forall x, \;\; \forall \varepsilon > 0.
\end{aligned}
$$

310   JEAN-LOUIS GOFFIN AND ALAN J. HOFFMAN

We have

$$x^T(H-K)x \leqslant x^T\big[\delta^2(1+\varepsilon^{-1})I+\varepsilon K\big]x \qquad \forall x, \quad \forall \varepsilon > 0,$$

and similarly, reversing the argument,

$$x^T(H-K)x \geqslant -x^T\big[\delta^2(1+\eta^{-1})I+\eta H\big]x \qquad \forall x, \quad \forall \eta > 0.$$

These two equations imply that

$$\|H-K\| \leqslant \max\{\varepsilon\|K\|+\delta^2(1+\varepsilon^{-1}), \eta\|H\|+\delta^2(1+\eta^{-1})\} \qquad \forall \varepsilon > 0, \quad \forall \eta > 0;$$

taking $\varepsilon = \delta/\|K\|^{1/2}$ and $\eta = \delta/\|H\|^{1/2}$, one gets

$$\|H-K\| \leqslant \delta^2 + 2\delta\max(\|H\|^{1/2}, \|K\|^{1/2}).$$

For the second part, let $\Delta^2 = \|H-K\|$, thus

$$\big|x^T(H-K)x\big| \leqslant \Delta^2\|x\|^2 \qquad \forall x;$$

using the inequality

$$|a-b| \leqslant \sqrt{|a^2-b^2|} \qquad (a,b \geqslant 0)$$

one gets

$$\big|(x^THx)^{1/2} - (x^TKx)^{1/2}\big| \leqslant \Delta\|x\| \qquad \forall x,$$

and

$$\delta(E,F) = \sup\{\big|(x^THx)^{1/2} - (x^TKx)^{1/2}\big| : \|x\| = 1\}$$

$$\leqslant \Delta = \|H-K\|^{1/2}. \qquad\qquad \blacksquare$$

Theorems 2 and 3 can be combined to give a relationship between the distances d and Δ, which is a statement about square roots of matrices.

THEOREM 4. *Let H and K be two $n \times n$ positive semidefinite matrices, and $A = H^{1/2}$, $B = K^{1/2}$. Then*

$$l_n^{-1}\|A - B\| \leqslant \|H - K\|^{1/2} \leqslant \left[2\|A - B\|\max(\|A\|,\|B\|) + \|A - B\|^2\right]^{1/2},$$

or

$$\frac{\|H - K\|}{\left[\|H - K\| + \max(\|H\|,\|K\|)\right]^{1/2} + \left[\max(\|H\|,\|K\|)\right]^{1/2}} \leqslant \|A - B\| \leqslant l_n\|H - K\|^{1/2},$$

where $l_n = k_n = 2\sqrt{2}\, n(n+2)$.

This theorem means that the square root satisfies a Lipschitz condition on the cone of positive semidefinite matrices:

$$\|H^{1/2} - K^{1/2}\| \leqslant l_n\|H - K\|^{1/2} \qquad \forall H, K \in p(R^n),$$

where the Lipschitz constant depends only upon the dimension of R^n; l_n is bounded by a polynomial of degree 1 in the dimension of $p(R^n)$.

It is now a simple matter to extend Theorems 2, 3, and 4 to the case of ellipsoids not necessarily centered at the origin.

THEOREM 5. *Let $E = e + AS = e + H^{1/2}$ and $F = f + BS = f + K^{1/2}S$ be two ellipsoids in R^n, and A, B, H, and K be $n \times n$ positive semidefinite symmetric matrices. Denote $\delta = \delta(E, F)$, $d = d(E, F)$, $\Delta = \Delta(E, F)$, and $M = \max(\|A\|,\|B\|) = \max(\|H\|^{1/2},\|K\|^{1/2}) = \frac{1}{2}\max(D(E), D(F))$. Then the following inequalities are satisfied:*

$$(k_n + 1)^{-1}d \leqslant \delta \leqslant d \leqslant (k_n + 1)\delta,$$

$$l_n^{-1}d \leqslant \Delta \leqslant (d^2 + 2dM)^{1/2},$$

$$\frac{\Delta^2}{\sqrt{\Delta^2 + M^2} + M} \leqslant d \leqslant l_n\Delta,$$

$$\delta \leqslant \Delta \leqslant \delta + (\delta^2 + 2\delta M)^{1/2},$$

$$\frac{\Delta^2}{2(M + \Delta)} \leqslant \delta \leqslant \Delta$$

with $k_n = l_n = 2\sqrt{2}\, n(n+2)$.

312   JEAN-LOUIS GOFFIN AND ALAN J. HOFFMAN

Proof. Let $\varepsilon = \|e - f\|$, and δ_0, d_0, and Δ_0 be the distances between $E - e$ and $F - f$.

One has $d = d_0 + \varepsilon$, $\Delta = \Delta_0 + \varepsilon$ and, by Lemma 1, $\delta \leqslant \delta_0 + \varepsilon$, $\varepsilon \leqslant \delta$, and $\delta_0 \leqslant \delta$. Hence a slight difference appears in the proofs for the various cases.

For instance, Theorem 4 implies

$$\Delta_0 \leqslant \left(d_0^2 + 2d_0 M \right)^{1/2}.$$

Hence $\Delta = \Delta_0 + \varepsilon \leqslant (d_0^2 + 2d_0 M)^{1/2} + \varepsilon$; the maximum of the right-hand side (subject to $d_0 \geqslant 0$, $\varepsilon \geqslant 0$, and $\varepsilon + d_0 = d$) is attained for $\varepsilon = 0$ and $d_0 = d$, and thus

$$\Delta \leqslant \left(d^2 + 2dM \right)^{1/2},$$

or

$$d \geqslant \frac{\Delta^2}{\sqrt{\Delta^2 + M^2} + M}.$$

The equivalent result from Theorem 3 implies

$$\Delta_0 \leqslant \left(\delta_0^2 + 2\delta_0 M \right)^{1/2};$$

hence

$$\Delta = \Delta_0 + \varepsilon \leqslant \left(\delta_0^2 + 2\delta_0 M \right)^{1/2} + \varepsilon,$$

and the maximum of the right-hand side subject to $\varepsilon \leqslant \delta$ and $\delta_0 \leqslant \delta$ is clearly attained for $\varepsilon = \delta$ and $\delta_0 = \delta$, and thus

$$\Delta \leqslant \left(\delta^2 + 2\delta M \right)^{1/2} + \delta,$$

or

$$\delta \geqslant \frac{\Delta^2}{2(M + \Delta)}.$$

The other cases follow similarly.   ■

DISTANCE OF ELLIPSOIDS313

4. CONCLUSION

Three metrics on the space of ellipsoids have been shown to be linked by various inequalities, and hence the induced topologies are identical. Not only is the notion of convergence unique, but rates of convergence can be related. Similar results clearly hold if the Euclidean norm is replaced by any of the L_p norms.

If k_n and l_n were defined to be the smallest constants satisfying Theorems 2 and 4 (with $l_n \leqslant k_n$), it would be quite interesting to know whether or not they must depend on n, the dimension.

REFERENCES

1   T. Bonnensen and W. Fenchel, *Theorie der Konvexen Korper*, Chelsea, New York, 1948.
2   F. Hausdorff, *Set Theory*, 2nd ed., Chelsea, New York, 1962.

Received 24 August 1981; revised 8 April 1982

Bounds for the Spectrum of Normal Matrices

E. R. Barnes*
School of Industrial and Systems Engineering
Georgia Institute of Technology
Atlanta, Georgia 30332-0205

and

A. J. Hoffman
IBM T.J. Watson Research Center
Yorktown Heights, New York 10598

Dedicated to Marvin Marcus

Submitted by N. J. Higham

ABSTRACT

Let A be a normal matrix with eigenvalues $\lambda_1, \lambda_2, \ldots, \lambda_n$, and let Γ denote the smallest disc containing these eigenvalues. We give some inequalities relating the center and radius of Γ to the entries in A. When applied to Hermitian matrices our results give lower bounds on the spread $\max_{ij}(\lambda_i - \lambda_j)$ of A. When applied to positive definite Hermitian matrices they give lower bounds on the Kantorovich ratio $\max_{ij}(\lambda_i - \lambda_j)/(\lambda_i + \lambda_j)$.

1. INTRODUCTION

Let $A = (a_{ij})$ be an $n \times n$ normal matrix with eigenvalues $\lambda_1, \ldots, \lambda_n$. We can write $A = U\Lambda U^*$, where U is a unitary matrix and $\Lambda = \operatorname{diag}(\lambda_1, \ldots, \lambda_n)$. This representation of A shows that the diagonal entries of

* This author's work was supported by the National Science Foundation grant DDM-9014823.

A are convex combinations of the eigenvalues of A. It follows that the smallest disc containing the eigenvalues of A is at least as large as the smallest disc containing the diagonal entries of A. In particular, if A is Hermitian we have

$$\max_{i,j}\left(\lambda_i - \lambda_j\right) \geqslant \max_{i,j}\left(a_{ii} - a_{jj}\right).$$

An improvement of this result is given by Mirsky in [4]. He shows that

$$\max_{i,j}\left(\lambda_i - \lambda_j\right)^2 \geqslant \max_{i \neq j}\left\{\left(a_{ii} - a_{jj}\right)^2 + 4|a_{ij}|^2\right\}. \tag{1.1}$$

We are going to derive the sharper result

$$\max_{i,j}\left(\lambda_i - \lambda_j\right)^2 \geqslant \max_{i,j}\left\{\left(a_{ii} - a_{jj}\right)^2 + 2\sum_{k \neq i}|a_{ik}|^2 + 2\sum_{k \neq j}|a_{jk}|^2\right\}. \tag{1.2}$$

Several other authors have been interested in the spread of Hermitian matrices. See for example [4], [5], and [6]. An application to combinatorial optimization problems is given in [1].

When A is Hermitian and positive definite, it is also of interest to estimate the Kantorovich ratio $\max_{ij}(\lambda_i - \lambda_j)/(\lambda_i + \lambda_j)$ of A. This quantity governs the rate of convergence of certain iterative schemes for solving linear systems of equations of the form $Ax = b$. See, for example, [7, Chapter 4]. In this case it is easy to show that

$$\max_{i,j}\frac{\lambda_i - \lambda_j}{\lambda_i + \lambda_j} \geqslant \max_{i,j}\frac{a_{ii} - a_{jj}}{a_{ii} + a_{jj}} \tag{1.3}$$

using the fact that the diagonal entries of A are convex combinations of the eigenvalues of A. We are going to prove the stronger result

$$\max_{i,j}\left(\frac{\lambda_i - \lambda_j}{\lambda_i + \lambda_j}\right)^2 \geqslant \max_{i,j}\frac{\left(a_{ii} - a_{jj}\right)^2 + \sum_{k \neq i}|a_{ik}|^2 + \sum_{k \neq j}|a_{jk}|^2}{\left(a_{ii} + a_{jj}\right)^2 + \sum_{k \neq i}|a_{ik}|^2 + \sum_{k \neq j}|a_{jk}|^2}. \tag{1.4}$$

Actually we are going to prove stronger results than (1.2) and (1.4), and we will do so for normal matrices. First we must establish some notation.

SPECTRUM OF NORMAL MATRICES   81

Let $\lambda = \{\lambda_1, \ldots, \lambda_n\}$ be a set of complex numbers, and let $\mathscr{A}(\lambda)$ denote the set of normal matrices whose spectrum is λ. Let Γ denote the smallest disc containing λ, and let $D(\lambda)$ denote the diameter of Γ. Define

$$d(\lambda) = \min_{i \neq j} |\lambda_i - \lambda_j|.$$

For any normal matrix $A = (a_{ij})$ we define

$$B_{ij}(A) = |a_{ii} - a_{jj}|^2 + 2 \sum_{k \neq i} |a_{ik}|^2 + 2 \sum_{k \neq j} |a_{jk}|^2.$$

We will prove that, for any indices i and j,

$$d^2(\lambda) = \min_{A \in \mathscr{A}(\lambda)} B_{ij}(A), \tag{1.5}$$

and

$$D^2(\lambda) = \max_{A \in \mathscr{A}(\lambda)} B_{ij}(A). \tag{1.6}$$

For Hermitian matrices (1.6) clearly implies (1.2).

2. THE VARIATION OF $B_{ij}(A)$ FOR $A \in \mathscr{A}(\lambda)$

We keep the notation of the introduction, except that for simplicity we write $B(A)$ instead of $B_{ij}(A)$.

THEOREM 2.1. *For any indices i and j and for any $A \in \mathscr{A}(\lambda)$ we have*

$$d^2(\lambda) \leqslant B(A) \leqslant D^2(\lambda), \tag{2.1}$$

and for each bound, there is an $A \in \mathscr{A}(\lambda)$ which attains the bound.

Proof   We first show if $A \in \mathscr{A}(\lambda)$, $B(A) \leqslant D^2(\lambda)$. Let e_i and e_j denote the ith and jth unit coordinate vectors, $i \neq j$, and let $u_1, \ldots, u_n$ be a set of orthonormal eigenvectors for A corresponding to the eigenvalues $\lambda_1, \ldots, \lambda_n$. If we write

$$e_i = \sum_{k=1}^{n} \alpha_k u_k \quad \text{and} \quad e_j = \sum_{k=1}^{n} \beta_k u_k,$$

                                    E. R. BARNES AND A. J. HOFFMAN

the vectors $\alpha = (\alpha_1, \ldots, \alpha_n)$ and $\beta = (\beta_1, \ldots, \beta_n)$ are orthonormal because e_i and e_j are orthonormal.

Let r and c denote the radius and center of the smallest disc containing λ. It is easy to see that

$$|a_{ii} - c|^2 + \sum_{k \neq i} |a_{ik}|^2 = \|(A - cI)e_i\|^2 = \left\| \sum_{k=1}^{n} \alpha_k(\lambda_k - c)u_k \right\|^2$$

$$= \sum_{k=1}^{n} \alpha_k^2 |\lambda_k - c|^2 \leqslant \sum_{k=1}^{n} \alpha_k^2 r^2 = r^2.$$

Similarly, $|a_{jj} - c|^2 + \sum_{k \neq j} |a_{jk}|^2 \leqslant r^2$.

The expression $|a_{ii} - c|^2 + |a_{jj} - c|^2$ achieves its minimum value $\frac{1}{2}|a_{ii} - a_{jj}|^2$ with respect to c at $c = \frac{1}{2}(a_{ii} + a_{jj})$. We therefore have

$$\frac{1}{2}|a_{ii} - a_{jj}|^2 + \sum_{k \neq i} |a_{ik}|^2 + \sum_{k \neq j} |a_{jk}|^2$$

$$\leqslant |a_{ii} - c|^2 + |a_{jj} - c|^2 + \sum_{k \neq i} |a_{ik}|^2 + \sum_{k \neq j} |a_{jk}|^2 \leqslant 2r^2, \quad (2.2)$$

which implies that

$$B(A) \leqslant 4r^2 = (2r)^2 = D^2(\lambda).$$

We must prove that this bound is attained.

We can write

$$B(\lambda) = 2\left(\sum_{k=1}^{n} |a_{ik}|^2 + \sum_{k=1}^{n} |a_{jk}|^2 \right) - |a_{ii} + a_{jj}|^2$$

$$= 2\left(\|Ae_i\|^2 + \|Ae_j\|^2 \right) - |(e_i, Ae_i) + (e_j, Ae_j)|^2$$

$$= 2 \sum_{k=1}^{n} t_k |\lambda_k|^2 - \left| \sum_{k=1}^{n} t_k \lambda_k \right|^2 \quad (2.3)$$

where $t_k = \alpha_k^2 + \beta_k^2$. From the definitions of α and β we have

$$0 \leqslant t_k \leqslant 1, \quad k = 1, \ldots, n \quad \text{and} \quad \sum_{k=1}^{n} t_k = 2. \tag{2.4}$$

We first show there are numbers $t_1, \ldots, t_n$ satisfying (2.4) such that the value on the right in (2.3) is $(2r)^2 = D^2(\lambda)$. We will then show that these t's correspond to a matrix $A \in \mathscr{A}(\lambda)$.

There are two cases to consider.

Case 1. The smallest disc containing λ is determined by two points, say λ_1 and λ_2, in λ. In this case we have $|\lambda_1 - \lambda_2|^2 = D^2(\lambda)$ and we take $t_1 = t_2 = 1$, $t_k = 0$, $k = 3, \ldots, n$. Substituting these values in (2.3) gives

$$2 \sum_{k=1}^{n} t_k |\lambda_k|^2 - \left| \sum_{k=1}^{n} t_k \lambda_k \right|^2 = 2\left(|\lambda_1|^2 + |\lambda_2|^2 \right) - |\lambda_1 + \lambda_2|^2$$

$$= |\lambda_1 - \lambda_2|^2 = D^2(\lambda),$$

as desired.

Case 2. The smallest disc containing λ is determined by three points, say $\lambda_1, \lambda_2, \lambda_3$. Clearly c is in the convex hull of these points. Let $c = \tau_1 \lambda_1 + \tau_2 \lambda_2 + \tau_3 \lambda_3$ be the representation of c as a convex combination of λ_1, λ_2, and λ_3. Since we are not in case 1, we have $0 < \tau_k < 1$, $k = 1, 2, 3$. Define $t_k = 2\tau_k$ for $k = 1, 2, 3$, and $t_k = 0$ for all other values of k. Then

$$D^2(\lambda) = (2r)^2 = 2 \sum_{k=1}^{3} t_k r^2 = 2 \sum_{k=1}^{3} t_k |\lambda_k - c|^2$$

$$= 2 \sum_{k=1}^{3} t_k |\lambda_k|^2 - \left| \sum_{k=1}^{3} t_k \lambda_k \right|^2.$$

To complete our analysis of case 2 we must show that $0 \leqslant t_k \leqslant 1$, $k = 1, 2, 3$. Equivalently, we must show that $0 \leqslant \tau_k \leqslant \frac{1}{2}$, $k = 1, 2, 3$. We will show that $0 \leqslant \tau_1 \leqslant \frac{1}{2}$. τ_2 and τ_3 can be treated similarly. The points

$$\lambda_1 \quad \text{and} \quad \frac{\tau_2}{\tau_2 + \tau_3} \lambda_2 + \frac{\tau_3}{\tau_2 + \tau_3} \lambda_3$$

lie in Γ. They are therefore separated by a distance of at most $2r$ units. But

$$\left| \lambda_1 - \left(\frac{\tau_2}{1 - \tau_1} \lambda_2 + \frac{\tau_3}{1 - \tau_1} \lambda_3 \right) \right| = \frac{|\lambda_1 - c|}{1 - \tau_1} = \frac{r}{1 - \tau_1},$$

and therefore $1/(1 - \tau_1) \leqslant 2$, or $\tau_1 \leqslant \frac{1}{2}$, as claimed.

We next show that for any t's satisfying (2.4), there exist real orthonormal vectors $u_1, \ldots, u_n$ such that

$$\left(e_i, u_k \right)^2 + \left(e_j, u_k \right)^2 = t_k, \qquad k = 1, \ldots, n. \tag{2.5}$$

To prove this we invoke a theorem of Horn [2] which says the following: If f and g are column vectors and T a doubly stochastic matrix satisfying $g' = f'T$, then there exists a real orthogonal matrix $U = (u_{ij})$ such that the doubly stochastic matrix $S = (u_{ij}^2)$ also satisfies $g' = f'S$.

For $t_1, \ldots, t_n$ satisfying (2.4) let T be a matrix whose ith and jth rows contain the vector $\frac{1}{2}(t_1, \ldots, t_n)$. Let every other row of T contain the vector $[1/(n - 2)](1 - t_1, \ldots, 1 - t_n)$. Then T is doubly stochastic and

$$(e_i + e_j)'T = (t_1, \ldots, t_n).$$

By Horn's theorem there exists a real orthogonal matrix $U = (u_{ij})$ such that

$$u_{ik}^2 + u_{jk}^2 = t_k, \qquad k = 1, \ldots, n.$$

The matrix $A = U \Lambda U^T$, where $\Lambda = \mathrm{diag}(\lambda_1, \ldots, \lambda_n)$, realizes the equality on the right in (2.1).

The left side of (2.1) is easy to treat. Consider the problem of minimizing the expression (2.3) over all values of t satisfying (2.4). Choose a $\bar{t} = (\bar{t}_1, \ldots, \bar{t}_n)$ which minimizes this expression and which has a minimum number of components satisfying $0 < t_j < 1$. Suppose some $\bar{t}_k$, say $\bar{t}_1$, satisfies $0 < \bar{t}_1 < 1$. Since $\sum_{k=1}^n \bar{t}_k = 1$, some other $\bar{t}_k$, say $\bar{t}_2$, is also strictly between 0 and 1. Let $t(\theta) = (\bar{t}_1 + \theta, \bar{t}_2 - \theta, \bar{t}_3, \ldots, \bar{t}_n)$. Then $t(\theta)$ satisfies (2.4) for an interval of values of θ. But (2.3) is a quadratic function of θ with leading coefficient negative if $\lambda_1 \neq \lambda_2$, or a linear function of θ if $\lambda_1 = \lambda_2$. In either case the minimum of (2.3), as a function of θ, will occur at an endpoint of the allowable interval, proving that $\bar{t}$ either does not give the minimum of (2.3), or does not have the minimum number of components satisfying $0 < t_j < 1$. But this contradicts our assumptions about $\bar{t}$. So $\bar{t}$ must have two components equal to 1, say $\bar{t}_i$ and $\bar{t}_j$, and the remainder equal to 0. For this t the expression (2.3) becomes

$$2\left(|\lambda_i|^2 + |\lambda_j|^2\right) - |\lambda_i + \lambda_j|^2 = |\lambda_i - \lambda_j|^2.$$

The minimum that this can be is $d^2(\lambda)$, and is obviously attainable. This completes the proof of Theorem 2.1. ∎

COROLLARY. *If $A = (a_{ij})$ is a Hermitian matrix with eigenvalues $\lambda_1 \geqslant \cdots \geqslant \lambda_n$, then*

$$(\lambda_1 - \lambda_n)^2 \geqslant \max_{i,j} \left\{ (a_{ii} - a_{jj})^2 + 2\sum_{k \neq i} |a_{ik}|^2 + 2\sum_{k \neq j} |a_{jk}|^2 \right\}. \quad (2.6)$$

Proof. This result follows immediately from (2.2). Note that the derivation of (2.2) does not assume that $i \neq j$, so i and j need not be distinct in (2.6). ∎

3. ERROR BOUNDS FOR THE SPREAD

In this section we derive an upper bound for the maximum distance between two eigenvalues of an arbitrary matrix. For sparse Hermitian matrices the upper bound is a small multiple of the lower bound given in (2.6).

THEOREM 3.1. *Let $A = (a_{ij})$ be an arbitrary $n \times n$ matrix with eigenvalues $\lambda_1, \ldots, \lambda_n$, and having at most $K - 1$ nonzero off-diagonal elements per row. Then*

$$\max_{i,j} |\lambda_i - \lambda_j| \leqslant \sqrt{K} \max_{i,j} \left(|a_{ii} - a_{jj}|^2 + 2\sum_{k \neq i} |a_{ik}|^2 + 2\sum_{k \neq j} |a_{jk}|^2 \right)^{1/2}. \quad (3.1)$$

Proof. By the Gerschgorin circle theorem, for any two eigenvalues λ_p and λ_q of A, there exist indices i and j such that

$$|\lambda_p - a_{ii}| \leqslant \sum_{k \neq i} |a_{ik}| \quad \text{and} \quad |\lambda_q - a_{jj}| \leqslant \sum_{k \neq j} |a_{jk}|.$$

                     E. R. BARNES AND A. J. HOFFMAN

It follows that

$$|\lambda_p - \lambda_q| \le |\lambda_p - a_{ii}| + |a_{ii} - a_{jj}| + |a_{jj} - \lambda_q|$$

$$\le |a_{ii} - a_{jj}| + \sum_{k \ne i} \frac{1}{\sqrt{2}}\left(\sqrt{2}|a_{ik}|\right) + \sum_{k \ne j} \frac{1}{\sqrt{2}}\left(\sqrt{2}|a_{jk}|\right)$$

$$\le \left(1 + \frac{K-1}{2} + \frac{K-1}{2}\right)^{1/2}$$

$$\times \left(|a_{ij} - a_{jj}|^2 + 2\sum_{k \ne i}|a_{ik}|^2 + 2\sum_{k \ne j}|a_{jk}|\right)^{1/2}. \tag{3.2}$$

The conclusion of the theorem follows immediately from this inequality. ∎

The next corollary gives a simple proof of a result due to Scott [5]. It shows that the Gerschgorin upper bound on the spread of a Hermitian matrix is a small multiple of the actual spread for sparse matrices.

COROLLARY (Scott).   *For a Hermitian matrix $A = (a_{ij})$ define*

$$G(A) = \max_{i,j}\left\{|a_{ii} - a_{jj}| + \sum_{k \ne i}|a_{ik}| + \sum_{k \ne j}|a_{jk}|\right\}.$$

Let $\lambda_1 \ge \lambda_2 \ge \cdots \ge \lambda_n$ denote the eigenvalues of A. If A has at most $K - 1$ nonzero off-diagonal elements per row, then

$$\frac{1}{\sqrt{K}}G(A) \le \lambda_1 - \lambda_n \le G(A). \tag{3.3}$$

Proof.   From the last inequality in (3.2) and (2.6) we have

$$G(A) \le \sqrt{K}\left\{\max_{i,j} B_{ij}(A)\right\}^{1/2} \le \sqrt{K}(\lambda_1 - \lambda_n).$$

This establishes the left side of (3.3). The right side follows from the first two inequalities in (3.2). ∎

SPECTRUM OF NORMAL MATRICES   87

In [3] Mirsky gives an upper bound of the spread of an arbitrary $n \times n$ matrix $A(n \geqslant 3)$. Denote the eigenvalues of A by $\lambda_1, \ldots, \lambda_n$. Mirsky shows that

$$\max_{i,j} |\lambda_i - \lambda_j| \leqslant \sqrt{2\|A\|^2 - \frac{2}{n}|\mathrm{tr}\, A|^2} \,. \tag{3.4}$$

The following theorem gives a bound on the maximum error in Mirsky's upper bound for Hermitian matrices.

THEOREM 3.2. *Let A be an $n \times n$ Hermitian matrix $(n \geqslant 3)$ with eigenvalues $\lambda_1 \geqslant \cdots \geqslant \lambda_n$. Define $M(A) = \{2\|A\|^2 - (2/n)(\mathrm{tr}\, A)^2\}^{1/2}$. Then*

$$\sqrt{\frac{2}{n}}\, M(A) \leqslant \lambda_1 - \lambda_n \leqslant M(A). \tag{3.5}$$

Proof. The second inequality follows from (3.4). To prove the first inequality, let r and c denote the radius and center, respectively, of the smallest disc containing the eigenvalues of A. In the proof of Theorem 2.1 we showed that

$$r^2 \geqslant |a_{ii} - c|^2 + \sum_{k \neq i} |a_{ik}|^2, \qquad i = 1, \ldots, n.$$

It follows that

$$nr^2 \geqslant \sum_{i=1}^{n} \left(|a_{ii} - c|^2 + \sum_{k \neq 1} |a_{ik}|^2 \right)$$

The expression on the right in this inequality assumes its minimum value with respect to c at $c = (\sum_{i=1}^{n} a_{ii})/n$. This minimum value is $\|A\|^2 - (1/n)(\mathrm{tr}\, A)^2$. It follows that

$$r^2 = \left(\frac{\lambda_1 - \lambda_n}{2} \right)^2 \geqslant \frac{1}{n} \left\{ \|A\|^2 - \frac{1}{n}(\mathrm{tr}\, A)^2 \right\},$$

and this is equivalent to the first inequality in (3.5). The proof is complete. ∎

88

4. THE KANTOROVICH RATIO

In this section we assume A is Hermitian positive definite and label the eigenvalues so that $\lambda_1 \geqslant \cdots \geqslant \lambda_n$. We will obtain two lower bounds for the Kantorovich ratio $(\lambda_1 - \lambda_n)/(\lambda_1 + \lambda_n)$.

THEOREM 4.1. *If* $A = (a_{ij})$ *is Hermitian and positive definite, then for any* i *and* j *we have*

$$\left(\frac{\lambda_1 - \lambda_n}{\lambda_1 + \lambda_n} \right)^2 \geqslant \frac{(a_{ii} - a_{jj})^2 + \Sigma_{k \neq i}|a_{ik}|^2 + \Sigma_{k \neq j}|a_{jk}|^2}{(a_{ii} + a_{jj})^2 + \Sigma_{k \neq i}|a_{ik}|^2 + \Sigma_{k \neq j}|a_{jk}|^2}. \quad (4.1)$$

Proof. If we square out all bracketed terms in (4.1) and cross-multiply and simplify, we see that this inequality is equivalent to

$$\frac{\Sigma_{k=1}^n |a_{ik}|^2 + \Sigma_{k=1}^n |a_{jk}|^2}{4 a_{ii} a_{jj}} \leqslant \frac{\lambda_1^2 + \lambda_n^2}{4 \lambda_1 \lambda_n}. \quad (4.2)$$

Using the notation introduced in the proof of Theorem 2.1, we have

$$a_{ii} = e_i^T A e_i = \sum_{k=1}^n \alpha_k^2 \lambda_k, \qquad \sum_{k=1}^n |a_{ik}|^2 = \| A e_i \|^2 = \sum_{k=1}^n \alpha_k^2 \lambda_k^2,$$

where $\alpha = (\alpha_1, \ldots, \alpha_n)$ is a unit vector. Since $\lambda_1 \geqslant \cdots \geqslant \lambda_n$ we have

$$\lambda_k^2 - \lambda_n^2 = (\lambda_k + \lambda_n)(\lambda_k - \lambda_n) \leqslant (\lambda_1 + \lambda_n)(\lambda_k - \lambda_n).$$

If we multiply this inequality by α_k^2 and sum over k, we obtain

$$\sum_{k=1}^n |a_{ik}|^2 \leqslant \lambda_n^2 + (\lambda_1 + \lambda_n)(a_{ii} - \lambda_n). \quad (4.3)$$

Similarly

$$\sum_{k=1}^n |a_{jk}|^2 \leqslant \lambda_n^2 + (\lambda_1 + \lambda_n)(a_{jj} - \lambda_n). \quad (4.4)$$

SPECTRUM OF NORMAL MATRICES   89

From these inequalities we have

$$\frac{\sum_{k=1}^{n}|a_{ik}|^2 + \sum_{k=1}^{n}|a_{jk}|^2}{4a_{ii}a_{jj}} \leqslant \frac{(\lambda_1 + \lambda_n)(a_{ii} + a_{jj}) - 2\lambda_1\lambda_n}{4a_{ii}a_{jj}}$$

$$= \frac{\lambda_1\lambda_n}{2}\left\{\frac{\lambda_1 + \lambda_n}{2\lambda_1\lambda_n}(x + y) - xy\right\}, \quad (4.5)$$

where $x = 1/a_{ii}$ and $y = 1/a_{jj}$. Since the diagonal entries of A lie in the interval $[\lambda_n, \lambda_1]$, we have $x, y \in [1/\lambda_1, 1/\lambda_n]$. We will maximize (4.5) subject to these restrictions on x and y.

For $y < \frac{1}{2}(1/\lambda_n + 1/\lambda_1)$ the expression in (4.5) is an increasing function of x, so we can increase it by taking $x = 1/\lambda_n$. For this value of x, (4.5) is a decreasing function of y, so we can increase it by taking $y = 1/\lambda_1$. For these values of x and y, (4.5) assumes the value $(\lambda_1^2 + \lambda_n^2)/4\lambda_1\lambda_n$ and (4.2) holds. Similarly, if $y > \frac{1}{2}(1/\lambda_n + 1/\lambda_1)$, (4.2) holds. Finally, if $y = \frac{1}{2}(1/\lambda_n + 1/\lambda_1)$, the expression (4.5) has the value

$$\frac{(\lambda_1 + \lambda_n)^2}{8\lambda_1\lambda_n} \leqslant \frac{2(\lambda_1^2 + \lambda_n^2)}{8\lambda_1\lambda_n} = \frac{\lambda_1^2 + \lambda_n^2}{4\lambda_1\lambda_n}.$$

Thus (4.2) holds also in this case. This completes the proof of Theorem 4.1.
∎

An examination of the proof of Theorem 4.1 shows that the inequality (4.1) cannot be tight if $(a_{ii} - a_{jj})^2$ is small compared to $(\lambda_1 - \lambda_n)^2$. The following theorem gives a lower bound for the Kantovorich ratio which is better than (4.1) for sufficiently small values of $(a_{ii} - a_{jj})^2$.

THEOREM 4.2.   *If $A = (a_{ij})$ is Hermitian and positive definite, then for any i and j we have*

$$\left(\frac{\lambda_1 - \lambda_n}{\lambda_1 + \lambda_n}\right)^2 \geqslant \frac{B_{ij}(A)}{(a_{ii} + a_{jj})^2 + B_{ij}(A)}. \quad (4.6)$$

Proof.   Let $r = \frac{1}{2}(\lambda_1 - \lambda_n)$ and $c = \frac{1}{2}(\lambda_1 + \lambda_n)$. From (2.2) we see that

$$2r^2 \geqslant |a_{ii} - c|^2 + |a_{jj} - c|^2 + \sum_{k \neq i}|a_{ik}|^2 + \sum_{k \neq j}|a_{jk}|^2.$$

90   E. R. BARNES AND A. J. HOFFMAN

It follows that

$$\left(\frac{\lambda_1 - \lambda_n}{\lambda_1 + \lambda_n}\right)^2 = \frac{2r^2}{2c^2} \geq \frac{|a_{ii} - c|^2 + |a_{jj} - c|^2 + \Sigma_{k \neq i}|a_{ik}|^2 + \Sigma_{k \neq j}|a_{jk}|^2}{2c^2}.$$

$$(4.7)$$

The right side of this inequality assumes its minimum value with respect to c at

$$c = \frac{\Sigma_{k=1}^{n}\left(|a_{ik}|^2 + |a_{jk}|^2\right)}{a_{ii} + a_{jj}}.$$

Substituting this value of c in (4.7) gives

$$\left(\frac{\lambda_1 - \lambda_n}{\lambda_1 + \lambda_n}\right)^2 \geq \frac{(a_{ii} - a_{jj})^2 + 2\left(\Sigma_{k \neq i}|a_{ik}|^2 + \Sigma_{k \neq j}|a_{jk}|^2\right)}{(a_{ii} + a_{jj})^2 + 2\left(\Sigma_{k \neq i}|a_{ik}|^2 + \Sigma_{k \neq j}|a_{jk}|^2\right) + (a_{ii} - a_{jj})^2},$$

$$(4.8)$$

which agrees with (4.6). This completes the proof of Theorem 4.1.   ∎

Clearly the inequality (4.8) is sharper than (4.1) for $(a_{ii} - a_{jj})^2$ sufficiently small.

REFERENCES

1 G. Finke, R. E. Burkard, and F. Rendl, Quadratic assignment problems, *Ann. Discrete Math.* 31:61–82 (1987).
2 A. Horn, Doubly stochastic matrices and the diagonal of a rotation matrix, *Amer. J. Math.* 76:620–630 (1954).
3 L. Mirsky, The spread of a matrix, *Mathematika* 3:127–130 (1956).
4 L. Mirsky, Inequalities for normal and Hermitian matrices, *Duke Math. J.* 24:591–598 (1957).
5 D. S. Scott, On the accuracy of the Gerschgorin circle theorem for bounding the spread of a real symmetric matrix, *Linear Algebra Appl.* 65:147–155 (1985).
6 R. A. Smith and L. Mirsky, The areal spread of matrices, *Linear Algebra Appl.* 2:127–129 (1969).
7 D. G. Luenberger, *Introduction to Linear and Nonlinear Programming*, Addison-Wesley, 1973.

Received 10 December 1990; final manuscript accepted 3 November 1992

Linear Inequalities and Linear Programming

1. On approximate solutions of systems of linear inequalities

I knew that if y is a vector which does not satisfy $Ax = b$, and A is nonsingular, then the distance of y to a solution x is bounded from above by the product of the norm of $b-Ay$ and the inverse of the smallest singular value of A. It seemed to me that there ought to be a theorem covering similar territory for a system of linear inequalities. This is it. It was unnoticed for many years, but eventually became known, and (along with various generalizations) widely used in the analysis of algorithms. I wish I were versed in that line of research. I also wish that this paper were more legible. A few years ago I tried to follow my arguments and could not succeed (and so wrote a different proof with the help of Güler and Rothblum. There are also other proofs in the literature). The arguments I used in 1952 were adapted (i.e. stolen shamelessly, albeit with acknowledgment) from S. Agmon's analysis of the relaxation method for solving systems of linear inequalities.

2. Cycling in the simplex algorithm

I have told in the Autobiographical Notes of the good fortune which brought me to the Applied Mathematics Divison of the National Bureau of Standards in 1951, and involved me in the project supporting the linear programming activities at the U.S. Air Force.

The first problem George Dantzig (the father of linear programming) gave me was to find whether the simplex method could cycle if no special degeneracy-avoiding prescription was in the code. On Mondays, Wednesdays and Fridays I thought it could; on Tuesdays, Thursdays and Saturdays I thought it couldn't. Finally I found this example, which showed it could. George thought I had done something very clever, like inventing the zipper. A few years later, I wrote an NBS report giving the example, and it has appeared in several books. But I was never able, despite numerous requests, to explain what was in my mind when I conceived the example. Jon Lee, in an article in *SIAM Rev.* **39** (1997), pp. 98–105, proposed an explanation that I think is correct. He also wrote a computer program to generate the steps of the cycle, and this program inspired the drawing on the cover of a whirling pentagon with blades and strings.

3. Computational experience in solving linear programs

A principal goal of our Air Force project was to compare methods for solving linear programming problems, so we undertook the experiment described here. Although (maybe because) it contained no theorems, it had some influence on the development of linear programming. We received more than 200 requests for reprints, an astonishingly large number even for those days before duplicating machines (Harold Kuhn attributed most of those requests to new computing centers looking for some verified answers to some specified linear programming problems in order to test simplex codes they had written). Second, the experiment and its description have been cited by the Mathematical Programming Society as models for the conduct and reporting of computational experiments. But candor compels me to say that, given our limited computational resources, we could only do a little bit of work, hence it was not difficult to describe our work in detail.

4. On abstract dual linear programs

I think this was the first paper which looked at the concepts of linear programming, especially the duality theorem, from the viewpoint of abstract algebraic structures. This viewpoint was also adopted, mutatis mutandis, by Edmonds, Fulkerson, Burkhard, Zimmerman and others. The principal result in this paper, which looks at duality from the standpoint of MAX algebra, can be proved from ordinary duality by using exponentials and going to a limit. Most theorems I know in MAX algebra yield to this approach, except the theorem (I have given in class for decades but never published) that a weak form of Cramer's rule is valid in solving systems of linear equations in MAX.

Very recently, I have been looking at a more concrete generalization of linear programming, in which (of the classic (A, b, c) triplet of linear programming A is real, but b or c (not both) has elements taken from a totally ordered abelian group (toag). There are many marvelous mathematical questions arising when one examines linear inequalities, and convexity in general, in toag world, and I think this line of research will flourish.

5. A proof of the convexity of the range of a nonatomic vector measure using linear inequalities

Lyapunov's theorem has been proved several times, not always correctly. Not knowing any measure theory, but wanting to understand the theorem because it is fundamental to methods for "fair division", a concept beloved by game theorists, we thought of this proof, based on (what else?) linear programming. In fact, the paper explains that the basic idea goes back to a paper of Dzielinski and Gomory on a linear programming approach to production scheduling.

6. A nonlinear allocation problem

First we consider a problem on allocation of manpower to tasks in a project (Pert) network, and prove a conjecture that had circulated for a few years in a small circle of aficionados. But the proof is applicable more generally. We prove under mild assumptions that, given an n-vector of proposed target (positive) revenues for n nonnegative activities constrained by linear inequalities, this n-vector is the optimum set of revenues in a maximizing problem for a suitably chosen set of unit profits. So any n-vector of target revenues is "best"; hence, we call our result a "Pangloss theorem" in allusion to Voltaire's Candide. I nurse fantasies that our Pangloss theorems will one day be recognized as a fabulous insight by an admiring coterie of Swedish economists.

Journal of Research of the National Bureau of Standards   Vol. 49, No. 4, October 1952   Research Paper 2362

On Approximate Solutions of Systems of Linear Inequalities[*]

Alan J. Hoffman

Let $Ax \leqq b$ be a consistent system of linear inequalities. The principal result is a quantitative formulation of the fact that if x "almost" satisfies the inequalities, then x is "close" to a solution. It is further shown how it is possible in certain cases to estimate the size of the vector joining x to the nearest solution from the magnitude of the positive coordinates of $Ax - b$.

1. Introduction

In many computational schemes for solving a system of linear inequalities

$$A_1 \cdot x = a_{11}x_1 + \ldots + a_{1n}x_n \leqq b_1$$

$$\vdots \tag{1}$$

$$A_m \cdot x = a_{m1}x_1 + \ldots + a_{mn}x_n \leqq b_m$$

(briefly, $Ax \leqq b$), one arrives at a vector $\bar{x}$ that "almost" satisfies (1). It is almost obvious geometrically that, if (1) is consistent, one can infer that there is a solution x_0 of (1) "close" to $\bar{x}$. The purpose of this report is to justify and formulate precisely this assertion.[1] We shall use fairly general definitions of functions that measure the size of vectors, since it may be possible to obtain better estimates of the constant c (whose important role is described in the statement of the theorem) for some measuring functions than for others. We shall make a few remarks on the estimation of c after completing the proof of the main theorem.

2. The Main Theorem

We require the following
Definitions:
For any real number a, we define
$$a^+ = \begin{cases} a & \text{if } a \geqq 0 \\ 0 & \text{if } a < 0. \end{cases}$$

For any vector $y = (y_1, \ldots, y_k)$, we define
$$y^+ = (y_1^+, \ldots, y_k^+). \tag{2}$$

A positive homogeneous function F_k defined on k-space is a real continuous function satisfying
(i) $\quad F_k(x) \geqq 0, \quad F_k(x) = 0$ if, and only if, $x = 0$

(ii) $\quad \alpha \geqq 0$ implies $F_k(\alpha x) = \alpha F_k(x) \tag{3}$

Theorem: Let (1) be a consistent system of inequalities and let F_n and F_m each satisfy (3). Then there exists a constant $c > 0$ such that for any x there exists a solution x_0 of (1) with

$$F_n(x - x_0) \leqq c F_m \, (Ax - b)^+).$$

The proof is essentially contained in two lemmas (2 and 3 below) given by Shmuel Agmon.[2]

Lemma 1. If F_m satisfies (3), there exists an $e > 0$ such that for every y and every subset S of the half spaces (1)

$$F_m(\bar{y}) \leqq e F_m(y)$$

where $y = (y_1, \ldots, y_m)$, $\bar{y} = (\bar{y}_1, \ldots, \bar{y}_m)$, and

$$\bar{y}_i = \begin{cases} y_i & \text{if the } i\text{th half space belongs to } S \\ 0 & \text{otherwise.} \end{cases}$$

Proof. It is clear from (3) (i) that any e will suffice for $y = 0$. By (3) (ii), we need only consider the ratio $F_m(\bar{y})/F_m(y)$ for y such that $\check{F}(y) = 1$, a compact set. Hence, for each subset S, $F_m(\bar{y})/F_m(y)$ has a maximum e_S. Set $e = \max e_S$.

Lemma 2. Let Ω be the set of solutions of (1), let x be a point exterior to Ω, and let y be the point in Ω nearest to x. Let S be the subset of the half spaces (1), each of which contains y in its bounding hyperplane, and let Ω_S be the intersection of these half spaces.

Then x is exterior to Ω_S and y is the nearest point of Ω_S to x.

Lemma 3. Let M be an $m \times n$ matrix obtained from A by substituting 0 for some of the rows of A, and let Ω be the cone of solutions of $Mz \leqq 0$. Let E be the set of all points x such that (i) x is exterior to Ω, and (ii) the origin is the point of Ω nearest to x.

Then there exists a $d_S > 0$ such that $x \epsilon E$ implies

$$F_m((Mx)^+) \geqq d_S F_n(x).$$

Proof of the theorem. Let x be any vector exterior to the solutions Ω of (1), let x_0 be the point of Ω nearest to x, and let S be defined as in lemma 2.

Let M be the matrix obtained from A by substitut-

[*] This work was sponsored (in part) by the Office of Scientific Research, USAF.
[1] A. M. Ostrowski has kindly pointed out that part of the results given below is implied by the fact that if K and L are two convex bodies each of which is in a small neighborhood of the other, then their associated gauge functions differ slightly.

[2] S. Agmon, The relaxation method for linear inequalities, National Applied Mathematics Laboratory Report 52-27, NBS (prepublication copy).

ing 0 for the rows not in S, and let $\bar{b}$ be the vector obtained from b by substituting 0 for the components not contained in S.

Then lemma 2 says that x_0 satisfies $Mz=\bar{b}$, x is exterior to the solutions of $Mz \leq \bar{b}$, and x_0 is the solution of $Mz \leq \bar{b}$ nearest to x. Perform the translation $z'=z-x_0$. Then $Mz \leq \bar{b}$ if, and only if,

$$Mz'=Mz-Mx_0=Mz-\bar{b} \leq 0.$$

Thus $x-x_0$ belongs to the set E of lemma 3, and

$$F_m((Mx-\bar{b})^+)=F_m((M(x-x_0))^+) \geq d_S F_n(x-x_0) \geq dF_n(x-x_0),$$

where $d=\min\limits_{S} d_S$.

Thus,

$$F_n(x-x_0) \leq \frac{1}{d} F_m(Mx-\bar{b}) \leq \frac{e}{d} F_m((Ax-b)^+),$$

using lemma 1. Setting $c=e/d$ completes the proof of the theorem.

3. Estimates of c for various norms

None of the estimates to be obtained is satisfactory, since each requires an inordinate amount of computation except in special cases. It is worth remarking, however, that even without knowledge of the size of c, the theorem is of use in insuring that any computation scheme that makes $(Ax-b)^+$ approach 0 will bring x close to the set of solutions of (1). This guarantees, for instance, that in Brown's method for solving games the computed strategy vector is approaching the set of optimal strategy vectors.

In what follows let

$|x|$ =maximum of the absolute values of the coordinates of x;

$||x||$ =sum of the absolute values of the coordinates of x;

$|||x|||$ =the square root of the sum of the squares of the coordinates of x.

Note that if F_m is any one of these norms, then $e=1$. We consider these cases:

Case I. $F_n=||| \quad |||, F_m=| \quad |$. If $C=(c_{ij})$ is a square matrix of rth order, let

$$\Gamma(C)=\left[\sum_{j=1}^{r} \left(\sum_{i=1}^{r} |C_{ij}| \right)^2 \right]^{\frac{1}{2}},$$

where the C_{ij}'s are the cofactors of the elements of c_{ij}. Using this notation, and assuming that the rows of (1) are normalized so that $\sum\limits_{j=1}^{n} a_{ij}{}^2=1$, Agmon (see p. 9 of reference in footnote 2) has shown that if

A is of rank r, then

$$C \leq \frac{\left[\sum\limits_{1 \leq j_1 < \ldots < j_r \leq n} (\Gamma_{i_1, \ldots, i_r}^{j_1, \ldots, j_r})^2 \right]^{\frac{1}{2}}}{\left[\sum\limits_{1 \leq j_1 < \ldots < j_r \leq n} |A_{i_1, \ldots, i_r}^{j_1, \ldots, j_r}|^2 \right]^{\frac{1}{2}}},$$

where $i_1, \ldots, i_r$ are r (fixed) linearly independent rows of (a_{ij}); $A_{i_1, \ldots, i_r}^{j_1, \ldots, j_r}$ is the $r \times r$ submatrix formed by the fixed rows and indicated columns, and where the summation is performed over all different combinations of the j's.

Case II. $F_n=| \quad |, F_m=| \quad |$.

Case III. $F_n=| \quad |, F_m=|| \quad ||$.

For cases II and III, it is convenient to have a description of E alternative to that contained in lemma 3. We shall use the notation of lemma 3.

Lemma 4. Let $K'=$ the cone spanned by the row vectors of M, with the origin deleted. Then $K'=E$.

Proof. Let $M_1, \ldots, M_m$ be the row vectors of M, and let $x=\lambda_1 M_1 + \ldots + \lambda_m M_m$, where $x \neq 0$, $\lambda_i \geq 0$, $i=1, \ldots, m$. Then x is exterior to Ω, and the origin is the point of Ω closest to x; that is, $z \in \Omega$ implies $(x-z) \cdot (x-z) - x \cdot x \geq 0$. For $z \cdot x = z \cdot (\lambda_1 M_1 + \ldots + \lambda_m M_m) = \lambda_1 z \cdot M_1 + \ldots + \lambda_m z \cdot M_m \leq 0$. Hence $(x-z) \cdot (x-z) - x \cdot x = z \cdot z - 2z \cdot x \geq 0$. This shows that $K' \subset E$.

We now prove that $E \subset K'$. Assume $x \in E$. Then,

$$z \in \Omega \text{ implies } z \cdot x \leq 0 \tag{4}$$

(otherwise let w be a sufficiently small positive number; then $w z \in \Omega$ and $wz \cdot wz - 2wz \cdot x < 0$). Consider z as the coordinates of a half space whose bounding plane contains the origin. Then (4) says that all half spaces ("through" the origin) containing the row vectors of M also contain x. It is a fundamental result in the theory of linear inequalities [3] that this later statement implies that x is in the cone generated by the rows of M. Hence $E \subset K'$.

4. Case II

It is clear from the proof of the theorem that all we need is to calculate $\min\limits_{x \in E} (|Mx|^+/|x|)$, for each M corresponding to a subset S of the vectors $A_1, \ldots, A_m$. Let $A_1, \ldots, A_k$ (say) be the vectors of the subset S. Then by lemma 4, $x \in E$ implies that there exist $\lambda_1, \ldots, \lambda_k$ with $\lambda_j \geq 0$ such that

$$x=\lambda_1 A_1 + \ldots + \lambda_k A_k. \tag{5}$$

Hence,

$$|x| \leq a_S \sum_{j=1}^{k} \lambda_j, \tag{6}$$

where a_S is the largest absolute value of the co-ordinates of $A_1, \ldots, A_k$.

It follows from the homogeneity of $|(Mx)^+|/|x|$ that

---

[3] T. S. Motzkin, Beiträge zur Theorie der Linearen Ungleichungen. Jerusalem, 1936, with references to proofs by Minkowski and Weyl.

we need only consider $x \epsilon E$ such that if x is expressed as in (5), $\sum_{j=1}^{k} \lambda_j = 1$.

Then

$$|(Mx)^+| = \max_i (A_i \cdot x)^+ = \max_i (A_i \cdot x) =$$
$$\max_i \left(A_i \cdot \sum_{j=1}^{k} \lambda_j A_j \right) = \max_i \sum_{j=1}^{k} g_{ij} \lambda_j,$$

where $g_{ij} = A_i \cdot A_j, \lambda_j \geq 0, \sum \lambda_j = 1$.

Hence,

$$\min |(Mx)^+| = \min_\lambda \max_i \sum_{j=1}^{k} g_{ij} \lambda_j = v_S, \qquad (7)$$

where v_S is the value of zero sum two person game whose matrix is g_{ij}.

Therefore, from (6) and (7)

$$\min_{x \epsilon E} \frac{|(Mx)^+|}{|x|} \geq \frac{v_S}{a_S}.$$

Can $v_S = 0$? Clearly, if, and only if, the origin is in convex body spanned by $A_1, \ldots, A_k$. But this would imply that the set E is the entire space (except for the origin). And it follows from the proof of the main theorem that this can occur only for a subset S that would never arise in lemma 2.

Therefore, using the language of the theorem

$$|x - x_0| \leq c|(Ax - b)^+|, \qquad (8)$$

where

$$c = \max_{v_S > 0} \frac{a_S}{v_S}.$$

A special case occurs when all $A_i \cdot A_j > 0$. Let $v = \min_{i,j} A_i \cdot A_j, a = \max_{i,j} |a_{ij}|$. Then,

$$|x - x_0| \leq \frac{a}{v} |(Ax - b)^+|. \qquad (9)$$

5. Case III

Reasoning, along the lines of case II, we need only estimate $\min_\lambda || \left(\sum_{j=1}^{k} g_{ij} \lambda_j \right)^+ ||$, and it is possible to derive from it an expression analogous to (8), which unfortunately does not seem to have a neat statement in terms of games or any other familiar object. An interesting special case occurs, however, if the matrix g_{ij} (for S all the rows of A), has the property that

$$w = \min_i \left(g_{ii} + \sum_{\substack{j=1 \\ g_{ij} < 0}}^{n} g_{ij} \right) > 0.$$

Then

$$\min || \left(\sum_{j=1}^{k} g_{ij} \lambda_j \right)^+ || \geq \min_\lambda \sum_{i=1}^{k} \sum_{j=1}^{k} g_{ij} \lambda_j$$
$$= \min_\lambda \sum_{j=1}^{k} \lambda_j \sum_{i=1}^{k} g_{ij} \geq \min_\lambda \sum_{j=1}^{k} \lambda_j w = w.$$

Then we obtain, with a having the same meaning as in (9)

$$|x - x_0| \leq \frac{a}{w} ||(Ax - b)^+||. \qquad (10)$$

Washington, June 5, 1952.

Cycling in the Simplex Algorithm*

A. J. Hoffman

1. Introduction

About two years ago, stimulated by conversation with G. Dantzig and A. Orden (both then with the U.S. Air Force), the author discovered an example to illustrate the fact that, if the so-called nondegeneracy assumption is not satisfied, then the simplex algorithm for solving linear programs may fail in the sense that a basis may recur.

Subsequent developments have diminished the importance of the example. First, modifications of the original simplex algorithm have been discovered ([1], [2]) which always work, even when the nondegeneracy assumption is not satisfied. Secondly, in the many degenerate cases which have been computed on the SEAC (National Bureau of Standards Eastern Automatic Computer), the original simplex method *has* worked. So far as the author knows, this has been universally the case in other simplex computations. Thus, it appears that in practice, the phenomenon illustrated by the example has failed to occur; moreover, with only a slight modification of the original algorithm one may be assured that the phenomenon will not occur.

Despite these developments, however, (or perhaps because of them), several mathematicians have requested that the example be made available for study, and the purpose of this report is to fulfill this request. The "cycling" or recurrence of bases, although "solved" as indicated in the preceding paragraph, is certainly not completely understood. It is hoped that this report, which presents what is probably the only existing example of cycling, will help in the investigation.

Since what follows is of interest only to those who are already familiar with the simplex algorithm for minimizing a linear form of non-negative variables subject to linear equations, we will presuppose familiarity not only with the ideas but also with the usual terminology of the method.

*This work was sponsored (in part) by the Office of Scientific Research, USAF.
Reprinted from National Bureau of Standards Report 2974 (1953), National Bureau of Standards, Washington.

2. The Definition of Cycling

Changes of bases occur in the simplex algorithm by means of certain rules: A vector to enter the basis is chosen by means of a certain process (called I), the vector it replaces is chosen by a process (called II). In Process I, one selects H_j to enter the basis if $z_k - c_k > 0$ and if $z_k - c_k = \max_j z_j - c_j$. In case of ties, one selects the smallest k of those that tie. In Process II, one decides that V_l leaves the basis if $h_{lj} > 0$ and if $h_{l0}/h_{lj} = \min_{h_{ij}>0} h_{i0}/h_{ij}$. In case of ties, one selects the smallest l.

The procedure continues until no k can be discovered (minimum attained) or no l can be discovered (there is no minimum).

If the given problem involves an $m \times n$ matrix, and if, given any $m - 2$ column vectors of the matrix, H_0 is *not* in the cone they span, then it is possible to show that, in a finite number of iterations, the simplex procedure terminates. Our example will be a case where H_0 does lie in such a cone and the procedure does *not* terminate. In fact, in our example, the simplex tableau, after eight iterations, is identical with its appearance at the start. Thus, the basis has cycled.

3. The Example

Let $\phi = \frac{2}{5}\pi$ and let w be any number greater than $(1 - \cos\phi)/(1 - 2\cos\phi)$. Let the original simplex tableau be

	H_1	H_2	H_3	$z_j^{(0)} - c_j^{(0)}$
H_0	1	0	0	0
H_1	1	0	0	0
H_2	0	1	0	0
H_3	0	0	1	0
H_4	0	$\cos\phi$	$\dfrac{\tan\phi\sin\phi}{w}$	$\dfrac{1-\cos\phi}{\cos\phi}$
H_5	0	$-w\cos\phi$	$\cos\phi$	$-w$
H_6	0	$\cos 2\phi$	$\dfrac{\tan\phi\sin 2\phi}{w}$	0
H_7	0	$-2w\cos^2\phi$	$\cos 2\phi$	$-2w$
H_8	0	$\cos 2\phi$	$-\dfrac{2\sin^2\phi}{w}$	$-4\sin^2\phi$
H_9	0	$2w\cos^2\phi$	$\cos 2\phi$	$w(4\cos^2\phi - 2)$
H_{10}	0	$\cos\phi$	$-\dfrac{\tan\phi\sin\phi}{w}$	$-4\sin^2\phi$
H_{11}	0	$w\cos\phi$	$\cos\phi$	$w(2\cos\phi)$

After one iteration, the tableau appears as follows:

	H_1	H_4	H_3	$z_j^{(1)} - c_j^{(1)}$
H_0	1	0	0	0
H_1	1	0	0	0
H_2	0	$\sec\phi$	$-\dfrac{\tan^2\phi}{w}$	$\dfrac{\cos\phi - 1}{\cos^2\phi}$
H_3	0	0	1	0
H_4	0	1	0	0
H_5	0	$-w$	$\sec\phi$	$\dfrac{w(1 - 2\cos\phi)}{\cos\phi}$
H_6	0	$4\cos^2\phi - 3$	$\dfrac{\tan^2\phi}{w}$	$2\sin\phi\tan\phi$
H_7	0	$-2w\cos\phi$	1	$-2w\cos\phi$
H_8	0	$4\cos^2\phi - 3$	$\dfrac{2\sin\phi\tan\phi}{w}$	$\dfrac{1 - \cos\phi}{\cos\phi}$
H_9	0	$2w\cos\phi$	$4\cos^2\phi - 3$	$-3w$
H_{10}	0	1	$-\dfrac{2\sin\phi\tan\phi}{w}$	$-2\sin\phi\tan\phi$
H_{11}	0	w	$4\cos^2\phi - 3$	$w(4\cos^2\phi - 3)$

After the next iteration, the tableau is:

	H_1	H_4	H_5	$z_j^{(2)} - c_j^{(2)}$
H_0	1	0	0	0
H_1	1	0	0	0
H_2	0	$\cos\phi$	$-\dfrac{\sin\phi\tan\phi}{w}$	$-4\sin^2\phi$
H_3	0	$w\cos\phi$	$\cos\phi$	$w(2\cos\phi - 1)$
H_4	0	1	0	0
H_5	0	0	1	0
H_6	0	$\cos\phi$	$\dfrac{\sin\phi\tan\phi}{w}$	$\dfrac{1 - \cos\phi}{\cos\phi}$
H_7	0	$-w\cos\phi$	$\cos\phi$	$-w$
H_8	0	$\cos 2\phi$	$\dfrac{\tan\phi\sin 2\phi}{w}$	0
H_9	0	$-2w\cos^2\phi$	$\cos 2\phi$	$-2w$
H_{10}	0	$\cos 2\phi$	$-\dfrac{2\sin^2\phi}{w}$	$-4\sin^2\phi$
H_{11}	0	$2w\cos^2\phi$	$\cos^2\phi$	$w(4\cos^2\phi - 2)$

Compare the original tableau with the present one. The present tableau is identical with the original, except that the column H_3–H_{11} have been cyclically permuted. Since there are no ties in process I (deciding which vector enters the bases), it follows that in six more iterations we will be back where we were at the beginning.

4. Remarks

It is easy to see that in this problem, we have, in any of the bases, achieved the minimum value of the function, namely 0, but the algorithm has not permitted us to discover it.

Let us add a column

$$H_{12} = \begin{pmatrix} 1 \\ 0 \\ 0 \\ \varepsilon \end{pmatrix},$$

where $0 < \varepsilon < (1 - \cos\phi)/\cos\phi$. Then the minimum is $-\varepsilon$, but the algorithm never discovers it. We still cycle in eight steps.

Now, add

$$H_{13} = \begin{pmatrix} -1 \\ 0 \\ 0 \\ \varepsilon \end{pmatrix}.$$

For this problem, there is no minimum, but the algorithm never discovers it, since we still cycle in eight steps.

A natural question to ask is: does this example depend very peculiarly on properties of $\phi = \frac{2}{5}\pi$? The answer is no. It is easy to see that all the nonzeo coefficients of the original tableau may be altered slightly without changing the decisions of process I or process II in any of the eight steps. To be sure, the third tableau will not be a permutation of the first tableau; but the ninth will nevertheless be identical with the first.

Finally, we regret that we are unable at this date to recall any details of the considerations that led to construction of the example, beyond the fact that the geometric meaning of processes I and II in the degenerate case was very much in the foreground. W. Jacobs and S. Gass of the U.S. Air Force have pointed out that if we let A denote the 2×2 matrix which is the intersection of the 2nd and 3rd rows of the first tableau with columns H_4 and H_5, then $A^5 = I$, and the other 2×2 matrices obtained from H_6 and H_7, etc., are A^2, A^3, A^4. One feels intuitively that a judicious use of matrices of finite order (with special care needed for process I of the simplex procedure) may produce other examples of cycling.

Bibliography

[1] A. Charnes, "Optimality and Degeneracy in Linear Programming," Econometrica, April 1952.

[2] G. B. Dantzig, A. Orden and P. Wolfe, "Notes on Linear Programming: Part I: The Generalized Simplex Method for Minimizing a Linear Form under Linear Inequality Restraints", Rand Corporation P-392, April 1953.

COMPUTATIONAL EXPERIENCE
IN SOLVING LINEAR PROGRAMS*

A. HOFFMAN, M. MANNOS, D. SOKOLOWSKY
and N. WIEGMANN

1. Introduction. This paper is a discussion of three methods which have been employed to solve problems in linear programming, and a comparison of results which have been yielded by their use on the Standards Eastern Automatic Computer (SEAC) at the National Bureau of Standards.

A linear program is essentially a scheme to run an organization or effect a plan efficiently, i.e., it is a technique of management which serves to minimize costs, maximize returns or achieve other ends of a similar nature. To illustrate the kind of "life situation" to which linear programming is applicable, and the technique of formulating the circumstances mathematically, let us examine a particular problem. For this purpose, we choose a simplification of the so-called "caterer problem" of W. Jacobs.

A caterer knows that in connection with the meals he has arranged to serve during the next n days, he will need $r_j (\geq 0)$ fresh napkins on the j-th day, $j = 1, 2, \ldots, n$. Laundering takes p days; that is, a soiled napkin sent for laundering at the end of the j-th day is returned in time to be used again, on the $(j + p)$th day. Having no usable napkins on hand or in the laundry, the caterer will meet his early needs by purchasing napkins at a cents each. Laundering costs b cents per napkin. How does he arrange matters to meet his needs and minimize his outlays for the n days?

Before expressing the caterer's problem algebraically, two conventions of notation will be stated. The subscript j throughout has the range $1, 2, \ldots, n$; every equation involving j is to hold for the entire range of values. Quantities with subscripts outside this range are always zero.

Received 3-16-53
SIAM Journal 1 (1953) 1-33

*This work was supported (in part) by the Office of Scientific Research, USAF. The coding and operation of the methods was performed by R. Bryce, I. Diehm, L. Gainen, B. Handy, Jr., B. Heindish, N. Levine, F. Meek, S. Pollack, R. Shepherd and O. Steiner.

18   A. HOFFMAN, M. MANNOS, D. SOKOLOWSKY AND N. WIEGMANN

Let x_j represent the napkins purchased for use on the j-th day; the remaining requirements, if any, are supplied by laundered napkins. Of the r_j napkins which have been used on that day plus any other soiled napkins on hand, let y_j be the number sent to the laundry and s_j the stock of soiled napkins left. Consequently

$$(1) \qquad y_j + s_j - s_{j-1} = r_j .$$

The stock of fresh napkins on hand on the j-th day must be at least as great as the need. Thus

$$(2) \qquad \sum_{i=1}^{j} x_i + \sum_{i=1}^{j} y_{i-p} \geq \sum_{i=1}^{j} r_i .$$

The total cost to be minimized, subject to the constraints (1) and (2) on the nonnegative variables x_j, y_j, s_j, is

$$\sum_{j=1}^{n} (ax_j + by_j) .$$

This is a mathematical formulation of the problem the caterer wishes to solve.

If desired, the equations (1) can be changed into inequalities. For example, (1) is equivalent to the pair of inequalities

$$y_j + s_j - s_{j-1} \geq r_j$$

$$-y_j - s_j + s_{j-1} \geq -r_j .$$

If we make this change, then it is clear that the problem just described is, mathematically, a special case of the following: let $A = (a_{ij})$ be a given $m \times n$ matrix; $b = (b_1,\ldots,b_m)$ an m-dimensional vector, $c = (c_1,\ldots,c_n)$ an n-dimensional vector. For all vectors $x = (x_1,\ldots,x_n)$ satisfying

$$(3) \qquad x_j \geq 0 \quad \text{for } j = 1,\ldots,n$$

$$
\begin{aligned}
a_{11}x_1 + \cdots + a_{1n}\,x_n &\geqq b_1 \\
a_{21}x_1 + \cdots + a_{2n}\,x_n &\geqq b_2 \\
\cdots \qquad \cdots \quad \cdots \quad \cdots & \\
a_{m1}x_1 + \cdots + a_{mn}\,x_n &\geqq b_n \,,
\end{aligned}
$$

(4)

(briefly, $Ax \geqq b$),

minimize $(c,x) = c_1 x_1 + \cdots + c_n x_n.$

The foregoing is the mathematical statement of the general linear programming problem. In geometric language, it is to find a point on a convex polyhedron (the region satisfying (3) and (4)) at which a given linear form $(c_1 x_1 + \cdots + c_n x_n)$ is a minimum.

The Computation Laboratory of the National Bureau of Standards, with the sponsorship and close cooperation of the Planning and Research Division of the Office of the Air Comptroller, U. S. Air Force, has been engaged in the task of discovering and evaluating methods for computational attack on this problem, and this paper is in a sense a progress report on a part of this work (see also Orden [10]).

The three techniques that have received most attention so far are (a) the "Simplex" method, devised by George Dantzig, (b) the Fictitious Play method of George Brown and (c) the Relaxation Method of T. S. Motzkin. Each will be described in more detail in the next section, but for the present it is appropriate to remark that the simplex method is a finite algorithm, and the other two are infinite processes. Further, the other two methods are designed to solve not the linear programming problem *per se* but two related problems: fictitious play finds a solution to a matrix game - i.e., a zero-sum 2-person game in normalized form - and relaxation finds an x satisfying (4) - i.e., solves a system of linear inequalities. However, it is known (see Gale, Kuhn and Tucker [6], Dantzig [4], Orden [9]) that the three problems (i) solving a linear program, (ii) solving a matrix game, (iii) solving a system of linear inequalities are in general equivalent in that each of (i), (ii) and (iii) can be so formulated that it becomes either of the other two.

For purposes of comparison, the following experiment was undertaken. Several symmetric matrix games (i.e., games whose matrices were skew-symmetric) were attacked by each method in turn and the results studied with respect to the accuracy achieved and the time required to obtain this accuracy. Many conjectures about the relative

     A. HOFFMAN, M. MANNOS, D. SOKOLOWSKY AND N. WIEGMANN

merits of the three methods by various criteria could only be verified by actual trial. Apart from the descriptions of the methods,
the paper is concerned principally with the results of the experiment,
but some other aspects of the comparison, revealed more strikingly by
other computations, will also be mentioned.

The games in question have as payoff matrices the submatrices
of order $5,6,7,8,9,10$ obtained from the following 10×10 array by deleting the last five rows and columns, the last four rows and columns,
etc.

$$
\begin{pmatrix}
0 & 1 & -2 & -1 & 3 & -2 & -1 & -4 & 1 & -2 \\
-1 & 0 & -1 & 1 & 2 & -2 & 1 & 1 & -1 & -1 \\
2 & 1 & 0 & -3 & 1 & 1 & 3 & -3 & 1 & -1 \\
1 & -1 & 3 & 0 & 1 & -1 & 4 & 2 & -1 & 5 \\
-3 & -2 & -1 & -1 & 0 & -1 & -5 & 6 & 1 & 6 \\
2 & 2 & -1 & 1 & 1 & 0 & -2 & -1 & -1 & 3 \\
1 & -1 & -3 & -4 & 5 & 2 & 0 & 2 & 1 & -4 \\
4 & -1 & 3 & -2 & -6 & 1 & -2 & 0 & 1 & -1 \\
-1 & 1 & -1 & 1 & -1 & 1 & -1 & -1 & 0 & -5 \\
2 & 1 & 1 & -5 & -6 & -3 & 4 & 1 & 5 & 0
\end{pmatrix}
$$

2. Description of the Methods

(a) *The Simplex Method*

The simplex method solves the problem: minimize $c_1 x_1 + \cdots + c_n x_n$
for all vectors $x = (x_1, \ldots, x_n)$ satisfying

$$x_j \geqq 0 \quad \text{for } j = 1, \ldots, n$$

$$Ax = b$$

where $A = (a_{ij})$ is an $m \times n$ matrix and $b = (b_1, \ldots, b_m)$ is m-dimensional
vector.

This differs slightly from the formulation of the general linear programming problem in which $Ax \geq b$, but the inequalities can be
made into equations by appending dummy nonnegative variables.

An algebraic description of the process* is given in Dantzig
[5] and Orden [9], and will not be repeated here. While it is not

---

*Interesting variations suggested by Charnes [3] and Wolfe [12] have not yet
been tested.

excessively complicated it is somewhat lengthy, and we merely remark now that it very much resembles elimination methods for solving equations. Even for those familiar with the algebra, however, (as well as for novices), the following geometric interpretation is illuminating. First, to state the problem in geometric language: if $A_1,\ldots,A_n$ are the m-dimensional column vectors of A, let $A'_1,\ldots,A'_n$ be the $(m+1)$-dimensional vectors obtained from $A_1,\ldots,A_n$ by appending $c_1,\ldots,c_n$ respectively as the $(m+1)$st coordinates. Let C be the convex cone in $(m+1)$-space spanned by these vectors. Let B be the line in $(m+1)$-space consisting of all points whose first m coordinates are $b_1,\ldots,b_m$. The object of the computation is to find the lowest point of B which is also in C, i.e., the point of B whose $(m+1)$th coordinate is a minimum.

The computation proceeds in the following way: assume that m of the vectors $A'_1,\ldots,A'_n$, say $A'_{\mu_1},\ldots,A'_{\mu_m}$ are given which are linearly independent and have the property that the m-dimensional cone D they span contains a point of B. (Such a set of vectors may have to be given initially by an artificial device which we shall not describe here). Of all the remaining vectors A'_i, let us look at the subset of those which are on the side of the hyperplane containing D that does not contain the positive $(m+1)$st coordinate axis. These vectors are all "lower" than D. Each of these vectors can be joined to the hyperplane containing D by a line segment parallel to the $(m+1)$st coordinate axis. Let A'_i be the vector with the property that this line segment has maximal length - i.e., A'_i is the "lowest" of the low vectors. Then A'_i and a certain set of $m-1$ of the vectors $A'_{\mu_1},\ldots,A'_{\mu_m}$ have the property that the m-dimensional cone they span contains a point of B, and this point will be lower than the intersection of D with B. We replace the discarded vector of the set $A'_{\mu_1},\ldots,A'_{\mu_m}$ with A'_i and proceed. This replacement process is an iteration in the simplex method, and clearly the computation must stop after a finite number of iterations with the desired lowest point of the intersection of C and B.

The symmetric games were formulated for the simplex method as follows:

Let $A = (a_{ij})$ be the $n \times n$ game matrix. We wish to find an $x = (a_1,\ldots,x_n)$ such that

$$Ax \leqq 0$$

22   A. HOFFMAN, M. MANNOS, D. SOKOLOWSKY AND N. WIEGMANN

$$x_i \geqq 0 \quad \text{for } i = 1,\ldots,n$$

$$\sum_{i=1}^{n} x_i = 1.$$

This is equivalent to: minimize $-(x_1 + x_2 + \cdots + x_n)$ subject to

$$(a_{11} + 1)\, x_1 + (a_{12} + 1)\, x_2 + \cdots + (a_{1n} + 1)\, x_n + w_1 = 1$$

$$\cdots \qquad \cdots \qquad \cdots \qquad \cdots \qquad \cdots$$

$$(a_{n1} + 1)\, x_1 + (a_{n2} + 1)\, x_2 + \cdots + (a_{nn} + 1)\, x_n + w_n = 1$$

$$x_i \geqq 0 \quad \text{for } i = 1,\ldots,n$$

$$w_i \geqq 0 \quad \text{for } i = 1,\ldots,n.$$

This latter problem is suitable for simplex computation. We omit
the proof of the equivalence as well as the justification for choosing this particular way of formulating the game as a simplex computation, since both depend on technical reasons irrelevant to the main
purpose of the paper.

    (b) *Fictitious Play*

It is well known (see Dantzig [4]) that a computational scheme
that will solve symmetric games can be adapted to the solution of
linear programming problems. The fictitious play method, devised by
G. Brown [2] and proved valid by J. Robinson [11], is a procedure
for solving an arbitrary matrix game, but the computation is simpler
(particularly from the standpoint of storage) if the game is symmetric. And since our primary interest is in linear programs rather
than games *per se*, we have confined our attention to the symmetric
case.

If $A = (a_{ij})$ is skew-symmetric, our object is to solve the
system of linear inequalities

$$\begin{aligned}
a_{11}x_1 + a_{21}x_2 + \cdots + a_{n1}x_n &\geqq 0\\
a_{12}x_1 + a_{22}x_2 + \cdots + a_{n2}x_n &\geqq 0\\
\cdots \quad \cdots \quad \cdots \quad \cdots \quad \cdots\\
a_{1n}x_1 + a_{2n}x_2 + \cdots + a_{nn}x_n &\geqq 0
\end{aligned}$$

(5)

subject to

$$x_i \geqq 0 \quad \text{for } i = 1, \ldots, n$$

$$\sum_{i=1}^{n} x_i = 1.$$

The fictitious play method consists of forming two sequences of n-dimensional vectors, $V(0)$, $V(1)$, $V(2), \ldots$, and $T(0)$, $T(1)$, $T(2)$, $\ldots$, with $V(0) = T(0) = 0$. $V(N)$ $(N = 1, 2, \ldots)$ is obtained by adding the r-th row of A to $V(N\text{-}1)$, where the r-th coordinate of $V(N\text{-}1)$ is the minimum of the coordinates of $V(N\text{-}1)$ and of smallest index among all the coordinates equal to the minimum. The smallest index criterion is used in order to be specific but is in no way essential. $T(N)$ is the same as $T(N\text{-}1)$ except for the r-th coordinate which is larger by 1. In effect, the j-th component $T_j(N)$ of $T(N)$ represents the number of times the j-th row of the matrix A has been selected by the above criterion in forming $V(N)$. Hence, if $V_j(N)$ denotes the j-th component of $V(N)$, we have

$$V_j(N) = a_{1j} T_1(N) + a_{2j} T_2(N) + \cdots + a_{nj} T_j(N) = \sum_{i=1}^{N} a_{ij} T_i(N).$$

Upon setting $x_i(N) = T_i(N)/N$, this is equivalent to

$$(6) \quad \frac{V_j(N)}{N} = a_{ij} x_1(N) + a_{2j} x_2(N) + \cdots + a_{nj} x_n(N) = \sum_{i=1}^{N} a_{ij} x_i(N).$$

Since $T_i(N) \geqq 0$, and $T_1(N) + T_2(N) + \cdots + T_n(N) = N$, it follows that $x_i(N) \geqq 0$ and $x_1(N) + x_2(N) + \cdots + x_n(N) = 1$. If the first player follows the strategy $(x_1(N), x_2(N), \ldots, x_n(N))$ his expectation is the least of the expressions (6), and except in the trivial case that the first row of A consists of nonegative elements, $\min_{j} V_j(N)/N$ will be negative, which, by (6), implies that the inequalities (5) will not be satisfied. It is, however, the main result of J. Robinson's paper that

$$\lim_{N \to \infty} \frac{\min_{j} V_j(N)}{N} = 0 .$$

This implies (see Hoffman [7], McKinsey [8]), that the vector $x(N) = (x_1(N), \ldots, x_n(N))$ approaches the convex set of all solutions to (5) as N increases indefinitely, though it does *not* imply (indeed, it is

24   A. HOFFMAN, M. MANNOS, D. SOKOLOWSKY AND N. WIEGMANN

not always true) that $x(N)$ converges. Of course, if the first player
is willing to follow a strategy such that his expected loss is no
greater than ϵ, he may follow the strategy $x(N)$, where $\min_j V_j(N)/N$
$\geq -\epsilon$.

These considerations suggest that the speed of convergence of
the fictitious play method should be determined by deciding how "long"
it takes for the process to arrive at a vector $x(N)$ such that the corresponding expected loss is no greater than ϵ for a given decreasing
sequence of positive numbers ϵ.

At least two criteria are relevant in measuring how "long" it
takes to attain an ϵ: the size of N and the time consumed on the
computer. Both are given in the table of results. A third criterion
is suggested by the manner in which the procedure was coded. It is
apparent from J. Robinson's proof and readily verifiable by experience that as the computation proceeds the same row vector of A will
be added to $V(N)$ for a large stretch of successive values of N.
Hence, the code picks out, not only the row to be added to $V(N)$, say
A_r, but decides how many times A_r is to be added to $V(N)$ before some
other row is added, i.e. it determines a number $S(N)$ such that

$$V(N+1) - V(N) = \cdots = V[N+S(N)] - V[N+S(N)-1] = A_r$$

but

$$V(N+S(N)+1) - V[N+S(N)] \neq A_r.$$

The number $S(N)$ is the least positive integer not less than

$$\min_{a_{rj} < 0} \frac{V_r(N) - V_j(N)}{a_{rj}}.$$

Then

$$V[N+S(N)] = V(N) + S(N) \cdot A_r,$$

$$T[N+S(N)] = T(N) + S(N) \cdot \delta_r,$$

(where δ_r is the r-th unit vector).

The foregoing computation and subsequent changes in $V(N)$ and
$T(N)$ are essentially computational "steps" in the SEAC code and
therefore the number of such steps it takes to attain a given ϵ is a

third proper criterion by which to measure the convergence rate of the process.

(c) *Relaxation Method*

There have been several proposed versions of the relaxation method and what we call the relaxation method here might more properly be termed the "furthest hyperplane" method.

The object of the computation is to find a point which satisfies a finite system of linear inequalities

$$\sum_{j=1}^{n} a_{ij}x_j + b_i \geq 0 \quad \text{for } i = 1,\ldots,m.$$

The set of points which satisfy one of these m inequalities is called a *half-space* and the set of points which satisfy the corresponding equation is called the *bounding hyperplane* of the half-space. A point satisfies the entire system of inequalities (i.e. is a solution) if and only if it lies in the intersection of the m half-spaces.

The procedure is inductive, producing an infinite sequence of points x^0, x^1, x^2,... which converges to a solution (see Agmon [1]), provided one exists. x^0 is arbitrary. Assuming we have x^k, ($k = 0,1,2,\cdots$),x^{k+1} is obtained as follows:

If x^k is a solution, $x^{k+1} = x^k$.

If x^k is not a solution, there are one or more of the given half spaces which do not contain it. Among the bounding hyperplanes of these half-spaces, let α be one at a maximum distance from x^k and let p be the point of α nearest x^k. Then

$$x^{k+1} = x^k + t(p\text{-}x^k) \quad \text{where } 0 < t < 2.$$

Three values of t were tried, namely $t = 3/4$ (Undershoot), $t = 1$ (normal--here $x^{k+1} = p$), and $t = 3/4$ (overshoot).

We now describe the process algebraically along with a summary of the machine procedure. The code does not use the algebraic formulation which would yield the fastest computational procedure (the reader can easily concoct such a procedure using the matrix AA^T), for the naive method followed required less internal storage.

26   A. HOFFMAN, M. MANNOS, D. SOKOLOWSKY AND N. WIEGMANN

Let the set of inequalities be normalized so that

$$\sum_{j=1}^{n} a_{ij}^2 = 1 \quad \text{for } i = 1, \ldots, m.$$

Let $y_i = \sum_{j=1}^{n} a_{ij}x_j + b_i$, choose an initial set of values for $x : x_1^{(0)}$, $\ldots, x_n^{(0)}$ and obtain a corresponding set of $y : y_1^{(0)}, \ldots, y_n^{(0)}$. If all the $y_i^{(0)}$ are non-negative, the $x_j^{(0)}$ form a solution. If not choose the largest (in absolute value) of the negative $y_i^{(0)}$, call this $y_k^{(0)}$, and form new values x_j according to $x_j^{(1)} = x_j^{(0)} - t \, a_{kj} \, y_k^{(0)}$ where $0 < t < 2$. Substitute the $x_j^{(1)}$ into the system and obtain a new system of $y : y_i^{(1)}$. Continue in this way to form a sequence $x_1^{(i)}, x_2^{(i)}, \ldots, x_n^{(i)}$ for $i = 1, 2, \ldots$.

The machine procedure has been as follows: Given the $m \times n$ matrix $A = (a_{ij})$, the problem is scaled so that $|\max a_{ij}| \leq 1$. Each inequality is multiplied by 10^{-4} so that the b_i are properly scaled. This scaling is to be kept in mind in interpreting the final results. Next conversion is made from the decimal to the binary system and the system is normalized. The matrix A and the b_i's are stored in the machine. The initial choice is $x_j^{(0)} = 0$ for $j = 1, 2, \ldots, n$ in all the problem work.

In order to measure convergence, it is reasonable to compare the size of the negative $y_i^{(k)}$ relative to k itself. In this case the following procedure was adopted: an $\epsilon > 0$ was chosen and a solution was considered as obtained when the minimum $y_i \geq -\epsilon$. The ϵ was made progressively smaller and, on the basis of previous experience was taken successively as 2^{-2}, 2^{-6}, 2^{-10}, 2^{-11}, $2^{-12}, \ldots, 2^{-22}$. The time required to satisfy a given ϵ was also noted in each instance.

The game problem was transformed into a pure inequality problem in the following manner: If the $m \times n$ matrix is $A = (a_{ij})$, the expected payoff for player one, if he engages in a mixed strategy $x = (x_1, \ldots, x_n)$ is given by

$$M = \min_{j} \sum_{i+1}^{n} a_{ij}x_i,$$

where $\sum x_i = 1$, $x_i \geq 0$.

Since the game is symmetric, the value is zero so that the problem reduces to solving

$$-Ax \geq 0, \quad x \geq 0, \quad \text{and} \quad \sum_{i=1}^{m} x_i = 1,$$

COMPUTATIONAL EXPERIENCE IN SOLVING LINEAR PROGRAMS   27

or to solving the $(2m + 2) \times m$ system of inequalities

$$-Ax \geqq 0, \quad x \geqq 0, \quad \sum_{i=1}^{m} x_i \geqq 1, \quad - \sum_{i=1}^{m} x_i \geqq -1.$$

3. Numerical Results

(a) *Simplex Method*

Table I gives the answers obtained, the number of iterations required and the time consumed on the machine by the computation.

TABLE I

	5×5	6×6	7×7	8×8	9×9	10×10
x_1	0.00000	0.00000	0.00000	0.00000	0.12341	0.03690
x_2	0.59999	0.00000	0.00000	0.00000	0.00000	0.00000
x_3	0.19999	0.20000	0.19999	0.04761	0.00000	0.08487
x_4	0.19999	0.20000	0.20000	0.26190	0.25949	0.22509
x_5	0.00000	0.00000	0.00000	0.00000	0.06012	0.10332
x_6		0.60000	0.59999	0.57142	0.02531	0.12915
x_7			0.00000	0.09523	0.03481	0.00000
x_8				0.02380	0.06645	0.00000
x_9					0.43037	0.39483
x_{10}						0.02582
No. of iterations	6	4	5	6	7	11
Time (mins.)	10	8	8½	9	12	15

Note that the number of iterations is about n for each of these $n \times 2n$ linear programming problems. This is in accord with our general experience using the simplex method on $m \times n$ problems that a solution takes approximately m iterations unless the artificial device mentioned in the description of the simplex method given in **2(a)** is needed. In that case it takes about $2n$ iterations to reach a solution. These estimates are completely heuristic, but they are based on over fifty simplex computations of various sizes and are probably the right order of magnitude. The success that the simplex method has enjoyed is based largely on the fact that the number of iterations required has not been larger.

28 COMPUTATIONAL EXPERIENCE IN SOLVING LINEAR PROGRAMS

(b) *The fictitious play method*

For each of the problems many answers were printed as different ϵ's were attained.

Let us look in detail at the results of the 6×6 game, which illustrate the typical properties of the convergence rates. In table II are given the approximate solutions $x_i(N)$ corresponding to the various values of ϵ. The step in going from $V(N)$ to $V[N+S(N)]$ is called an S-step. The time is counted from the beginning of the computation, excluding the time taken to print out the answers.

TABLE II

$\epsilon = 2^{-2}$	$\epsilon = 2^{-6}$	$\epsilon = 2^{-10}$
$x_1 = 0.0169491525$	$x_1 = 0.0002732987$	$x_1 = 0.0000013102$
$x_2 = 0$	$x_2 = 0$	$x_2 = 0$
$x_3 = 0.1016949152$	$x_3 = 0.1866630226$	$x_3 = 0.1990320137$
$x_4 = 0.1864406779$	$x_4 = 0.1997813610$	$x_4 = 0.1999989518$
$x_5 = 0$	$x_5 = 0$	$x_5 = 0$
$x_6 = 0.6949152542$	$x_6 = 0.6132823175$	$x_6 = 0.6009677241$
TIME = 0:02 S-steps = 7 $N = 59$	TIME = 0:12 S-steps = 52 $N = 3659$	TIME = 2:04 S-steps = 742 $N = 763,234$

$\epsilon = 2^{-11}$	$\epsilon = 2^{-12}$
$x_1 = 0.0000003290$	$x_1 = 0.0000000824$
$x_2 = 0$	$x_2 = 0$
$x_3 = 0.1995141064$	$x_3 = 0.1997565760$
$x_4 = 0.1999997367$	$x_4 = 0.1999999340$
$x_5 = 0$	$x_5 = 0$
$x_6 = 0.6004858277$	$x_6 = 0.6002434074$
TIME = 3:50 S-steps = 1480 $N = 3,039,348$	TIME = 6:40 S-steps = 2956 $N = 12,130,279$

Observe that, for ϵ sufficiently small, the number of S-steps required to "attain the ϵ" doubled as ϵ was halved, while N quadrupled. This phenomenon held for all the games solved as part of the experiment and for others not part of the experiment that were solved by the Brown method.

No arithmetic relationship between the computing time (required to attain a given ϵ) and the size of the matrix could be determined.

(c) *The Relaxation Method*

For the same reasons as given above for the fictitious play method, we present below (table III) the results of the 6×6 game obtained using the three methods of relaxation. (N = number of iterations.)

TABLE III

$\epsilon = 2^{-2}$	Undershoot	Normal	Overshoot
$x_1 =$	0.125000	0.015152	-0.034309
$x_2 =$	0.125000	0.015152	0.217439
$x_3 =$	0.125000	0.242424	0.225306
$x_4 =$	0.125000	0.090909	0.262019
$x_5 =$	0.125000	0.090909	0.136145
$x_6 =$	0.125000	0.166667	0.558348
$N =$	2	3	5
$T =$	0:02	0:03	0:05

$\epsilon = 2^{-6}$	Undershoot	Normal	Overshoot
$x_1 =$	-0.014057	0.000000	-0.014191
$x_2 =$	0.054008	0.043413	-0.004277
$x_3 =$	0.217143	0.210530	0.203763
$x_4 =$	0.174677	0.177211	0.203559
$x_5 =$	0.010341	-0.002738	0.029268
$x_6 =$	0.545075	0.546727	0.585016
$N =$	50	33	12
$T =$	0:56	0:37	0:13

$\epsilon = 2^{-10}$	Undershoot	Normal	Overshoot
$x_1 =$	-0.000697	-0.000923	-0.000269
$x_2 =$	0.003456	0.002975	0.000320
$x_3 =$	0.201578	0.201360	0.198920
$x_4 =$	0.198840	0.198293	0.200325
$x_5 =$	-0.000877	0.000000	-0.000288
$x_6 =$	0.596372	0.596994	0.599880
$N =$	198	121	37
$T =$	3:42	2:15	0:41

$\epsilon = 2^{-11}$	Undershoot	Normal	Overshoot
$x_1 =$	-0.000372	-0.000251	-0.000412
$x_2 =$	0.001661	0.001701	-0.000462
$x_3 =$	0.200672	0.200691	0.199648
$x_4 =$	0.199325	0.199442	0.200473
$x_5 =$	-0.000408	-0.000238	-0.000140
$x_6 =$	0.598102	0.598654	0.600318
$N =$	236	144	39
$T =$	4:24	2:41	0:44

$\epsilon = 2^{-12}$	Undershoot	Normal	Overshoot
$x_1 =$	-0.000179	-0.000216	-0.000219
$x_2 =$	0.000920	0.000866	-0.000194
$x_3 =$	0.200323	0.200202	0.199861
$x_4 =$	0.199665	0.199755	0.200260
$x_5 =$	-0.000067	-0.000172	0.000176
$x_6 =$	0.598970	0.599354	0.600318
$N =$	271	168	43
$T =$	5:03	3:08	0:48

$\epsilon = 2^{-13}$	Undershoot	Normal	Overshoot
$x_1 =$	-0.000099	-0.000093	0.000110
$x_2 =$	0.000436	0.000399	-0.000034
$x_3 =$	0.200179	0.200086	0.199827
$x_4 =$	0.199822	0.199877	0.200029
$x_5 =$	-0.000104	-0.000070	-0.000055
$x_6 =$	0.599505	0.599696	0.600185
$N =$	310	193	48
$T =$	5:47	3:36	0:54

(continued next page)

30   A. HOFFMAN, M. MANNOS, D. SOKOLOWSKY AND N. WIEGMANN

TABLE III (continued)

$\epsilon = 2^{-14}$	Undershoot	Normal	Overshoot
$x_1 =$	-0.000023	0.000000	-0.000041
$x_2 =$	0.000226	0.000177	0.000036
$x_3 =$	0.200085	0.200040	0.199914
$x_4 =$	0.199904	0.199916	0.200052
$x_5 =$	-0.000033	-0.000041	-0.000035
$x_6 =$	0.599760	0.599772	0.600149
$N =$	349	212	55
$T =$	6:31	3:57	1:02

$\epsilon = 2^{-15}$	Undershoot	Normal	Overshoot
$x_1 =$	-0.000026	-0.000017	-0.000027
$x_2 =$	0.000116	0.000110	0.000017
$x_3 =$	0.200047	0.200055	0.199932
$x_4 =$	0 199953	0.199957	0.200003
$x_5 =$	-0.000028	-0.000017	0.000021
$x_6 =$	0.599867	0.599912	0.600099
$N =$	383	233	66
$T =$	7:09	4:21	1:14

$\epsilon = 2^{-16}$	Undershoot	Normal	Overshoot
$x_1 =$	-0.000004	-0.000015	-0.000012
$x_2 =$	0.000056	0.000058	0.000003
$x_3 =$	0.200025	0.200011	0.199957
$x_4 =$	0.199974	0.199981	0.200028
$x_5 =$	-0.000014	-0.000007	-0.000015
$x_6 =$	0.599946	0.599953	0.600060
$N =$	427	256	78
$T =$	7:58	4:47	1:27

$\epsilon = 2^{-17}$	Undershoot	Normal	Overshoot
$x_1 =$	-0.000007	-0.000002	-0.000003
$x_2 =$	0.000030	0.000025	-0.000008
$x_3 =$	0.200013	0.200005	0.199984
$x_4 =$	0.199988	0.199989	0.200010
$x_5 =$	-0.000007	-0.000006	-0.000003
$x_6 =$	0.599965	0.599980	0.600023
$N =$	457	283	105
$T =$	8:31	5:17	1:58

$\epsilon = 2^{-18}$	Undershoot	Normal	Overshoot
$x_1 =$	-0.000003	-0.000002	-0.000001
$x_2 =$	0.000016	0.000013	0.000002
$x_3 =$	0.200006	0.200004	0.199993
$x_4 =$	0.199994	0.199995	0.200005
$x_5 =$	-0.000003	-0.000001	-0.000003
$x_6 =$	0.599986	0.599990	0.600013
$N =$	499	305	125
$T =$	9:19	5:42	2:20

$\epsilon = 2^{-19}$	Undershoot	Normal	Overshoot
$x_1 =$	-0.000001	-0.000002	-0.000001
$x_2 =$	0.000007	0.000006	0.000000
$x_3 =$	0.200003	0.200003	0.199996
$x_4 =$	0.199998	0.199998	0.200003
$x_5 =$	-0.000002	-0.000002	-0.000001
$x_6 =$	0.599993	0.599993	0.600005
$N =$	541	322	147
$T =$	10:05	6:01	2:45

$\epsilon = 2^{-20}$	Undershoot	Normal	Overshoot
$x_1 =$	-0.000001	0.000000	-0.000001
$x_2 =$	0.000003	0.000003	0.000000
$x_3 =$	0.200001	0.200001	0.199999
$x_4 =$	0.199999	0.199999	0.200000
$x_5 =$	-0.000001	-0.000001	0.000001
$x_6 =$	0.599996	0.599998	0.600004
$N =$	579	355	163
$T =$	10:48	6:38	3:03

$\epsilon = 2^{-21}$	Undershoot	Normal	Overshoot
$x_1 =$	0.000000	0.000000	0.000000
$x_2 =$	0.000002	0.000002	0.000000
$x_3 =$	0.200001	0.200001	0.199999
$x_4 =$	0.199999	0.200000	0.200001
$x_5 =$	0.000000	0.000000	0.000000
$x_6 =$	0.599998	0.599998	0.600001
$N =$	614	368	184
$T =$	11:28	6:52	3:26

(continued next page)

TABLE III (continued)

$\epsilon = 2^{-22}$	Undershoot	Normal	Overshoot
$x_1 =$	0.000000	0.000000	0.000000
$x_2 =$	0.000001	0.000001	0.000000
$x_3 =$	0.200000	0.200000	0.200000
$x_4 =$	0.200000	0.200000	0.200000
$x_5 =$	0.000000	0.000000	0.000000
$x_6 =$	0.599999	0.599999	0.600001
$N =$	653	388	201
$T =$	12:11	7:15	3:45

Observe that overshoot converged faster than normal, which in turn converged faster than undershoot. This held consistently for all the games.

Further, for ϵ sufficiently small, there is an approximately uniform increase in the number of iterations required to "attain a given ϵ" as ϵ is halved. For the 6×6 game, for example, from $\epsilon = 2^{-6}$ to $\epsilon = 2^{-20}$, the additional iterations required to go from $\epsilon = 2^{-i}$ to $\epsilon = 2^{-(i+1)}$ were approximately 38 (for undershoot), 24 (for normal), 12 (for overshoot).

The experiment did not reveal any arithmetic relationship between the size of the matrix and the computing time.

4. Conclusions. Any relative evaluation of proposed computation schemes requires specification of the size of the problem considered, the accuracy demanded and the amount of computation time reasonable to invest in obtaining this accuracy. Let us assume (in accordance with the requirements of most of the practical problems that have so far arisen in our work) that four or five decimal digits are required in the answer, and that the size of the matrix A is say, 7×7 or greater.

Then the simplex method is outstanding among the three. In the large size games considered in the experiment, the simplex method achieved answers to this precision in a third or a fourth of the time required by the most favorable of the others. This occurred despite the fact that simplex was coded to use magnetic tapes for storage of most of the numbers arising in the computation, whereas the other methods stored all the numbers within the high speed memory. It is

32   A. HOFFMAN, M. MANNOS, D. SOKOLOWSKY AND N. WIEGMANN

estimated that about 4/5 of the machine time required for simplex on
these games was spent in bringing the needed numbers from the tape
to the high speed memory and taking the numbers from the memory to
the tape. (Improvements in tape performance subsequent to these computations have reduced this ratio to about ½). The other methods
would be completely impractical if tape had to be used, and that is
why only the simplex method has solved moderately large problems
(where the matrix A is about 50×70). Even assuming a very large
memory so that, for instance, a large problem could be coded for relaxation in the most efficient way, then the fact that simplex could
be done internally would favor it even more. It is true, however,
that because simplex is more complicated algebraically, it is possible by clever coding to fit some problems into the high speed memory
when using fictitious play or relaxation that could not be so accom-
modated if simplex were employed. One such large problem arose in
our work: the computation matrix was 48×71 for the simplex method,
but formulated as a symmetric game and using ingenious coding devices,
it could be done within the high speed memory by fictitious play.
Nevertheless, simplex was completed in half the time that fictitious
play required to obtain the same accuracy.

Is there then an area of usefulness for the infinite methods?
The answer is yes, for problems satisfying the following conditions:
they are small enough to be done entirely within the memory, and the
precision demanded is very small or very large. Two objections to
the simplex method are: (i) in general, there is no reason to believe
that an answer from an early iteration has any meaning at all, so
there is no provision for doing less work if one is content with small
accuracy: and (ii) when the answers are finally obtained, there is no
way to improve them to obtain greater precision. (Wolfe's proposed
variation [12] of the simplex method will help (ii), but it is ques-
tionable that it would involve less work than the procedure suggested below).

If the purposes of the computation require only one or two deci-
mals in the answer, then one is perhaps better off using the infinite
methods. This is verified in the 6×6 problem, which, indeed, favors
fictitious play over relaxation for this purpose (which favorable
position held in general). If the purposes of the computation demand
greater precision than the simplex answers yield, then it is reasonable to use the simplex answers as a starting point for one of the
other methods. And here, the more favorable convergence rate for relaxation (see the comments in **3(b)** and **3(c)**) over fictitious play
favors the use of the former.

REFERENCES

Several of the references (indicated with *) are taken from *Activity Analysis of Production and Allocation,* edited by T. C. Koopmans, New York, 1951.

1. S. Agmon, *The Relaxation Method for Linear Inequalities,* (to be published).

2. G. W. Brown, *Iterative Solution of Games by Fictitious Play,* p. 379.

3. A. Charnes, *Optimality and Degeneracy in Linear Programming,* Econometrica, vol. 20, No. 2 (1952), p. 160.

***4.** G. B. Dantzig, *A Proof of the Equivalence of the Programming Problem and the Game Problem,* p. 330.

***5.** G. B. Dantzig, *Maximization of a Linear Function of Variables Subject to Linear Inequalities,* p. 339.

***6.** D. Gale, H. W. Kuhn, A. W. Tucker, *Linear Programming and the Theory of Games,* p. 317.

7. A. J. Hoffman, *On Approximate Solutions of Systems of Linear Inequalities,* Journal of Research of the National Bureau of Standards, vol. 49, No. 4 (1952), p. 263.

8. J. C. C. McKinsey, *Introduction to the Theory of Games,* New York, 1952, p. 94.

9. A. Orden, *Application of the Simplex Method to a Variety of Matrix Problems,* in Symposium on Linear Inequalities and Programming, edited by A. Orden and L. Goldstein, Washington, 1952, p. 28.

10. A. Orden, *Solution of Systems of Linear Inequalities on a Digital Computer,* Proceedings of the Association for Computing Machinery, May, 1952, Pittsburgh, Pa.

11. J. Robinson, *An Iterative Method of Solving a Game,* Annals of Mathematics, vol. 54 (1951), p. 296.

12. P. Wolfe, *The Epsilon-Technique and the Artifical Basis in the Simplex Solution of the Linear Programming Problem,* Planning Research Division, Office of Air Comptroller, United States Air Force, 1951 (mimeographed).

NATIONAL BUREAU OF STANDARDS

Reprinted from Naval Research Logistics Quarterly
Vol. 10 (1963), pp. 369–373

ON ABSTRACT DUAL LINEAR PROGRAMS*

A. J. Hoffman

*IBM Corporation
Thomas J. Watson Research Center
Yorktown Heights, New York*

INTRODUCTION

This article examines the duality theorem of linear programming in the context of a general algebraic setting. It is well known that, when the constants and variables of primal and dual programs are real numbers (or any ordered field), then (i) any value of the function to be maximized does not exceed any value of the function to be minimized, and (ii) max = min. Property (i) is a triviality, and property (ii) depends on the hyperplane separation theorem [3], the simplex method [2], or some other argument [4]. All of the arguments used to prove (ii), however, seem to depend on the properties of a field; the proof of (i), however, does not. In fact, its triviality will persist in the abstract setting described in the next section. We then formulate some questions, which it is the main purpose of this article to advertise. That these questions have some interest will be illustrated in the section entitled "Examples of Sets S for Which Duality Holds," where the duality theorem will be shown to hold in some unusual surroundings.

ABSTRACT FORMULATION OF LINEAR PROGRAMMING DUALITY

We shall be concerned only with that portion of the duality theorem which considers properties (i) and (ii) mentioned in the introduction.

We assume that we deal with a set S which contains all the constants and all possible values of our variables. Also, S admits the operations of addition (under which S is a commutative semi-group) and multiplication (under which S is a semi-group), and multiplication is distributive with respect to addition. Furthermore, S is partially ordered under a relation "$\leq$" satisfying $a \leq b$ implies $x + a \leq x + b$ for all $x \epsilon S$. Finally, S admits a subset $P \subset S$ such that $a \leq b$, $x \epsilon P$ implies $xa \leq xb$ and $ax \leq bx$.

We now formulate two dual linear programs: $A = (a_{ij})$ is an m by n matrix; $b = (b_1, \ldots, b_m)$ is a vector with m components; $c = (c_1, \ldots, c_n)$ is a vector with n components; all entries in S.

<u>Problem 1</u>: Choose n elements $x_1, \ldots, x_n$ of S so that

$$(1) \qquad \sum_j a_{ij} x_j \leq b_i \qquad (i = 1, \ldots, m)$$

$$(2) \qquad x_j \epsilon P \qquad (j = 1, \ldots, n)$$

---

*This research was supported in part by the Office of Naval Research under Contract No. Nonr 3775(00), NR 047040.

                                        A. J. HOFFMAN

in order to maximize

$$(3) \qquad \sum_j c_j x_j \, .$$

The meaning of (3) is that we seek elements $x_1^0, \ldots, x_n^0$ which satisfy (1) and (2) such that, if $x_1, \ldots, x_n$ are any elements satisfying (1) and (2), we have

$$(4) \qquad \sum_j c_j x_j \leq \sum_j c_j x_j^0 \, .$$

Problem 2: Choose m elements $y_1, \ldots, y_m$ satisfying

$$(5) \qquad \sum_i y_i a_{ij} \geq c_j \qquad (j = 1, \ldots, n)$$

$$(6) \qquad y_i \, \epsilon P \qquad (i = 1, \ldots, m)$$

in order to minimize

$$(7) \qquad \sum_i y_i b_i \, .$$

Remarks analogous to (4) explain the meaning to be attached to (7).

Before proving property (i), let us note that

$$(8) \qquad a_i \leq b_i \qquad i = 1, \ldots, k$$

implies

$$\sum_i a_i \leq \sum b_i \, .$$

To prove (8), it is clearly sufficient, by induction, to prove it in the case $k = 2$. But $a_1 \leq b_1$ implies $a_1 + a_2 \leq b_1 + a_2$. Also, $a_2 \leq b_2$ implies $b_1 + a_2 \leq b_1 + b_2$. Hence $a_1 + a_2 \leq b_1 + b_2$, by the transitivity of partial ordering.

To prove property (i), let $x_1, \ldots, x_n$ satisfy (1) and (2), $y_1, \ldots, y_m$ satisfy (5) and (6), and one sees that the usual proof applies. For, consider

$$(9) \qquad \sum_j \sum_i y_i a_{ij} x_j = \sum_i \sum_j y_i a_{ij} x_j \, .$$

The right-hand side of (9) is

$$\sum_i y_i \left(\sum_j a_{ij} x_j \right) \, .$$

Since

$$\sum_j a_{ij} x_j \leq b_i \,,$$

we have

$$y_i \sum_j a_{ij} x_j \leq y_i b_i \,,$$

since

$$y_i \in P \,;$$

and

$$\sum_i y_i \left(\sum_j a_{ij} x_j \right) \leq \sum_i y_i b_i \,,$$

by (8). Similarly, the left side of (9) is

$$\sum_j \left(\sum_i y_i a_{ij} \right) x_j \geq \sum_j c_j x_j \,.$$

By the transitivity of partial ordering,

$$\sum_j c_j x_j \leq \sum_i y_i b_i \,,$$

which is property (i).

We now pose the following problems:

1. Find all (some) sets S satisfying the postulates such that, if (1), (2), (5), and (6) have solutions, then the maximum of (3) and the minimum of (7) exist and are equal (i.e., duality holds).

Two examples of such sets S will be given in the next section.

2. If S is a set satisfying the postulates, for which duality fails, find all matrices A with the property: if $b_1, \ldots, b_m$ and $c_1, \ldots, c_n$ are taken so that (1), (2), (5) and (6) have solutions, then duality holds for this matrix A.

As an example of problem 1, let S be the set of integers, P the nonnegative integers, and multiplication, addition, and "$\leq$" have the usual meanings; then duality does not hold in general. The class of matrices A for which it does hold are the totally unimodular matrices [1,5].

EXAMPLES OF SETS S FOR WHICH DUALITY HOLDS

Example 1: Let U be a set, S any algebra of subsets of S (denote the complement of a by $\bar{a}$, interpret multiplication and addition as intersection and union respectively, "$\leq$" means "$\subset$", and P = S).

                                                 A. J. HOFFMAN

THEOREM 1: In example 1, duality holds.

PROOF: Observe that (1) and (2) always have solutions; trivially, we can set $x_j = \phi$ for every j. Also, (5) and (6) have solutions if, and only if, $\sum_i a_{ij} \geqq c_j$ for every j, which we shall assume. It is now straightforward to show that

$$x_j = \prod_i (\overline{a_{ij}} + b_i), \qquad (j = 1, \ldots, n)$$

and

$$y_i = \sum_j c_j \left(\overline{b_i} + a_{ij} \prod_k (\overline{a_{kj}} + b_k\, a_{kj}) \right) \qquad (i = 1, \ldots, m)$$

verify (1), (2), (5), (6) and the equality of (3) and (7).

  Example 2: Let S be the set of positive fractions. We shall say that a/b if (b/a) is an integer. Let multiplication in S be ordinary multiplication, addition in S be (g.c.d), "$\leqq$" mean "|", and $P = S$.

  Example 3: Let S be the set of all integers. Multiplication in S is ordinary addition, addition in S is min (i.e., $a + b = \min(a,b)$), "$\leqq$" in S is the ordinary inequality, and $P = S$.

THEOREM 2: In Examples 2 and 3, duality holds.

PROOF: We first remark that, by considering the exponents of each prime number present in each fraction, we see that duality for Example 2 will follow from Example 3, which we now treat.

  Clearly (1), (2), (5) and (6) have solutions. In Problem 1, we seek $\{x_j\}$ in order to maximize

$$(10) \qquad\qquad \min_j \{c_j + x_j\},$$

where

$$(11) \qquad\qquad \min_j \{a_{ij} + x_j\} \leqq b_i \qquad i = 1, \ldots, m.$$

  Let $j(i)$ be any mapping of $\{1, \ldots, m\}$ into $\{1, \ldots, n\}$,

and let

$$(12) \qquad \overline{x}_k = \begin{cases} \infty \text{ if } k \neq j(i) \text{ for any } i \\ \min\,(b_i - a_{ik}) \text{ over all } i \text{ such that } k = j(i) \end{cases}.$$

Clearly, $\{\overline{x}_k\}$ satisfy (11), and (10) becomes

$$\min_k \begin{cases} \infty \text{ if } k \neq j(i) \text{ for any } i \\ \min\,(c_k + b_i - a_{ik}), \text{ over all } i \text{ such that } k = j(i) \end{cases}.$$

Another way of stating this value of (3) is as follows: the mapping $j\,(i)$ picks out certain entries in the matrix $(c_j + b_i - a_{ij})$, and (10) is the least of those entries. In particular, we may select a mapping $j\,(i)$ so that

$$c_{j(i)} + b_i - a_{ij(i)} = \max_j \; c_j + b_j - a_{ij} \,, \qquad (i = 1, \ldots, m)\,.$$

Thus, we can obtain a value for (10) which is the minimum of the row maxima of $(c_j + b_i - a_{ij})$. For Problem 2, $y_i = \max_j \{c_j - a_{ij}\}$ satisfies (5) and (6), and (7) becomes $\min_i \max_j \{b_i + c_j - a_{ij}\}$. This is the same as the solution we found for Problem 1, proving the theorem.

REFERENCES

[1] Berge, C., "Théorie des Graphes et de ses Applications" (Dunod, Paris, 1958).

[2] Dantzig, G. B., "Inductive Proof of the Simplex Method," IBM Journal of Research and Development, $\underline{4}$, 505-506 (1960).

[3] Gale, D., Kuhn, H. W., and Tucker, A. W., "Linear Programming and the Theory of Games," chapter XIX, pp. 317-329 of Activity Analysis of Production and Allocation, Cowles Commission. Monograph No. 13, edited by T. C. Koopmans (John Wiley and Sons, New York, 1951).

[4] Goldman, A. J. and Tucker, A. W., "Theory of Linear Programming," Paper 4 of "Linear Inequalities and Related Systems," pp. 53-98, Annals of Mathematics Studies No. 38, edited by H. W. Kuhn and A. W. Tucker (Princeton, 1956).

[5] Hoffman, A. J. and Kruskal, J. B., "Integral Boundary Points of Convex Polyhedra," Paper 13 of "Linear Inequalities and Related Systems," pp. 223-246, Annals of Mathematics Studies No. 38, edited by H. W. Kuhn and A. W. Tucker (Princeton, 1956).

*　　*　　*

A Proof of the Convexity of the Range of a Nonatomic Vector Measure Using Linear Inequalities

Alan Hoffman
IBM Thomas J. Watson Research Center
P.O.B. 218
Yorktown Heights, New York 10598

and

Uriel G. Rothblum*
Faculty of Industrial Engineering and Management
Technion—Israel Institute of Technology
Haifa 32000, Israel
and
RUTCOR—Rutgers Center for Operations Research
Rutgers University
New Brunswick, New Jersey 08904

Submitted by Henry Wölkowicz

Dedicated to Ingram Olkin

ABSTRACT

This note shows how a standard result about linear inequality systems can be used to give a simple proof of the fact that the range of a nonatomic vector measure is convex, a result that is due to Liapounoff.

We denote the set of reals by R and the set of rationals by Q. Also, we let $\| \ \|_1$ be the l_1 norm on R^k, i.e., for every $a \in R^k$, $\|a\|_1 = \sum_{j=1}^k a_j$. A *measurable space* is a pair (X, Σ) where Σ is a subset of the power set $P(X)$ of X which contains the empty set and is closed under countable unions and under complements with respect to X. In particular, in this case the sets in Σ

*Research of this author was supported in part by ONR grant N00014-92-J1142.

LINEAR ALGEBRA AND ITS APPLICATIONS 199:373–379 (1994)   373

will be called *measurable*. A parametric family of measurable sets whose index set I is a subset of the reals, say $\{S_t : t \in I\}$, is called *increasing* if $S_{t'} \supseteq S_t$ for every $t, t' \in I$ with $t' \geq t$.

Throughout the remainder of this note let (X, Σ) be a given measurable space. A function $\mu : \Sigma \to R^k$ is called a *k-vector measure* on (X, Σ) if $\mu(\varnothing) = 0$ and for every countable collection of pairwise disjoint sets $S_1, S_2, \ldots$ in Σ one has $\mu(\bigcup_{i=1}^{\infty} S_i) = \sum_{i=1}^{\infty} \mu(S_i)$, where the series converges absolutely; in particular, in this case we call the integer k the *dimension* of the vector measure μ. A *scalar measure* is a vector measure with dimension 1. For a k-vector measure μ and $j \in \{1, \ldots, k\}$, we denote by μ_j the scalar measure defined for $S \in \Sigma$ by $\mu_j(S) = [\mu(S)]_j$. A vector measure μ is called *nonnegative* if $\mu(S) \geq 0$ for all measurable sets S, and *nonatomic* if every measurable set S with $\mu(S) \neq 0$ has a measurable subset T with $\mu(T) \neq 0$ and $\mu(T) \neq \mu(S)$.

The purpose of this note is to use a standard result about linear inequality systems to give a simple proof of the following theorem due to Liapounoff; see Liapounoff (1940), Halmos (1948) and Lindenstrauss (1966), for example.

THEOREM 1. *Let μ be a nonnegative, nonatomic vector measure. Then the set $\{\mu(S) : S \in \Sigma\}$ is convex.*

The following fact will be used in our proof. It can be established by a simple argument using Zorn's lemma. A more elementary proof that relies only on countable induction is given in the Appendix for the sake of completeness.

PROPOSITION 1. *Let μ be a nonnegative, nonatomic, scalar measure, and let S be a measurable set. Then there exists an increasing parametric family of measurable subsets of S, $\{S_t : t \in ([0, \mu(S)) \cap Q) \cup \{\mu(S)\}\}$, such that $\mu(S_t) = t$ for every $t \in ([0, \mu(S)) \cap Q) \cup \{\mu(S)\}$.*

Proof of Theorem 1. Suppose that S_0 and S_1 are measurable sets and $0 < \beta < 1$. We will show that for some measurable set T, $\mu(T) = (1 - \beta)\mu(S_0) + \beta\mu(S_1)$. We first note that it suffices to consider the case where S_0 and S_1 are disjoint, for otherwise let $S_0' \equiv S_0 \setminus (S_0 \cap S_1)$ and $S_1' \equiv S_1 \setminus (S_0 \cap S_1)$, and construct a set T' with $\mu(T') = (1 - \beta)\mu(S_0') + \beta\mu(S_1')$. Then $T \equiv T' \cup (S_0 \cap S_1)$ will satisfy $\mu(T) = (1 - \beta)\mu(S_0) + \beta\mu(S_1)$.

Let k be the dimension of μ, and let $\|\mu\|_1$ be the scalar measure defined by $\|\mu\|_1 \equiv \sum_{j=1}^{k} \mu_j$, i.e., for every measurable set U, $\|\mu\|_1(U) = \sum_{j=1}^{k} \mu_j(U) = \|\mu(U)\|_1$. Now, fix $i \in \{0, 1\}$ and let $I_i \equiv ([0, \|\mu\|_1(S_i)] \cap Q) \cup \{\|\mu\|_1(S_i)\}$. By applying Proposition 1 to $\|\mu\|_1$ and the set S_i we can construct an increasing parametric family of measurable subsets of S_i, say

$\{S_{it}: t \in I_i\}$, such that $\|\mu\|_1(S_{it}) = t$ for every $t \in I_i$. By taking set differences corresponding S_{it}'s we can define for each $p = 1, 2, \ldots$ *finite* partitions $\Pi_i^{(p)}$ of S_i into measurable sets such that $\|\mu\|_1 \leqslant 2^{-p}$ for every $U \in \Pi_i^{(p)}$; further, if $p' > p$ then $\Pi_i^{(p')}$ is a refinement of $\Pi_i^{(p)}$, i.e., all sets in $\Pi^{(p')}$ are subsets of sets in $\Pi^{(p)}$. Let $\Pi^{(p)} \equiv \Pi_0^{(p)} \cup \Pi_1^{(p)}$. In particular, $\Pi^{(p)}$ is a partition of $S_0 \cup S_1$.

Consider linear inequality systems with variables $\{x_U : U \in \Pi^{(1)}\}$ given by

$$\sum_{U \in \Pi^{(1)}} \mu(U) x_U = (1 - \beta)\mu(S_0) + \beta\mu(S_1), \tag{1}$$

$$0 \leqslant x_U \leqslant 1 \qquad \text{for all} \quad U \in \Pi^{(1)} \tag{2}$$

Let $a'^{(1)}$ be the vector in $R^{\Pi^{(1)}}$ defined by setting $a_U'^{(1)} = 1 - \beta$ for the sets $U \in \Pi^{(1)}$ that are included in S_0, and $a_U'^{(1)} = \beta$ for (the distinct class of) sets $U \in \Pi^{(1)}$ that are included in S_1. Evidently, the vector $a'^{(1)}$ satisfies (1)–(2); hence, this system is feasible. It now follows from a standard result about linear inequalities (see Chavatal, 1983, Theorem 3.4, p. 42) that there exists a solution $a^{(1)}$ of (1)–(2) such that at most k of the $a_U^{(1)}$'s are neither 0 nor 1.

For $p = 1, 2, \ldots$, we inductively consider linear inequality systems with variables $\{x_U : U \in \Pi^{(p)}\}$ and construct special solutions $a^{(p)} \in R^{\Pi^{(p)}}$ of these systems having the property that at most k of the $a_U^{(p)}$'s are neither 0 nor 1. The first system is given by (1)–(2), and its special solution $a^{(1)}$ was constructed in the above paragraph. Next assume that for some $p \in \{2, 3, \ldots\}$, $a^{(p-1)} \in R^{\Pi^{(p-1)}}$ was constructed, and consider the pth system consisting of

$$\sum_{U \in \Pi^{(p)}} \mu(U) x_U = (1 - \beta)\mu(S_0) + \beta\mu(S_1), \tag{3}$$

$$0 \leqslant x_U \leqslant 1 \qquad \text{for all} \quad U \in \Pi^{(p)}, \tag{4}$$

$$x_U = 0 \qquad \text{for} \quad U \in \Pi^{(p)} \quad \text{for which}$$

$$\text{the unique set } V \in \Pi^{(p-1)} \text{ containing } U$$

$$\text{has } a_V^{(p-1)} = 0, \tag{5}$$

$$x_U = 1 \qquad \text{for} \quad U \in \Pi^{(p)} \quad \text{for which}$$

$$\text{the unique set } V \in \Pi^{(p-1)} \text{ containing } U$$

$$\text{has } a_U^{(p-1)} = 1. \tag{6}$$

 ALAN HOFFMAN AND URIEL G. ROTHBLUM

Consider the vector $a'^{(p)} \in R^{\Pi^{(p)}}$ where for each set $U \in \Pi^{(p)}$ we let $a'^{(p)}_U = a^{(p-1)}V$ for the unique set $V \in R^{\Pi^{(p-1)}}$ which contains U. Evidently, $a'^{(p)}$ satisfies (3)–(6), and therefore this system is feasible. Another application of the standard result about linear inequalities shows that there exists a solution $a^{(p)}$ of (3)–(6) such that at most k of the $a^{(p)}_U$'s are neither 0 nor 1, completing our inductive construction.

For $p = 1, 2, \ldots$ let $T^{(p)} \equiv \bigcup \{U : U \in \Pi^{(p)}$ and $a^{(p)}_u = 1\}$. Then (6) assures that $T^{(1)}, T^{(2)}, \ldots$, is an increasing sequence of sets. Further, for $p = 1, 2, \ldots$, from Equations (3)–(6), the fact that at most k of the $a^{(p)}_U$'s are neither 0 nor 1, and the fact that $\| \mu \|_1(U) \leqslant 2^{-p}$ for every $U \in \Pi^{(p)}$ we see that

$$k\, 2^{-p} \geqslant \left\| \sum_{U \in \Pi^{(p)}} \mu(U) x_U - \mu(T^{(p)}) \right\|_1$$

$$= \|(1 - \beta)\mu(S_0) + \beta\mu(S_1) - \mu(T^{(p)})\|_1. \tag{7}$$

Let $T \equiv \bigcup_{p=1}^{\infty} T^{(p)}$. Then T is a measurable set, and (7) shows that $(1 - \beta)\mu(S_0) + \beta\mu(S_1) = \lim_{p \to \infty} \mu(T^{(p)}) = \mu(T)$, completing the proof. ∎

Our construction has some resemblance to the approach of Arstein (1980). But we obtain underlying extreme points from elementary arguments about linear inequalities over finite dimensional spaces, whereas he uses analytical arguments over finite dimensional spaces.

APPENDIX

The purpose of this appendix is to provide a proof of Proposition 1 that relies only on countable induction. We note that a simpler proof is available, establishing a stronger variant of the asserted result, by using Zorn's lemma. We first establish two elementary lemmas.

LEMMA 1. *Let μ be a nonnegative, nonatomic, scalar measure, and let S be a measurable set with $\mu(S) > 0$. Then for every $\varepsilon > 0$ there exists a measurable subset T of S with $0 < \mu(T) < \varepsilon$.*

Proof. The nonatomicity of μ implies that S has a measurable subset T' with $0 \neq \mu(T)$ and $\mu(T') \neq \mu(S)$. Let T_1 be the set with smaller μ measure among T' and $S \setminus T'$. Then T_1 is a measurable subset of S with $0 <

$\mu(T_1) \leqslant 2^{-1}\mu(S)$. By recursively iterating this procedure we can construct a sequence $T_1, T_2, \ldots$ of measurable subsets of S such that for each $k = 1, 2, \ldots$ we have $0 < \mu(T_k) \leqslant 2^{-1}\mu(T_{k-1}) \leqslant 2^{-k}\mu(S)$. The conclusion of the lemma now follows by selecting $T = T_k$ for any positive integer k with $2^{-k}\mu(S) \leqslant \varepsilon$. ∎

LEMMA 2. *Let μ be a nonnegative, nonatomic, scalar measure, and let S be a measurable set with $\mu(S) \geqslant 0$. Then for each $0 \leqslant \alpha \leqslant \mu(S)$ there exists a measurable subset T of S with $\mu(T) = \alpha$.*

Proof. The conclusion of our lemma is trivial if $\alpha = 0$ or if $\alpha = \mu(S)$, by selecting $T = \varnothing$ or $T = S$, respectively. Next assume that $0 < \alpha < \mu(S)$. Let

$$\alpha_1 \equiv \sup\{\, \mu(U) : U \text{ is a measurable subset of } S \text{ and } \mu(U) \leqslant \alpha \}; \quad (8)$$

in particular, Lemma 1 shows that $\alpha_1 > 0$. The definition of α_1 assures that one can select a measurable subset U_1 of S satisfying

$$2^{-1}\alpha_1 \leqslant \mu(U_1) \leqslant \alpha. \tag{9}$$

We continue by inductively selecting scalars $\alpha_2, \alpha_3, \ldots$ and measurable subsets $U_2, U_3, \ldots$ of S such that

$$\alpha_k \equiv \sup\Bigg\{ \mu(U) : U \text{ is a measurable subset of } S,$$

$$U \cap \left(\bigcup_{j=1}^{k-1} U_j = \varnothing \right), \text{ and } \mu(U) + \sum_{j=1}^{k-1} \mu(U_j) \leqslant \alpha \Bigg\} \tag{10}$$

and

$$U_k \cap \left(\bigcup_{j=1}^{k-1} U_j \right) = \varnothing, \qquad \mu(U_k) + \sum_{j=1}^{k-1} \mu(U_j) \leqslant \alpha, \quad \text{and}$$

$$\mu(U_k) \geqslant 2^{-1}\alpha_k. \tag{11}$$

We note that this inductive construction is possible because the selection of U_k in the kth step assures that $\Sigma_{j=1}^{k} \mu(U_j) \leqslant \alpha$; hence, $\mu(\varnothing) = 0$ is in the set over which the supremum in (10) in the $(k + 1)$st step is taken. Let $T \equiv \bigcup_{j=1}^{\infty} U_j$. Then T is a measurable subset of S and, as the U_k's are pairwise disjoint, $\mu(T) = \Sigma_{j=1}^{\infty} \mu(U_j) \leqslant \alpha$.

We will next show that $\mu(T) = \alpha$. Suppose that $\mu(T) \neq \alpha$, i.e., $\varepsilon \equiv \alpha - \mu(T) > 0$. Then $\mu(S \setminus T) = \mu(S) - \mu(T) \geqslant \alpha - \mu(T) > 0$; hence, by Lemma 1, there is a measurable subset U of $S \setminus T$ with $0 < \mu(U) < \varepsilon$. Now, for each $k = 1, 2, \ldots,$

$$\varnothing = U \cap T \supseteq U \cap \left(\bigcup_{j=1}^{k-1} U_j \right) \quad \text{and}$$

$$\mu(U) + \sum_{j=1}^{k-1} \mu(U_j) \leqslant \mu(U) + \mu(T) < \varepsilon + \mu(T) = \alpha; \qquad (12)$$

hence, $\mu(U)$ is an element in the set over which the supremum in (10) is taken, implying that $\alpha_k \geqslant \mu(U)$ and therefore $\mu(U_k) \geqslant 2^{-1}\alpha_k \geqslant 2^{-1}\mu(U) > 0$. Thus, we get a contradiction to the absolute convergence of $\Sigma_{j=1}^{\infty} \mu(U_j)$ which proves that, indeed, $\mu(T) = \alpha$. ∎

Proof of Proposition 1

We start by arbitrarily ordering the rationals in the interval $[0, \mu(S))$, say $q(0), q(1), \ldots,$ where $q(0) = 0$. Also, let $S_0 \neq \varnothing$ and $S_{\mu(S)} \equiv S$. We will use an inductive argument for our construction. Suppose that $S_{q(0)}, S_{q(1)}, \ldots, S_{q(k)}$ have been selected such that $\{S_{q(i)} : i \in \{0, 1, \ldots, k\}\} \cup \{S_{\mu(S)}\}$ is an increasing family of measurable subsets of S and $\mu(S_{q(i)}) = q(i)$ for $i \in \{0, 1, \ldots, k\}$. Let $q_* \equiv \max\{q(i) : i = 0, 1, \ldots, k,\ q(i) < q(k + 1)\}$ [the set over which this max is taken is nonempty because it contains $q(0) = 0$], and let $q^* \equiv \min\{\{q(i) : i = 0, 1, \ldots, k \ \text{and} \ q(i) > q(k + 1)\} \cup \{\mu(S)\}\}$. Then $q_* < q(k + 1) < q^*$ and $\mu(S_{q^*} \setminus S_{q_*}) = q^* - q_*$. Thus, $0 < q(k + 1) - q_* < q^* - q_*$, and Lemma 2 can be applied to select a measurable subset U of $S_{q^*} \setminus S_{q_*}$ such that $\mu(U) = q(k + 1) - q_*$. Letting $S(k + 1) \equiv S(q_*) \cup U$, we have that $\mu(S_{q_*} \cup U) = \mu(S_{q_*}) + \mu(U) = q_* + \mu(U) = q(k + 1)$. Further, we have that $\{s_{q(i)} : i \equiv \{0, 1, \ldots, k + 1\}\} \cup \{S_{\mu(S)}\}$ is an increasing family of measurable sets. The above inductive construction establishes the conclusion of Proposition 1. ∎

CONVEXITY OF VECTOR MEASURE RANGE 379

The authors would like to thank David Blackwell for interesting comments and Roger Wets for pointing out to them the related work of Z. Arstein. Special thanks are also due to Don Coppersmith for several useful comments. In particular, the idea for the proof of Proposition 1 as provided in the appendix is due to him; his arguments replaced our earlier proof which relied on Zorn's lemma.

REFERENCES

Arstein, Z. 1980. Discrete and continuous bang-bang and facial spaces or: Look for the extreme points, *SIAM Rev.* 22:172–184.

Chvatal, V. 1983. *Linear Programming*, Freeman, New York.

Halmos, P. R. 1948. The range of a vector measure, *Bull. Amer. Math. Soc.* 54:416–421.

Liapounoff, A. A. 1940. Sur les fonctions-vecteurs complètement additives, *Bull. Acad. Sci. URSS Ser. Math.* 4:465–478.

Lindenstrauss, J. 1966. A short proof of Liapounoff's convexity theorem, *J. Math. and Mech.* 15:971–972.

Received 19 October 1993; final manuscript accepted 28 June 1993

A nonlinear allocation problem

by E. V. Denardo
A. J. Hoffman
T. Mackenzie
W. R. Pulleyblank

We consider the problem of deploying work force to tasks in a project network for which the time required to perform each task depends on the assignment of work force to the task, for the purpose of minimizing the time to complete the project. The rules governing the deployment of work force and the resulting changes in task times of our problem are discussed in the contexts of a) related work on project networks and b) more general allocation problems on polytopes. We prove that, for these problems, the obvious lower bound for project completion time is attainable.

1. Introduction

A PERT network is an approach to organizing a project that consists of individual tasks satisfying precedence relations. The project is modeled as an acyclic directed graph with initial node s, terminal node t, and tasks $\{T_e\}$ corresponding to edges $\{e = (u, v)\}$ of the graph. For each T_e, we are given $d_e > 0$, the time to perform the task. $T_{(u,v)}$ cannot be started until all tasks $T_{(w,u)}$ are completed. Because of this, the earliest possible completion time y_t of the project is the length of a longest st-path with respect to the edge lengths d_e. This value represents the most time-consuming sequence of activities that must be performed in completing the project.

There is interest in studying how y_t can be altered if the set $\{d_e\}$ can be changed by actions of the project planner. The allocation of extra resources to critical tasks in a project so as to reduce their duration is commonly called "crashing." We cite two examples from the literature.

Example 1.1 [1]   For each T_e, d_e is changed to $d_e - m_e z_e$, where each m_e is a prescribed positive number, each z_e is a nonnegative variable bounded from above, and Σz_e equals a prescribed value.

Example 1.2 [2]   For each T_e, d_e is changed to d_e/z_e, where each z_e is a positive variable and Σz_e equals a prescribed value.

In both these cases, the general objective is to allocate the resources in such a way that the length of a longest path is minimized.

In this paper we consider a problem similar to Example 1.2, but one in which the resources may be reused in the manner described below. Again, d_e is replaced by d_e/z_e, but we impose conditions different from those of Example 1.2 on the positive work force variables z_e, which we do not require to be integral. As in Example 1.2, d_e represents the time to perform T_e with one unit of work force, and d_e/z_e the time with z_e units. When a task is completed, in the variation we consider, the work force assigned to that task may then be assigned to different tasks. As usual, a new task may start when all of its predecessor tasks have been completed.

A total of W units of work force are available. These units can be shifted from task to task, but at no time can more than W units be active. How shall the work force be assigned to minimize the completion time of the total project? We put some restrictions on this allocation below, but first we make some preliminary remarks.

301

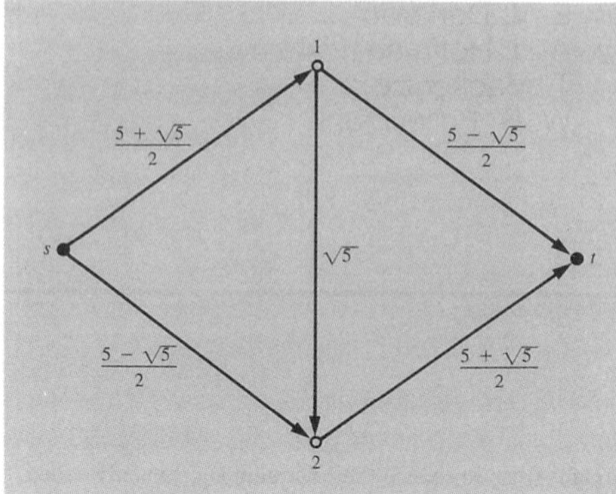

Figure 1

Sample PERT network. Labels on the edges are values of z_e for the optimal assignment of work force satisfying Condition 1.3 when all $d_e = 1$ and $W = 5$.

If work force can be reassigned at any time to any task that is available, given the precedence restrictions of the PERT network, it is easy to prove that *every* allocation that keeps the work force busy will complete the project in time $\Sigma d_e/W$.

If the work force assigned to a task cannot be changed during the performance of the task, then $\Sigma d_e/W$ is a lower bound on the completion time, and can be attained in many ways (for instance, by using the entire work force on each task as it becomes eligible).

The restriction we consider on the allocations was suggested to us by some managers of software projects. They pointed out that it is desirable to assign work force to tasks that grow from and relate to tasks that they have just performed. A rough attempt to model such a desideratum would be the requirement that the work force satisfy the following "flow" condition.

Condition 1.3 For each node $v \neq s$, t, the total work force assigned to tasks with terminal node v equals the total work force assigned to tasks with initial node v. The total work force assigned to tasks with initial node s equals W, which equals the total work force assigned to tasks with terminal node t.

Figure 1 gives an example of an allocation satisfying Condition 1.3, with all $d_e = 1$ and $W = 5$. The project is completed in time $\Sigma d_e/W = 1$. In view of the foregoing, this is optimum. Moreover, it is the *unique* optimum satisfying Condition 1.3. It is also the only allocation for which all three st-paths have the identical length, 1.

We show below that, for every problem, the minimum completion time is $\Sigma d_e/W$, and is attained by a unique allocation. In Section 2, we describe a general mathematical programming model that encompasses the above problem. This involves the concepts of *positive polytope* and *flat positive polytope*, for which we offer two *Pangloss** theorems. The first is proved in Section 3 for positive polytopes. A strengthening for flat positive polytopes is established in Section 4. In Section 5, we explore the relation between these two theorems, and in Section 6 we present concluding remarks.

For convenience, in the rest of this paper, we take $W = 1$. Because we impose no integrality conditions on z_e, we lose no generality.

2. Allocation problems on positive polytopes

A polytope is a bounded polyhedron. A point $\mathbf{x}$ is called *positive* (written $\mathbf{x} > \mathbf{0}$) if $x_i > 0$ for all i. We call a polytope *positive* if it is entirely contained in the nonnegative orthant and contains at least one positive point. We call a positive polytope *flat* if it is the set of nonnegative solutions to a nontrivial system of linear equations.

For a PERT network, let P be the set of all allocations satisfying Condition 1.3. (Recall that we have set $W = 1$.) Then P is a polyhedron, and, since PERT networks are acyclic, P is bounded. Each extreme point of P corresponds to an assignment of one unit to the edges in an st-path. The polytope P is positive because each edge is part of some st-path.

Let $\mathbf{d} = (d_1, \cdots, d_n) > \mathbf{0}$ be any positive vector indexed by the edges of our network. Consider the problem

$$\max_{\mathbf{x} \in P} \sum d_j x_j .\tag{2.1}$$

Since each extreme point of P assigns one unit to the edges belonging to a particular st-path, (2.1) computes the length of a longest st-path, with respect to edge lengths d_e.

We consider the PERT problem in which the time needed for task T_j equals d_j/z_j, where z_j is the amount of work force assigned to task T_j. Condition 1.3 states that $\mathbf{z} \in P$; that is, the assignment of work force must be an st flow of one unit. Hence, our PERT problem is this variation of (2.1):

$$\min_{\mathbf{z} \in P} \max_{\mathbf{x} \in P} \sum (d_j/z_j)x_j .\tag{2.2}$$

Thus, (2.2) allocates the work force so as to minimize the length of a longest st-path, where the length of edge j is d_j/z_j.

---

*In Voltaire's *Candide*, Dr. Pangloss says, "All is for the best . . . in this best of all possible worlds."

We actually study (2.1) for every positive polytope P and every positive vector $\mathbf{d}$, not just for those arising from PERT problems.

Note that for any positive $\mathbf{z} \in P$, if we define $c_j = d_j/z_j$ for all $j = 1, \cdots, n$, then

$$\max_{x \in P} \sum c_j x_j \geq \sum c_j z_j = \sum d_j . \tag{2.3}$$

One principal result is the following.

- **Theorem 2.1** *(Pangloss theorem for positive polytopes)*
Let $P \subset R^n$ be a positive polytope and let
$\mathbf{d} = (d_1, d_2, \cdots, d_n) > \mathbf{0}$. *Then there exists a*
unique $\mathbf{c} = (c_1, \cdots, c_n) > \mathbf{0}$ such that

there exists positive $\mathbf{z} \in P$ such that $c_j z_j = d_j$ for all j

and

$$\max_{x \in P} \sum c_j x_j = \sum d_j = \sum c_j z_j . \tag{2.4}$$

This can be restated as follows: Interpret each $\mathbf{x} \in P$ as a vector of feasible activities. Suppose we are given a vector $\mathbf{d}$ of target revenues for our activities. Then there exists a unique unit profit for each activity with the following property: If we allocate our resources so as to maximize the total revenue with respect to these unit profits, each activity in the optimum solution generates exactly its prescribed target revenue. More simply, every vector of proposed target revenues is the optimum set of revenues for a suitably chosen set of unit proofs. This is why we refer to this as a "Pangloss" theorem.

Theorem 2.1 is the same as

$$\min_{z \in P} \max_{x \in P} \sum (d_j/z_j)x_j = \sum d_j , \tag{2.5}$$

which we prove in the next section.

We have the following consequence.

- **Corollary 2.2**
Let z be the optimal solution to (2.5). Let S be the set of extreme points of P. Let $T \subseteq S$ be such that z can be expressed as a convex combination of $\{x : x \in T\}$ with positive weights. Then

$$\sum d_j = \sum (d_j/z_j)x_j \quad \text{for each } x \in T. \tag{2.6}$$

Let us interpret this corollary for the case of PERT networks. With P defined as in the second paragraph of this section, it is well known that each positive vector $\mathbf{x}$ in P is a positive convex combination of all extreme points of P. Hence, for a PERT network, Corollary 2.2 shows that every st-path has length Σd_j, the length of edge j being d_j/z_j. For a more general positive polytope P, however, there can exist extreme points $\mathbf{x}$, which we cannot use to express $\mathbf{z}$, for which $\Sigma(d_j/z_j)x_j < \Sigma d_j$.

We now describe an example of a polyhedron P that is positive but not flat. Consider a graph with node set $\{s, t, 1, 2, 3, 4, 5\}$ containing the following three st-paths:

$$p_1 = \{(s, 1), (1, 3), (3, 4), (4, t)\},$$

$$p_2 = \{(s, 1), (1, 3), (3, 5), (5, t)\},$$

$$p_3 = \{(s, 2), (2, 3), (3, 5), (5, t)\}.$$

Let $\mathbf{v}(p_i)$ be the path-edge incidence vector of path p_i, that is, the n-dimensional vector whose jth component equals 1 if edge j is in path p_i, and equals 0 otherwise. Define P to be the convex hull of $\{\mathbf{v}(p_1), \mathbf{v}(p_2), \mathbf{v}(p_3)\}$. Any hyperplane containing P must contain the incidence vector $\mathbf{v}(p_4)$ of $p_4 \equiv \{(s, 2), (2, 3), (3, 4), (4, t)\}$, since $\mathbf{v}(p_1) + \mathbf{v}(p_3) - \mathbf{v}(p_2) = \mathbf{v}(p_4)$. The positive polytope P is not flat because it does not contain $\mathbf{v}(p_4)$.

We return to flat positive polytopes in Section 4.

3. Pangloss theorem for positive polytopes

It is possible that this is a folk theorem in utility theory, but we do not know of any reference. We offer two proofs for Theorem 2.1. For the first, consider the optimization problem:

$$\min_{z \in P} F(\mathbf{z}) \equiv \sum -d_j \ln z_j . \tag{3.1}$$

- **Theorem 3.1**
Let P be a positive polytope and let $\mathbf{d} > 0$. Then (3.1) has a unique optimal solution z, which also solves (2.5).

Proof Since P contains a positive vector, a standard compactness argument shows that (3.1) has an optimal solution $\mathbf{z}$, all of whose components are strictly positive. Strict convexity of $-\ln z_j$ shows that $\mathbf{z}$ is unique.

To show that this $\mathbf{z}$ solves (2.5), we select any $\mathbf{x} \in P$ different from $\mathbf{z}$ and perturb $\mathbf{z}$ in the feasible direction $\mathbf{x} - \mathbf{z}$. Specifically, we consider the function $g(t) = F[\mathbf{z} + t(\mathbf{x} - \mathbf{z})]$ for $0 \leq t \leq 1$, with $F(\mathbf{z})$ defined as in (3.1). The function $g(t)$ is convex, differentiable, and nondecreasing in t—the last because $\mathbf{z}$ is optimal. Thus,

$$0 \leq g'(0) = \sum (-d_j/z_j)(x_j - z_j), \text{ for all } \mathbf{x} \in P. \tag{3.2}$$

Expression (3.2) simplifies to $\Sigma d_j \geq \Sigma(d_j/z_j)x_j$ for all $\mathbf{x} \in P$, which, when combined with (2.3), yields (2.5).

A more direct proof proceeds as follows:

First we establish uniqueness. Let $\mathbf{z}$ and $\bar{\mathbf{z}}$ be two minimizers in (2.5). Since $\mathbf{z} \in P$, and $\bar{\mathbf{z}}$ is a minimizer in (2.5),

$$\sum \frac{d_j}{\bar{z}_j} z_j \leq \sum d_j . \tag{3.3}$$

303

Similarly,

$$\sum \frac{d_j}{z_j}\bar{z}_j \leqq \sum d_j. \tag{3.4}$$

Add the inequalities (3.3) and (3.4). Since each $d_j > 0$ and $z_j/\bar{z}_j + \bar{z}_j/z_j \geqq 2$ and we have strict inequality unless $z_j = \bar{z}_j$, the sum of the left sides is at least $2\Sigma d_j$. The sum is $2\Sigma d_j$ if and only if $z_j = \bar{z}_j$ for all j. The sum of the right sides is $2\Sigma d_j$, so $\mathbf{z} = \bar{\mathbf{z}}$.

As noted in (2.3), for any positive $\mathbf{z} \in P$, $\max_{\mathbf{x}\in P} \Sigma(d_j/z_j)x_j \geqq \Sigma d_j$. We give two different proofs of the reverse inequality.

Let $\mathbf{M}$ be a matrix whose rows are the vertices of P, so that P is the convex hull $K(\mathbf{M})$ of the rows of $\mathbf{M}$. We must prove that

there exists $\mathbf{z} \in K(\mathbf{M})$ such that $\displaystyle\sum_j \frac{m_{ij}d_j}{z_j} \leqq \sum d_j$ for all i.
$$\tag{3.5}$$

Note that for every $\mathbf{z} \in K(\mathbf{M})$, we have $\Sigma_j m_{ij}d_j/z_j \geqq \Sigma d_j$ for at least one i. It is convenient to have $\mathbf{M}$ positive. To that end, let $\mathbf{J}$ be the matrix of 1's, $\epsilon > 0$. We prove that there exists $\mathbf{z}(\epsilon) \in K(\mathbf{M} + \epsilon\mathbf{J})$ such that

$$\sum_j \frac{m_{ij} + \epsilon}{z_j(\epsilon)} \leqq \sum d_j \text{ for all } i. \tag{3.6}$$

Now $\mathbf{z}(\epsilon)$ is in a compact region, so there exists a sequence of ϵ's converging to 0 such that the corresponding $\mathbf{z}(\epsilon)$'s converge to some $\mathbf{z} \in K(\mathbf{M})$. This $\mathbf{z}$ cannot have any coordinate 0; otherwise, since each column of $\mathbf{M}$ contains at least one positive entry (say in row i), (3.6) would be violated for row i and some ϵ. So, returning to (3.5), we assume that all entries in $\mathbf{M}$ are positive.

Here is an elementary proof of (3.5). Let $\mathbf{z}$ be a minimizer in (2.5), and assume that (3.5) is false, so that

$$\max_i \sum_j (d_j/z_j)m_{ij} = D^* > D \equiv \sum d_j.$$

It can be seen that the maximum cannot be attained for all i, so if $I^* \equiv \{i : \Sigma_j(d_j/z_j)m_{ij} = D^*\}$, then $|I^*| < m$. Now apply induction on the number of rows of $\mathbf{M}$, since (3.5) clearly holds if $\mathbf{M}$ has one row. Let $\mathbf{M}^*$ be the submatrix of $\mathbf{M}$ formed by rows in I^*. Then, by induction, there exists $\bar{\mathbf{z}} \in K(\mathbf{M}^*)$ with $\Sigma_j m_{ij}d_j/\bar{z}_j \leqq D$. If $\epsilon > 0$ is small, then setting $\hat{\mathbf{z}} = \epsilon\bar{\mathbf{z}} + (1 - \epsilon)\mathbf{z}$ yields a value of $\max_i \Sigma_j m_{ij}d_j/\hat{z}_j$ strictly smaller than D^*, contradicting the definition of $\mathbf{z}$. Here we use the fact that because $\mathbf{M}$ is strictly positive, $\Sigma_j m_{ij}d_j/w_j$ is a strictly convex function on positive $\mathbf{w}$, smaller when $\mathbf{w} = \bar{\mathbf{z}}$ than when $\mathbf{w} = \mathbf{z}$.

Another proof of (3.5) uses Brouwer's fixed-point theorem. Assume that $\mathbf{M}$ has m rows, and let Λ be the

simplex $\{\boldsymbol{\lambda} : \boldsymbol{\lambda} \geqq \mathbf{0}, \Sigma\lambda_i = 1\}$. With $\mathbf{z} = \boldsymbol{\lambda}\mathbf{M}$ and $D = \Sigma d_j$, consider the continuous mapping of Λ into itself:

$$\lambda_i' = \frac{\lambda_i + \left(\displaystyle\sum_j \frac{m_{ij}d_j}{z_j} - D\right)_+}{1 + \displaystyle\sum_k \left(\displaystyle\sum_j \frac{m_{kj}d_j}{z_j} - D\right)_+},$$

where $a_+ \equiv \max(a, 0)$. Let λ be a fixed point of this map, $\mathbf{z} = \boldsymbol{\lambda}\mathbf{M}$. Then

$$\lambda_i \sum_k \left(\sum_j m_{kj}\frac{d_j}{z_j} - D\right)_+ = \left(\sum_j m_{ij}\frac{d_j}{z_j} - D\right)_+. \tag{3.7}$$

We must show that for every i, the right side of (3.7) is 0. This is surely so if $\lambda_i = 0$. Further, if it is false, the left side of (3.7) would be positive for $\lambda_i > 0$, so that the right side would be positive.

Let $\Lambda^* = \{i : \lambda_i > 0\}$. For each $i \in \Lambda^*$, we would have $\Sigma_j m_{ij}d_j/z_j > D$, so

$$\sum_{i\in\Lambda^*} \lambda_i \sum_j \frac{m_{ij}d_j}{z_j} > D \sum_{i\in\Lambda^*} \lambda_i = D. \tag{3.8}$$

But (3.8) can be rewritten

$$D < \sum_j \frac{d_j}{z_j} \sum_{i\in\Lambda^*} \lambda_i m_{ij} = \sum_j \frac{d_j}{z_j} \sum_i \lambda_i m_{ij} = \sum d_j = D,$$

which is a contradiction.

4. Pangloss theorem for flat positive polytopes

• *Theorem 4.1 (Pangloss theorem for flat positive polytopes)*
Let P be a flat positive polytope, and let $\mathbf{d} = (d_1, \cdots, d_n) > \mathbf{0}$. Then there exists a unique $\mathbf{c} = (c_1, \cdots, c_n) > \mathbf{0}$ such that

there exists positive $\mathbf{z} \in P$ such that $c_j z_j = d_j$ for all j

and

for all $\mathbf{x} \in P$, $\displaystyle\sum c_j x_j = \sum d_j = D$. $\tag{4.1}$

This theorem asserts that any positive vector $\mathbf{d}$ is both the "best" and "worst" vector of revenues for some linear objective function $\mathbf{c}$, maximized over P.

Let us deduce Theorem 4.1 from Corollary 2.2. By hypothesis, P is flat and $\mathbf{z}$ is positive. It is easy to show that $\mathbf{z}$ can be written as a positive convex combination of all extreme points of P. Hence (2.6) holds for each extreme point of P, and (4.1) follows.

A more insightful proof of (4.1) comes from the concept of entropy (see [3] for other uses of entropy in combinatorial optimization). First, we present some preliminaries. Let $P \equiv \{x : Ax = b, x \geqq 0\}$ be the given flat positive polytope. Let A_j denote the jth column of A. The following optimization problem is motivated by the PERT problem:

$$
\begin{aligned}
&\text{Minimize } \theta = yb \quad \text{subject to} \\
p : \qquad &\qquad Az = b \\
&\qquad z \geqq 0 \\
w_j : \qquad &yA_j - d_j/z_j \geqq 0 \qquad \text{for all } j.
\end{aligned} \tag{4.2}
$$

Clearly, (4.2) is a convex program. Multipliers p and w are assigned to particular constraints. No multipliers are assigned to the constraints $z \geqq 0$, because an optimal solution to (4.2) would have $z > 0$; hence, corresponding multipliers would all be zero. Because we assume $z > 0$, the Karush–Kuhn–Tucker (KKT) optimality conditions for (4.2) are

$$
\begin{aligned}
&\qquad w \geqq 0, \\
y : \qquad &\qquad Aw = b \\
z_j : \qquad &pA_j - w_j d_j/(z_j)^2 = 0, \qquad \text{for all } j \\
&w_j(d_j/z_j - yA_j) = 0, \qquad \text{for all } j.
\end{aligned} \tag{4.3}
$$

We shall see that an optimal solution to (4.2) and its KKT multipliers can be obtained from the familiar program

$$
\min \sum -d_j \ln z_j : Az = b, z \geqq 0. \tag{4.4}
$$

• *Theorem 4.2*
Let z be an optimum for (4.4), and let y be its KKT multipliers for the constraints $Az = b$. Then $(y; z)$ is optimal for (4.2); that program's KKT multipliers are $p = y$ and $w = z$; and the optimal value θ^* is Σd_j.

Proof The KKT conditions for (4.4) are

$$
d_j/z_j = yA_j \qquad \text{for all } j. \tag{4.5}
$$

The pair $(y; z)$ satisfies the constraints in (4.2). To satisfy the optimality conditions in (4.3), we take $p = y$ and $w = z$. Finally, we multiply (4.5) by z_j and then sum, to obtain $\Sigma d_j = yAz = yb = \theta^*$.

Theorem 4.2 establishes a "self-dual" property of (4.2). Its optimal solution and its KKT multipliers equal each other.

5. Relation between the two Pangloss theorems

The contrast between Theorems 2.1 and 4.1 suggests that (4.1) characterizes flat positive polytopes.

• *Theorem 5.1*
Let P be a positive polytope. Then P is flat if and only if, for each $d > 0$, there exists positive $z \in P$ such that

$$
\sum d_j = \sum (d_j/z_j)x_j \text{ for all } x \in P. \tag{5.1}
$$

Proof The necessity is just Theorem 4.1. We prove the sufficiency. Let $d > 0$. By hypothesis, there exists $z \in P$ for which (5.1) holds. Let $c_j = d_j/z_j$ for each j. Equation (5.1) becomes $\Sigma c_j x_j = \Sigma d_j$ for each $x \in P$. Add sufficiently large multiples of this equation to the equations and inequalities of a minimal representation of P, to cause the representation to have the form $\{x : A_1 x = b^1, A_2 x \leqq b^2, x \geqq 0\}$, where all coefficients in A_1 and A_2 are positive, and every inequality is essential. Thus, at least one of these inequalities, which we write as $\Sigma a_j x_j \equiv (a, x) \leqq b$, has the following properties:

$$
\text{each } a_j > 0; \tag{5.2}
$$

$$
\text{at least one positive } z \in P \text{ satisfies } (a, z) = b; \tag{5.3}
$$

$$
\text{at least one } x \in P \text{ satisfies } (a, x) < b. \tag{5.4}
$$

Let $d = (a_1 z_1, \cdots, a_n z_n)$ and $c = (a_1, \cdots, a_n)$. Then, by (5.2) and (5.3), c and z satisfy the Pangloss theorem for positive polytopes, and they are unique. By (5.1), we must have $\Sigma a_j x_j = b$ for all $x \in P$, which contradicts (5.4). Hence P is flat.

6. Remarks

In closing, we mention three generalizations. First, PERT networks may require "dummy" edges that are not associated with any real task of the project, but are used to impose precedence constraints on the other tasks. A dummy edge e normally has $d_e = 0$. There is no difficulty in extending our previous results to this case, but uniqueness of the optimal solution value z_e for those e with $d_e = 0$ is lost. An alternative to the introduction of dummy edges is to let the nodes of an acyclic graph correspond to the tasks of a project, and to use the edges simply to indicate precedence. Results analogous to those presented in this paper hold in this framework.

Second, we can accommodate "nonconcurrence conditions," that is, requirements that certain pairs of tasks cannot be performed simultaneously, even if neither is a predecessor (direct or indirect) of the other in the acyclic graph. To do so, we consider each such pair in turn, adding an edge from the terminal node of one of the task edges to the initial node of the other if no edges previously added have made either one a predecessor of the other.

Third, our theorems about positive polytopes hold for any compact convex subset P of the nonnegative orthant that contains a positive vector—not just for polytopes.

216

Nimrod Megiddo has pointed out to us that problem (3.1) is an instance of finding the weighted analytic center of a polytope. (The word "center" is used as it is in barrier methods for linear programming.) Thus, efficient algorithms for finding the center (see [4]) are adaptable.

Acknowledgments
We thank Nimrod Megiddo, Don Coppersmith, Greg Glockner, Rolf Möhring, Michael Powell, David Jensen, Uri Rothblum, and Pete Veinott for helpful discussions during the course of this project.

References
1. J. E. Kelley, Jr., "Critical Path Planning and Scheduling: Mathematical Basis," *Oper. Res.* **9**, 296–320 (1961).
2. C. L. Monma, A. Schrijver, M. J. Todd, and V. K. Wei, "Convex Resource Allocation Problems on Directed Acyclic Graphs: Duality, Complexity, Special Cases, and Extensions," *Math. Oper. Res.* **15**, 736–748 (1990).
3. I. Csiszar, J. Körner, L. Lovász, K. Marton, and G. Simonyi, "Entropy Splitting for Antiblocking Pairs and Perfect Graphs," *Combinatorics* **10**, 27–40 (1990).
4. P. M. Vaidya, "A Locally Well-Behaved Potential Function and a Simple Newton-Type Method for Finding the Center of a Polytope," *Progress in Mathematical Programming*, N. Megiddo, Ed., Springer-Verlag, New York, 1989, pp. 79–90.

Received August 3, 1993; accepted for publication April 11, 1994

Eric V. Denardo *Center for Systems Science, Yale University, P.O. Box 208267, New Haven, Connecticut 06520.* Dr. Denardo has been at Yale University since 1968, in the Department of Administrative Sciences, in the School of Management and in the Department of Operations Research, prior to his present affiliation. He graduated from Princeton University in 1958, with a B.S. degree in engineering, and worked for Western Electric's Engineering Research Center until 1962, primarily on industrial uses of digital computers. From 1962 to 1965, he was a Ph.D. student at Northwestern University and a consultant to the RAND Corporation. At RAND (1965–1968), he worked on dynamic programming and management information systems. Dr. Denardo is perhaps best known for his thesis, papers, and monograph on dynamic programming. His more recent work is on uncertainty in manufacturing and in telecommunications. He has served on the editorial boards of *Management Science* and *Mathematics of Operations Research.*

Alan J. Hoffman *IBM Research Division, Thomas J. Watson Research Center, P.O. Box 218, Yorktown Heights, New York 10598 (HOFFA at YKTVMV, hoffa@watson.ibm.com).* Dr. Hoffman joined IBM in 1961 as a Research Staff Member in the Department of Mathematical Sciences at the IBM Thomas J. Watson Research Center; he was appointed an IBM Fellow in 1977. He received A.B. (1947) and Ph.D. (1950) degrees from Columbia University and worked at the Institute for Advanced Study (Princeton), National Bureau of Standards (Washington), Office of Naval Research (London) and General Electric Company (New York) prior to joining IBM. Dr. Hoffman has been adjunct or visiting professor at various universities and has supervised fifteen doctoral theses in mathematics and operations research. He is currently serving or has served on the editorial boards of *Linear Algebra and Its Applications* (founding editor) and ten other journals in applied mathematics, combinatorics, and operations research. Dr. Hoffman holds an honorary doctorate from the Israel Institute of Technology (Technion); he was a co-winner in 1992 (with Philip Wolfe of the Mathematical Sciences Department) of the von Neumann Prize of the Operations Research Society and the Institute of Management Science.

Todd Mackenzie *Department of Statistics, McGill University, Montreal, Quebec, H3A 2K6 Canada.* Mr. Mackenzie received his B.Sc. degree from Dalhousie University in 1990 and his M.Sc. degree from McGill University in 1993. He has worked as a research assistant in the Division of Clinical Epidemiology, Montreal General Hospital, since 1989 and is currently a Ph.D. student in the Department of Statistics at McGill University.

William R. Pulleyblank *IBM Research Division, Thomas J. Watson Research Center, P.O. Box 218, Yorktown Heights, New York 10598 (WRP at YKTVMV, wrp@watson.ibm.com).* Dr. Pulleyblank was a systems engineer with IBM Canada, Ltd. from 1969 through 1974. During this period, he also completed his doctoral degree at the University of Waterloo. From 1974 to 1990, he was a faculty member, first at the University of Calgary, later at the University of Waterloo. He spent four of these years working at research centers in Belgium, France, and Germany. His main research activities have been in the areas of algorithmic graph theory, combinatorial optimization, and polyhedral combinatorics. He has also worked on various applied problems. In addition to writing a large number of research papers and book chapters, Dr. Pulleyblank is a coauthor of TRAVEL, an interactive, graphics-based system for solving traveling salesman problems. He is currently involved in writing a student textbook on combinatorial optimization. He is Editor-in-Chief of *Mathematical Programming Series B* and also serves on several other editorial boards. Since August of 1990, he has been a member of the Mathematical Sciences Department at the IBM Thomas J. Watson Research Center in Yorktown Heights, NY. Dr. Pulleyblank is currently manager of the Optimization and Statistics Center.

306

Combinatorial Optimization

1. Integral boundary points of convex polyhedra

In this paper the concept (not the name) of total unimodularity was shown to be a neat explanation (via Cramer's rule) of the fact that some linear programming problems have all their vertices integral. I do not think that this paper would have ever been accepted for publication if we had not fancied it up with a soupçon of generalization: the main idea is too obvious and folklorish. And we also thought we introduced a new class of matrices with the "unimodular property", but Jack Edmonds later found that our new class wasn't really new after all. It is nevertheless true that totally unimodular matrices (as Berge christened them), and unimodular matrices generally, are key to understanding how linear programming duality underlies a wide variety of extremal combinatorial analysis. Incidentally, Joe Kruskal, the author of fundamental papers in combinatorics, combinatorial optimization and the theory of ordered sets is now a statistician.

2. Some recent applications of the theory of linear inequalities to extremal combinatorial analysis

Most if not all of this work was done while I was working at the London branch of the Office of Naval Research. As I explain in Autobiographical Notes, publication of this material in this form was due to the kind intervention of Al Tucker. Basically, it explores some of the consequences of total unimodularity, and some of the ways you can avoid total unimodularity yet get the same results.

This paper introduced the concept of circulation in directed graphs, which I created because s–t flow made some nodes look special. In a certain sense, the feasible circulation theorem and the max flow min cut theorem are equivalent (each is easily derivable from the other). But in operation, s–t flows and circulations look at different phenomena. I am pleased that (1) the idea of considering linear programs as canonically described by equations on variables with lower and upper bounds has become one of the standard representations, and (2) the word "circulation" is now used in many contexts apart from linear programming (I stole it, of course, from Harvey's work on blood).

3. Finding all shortest distances in a directed network

The problem is well-known in combinatorial optimization. I first encountered it while working for the government, when the calculation was required periodically in order to know the shortest rail distance between any two points on the network, for the purpose of setting freight rates. A basic step in the calculation is the multiplication of two matrices in the algebra MIN. Winograd had found a nifty way to multiply two matrices in ordinary algebra. We tuned his trick to matrix multiplication in MIN, added more tricks (Warshall, Hu), and voilà. I am disappointed that this paper is widely unnoticed.

4. On balanced matrices

We did not understand one of the papers Claude Berge wrote about one of his creations, balanced matrices. So we produced this study, which clarified the situation (for us) and actually proved a conjecture Berge had stated in the paper we didn't understand. The keys to our success were realizing what was going to be different from the case of perfect matrices; and realizing that the key to studying vertices of polyhedra described by nonnegative variables and linear inequalities was to understand the case where the inequalities were equations.

5. A generalization of max flow-min cut

I knew several proofs of the max flow-min cut theorem where the flow is carried on the edges of a directed graph; I knew tricks to reduce other versions (flow is on edges of undirected graph, flow is on nodes of a directed graph, etc.); but I thought it would be more aesthetic to prove a general theorem of which of all these max flow-min cut theorems would be special cases. Then in one proof you do them all. I found the right formulation using the concept of a set of paths closed with respect to "switching". It worked, although the proof invoked more machinery than I thought it would. As a consequence, it made the problem of calculating the maximum flow in this general situation interesting. Tom McCormick has made good progress on the theory of such calculations.

An important conseqence of this paper is that it begins by introducing the concept (not the term) of total dual integrality. I thought (incorrectly) that the concept was imbedded in linear programming folklore, and I proved the basic theorem just to be complete. But I was wrong about the folklore, and Edmonds and Giles did a great service by giving a name to the concept, and pointing out that total dual integrality is the idea behind most uses of linear programming to prove extremal combinatorial theorems.

6. On lattice polyhedra III

Don Schwartz and I had a couple of years earlier introduced the concept of lattice polyhedron to generalize various polyhedra (polymatroids, cut-set polyhedra, ...) where there were integer optimal solutions for integer data. The term was intended to be a pun: the rows of the relevant matrix corresponded to elements of a lattice (as in partially ordered sets) and answers occurred in the integer lattice. To my great disappointment, the pun was never appreciated, probably because it was never noticed.

The best part of "On lattice polyhedra III" starts by looking at the blocking relation between s–t paths of a directed graph and s–t cutsets. It then shows that this has a flawless generalization: a clutter on a set U of paths closed with respect to switching (see the discussion for the immediately preceding paper) has as its blocking clutter a set of subsets of U, whose incidence vectors form the rows of a certain lattice polyhedron, perfectly generalizing the s–t cutsets. Oh, the joys of axiomatization!

7. Local unimodularity in the matching polytope

Edmonds had shown (and I and others had followed him) that you could have interesting polyhedra with integer vertices even when the matrix giving the rows of inequalities defining the polyhedron was not totally unimodular. The notion of total dual integrality (Edmonds and Giles) explores this phenomenon in detail. Here we show that, if you define the generalized matching polytope of Edmonds in a certain way, a unimodular matrix (to be used in a Cramer's rule argument) arises in appropriate fashion. Later, Gerards and Sebö proved the much more general result (with less effort than we expended on a particular case) that total dual integrality always implies the existence of such a unimodular matrix.

8. A fast algorithm that makes matrices optimally sparse

Since most linear programs are sparse (and the simplex method needs to take advantage of that), it seemed to us desirable to see if we could rewrite the matrix as originally given to make it sparsest (or at least sparser). The sense in which (theoretically) we succeeded is described in the paper. And, beyond the theory, we performed some experiments, on real problems, which made us think at that time that our algorithm could be useful as well as interesting. But so far as we know, no one has incorporated any version of our algorithm in any production code.

Reprinted from
Linear Inequalities and Related Systems, eds. H. W. Kuhn and A. W. Tucker
© 1956 Princeton Univ. Press, pp. 223–246

INTEGRAL BOUNDARY POINTS OF CONVEX POLYHEDRA

A. J. Hoffman and J. B. Kruskal

INTRODUCTION

Suppose every vertex of a (convex) polyhedron in n-space has (all) integral coordinates. Then this polyhedron has the <u>integral property</u> (i.p.). Sections 1, 2, and 3 of this paper are concerned with such polyhedra.

Define two polyhedra[1]:

$$P(b) = \{x \mid Ax \geq b\} \, ,$$

$$Q(b, c) = \{x \mid Ax \geq b, \ x \geq c\} \, ,$$

where A, b, and c are integral and A is fixed. Theorem 1 states that $P(b)$ has the i.p. for every (integral) b if and only if the minors of A satisfy certain conditions. Theorem 2 states that $Q(b, c)$ has the i.p. for every (integral) b and c if and only if every minor of A equals 0, $+1$, or -1. Section 1 contains the exact statement of Theorems 1 and 2, and Sections 2 and 3 contain proofs.

A matrix A is said to have the <u>unimodular property</u> (u.p.) if it satisfies the condition of Theorem 2, namely if every minor determinant equals 0, $+1$, or -1. In Section 4 we give Theorem 3, a simple sufficient condition for a matrix to have the u.p. which is interesting in itself and necessary to the proof of Theorem 4. In Section 5 we state and prove — at length — Theorem 4, a very general sufficient condition for a matrix to have the u.p. Finally, in Section 6 we discuss how to recognize the unimodular property, and give two theorems, based on Theorem 4, for this purpose.

Our results include all situations known to the authors in which the polyhedron has the integral property <u>independently</u> of the "right-hand sides" of the inequalities (given that the "right-hand sides" are integral of course). In particular, the well-known "integrality" of transportation

---

[1] Unless otherwise stated, we assume throughout this paper that the inequalities defining polyhedra are consistent.

224   HOFFMAN AND KRUSKAL

type linear programs and their duals follows immediately from Theorems 2
and 4 as a special case.

1. DEFINITIONS AND THEOREMS

A point of n-space is an <u>integral point</u> if every coordinate is an
integer. A (convex) polyhedron in n-space is said to have the <u>integral
property</u> (i.p.) if <u>every</u> face (of every dimension) contains an integral
point. Of course, this is true if and only if every <u>minimal</u> face contains
an integral point. If the minimal faces happen to be vertices[2] (that is,
of dimension 0), then the integral property simply means that the vertices
of P are themselves all integral points.

Let A be an m by n matrix of integers; let b and b' be
m-tuples (vectors), and c and c' be n-tuples (vectors), whose components are integers or $\pm \infty$. We will let $\infty(-\infty)$ also represent a vector
all of whose components are $\infty(-\infty)$; this should cause no confusion. The
vector inequality $b < b'$ means that strict inequality holds at every
component. Let $P(b; b')$ and $Q(b; b'; c; c')$ be the polyhedra in n-space
defined by

$$P(b; b') = \{x \mid b \leq Ax \leq b'\} \, ,$$

$$Q(b; b'; c; c') = \{x \mid b \leq Ax \leq b' \quad \text{and} \quad c \leq x \leq c'\} \, .$$

Of course $Q(b; b'; -\infty, +\infty) = P(b; b')$. If S is any set of rows of A,
then define

$$\gcd(S) = \begin{cases} 0, & \text{if each minor determinant in } S \text{ which} \\ & \text{has as many rows as } S \text{ equals } 0, \\ \text{greatest common divisor (g.c.d.) of all} \\ \text{those minor determinants in } S \text{ which} \\ \text{have as many rows as } S, \text{ otherwise.} \end{cases}$$

THEOREM 1. The following conditions are equivalent:

(1.1) $P(b; b')$ has the i.p. for every b, b';

(1.2) $P(b; \infty)$ has the i.p. for every b;

(1.2') $P(-\infty; b')$ has the i.p. for every b';

(1.3)   if r is the rank of A, then for every
                 set S of r linearly independent rows
                 of A, gcd(S) = 1;

---

[2] It is well known (and, incidentally, is a by-product of our Lemma 1) that
all minimal faces of a convex polyhedron have the same dimension.

(1.4)   for every set S of rows of A, gcd(S) = 1 or 0.

The main value of this theorem lies in the fact that condition (1.3) implies
condition (1.1). However the converse implication is of esthetic interest.
If it is believed that (1.3) does not hold, (1.4) often offers the easiest
way to verify this, for it may suffice to examine <u>small</u> sets of rows.

A matrix (of integers) is said to have the <u>unimodular property</u>
(u.p.) if every minor determinant equals 0, + 1, or - 1. We see immediately that the entries in a matrix with the u.p. can only be 0, + 1, or - 1.

THEOREM 2. The following conditions are equivalent:

(1.5)
$\qquad$ $Q(b; b'; c; c')$ has the i.p. for every
$\qquad\qquad$ b, b', c, c';

(1.6)
$\qquad$ for some fixed c such that $- \infty < c < + \infty$,
$\qquad\qquad$ $Q(b, \infty; c; \infty)$ has the i.p. for every b;

(1.6')
$\qquad$ for some fixed c such that $- \infty < c < \infty$,
$\qquad\qquad$ $Q(- \infty; b'; c; \infty)$ has the i.p. for every b';

(1.6'')
$\qquad$ for some fixed c' such that $- \infty < c' < \infty$,
$\qquad\qquad$ $Q(b; \infty; - \infty; c')$ has the i.p. for every b;

(1.6''')
$\qquad$ for some fixed c' such that $- \infty < c' < \infty$,
$\qquad\qquad$ $Q(- \infty; b'; - \infty; c')$ has the i.p. for
$\qquad\qquad$ every b';

(1.7)   the matrix A has the unimodular property (u.p.).

The main value of this theorem for applications lies in the fact that condition (1.7) implies condition (1.5), a fact which can be proved directly
(with the aid of Cramer's rule) without difficulty. However the converse
implication is also of esthetic interest. The relationship between Theorems 1 and 2 is that Theorem 2 asserts the equivalence of stronger properties while Theorem 1 asserts the equivalence of weaker ones. Condition
(1.5) is clearly stronger than condition (1.1), and condition (1.7) is
clearly stronger than condition (1.3).

For A to have the unimodular property is the same thing as for
A transpose to have the unimodular property. Therefore if a linear program
has the matrix A with the u.p., both the "primal" and the dual programs
lead to polyhedra with the i.p. This can be very valuable when applying
the duality theorem to combinatorial problems (for examples, see several
other papers in this volume).

                          HOFFMAN AND KRUSKAL

2. PROOF OF THEOREM 1

We note that $(1.1) \implies (1.2)$ and $(1.2')$ trivially. Likewise $(1.4) \implies (1.3)$ trivially. To see that $(1.3) \implies (1.4)$, let S and S' be sets of rows of A. If $S \subset S'$, then the relevant determinants of S' are integral combinations of the relevant determinants of S. Hence $\gcd(S)$ divides $\gcd(S')$. From this we easily see that $(1.3) \implies (1.4)$.

As (1.2) and $(1.2')$ are completely parallel, we shall only treat the former in our proofs.

Let the rows of A be $A_1, \ldots, A_m$ and the components of b and b' be $b_1, \ldots, b_m$ and $b_1', \ldots, b_m'$. Suppose that we know that (1.3) for any matrix A_* implies (1.2) for the corresponding polyhedra $P_*(b; \infty)$. Also, suppose that (1.3) holds for the particular matrix A. Then setting

$$A_* = \begin{bmatrix} A \\ -A \end{bmatrix},$$

we see immediately that (1.3) holds for A_*. Consequently

$$P_*(b_1, \ldots, b_m, -b_1', \ldots, -b_m'; \infty)$$

has the i.p. But it is easy to see that this polyhedron is identical with $P(b; b')$; hence the latter also has the i.p. Therefore if for every matrix (1.3) implies (1.2), then (1.3) implies (1.1) for every matrix.

Let $P(b) = P(b; \infty)$ for convenience.

It only remains to prove that (1.2) is equivalent to (1.3)[3]. If S is any set of rows A_i of A, we define

$$F_S = F_S(b) = \{x \mid Ax \geq b \quad \text{and} \quad A_i x = b_i \text{ if } A_i \text{ in } S\},$$

$$G_S = \text{the subspace of } n\text{-space spanned by the rows } A_i \text{ in } S.$$

If $F_S(b)$ is not empty, it is the face of $P(b)$ corresponding to S. (We do not consider the empty set to be a face of a polyhedron.) We easily see that $F_S(b)$, if non-empty, corresponds to the usual notion of a face. Of course $F_\emptyset(b) = P(b)$, where $\emptyset$ is the empty set. We shall use the letter A to stand for the set of <u>all</u> rows of the matrix A. In general we will use the same letter to denote a set of rows and to denote the matrix formed by these rows. (This double meaning should cause no confusion.)

---

[3] The authors are indebted to Professor David Gale for this proof, which is much simpler than the original proof.

LEMMA 1. If $S \subset S'$, and if $F_S(b)$ and $F_{S'}(b)$ are faces (that is, not empty), then $F_{S'}(b)$ is a subface of $F_S(b)$. If $F_S(b)$ is a face, then it is a minimal face if and only if $G_S = G_A$, that is, if and only if S has rank r, where r is the rank of A.

PROOF. The first sentence of the lemma follows directly from the definitions. To prove the rest of the lemma, let S' be all rows of A which are in G_S. Then $G_S = G_{S'}$, and A_j is a linear combination of the A_i in S if and only if A_j is in S'. Clearly $G_S = G_A$ if and only if S' = A.

If $S' \neq A$, there is at least one row A_k in A - S'. Then there is a vector y such that $A_i y = 0$ for A_i in S, $A_k y < 0$. Let x be in F_S. As $A_k x \geq b_k$, there is a number $\lambda_k \geq 0$ for which $A_k(x + \lambda_k y) = b_k$. For every A_j in A - S' such that $A_j y < 0$, the equation $A_j(x + \lambda y) = b_j$ has a non-negative solution. Let λ_j be that solution. Define $\lambda = $ minimum λ_j, and let j' be a value such that $\lambda = \lambda_{j'}$. As λ_k exists, there is at least one λ_j, so λ exists. By the definition of λ,

$$A(x + \lambda y) \geq b \ ,$$
$$A_i(x + \lambda y) = b_i \qquad \text{for } A_i \text{ in } S,$$
$$A_{j'}(x + \lambda y) = b_{j'} \ .$$

Thus $F_{SUA_{j'}}$ is not empty, and is therefore a subface of F_S. Furthermore as $A_{j'}$ is not a linear combination of the A_i in S, $F_{SUA_{j'}}$ is a proper subface of F_S. Therefore F_S is not minimal.

On the other hand, if F_S is not minimal it has some proper subface F_{SUA_k}. Then there must be x_1 and x_2 in F_S such that $A_k x_1 = b_k$ and $A_k x_2 > b_k$. Therefore $A_k x$ varies as x ranges over F_S. But for A_i in S, $A_i x = b_i$ is constant as x varies over F_S. Hence A_k cannot be a linear combination of the A_i in S, so A_k is in A - S'. Hence $S' \neq A$. This proves the lemma.

If b, as usual, is an m-tuple and S is a set of r rows of A, then b_S is the "sub-vector" consisting of the r components of b which correspond to the rows of S. Let $\tilde{b}$ always represent an (integral) r-tuple. The components of $\tilde{b}$ and b_S will be indexed by the indices used for the rows of S, not by the integers from 1 to r. Let

$$L_S(\tilde{b}) = \{x \mid Sx = \tilde{b}\} \ .$$

LEMMA 2. Suppose S is a set of r linearly independent rows of A. Then for any $\tilde{b}$ there is a b such that

$$(2.1) \qquad b_S = \tilde{b} ;$$

$$(2.2) \qquad F_S(b) \text{ is a minimal face of } P(b).$$

PROOF. As S is a set of linearly independent rows, the equation $Sx = \tilde{b}$ has at least one solution: call it y. Define b as follows:

$$b_i = \begin{cases} \tilde{b}_i & \text{if } A_i \text{ in } S, \\ [A_i y] & \text{if } A_i \text{ not in } S . \end{cases}$$

Clearly $b_S = \tilde{b}$, so (2.1) is satisfied. Obviously b is integral. Furthermore y is seen to be in $F_S(b)$, so $F_S(b)$ is not empty, and hence is a face of $P(b)$. By Lemma 1, $F_S(b)$ is a minimal face, so (2.2) is satisfied.

LEMMA 3. Suppose S' is a set of rows of A of rank r, and $S \subset S'$ is a set of r linearly independent rows. For any b such that $F_{S'}(b)$ is a face (that is, not empty),

$$F_{S'}(b) = L_S(b_S).$$

PROOF. Let y be a fixed element in $F_{S'}(b)$, and let x be any element of $L_S(b)$. As $F_{S'}(b) \subset L_S(b_S)$ is trivial, we only need show the reverse inclusion. Thus it suffices to prove that x is in $F_{S'}(b)$.

As S has rank r, any row A_k in A can be expressed as a linear combination of the rows A_i in S:

$$A_k = \Sigma \alpha_{ki} A_i .$$

Then

$$A_i x = b_i = A_i y$$

for A_i in S, so

$$A_k x = \Sigma \alpha_{ki} A_i x = \Sigma \alpha_{ki} A_i y = A_k y .$$

Then as y is in $F_{S'}(b)$, x must be also. This completes the proof of the lemma.

LEMMA 4. Any minimal face of $P(b)$ can be expressed
in the form $F_S(b)$ where S is a set of r linearly
independent rows of A.

PROOF. Suppose the face is $F_{S'}(b)$. By Lemma 1, S' must have
rank r. Let S be a set of r linearly independent rows of S'. Then
by applying Lemma 3 to both $F_{S'}(b)$ and $F_S(b)$, we see that

$$F_{S'}(b) = L_S(b_S) = F_S(b).$$

This proves the lemma.

LEMMA 5. If S is a set of r linearly independent
rows of A, then the following two conditions are
equivalent:

(2.3) $L_S(\tilde{b})$ contains an integral point for every (integral) $\tilde{b}$;

(2.4) $\gcd(S) = 1.$

PROOF. We use a basic theorem of linear algebra, namely that any
integral matrix S which is r by n can be put into the form

$$S = UDV$$

where D is a (non-negative integral) diagonal matrix, and U and V are
(integral) unimodular matrices. (Of course U is r by r, V is
n by n, and D is r by n.) As U and V are unimodular, they have
integral inverses. Furthermore $\gcd(S) = \gcd(D)$. (For proofs of these
facts, see for example [3].)

Let the diagonal elements of D be d_{ii}. Clearly $\gcd(D) =$
$d_{11}d_{22} \cdots d_{rr}$. Therefore condition (2.4) is equivalent to the condition
that every $d_{ii} = 1$. Now we show that (2.3) is also equivalent to this
same condition.

Suppose that some diagonal element of D is greater than 1. For
convenience we may suppose that this element is $d_{11} = k > 1$. Let $\tilde{e}$ be
the r-tuple $(1, 0, \ldots, 0)$, and let $\tilde{b} = U\tilde{e}$. Then $L_S(\tilde{b})$ contains no
integral point. To see this, let x be in $L_S(\tilde{b})$. Then

$$Sx = UDVx = \tilde{b} = U\tilde{e} ,$$

so $DVx = \tilde{e}$. Clearly the first component of $y = Vx$ is $1/k$, so y is
not integral. Hence x cannot be integral. This shows that (2.3) cannot

                                HOFFMAN AND KRUSKAL

hold if some d_{ii} is greater than 1.

Suppose every $d_{ii} = 1$. Let x be in $L_S(\tilde{b})$ and set

$$Vx = (y_1, \ldots, y_r, y_{r+1}, \ldots, y_n).$$

Then

$$U^{-1}\tilde{b} = DVx = (y_1, \ldots, y_r),$$

and so $y_1, \ldots, y_r$ are integral. Let $y = (y_1, \ldots, y_r, 0, \ldots, 0)$.
Then $V^{-1}y$ is integral, and since $Dy = DVx$,

$$S(V^{-1}y) = UDV(V^{-1}y) = UDy = UDVx = \tilde{b}.$$

Thus $V^{-1}y$ is in $L_S(\tilde{b})$. This shows that (2.3) does hold if every $d_{ii} = 1$,
and completes the proof of the lemma.

Now it is easy to prove that (1.2) $\Longleftrightarrow$ (1.3). First we prove
$\Longrightarrow$. Let S be any set of r linearly independent rows of A. Let $\tilde{b}$
be any (integral) r-tuple. Choose a b which satisfies (2.1) and (2.2).
By (1.2), $F_S(b)$ must contain an integral point x. By Lemma 3 and (2.1),

$$F_S(b) = L_S(b_S) = L_S(\tilde{b}).$$

Hence $L_S(\tilde{b})$ contains x. Therefore (2.3) is satisfied, so by Lemma 5,
$\gcd(S) = 1$. This proves $\Longrightarrow$.

To prove $\Longleftarrow$, let $F_{S'}(b)$ be some minimal face of $P(b)$. By
Lemma 4 this face can be expressed as $F_S(b)$ where S consists of r
linearly independent rows of A. By Lemma 3, $F_S(b) = L_S(b_S)$. By (1.3),
$\gcd(S) = 1$, and by Lemma 5 $L_S(b_S)$ must contain an integral point x.
Hence $F_{S'}(b)$ contains the integral point x. Therefore every minimal
face of $P(b)$ contains an integral point, and hence also every face. This
proves $\Longleftarrow$, and completes the proof of Theorem 1.

3. PROOF OF THEOREM 2

The role of (1.6) and its primed analogues are exactly similar,
so we treat only the former in our proofs. For convenience we let

$$Q(b; c) = Q(b; \infty; c; \infty).$$

It is not hard to see that (1.7) $\Longrightarrow$ (1.5). For suppose that A
has the u.p. (that is, satisfies (1.7)). Then

$$A_* = \begin{bmatrix} A \\ -A \\ I \\ -I \end{bmatrix}$$

satisfies (1.3). By Theorem 1, the associated polyhedron

$$P_*(b_1, \ldots, b_m, -b_1', \ldots, -b_m', c_1, \ldots, c_n, -c_1', \ldots, -c_n')$$

has the i.p. But it is easy to see that this polyhedron is identical with $Q(b; b'; c; c')$. Therefore the latter has the i.p., so $(1.7) \Longrightarrow (1.5)$. (An alternate proof of this can easily be constructed using Cramer's Rule.)

Clearly $(1.5) \Longrightarrow (1.6)$. Hence it only remains to prove that $(1.6) \Longrightarrow (1.7)$. We shall prove[4] this by applying Theorem 1 to the matrix

$$A^* = \begin{bmatrix} I \\ A \end{bmatrix} .$$

Let d be any (integral) $(n+m)$-tuple, and let

$$c \cup b = (c_1, \ldots, c_n, b_1, \ldots, b_m) .$$

Then $P^*(c \cup b) = Q(b, c)$.

To verify condition (1.2) for A^*, we need to show that $P^*(d)$ has the i.p. for every d. Condition (1.6) yields only the fact that $P^*(d)$ has the i.p. for every d such that $d_I = c$. To fill this gap, note that A^* has rank n as it contains the n by n identity matrix, and let $F_{S'}^*(d)$ be any face of $P^*(d)$. This face contains some minimal face, which by Lemma 4 can be expressed as $F_S^*(d)$ where S consists of n linearly independent rows of A^*. By Lemma 3,

$$F_S^*(d) = L_S^*(d_S) \equiv \{x \mid Sx = d_S\} .$$

As S is an n by n matrix of rank n, $F_S^*(d)$ consists only of a single point. Call this point x. We shall show that x is integral.

Let I_1 be the rows of I in S, I_2 the rows of I not in S, A_1 the rows of A in S, and A_2 the rows of A not in S. We wish to pick an integral vector q such that

---

[4] The authors are indebted to Professor David Gale for this proof, which is much simpler than the original proof.

                    HOFFMAN AND KRUSKAL

(3.1) $$x + q \geqq c,$$

(3.2) $$(x + q)_{I_1} = c_{I_1}.$$

Let $q = c - d_I$. Then q satisfies these requirements, for

$$x_i + q_i \left\{ \begin{matrix} = \\ \geqq \end{matrix} \right\} d_i + (c_i - d_i) = c_i \left\{ \begin{matrix} \text{if the i-th row of} \\ I \text{ is in } I_1, \\ \text{otherwise.} \end{matrix} \right.$$

Define $d' = c \cup (d_A + Aq)$. Then d' is integral, and $d'_I = c$, so by (1.6) the polyhedron $P^*(d')$ has the i.p.

Now $F_S^*(d')$ is not empty because it contains $(x + q)$, as we may easily verify:

$$A^*(x + q) = I(x + q) \cup A(x + q) \geqq c \cup (d_A + Aq) = d',$$

$$S(x + q) = I_1(x + q) \cup A_1(x + q) = c_{I_1} \cup (d_{A_1} + A_{1q}) = d'_S.$$

Therefore $F_S^*(d')$ must contain an integral point. However $F_S^*(d')$ can contain only a single point for the same reasons that applied to $F_S^*(d)$. Hence $x + q$ must be that single point, so $x + q$ must itself be integral. As q is integral, x must be integral also. Thus $F_S^*(d)$, and a fortiori $F_{S'}^*(d)$, contains the integral point x. This verifies condition (1.2) for A^*.

By Theorem 1, (1.3) holds for A^*. As the rank of A^* is n, $\gcd(S) = |S| = 1$ for every set S of n linearly independent rows of A^*. From this we wish to show that A has the u.p. Suppose E is any non-singular square submatrix of A. Let the order of E be s. By choosing S to consist of the rows of A^* which contain E together with the proper set of $(n - s)$ rows of I, and by rearranging columns, we can easily insure that

$$S = \begin{bmatrix} I & 0 \\ F & E \end{bmatrix}$$

where I is the identity matrix of order $(n - s)$, F is some s by $(n - s)$ matrix, and 0 is the $(n - s)$ by s matrix of zeros. Then $|S| = |E| \neq 0$, so S is non-singular. Therefore S consists of n linearly independent rows, so

$$|E| = |S| = \gcd(S) = 1.$$

This completes the proof of Theorem 2.

4. A THEOREM BY HELLER AND TOMPKINS

In this and the remaining sections we give various sufficient conditions for a matrix to have the unimodular property.

> THEOREM 3. (Heller and Tompkins). Let A be an m by n matrix whose rows can be partitioned into two disjoint sets, T_1 and T_2 , such that A, T_1 , and T_2 have the following properties:

(4.1) every entry in A is $0, +1,$ or -1 ;

(4.2) every column contains at most two non-zero entries;

(4.3) if a column of A contains two non-zero entries, and both have the same sign, then one is in T_1 and one is in T_2 ;

(4.4) if a column of A contains two non-zero entries, and they are of opposite sign, then both are in T_1 or both in T_2 .

> Then A has the unimodular property.

This theorem is closely related to the central result of the paper by Heller and Tompkins in this Study. The theorem, as stated above, is given an independent proof in an appendix to their paper.

> COROLLARY.[5] If A is the incidence matrix of the vertices versus the edges of an ordinary linear graph G, then in order that A have the unimodular property it is necessary and sufficient that G have no loops with an odd number of vertices.

PROOF. To prove the sufficiency, recall the following. The condition that G have no odd loops is well-known to be equivalent to the property that the vertices of G can be partitioned into two classes so that each edge of G has one vertex in each class. If we partition the rows of A correspondingly, it is easy to verify the conditions (4.1)-(4.4). Therefore A has the u.p.

If A has an odd loop, let A' be the submatrix contained in the rows and columns corresponding to the vertices and edges of the loop. Then

___

[5] The authors are indebted to the referee for this result.

                          HOFFMAN AND KRUSKAL

it is not hard to see that $|A'| = \pm 2$. This proves the necessity.

5. A SUFFICIENT CONDITION FOR THE UNIMODULAR PROPERTY

We shall consider oriented graphs. For our purposes an oriented graph G is a graph (a) which has no circular edges, (b) which has at most one edge between any two given vertices, and (c) in which each edge has an orientation. Let V denote the set of vertices of G, and E the set of edges. If (r, s) is in E (that is, if (r, s) is an edge of G), then we shall call (s, r) an _inverse edge_. (Note that by (b), and inverse edge cannot be in E; thus an inverse edge cannot be an edge. This slight ambiguity in terminology should cause no confusion.) We shall often use the phrase _direct edge_ to denote an ordinary edge.

A _path_ is a sequence of distinct vertices $r_1, \ldots, r_k$, such that for each i, from 1 to $k - 1$, (r_i, r_{i+1}) is either a direct or an inverse edge. A path is directed if every edge is oriented forward, that is, if every edge (r_i, r_{i+1}) in the path is a direct edge. A path is _alternating_ if successive edges are oppositely oriented. More precisely, a path is alternating if its edges are alternately direct and inverse. An alternating path may be described as being $(++)$, $(+-)$, $(-+)$, or $(--)$. The first sign indicates the orientation of the first edge of the path, the second sign the orientation of the last edge of the path. A $+$ indicates a direct edge; a $-$ indicates an inverse edge. A _loop_ is a path which closes back on itself. More precisely, a loop is a sequence of vertices $r_1, \ldots, r_k$ in which $r_1 = r_k$ but which are otherwise distinct, and such that for each i (r_i, r_{i+1}) is either a direct or an inverse edge. A loop is alternating if successive edges are

Diagram 1

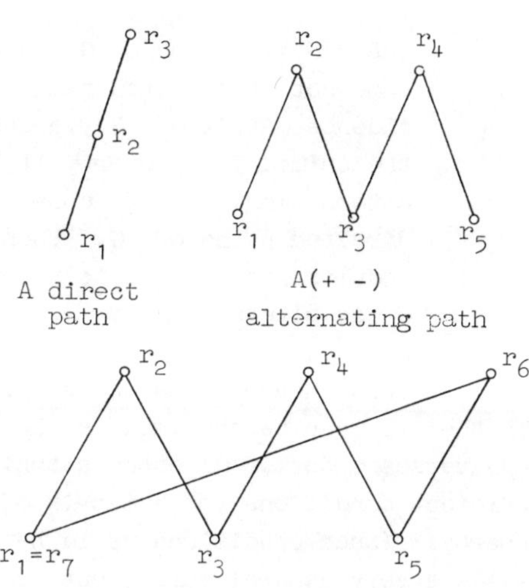

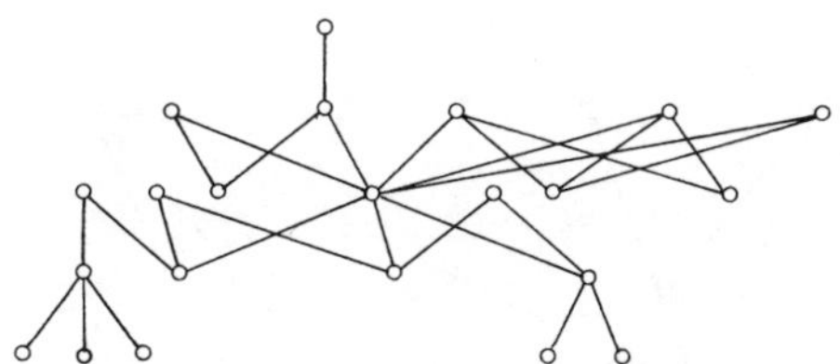

An alternating graph
(arrows omitted - all should be upward)

oppositely oriented and if the first and last edges are oppositely oriented.
An alternating loop must obviously contain an even number of edges.

A graph is <u>alternating</u> if every loop in it is alternating. Let
$V = \{v_1, \ldots, v_m\}$ be the vertices of G, and let $P = \{p_1, \ldots, p_n\}$ be
some set of directed paths in G. Then the incidence matrix $A = \|a_{ij}\|$
of G versus P is defined by

$$a_{ij} = \begin{cases} 1 & \text{if } v_i \text{ is in } p_j, \\ 0 & \text{if } v_i \text{ is not in } p_j. \end{cases}$$

We let A_v represent the row of A corresponding to the vertex v and
A^p represent the column of A corresponding to the path p. We often
write a_{vp} instead of a_{ij} for the entry common to A_v and A^p.

> THEOREM 4. Suppose G is an oriented graph, P is
> some set of directed paths in G, and A is the in-
> cidence matrix of G versus P. Then for A to have
> the unimodular property it is sufficient that G be
> alternating. If P consists of the set of <u>all</u>
> directed paths of G, then for A to have the uni-
> modular property it is necessary and sufficient that
> G be alternating.

This theorem does not state that every matrix of zeros and ones
with the u.p. can be obtained as the incidence matrix of an alternating
graph versus a set of directed paths. Nor does it give necessary and
sufficient conditions for a matrix of zeros and ones to have the unimodular
property. (Such conditions would be very interesting.) However it does
provide a very general sufficient condition. For example, the coefficient
matrix of the i by j transportation problem (or its transpose, depending on which way you write the matrix) is the incidence matrix of the
alternating graph versus the set of all directed paths. Hence this matrix
has the u.p., from which by Theorem 2 follows the well-known i.p. of
transportation problems and their duals. The extent to which alternating graphs
can be more general than the graph shown to the left is a measure of how
general Theorem 4 is.

So that the reader may follow our arguments more easily, we describe
here what alternating graphs look like.

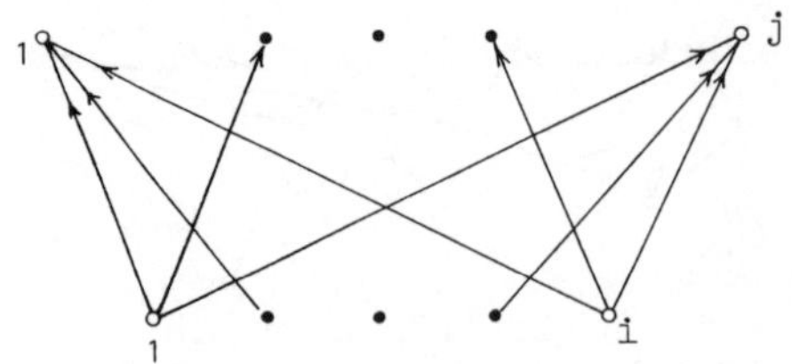

Diagram 2

                    HOFFMAN AND KRUSKAL

(As logically we do not need these facts and as the proofs are tedious, we
omit them.) An integral _height_ function h(v) may be defined in such a
way that (r, s) is a direct edge only when (but not necessarily when)
h(r) + 1 = h(s). If we define r ≦ s to mean that there is a directed path
from r to s, then ≦ is a partial order. Then (r, s) is a direct
edge if and only if both r < s and there is no element t such that
r < t < s.

PROOF OF NECESSITY. We consider here the case in which P is
the set of _all_ directed paths in G, and we prove that for A to have the
u.p. it is necessary that G be alternating. It is easy to verify that
the matrix (shown below) of odd order which has ones down the main diagonal
and sub-diagonal and in the upper right-hand corner, and zeros elsewhere,
has determinant + 2.

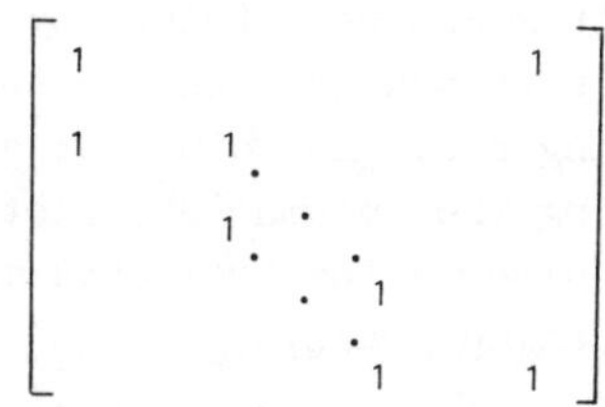

We shall show that if G is not alternating then it contains this matrix,
perhaps with rows and columns permuted, as a submatrix.

Let ℓ be a non-alternating loop in G. If ℓ has an odd number
of distinct vertices, consider the rows in A which correspond to these
vertices, and consider the columns in A which correspond to the one-edge
directed paths which correspond to the edges in ℓ. The submatrix contained in these rows and columns is clearly the matrix shown above, up to row
and column permutations. Hence in this case A does not have the u.p. If
ℓ has an even number of distinct vertices, then find in it three successive
vertices r, s, t such that (r, s) and (s, t) are both direct (or both
inverse) edges. (To find r, s, t it may be necessary to let s be the
initial-terminal vertex of ℓ, in which case r, s, t are successive only
in a cyclic sense.) Consider the rows of A which correspond to all the
vertices of ℓ except s. Consider the columns of A which correspond to
the following directed paths: the two-edge path r, s, t (or t, s, r) and
the one-edge paths using the other edges in ℓ. The submatrix contained in
these rows and columns is the square matrix of odd order shown above, up to
row and column permutations. Hence in this case also A does not have the
u.p. This completes the proof of necessity.

The proof of the sufficiency condition, when P may be <u>any</u> set of directed paths in G,   occupies the rest of this section. As this proof is long and complicated, it has been broken up into lemmas.

If $r_1, \ldots, r_k$ is a loop, then $r_1, \ldots, r_k, r_2, \ldots, r_1$ is called a <u>cyclic permutation</u> of the loop. Clearly a loop is alternating if and only if any cyclic permutation is alternating.

> LEMMA 6. Suppose A is the incidence matrix of an
> alternating graph G versus some set of directed
> paths P in G. For any submatrix A' of A, there
> is an alternating graph G' and a set of directed
> paths P' in G' such that A' is the incidence
> matrix of G' versus P'.

PROOF. Any submatrix can be obtained by a sequence of row and column deletions. Hence it suffices to consider the two cases in which A' is formed from A by deleting a single column or a single row. If A' is formed from A by deleting the column A^p, let G' = G, and P' = P - {p}. Then A' is clearly the incidence matrix of G' versus P', and G' is indeed an alternating graph.

Suppose now that A' is formed from A by deleting row A_t. Define

$$V' = V - \{t\},$$

$$E' = \{(v, w) \mid v, w \text{ in } V' \text{ and either}$$
$$(v, w) \text{ in } E \text{ or } (v, t)$$
$$\text{and } (t, w) \text{ in } E\},$$

$$G' = \text{the graph with vertices } V' \text{ and edges } E',$$

$$P' = \{p - \{t\} \mid p \text{ in } P\}.$$

Clearly A' is the incidence matrix of G' versus P'. We shall prove (a) that P' is a collection of directed paths and (b) that G' is alternating.

The proof of (a) is quite simple. Suppose v, w are successive vertices of p' = p - {t} in P'. It may or may not happen that p contains t. In either case, however, if v, w are successive vertices in p, then (v, w) is a

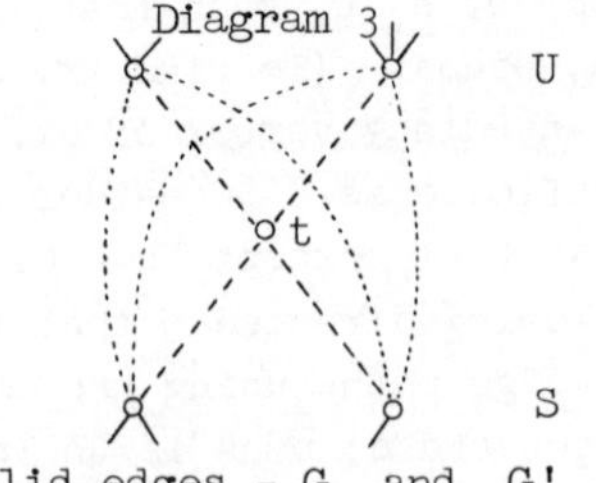

Solid edges - G and G'
Dashed edges - G only
Dotted edges - G' only

direct edge in G, so (v, w) is a direct edge in G'. If v, w are
not successive vertices in p, then necessarily v, t, w are successive
vertices in p. In this case (v, t) and (t, w) are direct edges in G,
so (v, w) is a direct edge in G'.

The proof of (b) is more extended. Define

$$S = \{s \mid (s, t) \text{ in } E\}$$
$$U = \{u \mid (t, u) \text{ in } E\} .$$

Then each "new" edge in E', that is, each edge of E' - E, is of the
form (s, u) with s in S and u in U. Let ℓ be any loop in G'.
If ℓ contains no new edge, then ℓ is also a loop in G and hence
alternating. If ℓ contains a new edge, it contains at least two vertices of S $\cup$ U. Hence the vertices of S $\cup$ U break ℓ up into pieces
which are paths of the form

$$p = v, r_1, \ldots, r_k, v'$$

where v and v' are in S $\cup$ U and the r's are not.

CASE (U, U): both v and v' belong to U. In this case

$$t, v, r_1, \ldots, r_k, v', t$$

is a loop in G, hence alternating. Therefore p is an alternating path.
As (t, v) is a direct edge and (v', t) is an inverse edge in G, p
must be a (- +) alternating path in G'.

CASE (S, S): both v and v' belong to S. In this case dual
argument to the above proves that p must be a (+ -) alternating path
in G'.

CASE (U, S): v belongs to U and v' belongs to S. In this
case p must be exactly the one-edge path v, v'. For if not, p consists solely of edges in E, so the loop which we may represent symbolically
v', t, p is a loop in G. But as (v', t) and (t, v) are both direct
edges in G this loop is not alternating, which is impossible. As (v, v')
is an inverse edge in G', p is a (- -) alternating path in G'.

CASE (S, U): v belongs to S and v' belongs to U. In this
case dual argument to the above proves that p must be exactly the one-
edge path v, v' and hence a (+ +) alternating path in G'.

Using these four cases, we easily see that the pieces of ℓ are
alternating and fit together in such a way that ℓ itself is alternating -
except for one technical difficulty, namely the requirement that the

initial and terminal edges of ℓ must have opposite orientations. However, if we form a cyclic permutation of ℓ and apply the reasoning above to this new loop, we obtain the necessary information to complete our proof that ℓ is alternating. This completes the proof of (b) and Lemma 6.

In view of Lemma 6, the sufficiency condition of Theorem 4 will be proved if we prove that every square incidence matrix of an alternating graph versus a set of directed paths has determinant, 0, + 1, or - 1. We prove this by a kind of induction on two new variables, $c(G)$ and $d(G)$, which we shall now define:

$c(G)$ = the number of unordered pairs $\{st\}$ of distinct vertices of G which satisfy

(5.1)
>there is a vertex u such that (s, u) and (t, u) are direct edges of G;

$d(G)$ = the number of unordered pairs $\{st\}$ of distinct vertices of G which satisfy

(5.2)
>there is no directed path from s to t nor any directed path from t to s.

Though not logically necessary the following information may help orient the reader to the significance of these two variables. Assume G is alternating. Then using the partial-order $\leqq$ introduced informally earlier, $d(G)$ is the number of pairs of vertices which are incomparable under $\leqq$. Any pair $\{st\}$ which satisfies (5.1) also satisfies (5.2), so $c(G) \leqq d(G)$. If $c(G) = 0$, then each vertex of G has at most one "predecessor", and G consists of a set of trees, each springing from a single vertex and oriented outward from that vertex. If $d(G) = 0$, then G is even more special: it consists of a single directed path.

>LEMMA 7. If G is alternating, and $\{st\}$ satisfies (5.1), then it also satisfies (5.2). Hence $c(G) \leqq d(G)$.

PROOF. Let u be a vertex such that (s, u) and (t, u) are direct edges of G. Suppose there is a directed path

$$s, \; r_1, \; \ldots, \; r_k, \; t \; .$$

If none of the r's is u, then

$$s, \; r_1, \; \ldots, \; r_k, \; t, \; u, \; s$$

is a loop, hence alternating. As (t, u) is a direct edge, (r_k, t) is an inverse edge, so the path is not directed, a contradiction. If one of the r's is u, take the piece from u to t. By renaming, we may call this directed path

$$u, \; r_1, \; \ldots, \; r_k, \; t \; .$$

Then

$$t, \; u, \; r_1, \; \ldots, \; r_k, \; t$$

is a loop, hence alternating. As (t, u) is a direct edge, (u, r_1) must be an inverse edge, so the path is not directed, a contradiction. Therefore, there can be no directed path from s to t. By symmetrical argument, there can be no directed path from t to s. Therefore $\{st\}$ satisfies (5.2). It follows trivially that $c(G) \leq d(G)$. This completes the proof of Lemma 7.

The induction proceeds in a slightly unusual manner. The "initial case" consists of all graphs G for which $c(G) = 0$. The inductive step consists of showing that the truth of the assertion for a graph G such that $c(G) > 0$ follows from the truth of the assertion for a graph G' for which $d(G') < d(G)$. It is easy to see that by using the inductive step repeatedly, we may reduce to a graph G^* for which either $c(G^*)$ or $d(G^*)$ is 0. But as $d(G^*) = 0$ implies $c(G^*) = 0$ by the inequality between c and d, we are down to the initial case either way.

We now treat the initial case.

LEMMA 8. Let A be the incidence matrix of an alternating graph G versus some set of directed paths P. Suppose that P contains as many directed paths as G contains vertices, so A is square. Suppose that $c(G) = 0$. Then $|A| = 0, + 1,$ or $- 1$.

PROOF. If (r, s) is a direct edge of G, we call r a predecessor of s and s a successor of r. The fact that $c(G) = 0$ means that each vertex of G has at most one predecessor. If V' is a subset of V, and r is in V' but has no predecessor in V', then r is called an initial vertex of V'.

Every non-empty subset V' of V has at least one initial vertex. For if V' has none, then we can form in V' a sequence $r_1, r_2, \ldots$ of vertices such that for every i, r_{i+1} is a predecessor of r_i. Let r_j be the first term in the sequence which is the same as a vertex picked earlier, and let r_i be the earlier name for this vertex. Then r_i, r_{i+1}, ..., r_j is a loop all of whose edges are inverse. As G is alternating, this is impossible.

Let $U(r) = \{s \mid s$ is a successor of $r\}$. Let r_1 be an initial vertex in V. Recursively, let r_i be an initial vertex of $V - \{r_1, r_2, ..., r_{i-1}\}$. Then define matrices $B(i)$ recursively:

$$B(0) = A,$$
$$B(i) = B(i - 1)$$

with the row $B_{r_i}(i - 1)$ replaced by

$$B_{r_i}(i - 1) - \sum_{s \text{ in } U(r_i)} B_s(i - 1) .$$

Let B be the final $B(i)$. We see immediately that $|A| = |B(1)| = \cdots = |B|$. Thus we only need show that $|B| = 0, +1,$ or -1.

We claim that each column B^p of B consists of zeros with either one or two exceptions: if w is the final vertex of the directed path p, then $b_{wp} = 1$, and if v is the unique predecessor to the initial vertex of p, then $b_{vp} = -1$. As the initial vertex of p may have no predecessor at all, the -1 may not occur.

We shall not prove in detail the assertions of the preceding paragraph. We content ourselves with considering the column corresponding to a fixed path p during the transition from $B(i - 1)$ to $B(i)$. Only one entry is altered, namely $b_{r_i p}(i - 1)$. There are four possible cases.

CASE (i): neither r_i nor any of its successors is in p.

CASE (ii): r_i is not in p but one of its successors is in p.

CASE (iii): both r_i and one of its successors is in p.

CASE (iv): r_i is in p but none of its successors is in p.

At most one successor of a vertex can be in a directed path because G is alternating, so these cases cover every possibility. In case (i), the entry we are considering starts as 0 and ends as 0. In case (ii), it starts as 0 and ends as -1. In case (iii), it starts as 1 and ends as 0. In case (iv), it starts as 1 and ends as 1. From these facts, it is not hard to see that B satisfies our assertions.

From our assertions about B it is trivial to check that B satisfies the hypotheses of Theorem 3. It is only necessary to partition the rows of B into two classes, one class being empty and the other class

containing every row. Then by Theorem 3, B has the u.p. Therefore,
$|B| = 0, + 1,$ or $- 1$. As $|A| = |B|$, this completes the proof of Lemma 8.

We now prove the inductive step.

LEMMA 9. Suppose that A is the square incidence matrix
of an alternating graph G versus a set of directed paths
P. Suppose that $c(G) > 0$. Then there is a square matrix
A' such that $|A'| = |A|$ and such that A' is the
square incidence matrix of an alternating graph G'
versus a set of directed paths P', where $d(G') < d(G)$.

PROOF. As $c(G) > 0$, G contains a vertex u which has at
least two distinct predecessors, s and t. Define

$$A' = A \text{ with row } A_t \text{ replaced by } A_s + A_t.$$

Clearly $|A'| = |A|$. Define

$$V' = V,$$
$$E_s = \{(s, w) \mid (s, w) \text{ in } E\},$$
$$E_t = \{(t, w) \mid (s, w) \text{ in } E\},$$
$$E' = E \cup E_t \cup \{(s, t)\} - E_s,$$
$$G' = \text{the graph with vertices } V' \text{ and edges } E',$$
$$p' = \begin{cases} p \text{ if } p \text{ does not contain } s, \\ p \text{ with } t \text{ inserted after } s \text{ if } p \text{ does} \\ \text{contain } s, \end{cases}$$
$$P' = \{p' \mid p \text{ in } P\}.$$

We shall prove (a) that G' is alternating, (b) that P' is a set of
directed paths of G', (c) that $d(G') < d(G)$, and (d) that A' is the
incidence matrix of G' versus P'.

Diagram 4

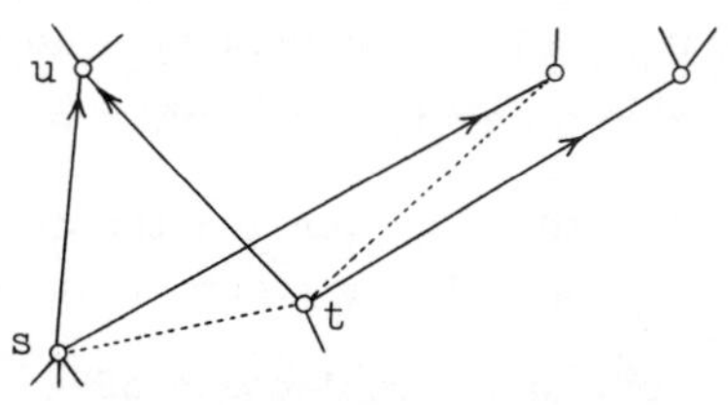
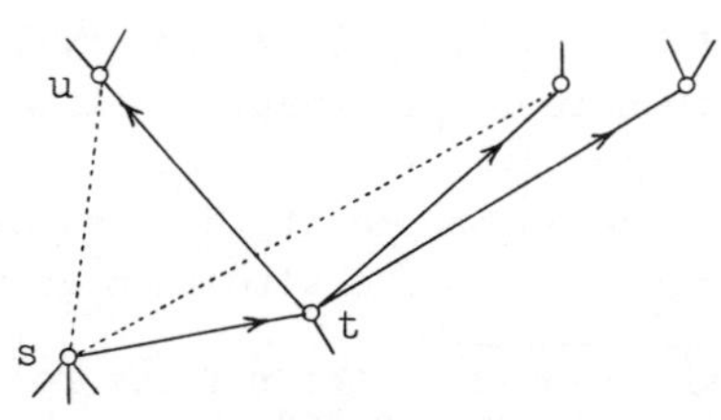

Graph G   Graph G'

The proof of (b) is simple. If p does not contain s, then every edge in p' = p is in E', so p' is a directed path in G'. If p does contain s, write p thus:

$$r_1, \ldots, r_i, s, r_{i+1}, \ldots, r_j .$$

Then p' is

$$r_1, \ldots, r_i, s, t, r_{i+1}, \ldots, r_j .$$

Each edge of p except (s, r_{i+1}) is also in E'. Hence to show that p' is a directed path in G', we only need show that (s, t) and (t, r_{i+1}) are in E'. The former is in E' by definition, and the latter is in E_t because (s, r_{i+1}) must be in E. This proves (b).

To prove (c), let r_1 and r_2 be any pair of vertices such that there is a directed path p from one to the other in G. Then p' is a directed path from one to the other in G'. Hence every pair of vertices which satisfies (5.2) in G' also satisfies (5.2) in G. Furthermore, {st} does not satisfy (5.2) in G' because (s, t) is in E', while {st} does satisfy (5.2) in G by Lemma 7. This proves that $d(G') < d(G)$.

To prove (d), we first show that A' consists entirely of zeros and ones. The only way in which this could fail to happen is if A_s and A_t both contained ones in the same column. But if this were the case, then the directed path corresponding to this column would contain both s and t, which cannot happen by Lemma 7. To see that A' is the desired incidence matrix, consider how P' differs from P. Each directed path which did not contain s remains unchanged; each directed path which did contain s has t inserted in it. Thus the change from A to A' should be the following. Each column which has a zero in row A_s should remain unchanged; each column which has a one in row A_s should have the zero in row A_t changed to a one. But adding row A_s to A_t accomplishes exactly this. Therefore (d) is true.

The proof of (a) is more complicated.[6] Define S' to be the set of successors of s in G which are not also successors of t. Note that every edge in G' which is not in G terminates either in t, or in a vertex of S'. Let ℓ be any loop of G'. If ℓ is already a loop of G, then it is alternating. If not, it must contain either the edge (s, t) or an edge (t, s') with s' in S'. (Of course, ℓ might contain the inverse of one of these edges instead. If so, reversing the order of ℓ brings us to the situation above.) Ignoring trivial loops, that is,

---

[6] We are indebted to the referee for this proof, which replaces a considerably more complicated one.

loops of the form aba (which are alternating trivially), ℓ must have
the form

$$\cdot \qquad str_1 \, \cdots \, r_k s$$

or

$$ts'r_1 \, \cdots \, r_k t, \quad \text{with}\quad s' \quad \text{in}\quad S'.$$

The first form is impossible. To prove this, first suppose that no r_i
is in S' U {u}. Then $sutr_1 \, \cdots \, r_k s$ is a loop of G, hence alternating.
Thus (r_k, s) is an inverse edge and belongs to both G and G', which
is impossible. Now suppose that some r_i is in S' U {u}, and let r_j
be the last such r_i. Then $sr_j \, \cdots \, r_k s$ is a loop of G, hence alternating. Hence (r_k, s) is inverse, which is impossible as before.

   We may now assume that ℓ is $ts'r_1 \, \cdots \, r_k t$. No r_i can be s.
Clearly r_1 cannot be s, and if $r_j = s$, j > 1, then $ss'r_1 \, \cdots \, r_{j-1} s$
is a loop of G, hence alternating, so (r_{j-1}, s) is inverse and belongs
to both G and G', which is impossible. Thus $r_1, \ldots, r_k$ are distinct
from s. Suppose that r_k is in S'. Then $ss'r_1 \, \cdots \, r_k s$ is a loop of
G, hence alternating. Consequently, so is ℓ. Suppose that r_k is not
in S' and that no r_i is u. Then $ss'r_1 \, \cdots \, r_k tus$ is a loop of G,
hence alternating. Thus $s'r_1 \, \cdots \, r_k t$ is a (- -) alternating path in G
and also in G'. Hence ℓ is alternating. Finally, suppose that r_k is
not in S' and that r_j is u. Then $ss'r_1 \, \cdots \, r_{j-1} us$ and $tur_{j+1} \, \cdots \, r_k t$
are loops of G, hence alternating. Thus $s'r_1 \, \cdots \, r_{j-1} u$ is a (- +)
alternating path, and $ur_{j+1} \, \cdots \, r_k t$ is a (- -) alternating path. Fitting
these paths together and adjoining t at the beginning, we see that ℓ
is alternating. This completes the proof of (a), of Lemma 9, and of the
sufficiency condition of Theorem 4.

6. HOW TO RECOGNIZE THE UNIMODULAR PROPERTY

   To apply Theorem 3 is easy, although even there one point is
important. To say that A has the unimodular property is the same thing
as to say that A^T, the transpose of A, has the unimodular property.
However the hypotheses of Theorem 3 or 4 may quite easily be satisfied for
A^T but not for A. Consequently it is desirable to examine both A and
A^T when using these theorems.

   To apply Theorem 4 is not so easy: how shall we recognize whether
matrix A (or matrix A^T) is the incidence matrix of an alternating graph
versus some set of directed paths? We point out that in actual applications
the graph G generally lies close at hand. For example, it was pointed
out in Section 5 that the coefficient matrix A of the i by j transportation problem is the incidence matrix of the alternating graph shown

in Diagram 2 (at the beginning of Section 5) versus all its directed paths.
This graph is no strange object - it portrays the i "producing points",
the j "consuming points", and the transportation routes between them.

In a given linear programming problem there will often be one
(or several) graphs which are naturally associated with the coefficient
matrix. Whenever the problem can be stated in terms of producers, con-
sumers, and intermediate handlers, this is the case. It may well be possible
in this situation to identify the matrix as a suitable incidence matrix.

However it is still useful to have criteria available which can
be applied directly to the matrix A and which guarantee that A can be
obtained as a suitable incidence matrix. The two following theorems give
such conditions. Each corresponds to a very special case of Theorem 4.
Theorem 5, historically, derives from the integrality of transportation-
type problems, and finds application in [2]; Theorem 6 from the integrality
of certain caterer-type problems (see [1]).

We shall write $A_i \geqq A_j$ to indicate that row A_i is component-
wise $\geqq$ row A_j.

> THEOREM 5. Suppose A is a matrix of 0's and 1's,
> and suppose that the rows of A can be partitioned
> into two disjoint classes V_1 and V_2 with this
> property: if A_i and A_j are both in V_1 or both in
> V_2, and if there is a column A^k in which both A_i
> and A_j have a 1, then either $A_i \leqq A_j$ or $A_i \geqq A_j$.
> Then A has the unimodular property.

This theorem corresponds to a generalized transportation situation, in which each upper vertex of the transportation graph has attached
an outward flowing tree and each lower vertex has attached an inward flowing tree. Only directed paths which have at least one vertex in the original transportation graph can be represented as columns of the matrix A.

PROOF. Briefly the proof is this: Let vertices v_i in V
correspond to the rows A_i of A. Define a partial-order $\leqq$ on the
vertices:

$$v_i \leqq v_j \quad \text{if } A_i \text{ in } V_1 \text{ and } A_j \text{ in } V_2$$
$$\text{or } A_i, A_j \text{ in } V_1 \text{ and } A_i \leqq A_j$$
$$\text{or } A_i, A_j \text{ in } V_2 \text{ and } A_i \geqq A_j.$$

Let G be the graph naturally associated with this partially-ordered set.
We leave to the reader verification of the fact that G is alternating,
and that the columns of A represent directed paths in P.

Say that two column vectors of the same size consisting of 0's and 1's are <u>in accord</u> if the portions of them between (in the inclusive sense) their lowest common 1 and the lower of their highest separate 1's are identical.

> THEOREM 6. Suppose A is a matrix of 0's and 1's,
> and suppose that.the rows of A can be rearranged in
> such a way that every pair of columns is in accord.
> Then A has the unimodular property.

This theorem corresponds to a situation in which $c(G) = 0$, that is, every vertex has at most one predecessor (or to the dual situation in which every vertex has at most one successor). The columns of A may represent any directed paths in the graph.

PROOF. Let vertices v_i in V correspond to the rows A_i in A. Assume that the rows are already arranged as described above. Define E as follows:

> (v_i, v_j) is in E if $i > j$ and if there is a
> column A^k of A such that a_{ik} and
> a_{jk} are both 1 while all intervening
> entries are 0's.

Let G be the graph with vertices V and edges E. We leave to the reader verification of the fact that G is an alternating graph in which every vertex has at most one successor, and that the columns of A represent directed paths in G.

BIBLIOGRAPHY

[1] GADDUM, J. W., HOFFMAN, A. J., and SOKOLOWSKY, D., "On the solution
     of the caterer problem," Naval Research Logistics Quarterly, Vol. 1,
     1954, pp. 223-227. See also JACOBS, W., "The caterer problem," <u>ibid</u>.,
     pp. 154-165.

[2] HOFFMAN, A. J., and KUHN, H. W., "On systems of distinct representa-
     tives," this Study.

[3] JACOBSON, NATHAN, Lectures in Abstract Algebra, Vol. II, (1953),
     D. Van Nostrand Co., pp. 88-92.

National Bureau of Standards   A. J. Hoffman
Princeton University   J. B. Kruskal

Reprinted from
Proc. Symp. Appl. Math., Vol. X (1960), pp. 113–127

SOME RECENT APPLICATIONS OF THE THEORY OF LINEAR INEQUALITIES TO EXTREMAL COMBINATORIAL ANALYSIS

BY

ALAN J. HOFFMAN

1. **Introduction.** The purpose of this talk is to give an account of some aspects of recent research on the interplay between the theory of linear inequalities and a certain class of combinatorial problems. The kind of problem to be considered can be illustrated by the following example:

Let $A = (a_{ij})$ be a square incidence matrix of order v such that every row contains $k > 0$ ones and every column contains $k > 0$ ones. All other entries of A are 0. Mann and Ryser [1] have observed that A can then be expressed in the form

$$(1.1) \qquad A = P_1 + \cdots + P_k,$$

where the P_i are permutation matrices. Obviously, an inductive argument will suffice to prove (1.1) if it can be shown that the hypotheses imply the existence of a permutation matrix $P = (p_{ij})$ such that

$$(1.2) \qquad p_{ij} = 1 \quad \text{only if} \quad a_{ij} = 1.$$

Mann and Ryser establish the existence of such a P by exploiting the Egervary-Hall-König (see [2; 3; 4] and the discussions below) theorem on systems of distinct representatives. An alternative approach to (1.2) is to consider the convex set of all vectors with kv components $X = (\cdots, x_{ij}, \cdots)$ (where the subscript "ij" appears if and only if $a_{ij} = 1$), satisfying

$$(1.3) \qquad \sum_i x_{ij} = \sum_j x_{ij} = 1, \quad x_{ij} \geqq 0.$$

This convex set is not empty, since the hypotheses on A imply that setting $x_{ij} = 1/k$ satisfies (1.3). The set is also bounded, so it admits a vertex. The co-ordinates of the vertex may be obtained by solving a certain set of equations contained in the system of equations and inequalities (1.3), and it is easy to show ([5; 6] and many other places) that the determinant associated with this set of equations is ± 1. Since the right hand side is integral, it follows from Cramer's rule that the co-ordinates of the vertex are integers. Conditions (1.3) imply that for each i, $x_{ij} = 0$ for all j with exactly one exception, for which $x_{ij} = 1$. Similarly, every column consists entirely of 0 entries except for exactly one entry which is 1. But this means that our vertex of (1.3) is the desired permutation P satisfying (1.2). (A slight generalization of the result is contained in [8].)

Observe that we have replaced a combinatorial argument—to wit, the

                           A. J. HOFFMAN

appeal to Egervary-Hall-König—by a quasi-geometric discussion involving
polyhedra and vertices, to prove the combinatorial result (1.1). This
suggests that invocation of concepts from the theory of linear inequalities
may be useful in studying certain kinds of combinatorial situations. As a
matter of fact, about a dozen mathematicians have chewed on this bone
during the past five or six years, and this talk will summarize their findings.

2. **Systems of representatives.** The first result in this direction of using
linear inequalities on combinatorial problems seems to be due to Rado [9],
but his paper was regrettably overlooked by later workers until 1956. The
more recent work began with the observation that the Hall theorem on
systems of distinct representatives has itself an easy proof through the
theory of linear inequalities. That theorem is:

Let R be a finite set with elements $P_1, \cdots, P_n$,

$$\mathscr{S} = \{S_1, \cdots, S_m\} \text{ a family of subsets of } R.$$

(2.1)

$$\text{A subset } S = \{P_{i_1}, \cdots, P_{i_m}\} \subset R$$

of m distinct elements is called a system of representatives of $\mathscr{S}$ if

(2.2)
$$P_{i_k} \in S_k, \quad k = 1, \cdots, m.$$

In order that $\mathscr{S}$ admit a system of distinct representatives, it is necessary
and sufficient that, for any $I \subset \{1, \cdots, m\}$,

(2.3)
$$\bar{I} \leq \overline{\bigcup_{i \in I} S_i}.$$

(Here and elsewhere $\bar{M} =$ the number of elements in the set M.)

Proof by linear inequalities: Let C be the $n \times m$ matrix given by:

$$c_{ij} = \begin{cases} 1 \text{ if } P_i \in S_j, \\ 0 \text{ if } P_i \notin S_j \end{cases}.$$

Consider the linear programming problem:

minimize $\sum_{ij} c_{ij} x_{ij}$, where (x_{ij}) varies over all n by m

matrices satisfying: $x_{ij} \geq 0$, $\sum_i x_{ij} \leq 1$, $\sum_j x_{ij} \leq 1$.

It is easy to prove, using the unimodular property (see [5]) exploited in §1,
that the maximum is m if and only if R contains a set S satisfying (2.2).
But the maximum of the primal linear program equals the minimum in the
dual program:

minimize $\sum_i u_i + \sum_j v_j$ $\qquad\qquad u_i \geq 0, \quad v_j \geq 0, \quad u_i + v_j \geq c_{ij}.$

Again by the unimodular property, it is sufficient to consider only the case where each u_i and each v_j is 0 or 1. So (2.2) holds for some S if and only if the smallest number of rows and columns collectively comprising all 1's in the matrix C is m, and it is easy to show that this condition is a consequence of (2.3). The necessity of (2.3) is, of course, trivial.

This method of proof, noted independently by several people (Motzkin [10], Kuhn [11], Hoffman [12], and probably others less vocal), while not as brief as the elegant induction of Halmos and Vaughan [13], has the redeeming feature that it fits the theorem into a larger context that enables us to know it better. We can now recognize it as a special case of the duality theorem of linear programming. Further, this recognition permits a facility in generalization.

One such direction was inspired by a result of Mann and Ryser (see [1; 14], also Hoffman and Kuhn [15; 16]). M. Tinsley and R. Rado have privately communicated alternative proofs of the main results of [1] and [15].

Let R and $\mathscr{S}$ be as in (2.1). Let $\mathscr{T} = \{T_1, \cdots, T_p\}$ be a partition of R; i.e., $\bigcup T_i = R$, $T_i \cap T_j = \emptyset$ if $i \neq j$:

Let $c_k \leq d_k$ $(k = 1, \cdots, m)$ be given integers. In order that there exist a set $S \subset R$ satisfying (2.2) and

$$(2.4) \qquad c_k \leq \overline{S \cap T_k} \leq d_k, \qquad\qquad k = 1, \cdots, m$$

it is necessary and sufficient that, for all $A \subset \{1, \cdots, m\}$ and $B \subset \{1, \cdots, p\}$.

$$(2.5) \qquad \overline{\left(\bigcup_{i \in A} S_i\right) \cap \left(\bigcup_{k \in B} T_k\right)} \geq \overline{A} - m + \sum_{k \in B} c_k$$

and

$$(2.6) \qquad \overline{\left(\bigcup_{i \in A} S_i\right) \cap \left(\bigcup_{k \notin B} T_k\right)} \geq \overline{A} - \sum_{k \in B} d_k.$$

As before, the necessity of these conditions is trivial. The proof of their sufficiency is given in [16].

Another generalization is given by Ford and Fulkerson [17]:

Let R and $\mathscr{S}$ be as in (2.1). Let $a_i \leq b_i$ $(i = 1, \cdots, n)$ be integers associated with the elements of R. A subset $S \subset R$ of not necessarily distinct elements satisfying (2.2) in which the number of times P_i is used is at least a_i and at most b_i is called a system of restricted representatives. Such a set S exists if and only if, for any $X \subset \{1, \cdots, n\}$, we have

$$(2.7) \qquad \overline{X} \leq \min\left(m - \sum_{\substack{P_i \notin \bigcup_{j \in X} S_j}} a_i, \ \sum_{\substack{P_i \in \bigcup_{j \in X} S_j}} b_i\right).$$

The proof given in [17] depends on a result on network flow called the min-cut max-flow theorem [18; 19], about which we will say more in the next section. But it is equally possible to prove it via the theory of linear

116   A. J. HOFFMAN

inequalities, or as a direct corollary of the theorem just quoted. Let $b = \max b_i$, and construct the set consisting of b copies of R, whose points are each P_i repeated b times. Then, if we let the kth summand of the partition $\mathscr{T} = \{T_1, \cdots, T_n\}$ be the b copies of P_k, we have a situation to which the previous hypotheses apply. Then (2.5) and (2.6) together are equivalent to (2.7).

Ford and Fulkerson have also considered the question of finding a subset S which is not only a system of restricted representatives for $\mathscr{S} = \{S_1, \cdots, S_m\}$, but also for another family $\mathscr{T} = \{T_1, \cdots, T_m\}$ (note $\mathscr{T}$ is generally *not* a partition). They have shown that such a system of restricted representatives exists if and only if, for every $X, Y \subset \{1, \cdots, m\}$, we have

$$\overline{X} + \overline{Y} \leqq m - \sum_{\substack{P_i \notin \bigcup_{j \in X} S_j;\, P_i \notin \bigcup_{j \in Y} T_j}} a_i + \sum_{\substack{P_i \in \bigcup_{j \in X} S_j;\, P_i \in \bigcup_{j \in Y} T_j}} b_i.$$

See [17] for the proof, which is based on consideration of flows in networks.

Another result [20] on two families of sets deals with the problem of choosing a subset S of the given set R such that the intersection of S with each set of each family has a number of elements lying within prescribed bounds. Note that the previous theorems deal with the *assignment* of elements to sets which contain them, ignoring the fact that an element assigned to one set may be contained in others. That consideration is not ignored in the present case, so it is not astonishing that fairly stringent conditions are imposed on the two families.

Let R be a set, $\mathscr{S} = \{S_1, \cdots, S_m\}$ and $\mathscr{T} = \{T_1, \cdots, T_n\}$ two families of subsets of R such that $S_i \cap S_j \neq \emptyset$ implies $S_i \subset S_j$ or $S_j \subset S_i$; $T_i \cap T_j \neq \emptyset$ implies $T_i \subset T_j$ or $T_j \subset T_i$. Let $a_i \leqq b_i$ $(i = 1, \cdots, m)$ and $c_j \leqq d_j$ $(j = 1, \cdots, n)$ be prescribed integers. In order that there exist a set $S \subset \boldsymbol{R}$ such that

$$a_i \leqq \overline{S \cap S_i} \leqq b_i \qquad\qquad i = 1, \cdots, m,$$

and

$$c_j \leqq \overline{S \cap T_j} \leqq d_j \qquad\qquad j = 1, \cdots, n,$$

it is necessary and sufficient that, for all $I \subset \{1, \cdots, m\}$ and $J \subset \{1, \cdots, n\}$, we have

$$\sum_{i \in I_o} a_i + \sum_{j \in J_e} c_j \leqq \sum_{i \in I_e} b_i + \sum_{j \in J_o} d_j + \overline{S^o \cap T^e},$$

and

$$\sum_{j \in J_o} c_j + \sum_{i \in I_e} a_i \leqq \sum_{j \in J_e} d_j + \sum_{i \in I_o} b_i + \overline{S^e \cap T^o}.$$

Now to explain the notation. By virtue of the conditions on S, it is possible to count the number of sets in $\{S_i\}_{i \in I}$ containing a given S_i $(i \in I)$. We count the set S_i itself, so that, for example, the number associated with a maximal S_i—maximal in the set $\{S_i\}_{i \in I}$—is 1. If the number associated with S_i is odd, we assign i to I_o; if even, we assign i to I_e. This explains the symbols I_o and I_e, and a similar discussion serves to define J_o and J_e.

Further, S^o is the set of all elements of R contained in an odd number of sets $\{S_i\}_{i \in I}$, S^e is the set of all elements of R contained in an even number of sets $\{S_i\}_{i \in I}$. T^o and T^e are defined similarly.

The original proof was a fairly elaborate deduction based on the fact that the incidence matrix of elements of R versus sets in both families had the unimodular property, and exploitation of the well-known result in the theory of linear inequalities that

$$(2.8) \qquad \begin{array}{l} Ax \leqq b \text{ is consistent if and only if} \\ y \geqq 0, \quad yA = 0 \text{ implies } (y,b) \geqq 0. \end{array}$$

But a simpler proof based on flow considerations was subsequently discovered.

3. **Flows in networks.** The discussion will concentrate on directed or oriented graphs, although most of what is said applies equally well to unoriented graphs.

The prototype of theorems of this type is the well-known result of Menger (see [21; 22]): If A and B are disjoint subsets of the nodes of a graph, the largest number of nodewise disjoint paths from A to B is the smallest number of nodes in any set of nodes intersecting each path. An analogous result is that the largest number of arcwise disjoint paths from A to B is the smallest number of arcs in any set of arcs intersecting each path. One can combine and generalize these two statements as follows:

Let G be a directed graph with capacities c_{ij} on arcs from node i to node j, and capacities c_{ii} on the nodes i. Let A and B be distinct nodes, designated source and sink respectively. A flow is an assignment of numbers x_{ij} to the arcs satisfying

$$0 \leqq x_{ij} \leqq c_{ij}, \qquad\qquad i \neq j,$$

$$\sum_j x_{ij} = \sum_j x_{ji} \leqq c_{ii}, \qquad\qquad i \neq A, B.$$

Then the maximal flow from A to B; i.e., $\max \sum_j x_{Aj} (= \max \sum_j x_{jB})$ equals the "minimal cut," the smallest sum of capacities of any collection of nodes and arcs which meets every path.

This result is due to Ford and Fulkerson [18]. A proof via the duality theorem of linear programming has been given by Dantzig and Fulkerson [19]. Note that the fact that "max flow $\leqq$ min cut" is trivial. The effort is to prove the equality. This is the analog of the situation in the previous section on systems of representatives where the necessity was trivial, and the only effort was required to prove the sufficiency.

Another result on flows, which does not specialize any of the nodes, is the "circulation theorem" [23]:

Let $a_{ij} \leqq b_{ij}$ $(i \neq j)$ be numbers associated with the arcs from i to j,

                            A. J. HOFFMAN

$c_i \leq d_i$ be numbers associated with the ith node. A flow in the network is an assignment x_{ij} $(i \neq j)$ satisfying

$$a_{ij} \leq x_{ij} \leq b_{ij}, \qquad\qquad i \neq j,$$
$$c_i \leq \sum_j x_{ji} - \sum_j x_{ij} \leq d_i, \qquad\qquad \text{all } i.$$

Such a flow exists if and only if, for any subset S of the nodes (with S' its complement), we have

$$\sum_{i \in S; j \in S'} b_{ji} \geq \sum_{i \in S} c_i + \sum_{i \in S; j \in S'} a_{ij}$$

and

$$\sum_{i \in S; j \in S'} a_{ji} \leq \sum_{i \in S} d_i + \sum_{i \in S; j \in S'} b_{ij}.$$

In the event that $c_i = d_i = 0$ for all i (i.e., what enters a node must leave it), the two conditions collapse into the single condition: for any subset S of the nodes,

$$(3.1) \qquad \sum_{i \in S; j \in S'} a_{ji} \leq \sum_{i \in S; j \in S'} b_{ij}.$$

This circulation theorem was originally proven through the theory of linear inequalities. It is closely related to Gale's "feasibility theorem" [24] for flows in undirected graphs. Gale has also shown [25] the equivalence of the circulation theorem and the min-cut max-flow theorem, and has further explored the relation (in theorems of this type) between the case of directed graphs and undirected graphs. Other remarks on this point are made in [19]. There has also been additional study of "dynamic flows" by Ford and Fulkerson [26], and by Gale [27].

We have thus seen that Hall's theorem can be generalized in various ways to results on systems of representatives and results on flows in networks. It is not at all difficult to show that any of the results already cited includes Hall's theorem as a special case. The tools for the generalization were the theory of linear inequalities and flow theorems—and although the latter can be discussed without invoking convex polyhedra and associated concepts, it is nevertheless quite natural to use them. Secondly [17], the demonstration that the flow theorems include the results on systems of representatives explicitly invokes the unimodular property in the manner of the discussion in the Introduction.

So it is in order for us to attribute combinatorial power to methods in the theory of linear inequalities, at least tentatively. But are these results actually stronger than Hall, or is it possible to deduce them as special cases of Hall's theorem? In short, while this trip has been fun, has it been entirely necessary?

At the present time, the answer appears to be yes and no. The next section will outline the way in which Hall's theorem may be twisted to yield all the results so far described. §5 will, on the other hand, present a result that seems at this time to be genuinely stronger than Hall's theorem.

4. **Deduction of previous results from Hall's theorem.** As has been remarked earlier, it is sufficient to show that the circulation theorem is a consequence of Hall. Actually, it is sufficient merely to deduce the special case (3.1) of the circulation theorem. Ford and Fulkerson [**17**] have noted that there is an alternative path—from Hall to the min-cut max-flow theorem—and it is likely that their method is closely related to the one outlined below.

Although it would be possible to proceed directly from Hall to the circulation theorem, the notation would be very cumbersome, so we shall accomplish our aim in three steps:

(a) Let $(a_1, \cdots, a_m)$, $(b_1, \cdots, b_n)$ be non–negative integers such that $\sum a_i = \sum b_j = S$. Let K be a subset of the set of all ordered pairs (i,j) $i = 1, \cdots, m, j = 1, \cdots, n$.

Then there exist integral x_{ij}, $i = 1, \cdots, m$, $j = 1, \cdots, n$ satisfying

$$x_{ij} \geqq 0,\ (i,j) \in K \text{ implies } x_{ij} = 0,$$

$$(4.2) \qquad \sum_j x_{ij} = a_i \qquad\qquad i = 1, \cdots, m,$$

$$\sum_j x_{ij} = b_j \qquad\qquad j = 1, \cdots, n,$$

if and only if, for every $I \subset \{1, \cdots, m\}$ $J \subset \{1, \cdots, n\}$ such that $I \times J \subset K$, we have

$$(4.3) \qquad 0 \geqq \sum_{i \in I} a_i + \sum_{j \in J} b_j - S.$$

Proof. The necessity being trivial, we shall only discuss the sufficiency. Let R be a set with S elements $\{P_1, \cdots, P_s\}$ $\mathscr{T} = \{T_1, \cdots, T_s\}$ a family of S subsets of R, defined as follows: For each $j = 1, \cdots, n$, we have b_j identical sets in $\mathscr{T}$. If $(1,j) \notin K$, then $P_1, \cdots, P_{a_1}$ are in each of these sets.

If $(1,j) \in K$, then $P_1, \cdots, P_{a_1}$ are not in any of these sets.

If $(2,j) \notin K$, then $P_{a_1+1}, \cdots, P_{a_2}$ are in each of these sets.

If $(2,j) \in K$, then $P_{a_1+1}, \cdots, P_{a_2}$ are not in any of these sets.

Continue in this fashion. Thus the sets in $\mathscr{T}$ are in classes corresponding to the b_j, and the elements of R are in classes corresponding to the a_i.

Now we assert that (4.3) implies the existence of a system of distinct representatives for $\mathscr{S}$. To prove this, we verify (2.3). Let $L \subset \{1, \cdots, S\}$. If we allow L to include all indices of sets belonging to any of the n classes of the sets in $\mathscr{S}$ of which it contains already at least one index, then the left side of (2.3) is increased, although the right side stays the same, so (2.3) is possibly harder to satisfy. Accordingly, we may assume that L is composed of all T_i's arising from some subset $J \subset \{1, \cdots, n\}$. Then

$$\overline{\bigcup_{l \in L} S_l} = \sum_{\substack{(i,j) \notin K \\ \text{for some } j \in J}} a_i.$$

120   A. J. HOFFMAN

Hence, verifying (2.3) is equivalent to verifying that

$$\bar{L} = \sum_{j \in J} b_j \leq \sum_{\substack{(i,j) \notin K \\ \text{for some } j \in J}} a_i.$$

Alternatively, we must show that if $I \subset \{1, \cdots, m\}$ is such that $I \times J \subset K$, then

$$\sum_{j \in J} b_j \leq S - \sum_{i \in I} a_i,$$

which is (4.3).

If we let x_{ij} be the number of elements of R in the system of distinct representatives which belong to the ith class of elements and represent the jth class of sets, then conditions (4.2) are obviously satisfied.

(b) Let $(a_1, \cdots, a_m)$, $(b_1, \cdots, b_n)$ be given non-negative integers such that $\sum a_i = \sum b_j = S$. Let c_{ij} be non-negative integers, $i = 1, \cdots, m, j = 1, \cdots, n$. Then in order that there exist integers x_{ij} satisfying

$$
\begin{aligned}
0 &\leq x_{ij} \leq c_{ij}, \\
\sum_j x_{ij} &= a_i && i = 1, \cdots, m, \\
\sum_j x_{ij} &= b_j && j = 1, \cdots, n,
\end{aligned}
$$

(4.4)

it is necessary and sufficient that, for all $I \subset \{1, \cdots, m\}$, $J \subset \{1, \cdots, n\}$, we have

(4.5)
$$\sum_{i \in I; j \in J} c_{ij} \geq \sum_{i \in I} a_i + \sum_{j \in J} b_j - S.$$

Proof. The necessity being easy, we treat only the sufficiency. This will be done by exploiting a device independently discovered by Kantorovich [28] and Dantzig [29]. An alternative device, which would have served equally well, has recently been discovered by Wagner [30].

Consider the vector $\alpha = (a_1, \cdots, a_m; c_{11}, \cdots, c_{mn})$ of $mn + m$ non-negative components, and the vector $\beta = (b_1, \cdots, b_n; c_{11}, \cdots, c_{mn})$ of $mn + n$ non-negative components. Observe that the sum of the co-ordinates of α is the same as the sum of the co-ordinates of β, namely $S + \sum_{i,j} c_{ij}$. We shall apply result (a) above, with the co-ordinates of α and β the respective row and column sums.

To specify the set K, first agree on the notation that co-ordinates of α are labeled either i or (i,j) and co-ordinates of β are labeled either j or (i,j). Then K consists of the following combination:

Row	Column
i	j
i	(k,j) if and only if $k \neq i$
(i,j)	k if and only if $k \neq j$
(i,j)	all (k,l) except when $k = i$, $l = j$.

To prove that (4.5) implies (4.4), we shall show that (4.5) implies (4.3) in the present situation, and it is clear that this will prove (4.4), with $x_{ij,j} = x_{i,ij} = x_{ij}$; $x_{ij,ij} = c_{ij} - x_{ij}$.

Let I be a subset of the row indices, J a subset of the column indices. Let $I = M \cup L$, where $M \subset \{1, \cdots, m\}$, $L \subset \{(1,1), \cdots, (m,n)\}$. Similarly, let $J = N \cup P$, where $N \subset \{1, \cdots, n\}$, $P \subset \{(1,1), \cdots, (m,n)\}$. Imagine M and N chosen. What are the possible choices for L and P so that $I \times J \subset K$?

Clearly,

$$L \subset \{(i,j) | j \notin N\},$$
$$P \subset \{(i,j) | i \notin M\},$$
$$L \cap P = \emptyset.$$

If we want to consider the choice of L and P to make (4.3) hardest to satisfy, we should (and may) choose them so that

$$L \cup P = \{(i,j), i \notin M \text{ or } j \notin N\}.$$

Making this choice, (4.3) becomes

$$0 \geqq \sum_{i \in M} a_i + \sum_{j \in N} b_j + \sum_{(i,j) \in L \cup P} c_{ij} - S - \sum_{i,j} c_{ij}$$
$$= \sum_{i \in M} a_i + \sum_{j \in N} b_j - S - \sum_{i \in M; j \in N} c_{ij},$$

which is (4.5).

(c) To prove the special case (3.1) of the circulation theorem, consider the following problem: find integral x_{ij} $(i,j = 1, \cdots, n)$ such that

$$0 \leq x_{ii} \leq \infty,$$
$$a_i \leq x_{ij} \leq b_i,$$
$$\sum_i x_{ij} = \sum_j x_{ij} = M,$$

where M is an arbitrarily large integer.

Clearly this problem has a solution if and only if there is a flow. This problem can be transformed, by the substitution

$$y_{ij} = x_{ij} - a_{ij}, y_{ii} = x_{ii}, \quad \text{into}$$
$$0 \leq y_{ii} \leq \infty,$$
$$0 \leq y_{ij} \leq b_{ij} - a_{ij},$$
$$\sum_j y_{ij} = M - \sum_j a_{ij}, \qquad\qquad i = 1, \cdots, n$$
$$\sum_i y_{ij} = M - \sum_i a_{ij}, \qquad\qquad J = 1, \cdots, n.$$

We can now apply (4.5). Obviously, since the upper bound on y_{ii} is ∞, we need only consider the case $I \cap J = \emptyset$. Secondly, since $S \stackrel{?}{=} nM + \text{other}$

122   A. J. HOFFMAN

terms, (4.5) will be trivially satisfied unless $\bar{I} + \bar{J} = n$. In short, we need only consider the case where $I \subset \{1, \cdots, n\}$ and J is the complement I' of I. In this case, (4.5) becomes

$$\sum_{i \in I; j \in I'} (b_{ij} - a_{ij}) \geq - \sum_{i \in I; \text{all} j} a_{ij} - \sum_{j \in I'; \text{all} i} a_{ij} + \sum_{i,j} a_{ij}.$$

With a little manipulation, this reduces to

$$\sum_{i \in I; j \in I'} b_{ij} \geq \sum_{i \in I; j \in I'} a_{ji}$$

which is (3.1).

5. **The "most general" theorem of the Hall type.** In the previous sections we have examined a number of combinatorial results on systems of representatives and network flow, and seen that they are all closely related—indeed, that the simplest implies all its more complicated generalizations.

This is a slight disappointment for anyone promulgating the thesis that linear inequalities are useful in discovering and proving theorems of this general character, though not a fatal disappointment, since it is a historical fact that some of these results were first reached that way. The case for linear inequalities can be made even stronger, however, by considering the following problem in systems of representatives, which appears to be quite general. Indeed, all problems considered thus far in this talk are subsumed in it.

Contemplation of the problem leads to a result [31] which appears in some sense to deserve the label of the "most general theorem" of this class.

Let R be a set with elements $\{P_1, \cdots, P_n\}$ and let $\mathscr{S} = \{S_1, \cdots, S_m\}$ be a family of subsets R. Let $a_i \leq b_i$ $(i = 1, \cdots, m)$ be given integers and $0 \leq w_j$ $(j = 1, \cdots, n)$ be given integers. Do there exist numbers x_j $(j = 1, \cdots, n)$ satisfying

$$0 \leq x_j \leq w_j \qquad (j = 1, \cdots, n),$$

(5.1) and

$$a_i \leq \sum_{P_j \in S_i} x_j \leq b_i \qquad (i = 1, \cdots, m)?$$

Define, for any $A \subset \{1, \cdots, m\}$,

$$n_j(A) = \overline{\{i \mid P_j \in S_i, i \in A\}}, \qquad j = 1, \cdots, n.$$

Then it is easy to show that a necessary condition for the existence of a solution to (5.1) is that

$$A, B \subset \{1, \cdots, m\}, A \cap B = \emptyset, \quad \text{and}$$

(5.2) $$|n_j(A) - n_j(B)| \leq 1 \quad (j = 1, \cdots, n) \quad \text{imply}$$

$$\sum_{i \in A} a_i \leq \sum_{i \in B} b_i + \sum_{n_j(A) > n_j(B)} w_j.$$

Condition (5.2) not only "sounds like" the condition (2.3) and all the other conditions we have met so far, but coincides with them when the combinatorial situations studied are put in such form that our "general problem" subsumes them. Then it is natural to inquire under what circumstances (5.2) is sufficient for the existence of a solution to (5.1). An answer is contained in the statement:

The following conditions are equivalent:

(5.3)   The m by n incidence matrix of sets S_i versus elements P_j has the unimodular property;

(5.4)   for every choice of integral $a_i \leq b_i$ $(i = 1, \cdots, m)$ and integral $w_j \geq 0$ $(j = 1, \cdots, n)$, (5.2) implies the existence of an integral solution to (5.1).

One can list other conditions, equivalent to the above two, which explore the relationship between real and integral solutions to (5.1), but from a combinatorial viewpoint, the principal interest is the equivalence of (5.3) and (5.4).

(5.3) implies (5.4) is the statement that if the incidence matrix has the unimodular property, then there is a theorem of the Hall type. (5.4) implies (5.3) is the statement that a theorem of the Hall type exists for all choices of boundary values *only* if the incidence matrix has the unimodular property. In summary, the unimodular property for the incidence matrix is a necessary and sufficient condition that (5.2) be a necessary and sufficient condition for the existence of an integral solution to (5.1). It is in this sense that the result may properly be regarded as the most general theorem of its kind.

While not difficult, the proof is long and will be published elsewhere. The key idea in the proof that (5.3) implies (5.4) is the consideration of dual sets of inequalities, where the unimodular property guarantees that if (5.1) is consistent, it has an integral solution, and further guarantees that examination of the extreme rays of the dual system is equivalent to checking (5.2). Ideas of the same general sort are involved in the proof that (5.4) implies (5.3).

In view of this theorem, it is of some interest to look for incidence matrices with the unimodular property. A special case of a result of Heller [**32**] shows that if all n columns of such a matrix are distinct and non-zero, and if there are n rows, then $n \leq m(m + 1)/2$. Further, this upper bound is attained if and only if the columns are all possible intervals of a simply ordered set of m points.

The most general sufficient condition known [**5**] for a matrix to have the unimodular property is that it be the incidence matrix of nodes versus directed paths in a directed graph all of whose loops are of even order with successive arcs alternating in direction. Until recently, every known incidence matrix with the unimodular property arose in this way, but the condition is

124   A. J. HOFFMAN

unaesthetically not symmetric with respect to rows and columns, although the unimodular property is. In fact, it is not difficult to give examples of incidence matrices where the rows represent nodes of an "alternating" graph, and the columns the directed paths, but no alternating graph can be found for which the columns represent nodes and the rows represent directed paths; for example, Heller [32],

$$\begin{matrix} 1 & 1 & 1 & 1 \\ 1 & 1 & 0 & 0 \\ 0 & 1 & 1 & 0 \\ 0 & 0 & 1 & 1 \\ 1 & 0 & 0 & 1 \end{matrix}$$

An example of an incidence matrix with the unimodular property that could not arise from an alternating graph whatever role be selected for rows and columns can be obtained by appending the column vector (1, 1, 1, 1, 1) to the above matrix.

6. Remarks.

a. It is interesting to see instances of matrices with the unimodular property arising in various contexts. It has already been noted [5] that, in linear programming, the transportation problem, the warehouse problem in the form discussed by Cahn [33] and by Charnes and Cooper [34], the caterer problem of Jacobs [35 and 36] and certain production scheduling problems involving fulfilling cumulative requirements all involve incidence matrices arising from alternating graphs.

One can also see the possibility of direct application of the result in §5 in work of Mirsky [37] offering an alternative proof of Horn's characterization [38] of the vector of diagonal elements of a hermitian matrix, and in Følner's discussion [39] of Banach mean values in groups. In neither of these cases does the author use (5.2) to prove (5.1) but he could have.

b. Although our emphasis has been using inequalities to prove combinatorial theorems, there is some interest in the reverse process. For example, Birkhoff [40] used Hall's theorem to prove that the vertices of the convex set of doubly stochastic matrices are the permutation matrices, a result for which many other proofs [41; 42; 43] have subsequently been offered.

c. It is also worth pointing out that, parallel to the theoretical interplay between linear inequalities and combinatorics, there has been a computational interplay. It was pointed out some time ago, see e.g. [44], that linear programming furnishes algorithms for certain combinatorial problems. So far, however, it has appeared from the Hungarian method of Kuhn [45] and its generalizations, modifications and extensions [46; 47; 48; 49] that the more fruitful relationship is the other way around: effective methods for choosing systems of distinct representatives yield algorithms for solving the transportation problem.

THE THEORY OF LINEAR INEQUALITIES

d. Several research questions are suggested by the material covered in this talk:

(1) The discovery of new classes of matrices with the unimodular property.

(2) The discovery of new ways of twisting problems so that matrices with the unimodular property appear. The theorem of Dilworth that the smallest number of chains whose union comprises all elements of a partially ordered set is the largest number of incomparable elements in the set (see Dilworth [49] and Fulkerson [50]) does not appear at first blush to be accessible by these methods since the matrix of elements versus chains does *not* have the unimodular property. But a method can be found to reformulate the problem [51] so that the unimodular property can be exploited. Perhaps the results of Ryser on term rank [52], and the graph theoretic theorems of Rabin and Norman [53], Berge [54], Tutte [55], [56], etc., which have at least a verbal similarity to the situations described in this talk, can be shown to be accessible by these methods. Thus far all attempts to derive them as corollaries of the main theorem of §5 have failed.

(3) The discovery of new applications of these results (which deal with finite sets) to infinite situations (through suitable finite approximation), and the discovery of non-trivial generalization—not accessible by finite approximation—to infinite combinatorial problems.

(4) There exist (see, e.g., Fan [57] and Duffin [58]) infinite-dimensional generalizations of the duality theorems in the theory of finite systems of linear inequalities; what use (if any) is the unimodular property in such circumstances, or what is the appropriate generalization of this property?

REFERENCES

1. H. B. Mann and H. J. Ryser, *Systems of distinct representatives*, Amer. Math. Monthly vol. 60 (1953) pp. 397–401.

2. E. Egervary, *Matrixok combinatorius tulajdonsagairol*, Mat. es Fiz. Lapok vol. 38 (1931) pp. 16–28 (translated as *On combinatorial properties of matrices*, by H. W. Kuhn, Office of Naval Research Logistics Project Report, Department of Mathematics, Princeton University, 1953).

3. P. Hall, *On representatives of subsets*, J. London Math. Soc. vol. 10 (1935) pp. 26–30.

4. D. König, *Theorie der endlichen und unendlichen Graphen*, New York, Chelsea Publishing Co., 1950.

5. A. J. Hoffman and J. B. Kruskal, *Integral boundary points of convex polyhedra*, in [7, pp. 223–246].

6. I. Heller and C. B. Tompkins, *An extension of a theorem of Dantzig's*, in [7, pp. 247–254].

7. H. W. Kuhn and A. W. Tucker, eds., *Linear inequalities and related systems*, Annals of Mathematics Studies, no. 38, Princeton, 1956.

8. A. J. Hoffman, *Generalization of a theorem of König*, J. Washington Acad. Sci. vol. 46 (1956) p. 211.

126 A. J. HOFFMAN

9. R. Rado, *Theorems on linear combinatorial topology and general measure*, Ann. of Math. vol. 44 (1943) pp. 228–270.

10. T. S. Motzkin, *The assignment problem*, Proceedings of the Sixth Symposium in Applied Mathematics, New York, McGraw-Hill, 1956.

11. H. W. Kuhn, *A combinatorial algorithm for the assignment problem*, Issue 11 of Logistics Papers, George Washington University Logistics Research Project, 1954.

12. A. J. Hoffman, *Linear programming*, Applied Mechanics Reviews (9), 1956.

13. P. R. Halmos and Herbert E. Vaughan, *The marriage problem*, Amer. J. Math. vol. 72 (1950) pp. 214–215.

14. H. J. Ryser, *Geometrics and incidence matrices*, Slaught Memorial Paper Contributions to Geometry, Amer. Math. Monthly vol. 62 (1955) pp. 25–31.

15. A. J. Hoffman and H. W. Kuhn, *Systems of distinct representatives and linear programming*, Amer. Math. Monthly vol. 63 (1956) pp. 455–460.

16. ———, *On systems of distinct representatives*, in [7, pp. 199–206].

17. L. R. Ford, Jr. and D. R. Fulkerson, *Network flow and systems of representatives*, Canad. J. Math. vol. 10 (1958) pp. 78–84.

18. ———, *Maximal flow through a network*, Canad. J. Math. vol. 8 (1956) pp. 399–404.

19. G. B. Dantzig and D. R. Fulkerson, *On the min-cut max-flow theorem of networks*, in [7, pp. 215–221].

20. A. J. Hoffman, 1955, Unpublished.

21. K. Menger, *Zur allgemeinen Kurventheorie*, Fund. Math. vol. 10 (1927) pp. 96–115.

22. G. Hajos, *Zum Mengerschen Graphensatz*, Acta Litterarum ac Scientiarum, Szeged vol. 7 (1934) pp. 44–47.

23. A. H. Hoffman, 1956, Unpublished.

24. D. Gale, *A theorem on flows in networks*, Pacific J. Math. vol. 7 (1957) pp. 1073–1082.

25. ———, 1956, Unpublished.

26. L. R. Ford, Jr. and D. R. Fulkerson, *Dynamic network flow*, 1957, Unpublished.

27. D. Gale, *Transient flows in networks*, 1958, Unpublished.

28. L. V. Kantorovich and M. K. Gavurin, *Problems of increasing the effectiveness of transport works*, AN USSR, 1949, pp. 110–138. I am indebted to G. B. Dantzig for this reference.

29. G. B. Dantzig, *Upper bounds, secondary constraints and block triangularity in linear programming*, Econometrica vol. 23 (1955) pp. 174–183.

30. H. M. Wagner, *On the capacitated Hitchcock problem*, 1958, Unpublished.

31. A. J. Hoffman, 1956, Unpublished.

32. I. Heller, *On linear systems with integral valued solutions*, Pacific J. Math. vol. 7 (1957) pp. 1351–1364.

33. A. S. Cahn, *The warehouse problem*, Bull. Amer. Math. Soc. vol. 54 (1948) p. 1073 (abstract).

34. A. Charnes and W. W. Cooper, *Generalizations of the warehousing model*, Operations Res. Q. vol. 6 (1955) pp. 131–172.

35. W. W. Jacobs, *The caterer problem*, Naval Res. Logist. Quart. vol. 1 (1954) pp. 154–165.

36. J. W. Gaddum, A. J. Hoffman and D. Sokolowsky, *On the solution of the caterer problem*, Naval Res. Logist. Quart. vol. 1 (1954) pp. 223–229.

37. L. Mirsky, *Matrices with prescribed characteristic roots and diagonal elements*, J. London Math. Soc. vol. 33 (1958) pp. 14–21.

38. A. Horn, *Doubly stochastic matrices and the diagonal of a rotation matrix*, Amer. J. Math. vol. 76 (1954) pp. 620–630.

39. E. Folner, *On groups with full Banach mean value*, Math. Scand. vol. 3 (1955) pp. 243–254.

40. G. Birkhoff, *Three observations on linear algebra*, Universidad Nacional de Tucuman, Revista Series A vol. 5 (1946) pp. 147–151.

41. A. J. Hoffman and H. W. Wielandt, *The variation of the spectrum of a normal matrix*, Duke Math. J. vol. 20 (1953) pp. 37–39.

42. J. von Neumann, *A certain zero-sum two-person game equivalent to the operational assignment problem*, in Contributions to the Theory of Games, vol. II, pp. 5–12 (edited by H. W. Kuhn and A. W. Tucker), Annals of Mathematics Studies, no. 28, Princeton, 1953.

43. J. Hammersley and W. Mauldon, *General principles of antithetic variates*, Proc. Cambridge Philos. Soc. vol. 52 (1956) pp. 476–481.

44. G. B. Dantzig, *Application of the simplex method to a transportation problem*, T. C. Koopmans, ed., Activity Analysis of Production and Allocation, Cowles Commission Monograph No. 13, New York, Wiley, 1951.

45. H. W. Kuhn, *The Hungarian method for solving the assignment problem*, Naval Res. Logist. Quart. vol. 2 (1955) pp. 83–97.

46. L. R. Ford, Jr. and D. R. Fulkerson, *A simple algorithm for finding maximal network flows and an application to the Hitchcock problem*, Canad. J. Math. vol. 9 (1957) pp. 210–218.

47. M. M. Flood, *The traveling salesman problem*, J. Operations Res. Soc. Amer. vol. 4 (1956) pp. 61–75.

48. J. R. Munkres, *Algorithms for the assignment and transportation problem*, J. Soc. Indust. Appl. Math. vol. 5 (1957) pp. 32–38.

49. R. P. Dilworth, *A decomposition theorem for partially ordered sets*, Ann. of Math. vol. 51 (1950) pp. 161–166.

50. D. R. Fulkerson, *Note on Dilworth's decomposition theorem for partially ordered sets*, Proc. Amer. Math. Soc. vol. 7 (1956) pp. 701–702.

51. G. B. Dantzig and A. J. Hoffman, *Dilworth's theorem on partially ordered sets*, in [7, pp. 207–214].

52. H. J. Ryser, *The term rank of a matrix*, Canad. J. Math. vol. 60 (1957) pp. 57–65.

53. M. O. Rabin and R. Z. Norman, *An algorithm for the minimum cover of a graph*, 1957, Unpublished.

54. C. Berge, *Two theorems in graph theory*, Proc. Nat. Acad. Sci. vol. 43 (1957) pp. 842–844.

55. W. T. Tutte, *The factorization of linear graphs*, J. London Math. Soc. vol. 22 (1947) pp. 107–111.

56. ———, *The factors of graphs*, Canad. J. Math. vol. 4 (1952) pp. 314–328.

57. K. Fan, *On systems of linear inequalities*, in [7, pp. 99–156].

58. R. J. Duffin, *Infinite programs*, in [7, pp. 157–170].

GENERAL ELECTRIC COMPANY,
NEW YORK, NEW YORK

A. J. Hoffman
S. Winograd

Finding All Shortest Distances in a Directed Network

Abstract: A new method is given for finding all shortest distances in a directed network. The amount of work (in performing additions, subtractions, and comparisons) is slightly more than half of that required in the best of previous methods.

Introduction

Let $D = (d_{ij})$ be a real square matrix of order n with 0 diagonal. We shall think of each of the numbers d_{ij} as respresenting the "length" of a link from vertex i to vertex j in a directed network. While we do not assume that all d_{ij} are nonnegative, we do assume that, if σ is any permutation of $N = \{1, \cdots, n\}$, then $\sum_i d_{i\sigma i} \geq 0$. This is equivalent to the customary assumption that the sum of the lengths around any cycle is nonnegative, an assumption generally made in shortest-distance problems.

Our problem is to calculate all "shortest distances" from i to j for all $i \neq j$. More formally, define a path P from i to j as an ordered sequence of distinct vertices $i = i_0, i_1, \cdots, i_k = j$, and define its length $L(P)$ by $L(P) = \sum_{r=0}^{k-1} d_{i_r i_{r+1}}$. Our problem is to calculate a square matrix $E = (e_{ij})$ of order n such that $e_{ij} = \min_P L(P)$, where P ranges over all paths from i to j.

To our knowledge, the most efficient method in the literature is due to Floyd [1] and Warshall [2], who showed that E can be calculated in n^3 additions and n^3 comparisons. (Here and elsewhere we suppress terms of lower order unless they are needed in the course of an argument.) The purpose of this paper is to announce an improved method.

• Theorem

If D is the matrix of link lengths, E the matrix of shortest distances of a directed network on n vertices, and if $\epsilon > 0$ is given, then E can be calculated from D in $(2 + \epsilon)n^{5/2}$ addition-subtractions and n^3 comparisons.

Proof

The proof of the theorem will consist in producing an algorithm and showing that it has the stated properties. Our algorithm borrows much from Shimbel [3], as well as from [1] and [2], but has two special features which we now outline briefly.

Let A be a $p \times q$ matrix, B a $q \times r$ matrix, and define $A \circ B = C = (c_{ij})$ to be the $p \times r$ matrix given by

$$c_{ij} = \min_k (a_{ik} + b_{kj}) .$$

A straightforward approach to calculating C would require pqr additions and $pr(q-1)$ comparisons. Our method, discussed in the following section, requires $pr(q-1)$ comparisons also, but fewer than $(q-1/2)$ $\sqrt{2pr(p+r)} + pr$ addition-subtractions.

The second special feature is that we suitably partition the vertices of our network into subsets of proper size and proceed to calculate E by a sequence of operations of the form $A \circ B$ and solutions of shortest-distance problems on the subsets. This part is a direct generalization of [1] and [2] in which the subsets consist of exactly one vertex. Hu [4] has also described a partitioning of D to take advantage of sparseness and geography, which is a different matter. Presumably our method could be modified to take similar advantages, but we do not pursue this point.

Pseudomultiplication of matrices

• *Lemma*
Let A and B be matrices of dimension $p \times q$ and $q \times r$ respectively. Define $(A \circ B)_{ij} = \min_k(a_{ik} + b_{kj})$. Then $A \circ B$ can be calculated in $pr(q - 1)$ comparisons and fewer than $(q - 1/2) \sqrt{2pr(p + r)} + pr$ additions.

Proof
Define, for any collection $M^1, M^2, \cdots, M^k$ of matrices of the same dimension,

$$M = \min(M^1, \cdots, M^k) = (m_{ij}) = \min(M^1_{ij}, \cdots, M^k_{ij}).$$

Partition the *columns* of A into nonempty subsets $S_1, \cdots, S_k$ of size $d_1, \cdots, d_k$ respectively. Partition the *rows* of B conformally. Let A_i, $i = 1, \cdots, k$, be the submatrix of A consisting of the columns in S_i. Let B_i' be the corresponding submatrix of rows of B. Clearly

$$A \circ B = \min(A_1 \circ B_1', \cdots, A_k \circ B_k').$$

Calculate $A_i \circ B_i'$, $i = 1, \cdots, k$ as follows: Form all p $d_i(d_i - 1)/2$ differences $a_{tj} - a_{tk}$, $t = 1, \cdots, p$, j, $k \in S_i$, $j < k$. Similarly, form all r $d_i(d_i - 1)/2$ differences $b_{ku} - b_{ju}$, $u = 1, \cdots, s$, $j, k \in S_i$, $j < k$. In order to find the (t,u)th entry of $A_i \circ B_i'$, we observe that $a_{tj} + b_{ju} \le a_{tk} + b_{ku}$ if and only if $a_{tj} - a_{tk} \le b_{ku} - b_{ju}$. Since we have already calculated these differences, it is clear that $(d_i - 1)$ comparisons will yield, for each (t,u), the index l such that $a_{tl} + b_{lu} = \min_{k \in S_i}(a_{tk} + b_{ku})$. Next, for each (t,u), we calculate $a_{tl} + b_{lu}$. Thus we have found $A_i \circ B_i'$ in $[(p + r)/2]d_i(d_i - 1)$ subtractions, pr additions and $pr(d_i - 1)$ comparisons.

It follows that $A \circ B = \min_i\{(A_i \circ B_i')\}$ can be calculated in

$$[(p + r)/2] \sum d_i^2 - [(p + r)/2]q + prk \tag{1}$$

addition-subtractions (here we have used $\sum d_i = q$), and

$$pr\left(\sum (d_i - 1) + k - 1\right) = pr(q - 1) \text{ comparisons.}$$

Let us study (1) further. Define m to be the smallest integer not less than $\sqrt{2pr/p + r}$. Thus

$$m = \sqrt{(2pr)/(p + r)} + \theta, \qquad 0 \le \theta < 1. \tag{2}$$

Write $q = am + b$, $0 \le b \le m - 1$.
Case 1. $b = 0$. Choose $k = a$, $d_1 = \cdots = d_k = m$. Then (1) becomes

$$[(p + r)/2]qm - [(p + r)/2]q + pr\,q/m, \tag{3}$$

which is easily seen to be less than the number specified

in the lemma. Here we use in our estimates $\sqrt{2pr(p + r)} \le m \le \sqrt{2pr(p + r)} + 1$, and $p + r \le 2pr$ for positive integers p and r.
Case 2. $b \ne 0$. Set $k = a + 1$, $d_1 \cdots = d_a = m$, $d_{a+1} = b$. Then

$$\sum d_i^2 = m^2a + b^2 = mq + b^2 - mb \le mq + 1 - m,$$

since $1 \le b \le m - 1$. Using this estimate, $\sqrt{2pr(p + r)} \le m \le \sqrt{2pr/(p + r)} + 1$, $k \le q/m + 1$, we obtain the estimate given in the statement of the lemma.

Finding all shortest distances (description and validation of algorithm)

Let N be partitioned into nonempty subsets $S_1 \cup \cdots \cup S_k$ of respective sizes $d_1, \cdots, d_k$. We shall proceed to modify the matrix D by successive steps so that the resulting matrix is E. In our description, the letter D will always stand for the current step of the modification of D. $D[S, T]$ will mean the submatrix of D formed by rows in S, columns in T, $D[S] = D[S, S]$. $\mathscr{E}D[S]$ stands for the shortest distance matrix computed from the submatrix $D[S]$. $\bar{S}$ means the complement of S. The expression $D[S, T] \leftarrow F$ means that in D, $D[S, T]$ gets replaced by F. All other entries of D are unchanged.

a)   Let $i = 1$

b) $D[S_i] \leftarrow \mathscr{E}D[S_i]$

c) $D[\bar{S}_i, S_i] \leftarrow D[\bar{S}_i, S_i] \circ D[S_i]$

$D[S_i, \bar{S}_i] \leftarrow D[S_i] \circ D[S_i, \bar{S}_i]$

d) $D[\bar{S}_i] \leftarrow \min\{D[\bar{S}_i], D[\bar{S}_i, S_i] \circ D[S_i, \bar{S}_i]\}$

e)   Increase i by 1. If $i = k$, stop. Otherwise, go to b.

After steps a) through e) are completed the first time, d_{ij} equals the shortest distance from i to j in which we are restricted to paths in which all intermediate vertices, if any, belong to S_1. This holds for all i, j. Manifestly, after we have completed a) through e) l times, d_{ij} equals the shortest distance from i to j in which we are restricted to paths where all intermediate vertices, if any, belong to $S_1 \cup \cdots \cup S_l$. Thus, by induction, the algorithm is easily seen to be valid.

We now show inductively that the number of comparisons $f(n)$ required by this algorithm is at most n^3. Examination of a) through e) shows that

$$f(n) = \sum f(d_i) + 2\sum (n - d_i)d_i(d_i - 1)$$
$$+ \sum (n - d_i)(n - d_i)(d_i - 1)$$
$$+ \sum (n - d_i)(n - d_i).$$

Assuming inductively that $f(d_i) \le d_i^3$, and using $\sum d_i = n$, we get $f(n) \le n^3$. Note that this does not depend on the magnitudes (d_i).

413

Count of addition-subtractions in the algorithm

If now we let $f(n)$ be the number of addition-subtractions required, we find from a) through e) that

$$f(n) < \sum f(d_i) + 2 \sum d_i \sqrt{2(n - d_i)nd_i}$$
$$+ 2 \sum (n - d_i)^2 + 2 \sum d_i(n - d_i) \sqrt{n - d_i} . \quad (4)$$

(Here we have suppressed the factor "$-1/2$" in the lemma.) In order to get an estimate of how n grows, let us tentatively assume $n = a^t$, $d_i = a^{t-1}$, $k = a$. Then we have

$$f(a^t) \leq af(a^{t-1}) + 2a\{a^{t-1}[2(a^t - a^{t-1})a^{2t-1}]^{1/2}$$
$$+ (a^t - a^{t-1})a^{t-1} + a^{t-1}(a^t - a^{t-1})^{3/2}\}$$
$$= af(a^{t-1}) + 2a \cdot a^{(5/2)t}\left\{\frac{1}{a}\left[2\left(1 - \frac{1}{a}\right)\frac{1}{a}\right]^{1/2}\right.$$
$$\left. + \frac{1}{a}\left(1 - \frac{1}{a}\right)^{3/2}\right\} + \mathrm{O}(a^{2t})$$
$$= af(a^{t-1})$$
$$+ 2a^{(5/2)/t}\left[\sqrt{1 - (1/a)} \sqrt{(2/a) + 1} - (1/a)\right]$$
$$+ \mathrm{O}(a^{2t}) . \quad (5)$$

Setting $f(a^t) = Aa^{(5/2)t}$, we find

$$Aa^{(5/2)t} \leq 1/(a^{3/2}) \, Aa^{(5/2)t}$$
$$+ 2[\sqrt{1 - (1/a)} \sqrt{(2/a) + 1} - (1/a)]a^{(5/2)t}$$
$$A \leq \frac{2\sqrt{1 - (1/a)} \sqrt{(2/a) + 1} - (1/a)}{1 - 1/(a^{3/2})} = 2 + \epsilon(a) ,$$

where $\epsilon(a) \to 0$ as $a \to \infty$.

But in order to establish this rigorously, we must proceed more carefully, without assuming that $n = a^t$. Because the details are tedious, we shall confine ourselves to an outline of the algorithm and proof.

First, an integer a is chosen. If $n < a$, the problem is solved by the method of [1] and [2]. If $n \geq a$, write $n = am + b$, $0 \leq b < a$. Then $n = b(m + 1) + (a - b)m$. Let $d_1, \cdots, d_b = m + 1$, $d_{b+1} = \cdots = d_a = m$. Partition n into subsets of size $d_1, \cdots, d_a$ and apply the algorithm given in this section. To prove that $f(n)$, the number of additions required, is at most $An^{5/2} +$ terms of lower order in n, the strategy is to assume inductively that $f(n) = An^{5/2} + P(a)n^2$, where P is a certain polynomial in a. Then using (4), replace m by $m + 1$ throughout. One finds that an auspicious choice of $P(a)$ makes $f(n) \leq f[a(m + 1)] \leq A(am)^{5/2} + P(a)(am)^2 \leq An^{3/2} + P(a)n^2$. This choice of $P(a)$ also makes the formula valid if $n \leq a$ and completes the proof.

Acknowledgment

This work was supported in part by the Office of Naval Research under contract numbers N0014-69-C-0023 and N0014-71-C-0112.

References

1. R. W. Floyd, "Algorithm 97, shortest paths," *Comm. of ACM* **5**, 345 (1962).
2. S. Warshall, "A theorem on Boolean matrices," *J. ACM* **9**, 11 (1962).
3. A. Shimbel, "Structure in communication nets," Proc. of the Symposium on Information Networks, (April 1954) 199–203, Polytechnic Institute of Brooklyn, New York (1955).
4. T. C. Hu, "A decomposition algorithm for shortest paths in a network," *Operations Research*, **16**, No. 1, 91–102 (January-February 1968).

Received November 15, 1971

The authors are located at the IBM Thomas J. Watson Research Center, Yorktown Heights, New York 10598.

Mathematical Programming Study 1 (1974) 120–132 North-Holland Publishing Company

ON BALANCED MATRICES

D.R. FULKERSON*

Cornell University, Ithaca, N.Y., U.S.A.

A.J. HOFFMAN**

IBM Thomas J. Watson Research Center, Yorktown Heights, N.Y., U.S.A.

and

Rosa OPPENHEIM

Rutgers University, New Brunswick, N.J., U.S.A.

Received 14 February 1974
Revised manuscript received 15 April 1974

*Dedicated to A.W. Tucker, as a token of our gratitude for
over 40 years of friendship and inspiration*

1. Introduction

In his interesting paper [2], Claude Berge directs our attention to two questions relevant to the use of linear programming in combinational problems. Let A be a (0, 1)-matrix, w and c nonnegative integral vectors, and define the polyhedra

$$P(A, w, c) = \{y : yA \geq w, 0 \leq y \leq c\}, \tag{1.1}$$
$$Q(A, w, c) = \{y : yA \leq w, 0 \leq y \leq c\}. \tag{1.2}$$

* The work of this author was supported in part by N.S.F. Grant GP-32316X and by O.N.R. Grant N00014-67A-0077-002F.
** The work of this author was supported in part by the U.S. Army under contract #DAHCO4-C-0023.

Let $1 = (1, \ldots, 1)$ denote the vector all of whose components are 1. The two questions are:

> If $P(A, w, c)$ is not empty, is the minimum value of $1 \cdot y$, taken over all $y \in P(A, w, c)$, achieved at an integral vector y? $\qquad$ (1.3)

> Is the maximum value of $1 \cdot y$, taken over all $y \in Q$ (A, w, c) achieved at an integral vector y? $\qquad$ (1.4)

Berge defines a $(0,1)$-matrix A to be balanced if A contains no square submatrix of odd order whose row and column sums are all two. He shows that the answer to (1.3) is affirmative for all $(0,1)$-vectors w and c if and only if A is balanced. He shows that the answer to (1.4) is affirmative for all w whose components are 1 or ∞ and for all $(0,1)$-vectors c if and only if A is balanced. Finally, he remarks that for all c whose components are 0 or ∞ and all w whose components are nonnegative integers, the Lovász-Fulkerson perfect graph theorem [4, 6, 7] implies that the answer to (1.3) is affirmative if and only if A is balanced.

In this paper we prove that if A is balanced, then the answers to (1.3) and (1.4) are affirmative for all nonnegative integral w and c. We do not use the perfect graph theorem as a lemma, nor the results of Berge in [2] or in earlier work on balanced matrices [1].

The above results and those of Berge are used to relate the theory of balanced matrices to those of blocking pairs of matrices and anti-blocking pairs of matrices [3, 4, 5]. We summarize below some pertinent aspects of these two geometric duality theories.

We first discuss briefly the blocking theory. Let A be a nonnegative m by n matrix, and consider the convex polyhedron

$$\{x : A x \geq 1, x \geq 0\}. \qquad (1.5)$$

A row vector a^i of matrix A is inessential (does not represent a facet of (1.5)) if and only if a^i is greater than or equal to a convex combination of other rows of A. The (nonnegative) matrix A is proper if none of its rows is inessential. Let A be proper with rows $a^1, \ldots, a^m$. Let B be the r by n matrix having rows $b^1, \ldots, b^r$, where $b^1, \ldots, b^r$ are the extreme points of (1.5). Then B is proper and the extreme points of the polyhedron

$$\{x : B x \geq 1, x \geq 0\} \qquad (1.6)$$

are $a^1, \ldots, a^m$. The matrix B is called the blocking matrix of A and vice-versa. Together, A and B constitute a blocking pair of matrices, and the polyhedra (1.5) and (1.6) they generate are called a blocking pair of polyhedra. (Thus for any blocking pair of polyhedra, the non-trivial facets of one and the extreme points of the other are represented by exactly the same vectors; trivial facets are those corresponding to the nonnegativity constraints.)

Let A be a nonnegative m by n matrix and consider the packing program

$$\begin{aligned} \text{maximize} \quad & 1 \cdot y, \\ \text{subject to} \quad & y A \leq w, \qquad y \geq 0, \end{aligned} \tag{1.7}$$

where w is nonnegative. Let B be an r by n nonnegative matrix having rows $b^1, \ldots, b^r$. The max–min equality is said to hold for the ordered pair A, B if, for every n-vector $w \geq 0$, the packing program (1.7) has a solution vector y such that

$$1 \cdot y = \min_{1 \leq j \leq r} b^j \cdot w. \tag{1.8}$$

One theorem about blocking pairs asserts that the max–min equality holds for the ordered pair of proper matrices A, B if and only if A and B are a blocking pair. Hence if the max–min equality holds for A, B, it also holds for B, A. (Note that the addition of inessential rows to either A or B does not affect the max–min equality.)

Now let A be a proper (0,1)-matrix, with blocking matrix B. The strong max–min equality is said to hold for A, B if, for any nonnegative integral vector w, the packing program (1.7) has an integral solution vector y, which of course satisfies (1.8). A necessary, but not sufficient, condition for the strong max–min equality to hold for A, B is that each row of B be a (0,1)-vector. To say that an m by n (0,1)-matrix A is proper is simply to say that A is the incidence matrix of m pairwise non-comparable subsets of an n-set, i.e., A is the incidence matrix of a clutter. If the strong max–min equality holds for A and its blocking matrix B, then B is the incidence matrix of the blocking clutter, i.e., B has as its rows all (0,1)-vectors that make inner product at least 1 with all rows of A, and that are minimal with respect to this property. If A and B are a blocking pair of (0,1)-matrices, the strong max–min equality may hold for A, B, but need not hold for B, A. This is in decided contrast with the similar situation for anti-blocking pairs of matrices, which we next briefly discuss.

Let A be an m by n nonnegative matrix with rows $a^1, \ldots, a^m$, having no zero columns, and consider the convex polyhedron

$$\{x : A x \le 1, x \ge 0\}. \tag{1.9}$$

(While a row vector a^i of A is inessential in (1.9) if and only if a^i is less than or equal to a convex combination of other rows of A, we shall not limit A to "proper" matrices in this discussion, as we did for blocking pairs, because there will not be a one–one correspondence between non-trivial facets of one member of a pair of anti-blocking polyhedra and the extreme points of the other.) Let D be the r by n matrix having rows $d^1, \ldots, d^r$, where $d^1, \ldots, d^r$ are the extreme points of (1.9). Then D is nonnegative, has no zero columns, and the extreme points of

$$\{x : D x \le 1, x \ge 0\} \tag{1.10}$$

are $a^1, \ldots, a^m$ and all projections of $a^1, \ldots, a^m$. D is called an anti-blocking matrix of A, and vice-versa. Together, A and D constitute an anti-blocking pair of matrices, and the polyhedra (1.9) and (1.10) are an anti-blocking pair of polyhedra.

Now consider the covering program

$$\begin{aligned} \text{minimize} \quad & 1 \cdot y, \\ \text{subject to} \quad & y A \ge w, \qquad y \ge 0, \end{aligned} \tag{1.11}$$

where w is nonnegative. Let D be an r by n nonnegative matrix having no zero columns with rows $d^1, \ldots, d^r$. The min–max equality is said to hold for the ordered pair A, D if, for every n-vector $w \ge 0$, the covering program (1.11) has a solution vector y satisfying

$$1 \cdot y = \max_{1 \le j \le r} d^j \cdot w. \tag{1.12}$$

Then the min–max equality holds for A, D if and only if A and D are an anti-blocking pair. Hence, if the min–max equality holds for A, D, it also holds for D, A.

 D.R. Fulkerson et al., On balanced matrices

Now let A be a (0,1)-matrix, with anti-blocking matrix D. The strong min–max equality is said to hold for A, D if, for every nonnegative integral vector w, the covering program (1.11) has an integral solution vector y; y of course satisfies (1.12). A necessary and sufficient condition for the strong min–max equality to hold for A, D is that all the essential rows of D be (0,1)-vectors. Hence, if the strong min–max equality holds for A, D, it also holds in the reverse direction D, A (where we may limit D to its essential rows). In this case, it can be shown that the essential (maximal) rows of A are the incidence vectors of the cliques of a graph G on n vertices, and the essential rows of D are the incidence vectors of the anti-cliques (maximal independent sets of vertices) of G. Graph G is thus pluperfect, or equivalently, perfect. The fact that the strong min–max equality for A, D implies the strong min–max equality for D, A is the essential content of the perfect graph theorem.

We shall show in Section 5 that the results described above and those of Berge imply: (a) If A is balanced and B is the blocking matrix of A, then the strong max–min equality holds for both A, B and B, A, and (b) if A is balanced and if D is an anti-blocking matrix of A, then the strong min–max equality holds for A, D (and hence for D, A).

2. Vertices of some polyhedra

We first state the lemmas of this section, and then give their proofs.

Lemma 2.1. *If A is balanced, and if $\{x : A x = 1, x \geq 0\}$ is not empty, then every vertex of this polyhedron has all coordinates 0 or 1.*

Lemma 2.2. *If A is balanced, and if $\{x : A x \geq 1, x \geq 0\}$ is not empty, then every vertex of this polyhedron has all coordinates 0 or 1.*

Lemma 2.3. *If A is balanced, and if $\{x, z : A x - z = 1, x \geq 0, z \geq 0\}$ is not empty, then every vertex of this polyhedron has all coordinates 0 or 1.*

Lemma 2.4. *If A is balanced, then every vertex of $\{x, z : A x - z \leq 1, x \geq 0, z \geq 0\}$ is integral. Hence if A is balanced, every vertex of $\{x : A x \leq 1, x \geq 0\}$ has coordinates 0 or 1.*

Note that Lemma 2.1 is a special case of Lemma 2.3, but it is convenient to separate the proofs.

Proof of Lemma 2.1. If A is balanced, then every submatrix of A is balanced. We shall prove Lemma 2.1 by induction on the number of rows of A. It is clearly equivalent to prove that if $x > 0$ satisfies $A x = 1$, then there exists a set of non-overlapping columns $a_{j_1}, \ldots, a_{j_k}$ of A (i.e., $a_{j_r} \cdot a_{j_s} = 0$ for $r \neq s$) whose sum is the vector 1. For any set S of non-overlapping columns, define $C(S)$, the "cover of S", to be the number of i such that $\sum_{j \in S} a_{ij} = 1$. Let S^* be a set of non-overlapping columns such that $C(S^*) \geq C(S)$ for any set S of non-overlapping columns. If $C(S^*) = m = $ number of rows of A, we are done, so assume $C(S^*) = k < m$, and, say, $\sum_{j \in S*} a_{ij} = 1$ for $i = 1, \ldots, k$. Let $\bar{A}$ be the submatrix of A formed by rows $1, \ldots, k$. We have $\bar{A} x = 1$, $x > 0$, so, by the induction hypothesis, any column of $\bar{A}$ is contained in a set T of non-overlapping columns of $\bar{A}$ such that $C(T) = k$. In particular, let j^* be a column index such that $a_{ij}{}^* = 1$ for some $i \in \{k + 1, \ldots, m\}$, and let the aforementioned T contain j^*. Now some column indices in T (possibly none) may coincide with some column indices in S^*. Let $V = T \backslash S^*$, $U = S^* \backslash T$, both non-empty. Define a graph $G(\bar{A})$ whose points are the indices in $V \cup U$, with j and ℓ adjacent if and only if $\bar{a}_j \cdot \bar{a}_\ell > 0$. Clearly, $G(A)$ is bipartite with parts U and V. Let W be the vertices of the connected component $W(\bar{A})$ of $G(\bar{A})$ containing j^* (W may be $V \cup U$). It follows that

$$\sum_{j \in U \cap W} a_{ij} = \sum_{j \in V \cap W} a_{ij} = 0 \text{ or } 1 \quad \text{for } i = 1, \ldots, k. \tag{2.1}$$

Suppose that, for each $i = k + 1, \ldots, m$,

$$\sum_{j \in V \cap W} a_{ij} \leq 1. \tag{2.2}$$

Since $j^* \in W$, it follows from (2.1) and (2.2) that the columns of A with indices in

$$(S^* \backslash (U \cap W)) \cup (V \cap W)$$

are a non-overlapping set of columns with cover $\geq k + 1$, contradicting the definition of S^*. Hence (2.2) is untenable. Now consider the graph $W(A)$ with point set W, where j and ℓ are adjacent if and only if $a_j \cdot a_\ell > 0$. Recall that $W(\bar{A})$ is connected and bipartite. The graphs $W(\bar{A})$ and $W(A)$ have the same point set W, but $W(A)$ has more edges. In particular, there exists at least one pair of points in $W \cap V$ which are adjacent in $W(A)$. Let j and ℓ be points in $W \cap V$ such that the shortest path P in $W(\bar{A})$ joining j and ℓ contains no points j' and ℓ' in $W \cap V$ adjacent in $W(A)$ other than j and ℓ. Clearly such a path exists and is of even length. Let this path be

$$j = j_1, i_1, j_2, i_2, \ldots, j_p, i_p, j_{p+1} = \ell,$$

where the first, third, fifth, ... indices are in V, the second, fourth, ... indices are in U. Let $r^* \in \{k + 1, \ldots, m\}$ satisfy $a_{r^* j_1} = a_{r^* j_{p+1}} = 1$ and choose $r_1, \ldots, r_p, s_1, \ldots, s_p$ such that

$$
\begin{aligned}
a_{r_t j_t} = a_{r_t i_t} &= 1, & t = 1, \ldots, p, \\
a_{s_t i_t} = a_{s_t j_{t+1}} &= 1, & t = 1, \ldots, p.
\end{aligned}
$$

That such indices exist follows from the construction of the path P. It is now clear that the submatrix of A formed by the columns $i_1, \ldots,$ $i_p, j_1, \ldots, j_{p+1}$ and rows $r^*, r_1, \ldots, r_p, s_1, \ldots, s_p$ violates the hypothesis that A is balanced. Thus $C(S^*) = m$, proving Lemma 2.1.

Proof of Lemma 2.2. If x is a vertex of $\{x : A x \geq 1, x \geq 0\}$, it is a vertex of the polyhedron obtained by deleting the inequalities of $A x \geq 1$ that are strict. By Lemma 2.1, every vertex of this polyhedron has all coordinates 0 or 1.

Proof of Lemma 2.3. If (x, z) is a vertex of $\{x, z : A x - z = 1, x \geq 0,$ $z \geq 0\}$, then x is a vertex of $\{x : A x \geq 1, x \geq 0\}$. Lemma 2.3 thus follows from Lemma 2.2.

Proof of Lemma 2.4. If (x, z) is a vertex of $\{x, z : A x - z \leq 1, x \geq 0,$ $z \geq 0\}$, it is a vertex of the polyhedron obtained by deleting the inequalities of $A x - z \leq 1$ that are strict. Thus Lemma 2.4 follows from Lemma 2.3.

3. Solution of Problem (1.3)

We first prove a lemma.

Lemma 3.1. *Let A be a $(0,1)$-matrix satisfying the condition: For all nonnegative integral vectors w and c such that $P(A, w, c)$ is not empty, the minimum value of $1 \cdot y$, $y \in P(A, w, c)$ is an integer. Then for all nonnegative integral vectors w and c such that $P(A, w, c)$ is not empty, there exists an integral vector y that minimizes $1 \cdot y$ over $y \in P(A, w, c)$.*

Proof. The lemma is true if $1 \cdot c = 0$, and so we argue by induction on $1 \cdot c$.

Assume $y = (y_1, y_2, \ldots, y_m)$ is a solution to the linear program

$$\begin{aligned} \text{minimize} \quad & 1 \cdot y, \\ \text{subject to} \quad & y \in P(A, w, c), \end{aligned} \tag{3.1}$$

with at least one component not integral, say $y_1 = r + \theta$, where $r \geq 0$ is an integer and $0 < \theta < 1$. Let $1 \cdot y = k$, where k is an integer. For any number z, define $z^+ = \max(0, z)$, and for any vector $z = (z_1, z_2, \ldots)$, define $z^+ = (z_1^+, z_2^+, \ldots)$. Let $\alpha = (r, y_2, \ldots, y_m)$, and note that $0 \leq \alpha \leq \tilde{c} = (c_1 - 1, c_2, \ldots, c_m)$. Let a^1 be the first row of A. Since $\alpha A \geq w - a^1$ and $\alpha A \geq 0$, we have $\alpha A \geq (w - a^1)^+$. Thus $\alpha \in P(A, (w - a^1)^+, \tilde{c})$, and $1 \cdot \alpha = k - \theta < k$. Now $1 \cdot \tilde{c} < 1 \cdot c$. Hence, by the induction assumption there exists an integral vector $\beta = (\beta_1, \ldots, \beta_m)$ such that $\beta A \geq (w - a^1)^+ \geq w - a^1, 0 \leq \beta \leq \tilde{c}$, and $1 \cdot \beta = \ell \leq k - \theta < k$, where ℓ is an integer. Therefore, the integral vector $\bar{\beta} = (\beta_1 + 1, \beta_2, \ldots, \beta_m) \in P(A, w, c)$, $1 \cdot \bar{\beta} = \ell + 1 \leq k$. But no solution to (3.1) can have value less than k, and hence $1 \cdot \bar{\beta} = k$. Thus $\bar{\beta}$ is an integral vector solving (3.1).

Theorem 3.2.[1] *Let A be balanced, and let w and c be nonnegative integral vectors such that $P(A, w, c)$ is not empty. Then the linear program (3.1) has an integral solution.*

Proof. Since $P(A, w, c)$ is not empty and bounded, (3.1) has a solution. Hence, by the duality theorem of linear programming, the dual program

$$\begin{aligned} \text{maximize} \quad & w \cdot x - c \cdot z, \\ \text{subject to} \quad & A x - z \leq 1, \quad x \geq 0, \quad z \geq 0, \end{aligned} \tag{3.2}$$

---

[1] Added in proof: A different (and earlier) demonstration of Theorem 3.2 was given by L. Lovász.

has a solution. One such must occur at a vector with integral coordinates, by Lemma 2.4, so the common value of (3.2) and of (3.1) is an integer. But this means that the hypothesis of Lemma 3.1 holds. Hence, the conclusion of Lemma 3.1 holds, proving the theorem.

Note that the theorem holds if all coordinates of the vector c are ∞, an observation we will need below.

4. Solution of Problem (1.4)

We devote this section to the proof of Theorem 4.1 below.

Theorem 4.1. *Let A be a balanced matrix, and let w and c be nonnegative integral vectors. Then the linear program*

$$\begin{aligned} \text{maximize} \quad & 1 \cdot y, \\ \text{subject to} \quad & y \in Q(A, w, c) \end{aligned} \tag{4.1}$$

has an integral solution vector y.

Proof. We first remark that if A is balanced, the matrix (A, I) is balanced. Thus it suffices to prove that if A is balanced and $w \geq 0$ is integral, then the linear program

$$\begin{aligned} \text{maximize} \quad & 1 \cdot y, \\ \text{subject to} \quad & y A \leq w, \qquad y \geq 0, \end{aligned} \tag{4.2}$$

has an integral solution vector y. We shall prove this by a double induction on the pair of integers $(1 \cdot w, m)$, where A has m rows. Note that the theorem clearly is valid for any $m \geq 1$ if $1 \cdot w = 0$; it is also valid for any nonnegative integer value of $1 \cdot w$, if $m = 1$ (i.e., if (4.2) is a problem in one variable.)

Let $y = (y_1, y_2, \ldots, y_m)$ be a fractional solution of (4.2). If at least one y_i is zero, we are in the situation described by the pair of integers $(1 \cdot w, m - 1)$, since any submatrix of A is balanced, and the induction hypothesis applies. Thus we suppose all $y_i > 0$. By Lemma 2.2 and the duality theorem of linear programming, we know that $1 \cdot y = k$, where k is an integer. Now suppose there is at least one j such that $y \cdot a_j < w_j$, where a_j is the jth column of A. Thus $w_j > 0$. If $y \cdot a_j \leq w_j - 1$, we consider the pair of integers $(1 \cdot w - 1, m)$. By the inductive hypothesis, there is an integral vector z such that $z A \leq 0$, $z \geq 0$, $1 \cdot z = 1 \cdot y = k$, and we are done. Thus we may assume that $y \cdot a_j = w_j - 1 + \theta$, where $0 < \theta < 1$. Hence $a_j \neq 0$. Then clearly we can find a vector z such that $z \geq 0$, $z A \leq (w_1, w_2, \ldots, w_j - 1, \ldots, w_n)$, $z \leq y$, and $1 \cdot z = k - \theta$. By the inductive hypothesis for the pair of integers $(1 \cdot w - 1, m)$, there is an integral vector α satisfying $\alpha \geq 0$, $\alpha A \leq (w_1, \ldots, w_j - 1, \ldots, w_n) \leq w$, $1 \cdot \alpha \geq k - \theta$, hence $1 \cdot \alpha = k$, and we are done.

Thus $y \cdot a_j = w_j$ for all j and $y_i > 0$ for all i. By the principle of complementary slackness, every optimal solution of the dual problem

$$\begin{aligned} \text{minimize} \quad & w \cdot x, \\ \text{subject to} \quad & A\,x \geq 1, \qquad x \geq 0 \end{aligned} \tag{4.3}$$

satisfies $A x = 1$, $x \geq 0$, $w \cdot x = k$. Select one such x. Then y and x are optimal solutions, respectively, of the dual programs

$$\begin{aligned} \text{minimize} \quad & 1 \cdot y, \\ \text{subject to} \quad & y A \geq w, \qquad y \geq 0, \end{aligned} \tag{4.4}$$

$$\begin{aligned} \text{maximize} \quad & w \cdot x, \\ \text{subject to} \quad & A x \leq 1, \qquad x \geq 0, \end{aligned} \tag{4.5}$$

with common value $1 \cdot y = w \cdot x = k$. By the remark at the end of the last section, there exists an integral vector α such that $\alpha \geq 0$, $\alpha A \geq w$, $1 \cdot \alpha = k$. If $\alpha A = w$, we are done. So assume $\alpha \cdot a_j > w_j$ for at least one j. Since $y_i > 0$ for all i, there is a number t, $0 < t < 1$, such that $y_i > (1 - t)\alpha_i$ for all i. Let vector z solve $y = (1 - t)\alpha + t z$, i.e., $z = (1/t) [y - (1 - t)\alpha]$. Thus $z \geq 0$ and $1 \cdot z = k$. Now, since $y A = w$ and $\alpha A \geq w$, it follows that $z A \leq w$. Moreover, since there is a j such that $\alpha \cdot a_j > w_j$, we have $z \cdot a_j < w_j$. Thus z is a solution to (4.1) with $z \cdot a_j < w_j$ for some j. However, as we have already seen, in this case the theorem is true by induction, and this completes the proof of Theorem 4.1.

5. Blocking pairs and anti-blocking pairs

Our purpose in this section is to prove the following theorems, which were mentioned in Section 1.

130   *D.R. Fulkerson et al., On balanced matrices*

Theorem 5.1. *Let A be balanced and let B be the blocking matrix of A. Then the strong max–min equality holds for both A, B and B, A.*

Theorem 5.2. *Let A be balanced with no zero columns and let D be an anti-blocking matrix of A. Then the strong min–max equality holds for both A, D and D, A.*

Note that we have not assumed the $(0,1)$-matrix A in the statement of Theorem 5.1 to be proper; it would be no restriction to do so, however; we could just consider the minimal (essential, in the blocking sense) rows of A.

Proof of Theorem 5.1. That the strong max–min equality holds for the ordered pair A, B follows from Theorem 4.1 by taking the components of the vector c in Theorem 4.1 all equal to ∞.

To show that the strong max–min equality holds in the reverse direction B, A, we first note that [2, Theorem 2] can be rephrased in blocking terminology as follows: Let A be balanced and let B have as its rows all $(0,1)$-vectors that make inner product at least 1 with every row of A and that are minimal with respect to this property (i.e., B is the incidence matrix of the blocking clutter of the clutter of minimal rows of A); then the linear program

$$\begin{aligned}
\text{maximize} \quad & 1 \cdot y, \\
\text{subject to} \quad & y\,B \le 1, \qquad y \ge 0,
\end{aligned} \qquad (5.1)$$

has a $(0,1)$ solution vector y satisfying $1 \cdot y = \min 1 \cdot a^i$, taken over all rows a^i of A. To get the strong max–min equality for B, A from this, we need to pass from the vector 1 on the right-hand side of $y\,B \le 1$ to a general nonnegative integral vector w. This transformation can be effected inductively by first observing that if A is balanced, and if we duplicate a column of A, the resulting matrix A' is balanced [2, Prop. 5]. Pictorially:

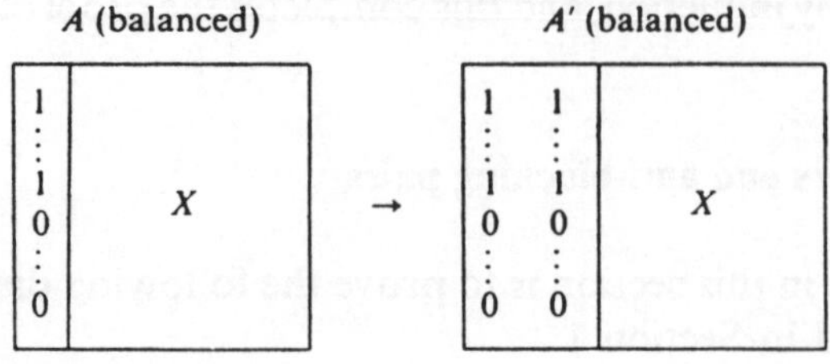

B (blocker of A)　　　　　　　B′ (blocker of A′)

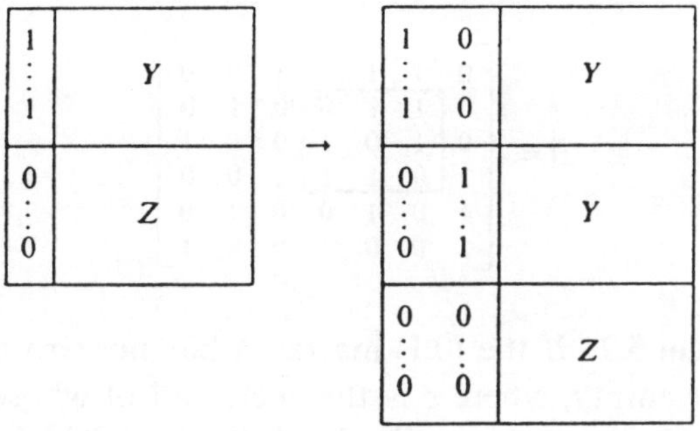

Thus, if the first component of w is 2 instead of 1, we can consider the linear program

$$\begin{aligned}
&\text{maximize} \quad 1 \cdot y, \\
&\text{subject to} \quad y B' \le 1, \qquad y \ge 0
\end{aligned} \tag{5.2}$$

instead of

$$\begin{aligned}
&\text{maximize} \quad 1 \cdot y, \\
&\text{subject to} \quad y B \le (2, 1, \ldots, 1), \qquad y \ge 0.
\end{aligned} \tag{5.3}$$

It follows that a general nonnegative integral vector w can be dealt with by deleting certain columns of A (those corresponding to zero components of w), replicating others, yielding a new balanced matrix, and making the appropriate transformations on the blocker B of A (a zero component of w means that we delete the corresponding column of B and also delete all rows of B that had a 1 in that column). In this way, one can deduce from [2, Theorem 2] that if A is balanced, the strong max–min equality holds for B, A.

In connection with Theorem 5.1 and its proof, we point out that the blocking matrix B of a balanced matrix A may not be balanced. For example, let

$$A = \begin{bmatrix}
1 & 1 & 0 & 0 & 0 & 0 & 0 \\
0 & 0 & 1 & 1 & 0 & 0 & 0 \\
0 & 0 & 1 & 0 & 0 & 0 & 1 \\
0 & 0 & 0 & 1 & 0 & 1 & 0 \\
0 & 0 & 0 & 0 & 1 & 1 & 1
\end{bmatrix}$$

132   *D.R. Fulkerson et al., On balanced matrices*

Matrix A is balanced, with blocking matrix

$$
B = \begin{bmatrix}
0 & 1 & 1 & 1 & 1 & 0 & 0 \\
0 & 1 & 1 & 0 & 0 & 1 & 0 \\
0 & 1 & 0 & 1 & 0 & 0 & 1 \\
1 & 0 & 1 & 1 & 1 & 0 & 0 \\
1 & 0 & 1 & 0 & 0 & 1 & 0 \\
1 & 0 & 0 & 1 & 0 & 0 & 1
\end{bmatrix}
$$

Proof of Theorem 5.2. If the $(0,1)$-matrix A has no zero columns, then $P(A, w, c)$ is not empty, where c is the vector all of whose components are ∞. The strong min–max equality for A, D, where D is an anti-blocking matrix of A, now follows from Theorem 3.2 and the discussion in Section 1 concerning anti-blocking pairs. Moreover, as noted in Section 1, the strong min–max equality for A, D implies the strong min–max equality for D, A.

Theorem 5.2 can be paraphrased as follows. The maximal (essential, in the anti-blocking sense) rows of a balanced matrix A are the incidence vectors of the cliques of a perfect graph G. Consequently, the essential rows of D are the incidence vectors of the anti-cliques of G.

References

[1] C. Berge, *Graphes et hypergraphes* (Dunod, Paris, 1970) ch. 20.
[2] C. Berge, "Balanced matrices", *Mathematical Programming* 2 (1972) 19–31.
[3] D.R. Fulkerson, "Blocking polyhedra", in: *Graph theory and its applications*, Ed. B. Harris (Academic Press, New York, 1970) pp. 93-112.
[4] D.R. Fulkerson, "Anti-blocking polyhedra", *Journal of Combinatorial Theory* 12 (1) (1972) 50–71.
[5] D.R. Fulkerson, "Blocking and anti-blocking pairs of polyhedra", *Mathematical Programming* 1 (1971) 168–194.
[6] D.R. Fulkerson, "On the perfect graph theorem", in: *Mathematical programming*, Eds. T.C. Hu and S.M. Robinson (Academic Press, New York, 1973) pp. 69–76.
[7] L. Lovász, "Normal hypergraphs and the perfect graph conjecture", *Discrete Mathematics* 2 (1972) 253–267.

Mathematical Programming 6 (1974) 352–359. North-Holland Publishing Company

A GENERALIZATION OF MAX FLOW–MIN CUT

A.J. HOFFMAN[*]

IBM Watson Research Center, Yorktown Heights, New York, U.S.A.

Received 20 November 1973
Revised manuscript received 17 April 1974

We present a theorem which generalizes the max flow–min cut theorem in various ways. In the first place, all versions of m.f.–m.c. (emphasizing nodes or arcs, with graphs directed or undirected, etc.) will be proved simultaneously. Secondly, instead of merely requiring that cuts block all paths, we shall require that our general cuts be weight assignments which meet paths in amounts satisfying certain lower bounds. As a consequence, our general theorem will not only include as a special case m.f.–m.c., but also includes the existence of integral solutions to a transportation problem (with upper bounds on row and column sums) and its dual.

1. Introduction

We shall present a theorem which generalizes the max flow–min cut theorem in various ways. In the first place, all versions of m.f.–m.c. (emphasizing nodes or arcs, with graphs directed or undirected, etc.) will be proved simultaneously. Secondly, instead of merely requiring that cuts block all paths, we shall require that our general cuts be weight assignments which meet paths in amounts satisfying certain lower bounds, and we follow Edmonds [2] in our choice of bounds. As a consequence, our general theorem will not only include as a special case m.f.–m.c., but also includes the existence of integral solutions to a transportation problem (with upper bounds on row and column sums) and its dual.

The viewpoint which dominates in this work is very close to the original paper by Ford and Fulkerson [3] on network flows; in fact, the present results can be conceived as an attempt to extract the essence of the arguments given in [3] in a somewhat more general setting.

We shall begin with a finite set U and a system $\mathcal{S} = \{S_0, S_1, ..., S_m\}$ of subsets $S_i \subset U$, $S_0 = \emptyset$. We also assume each non-empty set S_i is linearly

---

[*] This work was supported (in part) by the U.S. Army under contract #DAHC04-72-C-0023.

ordered, and write "$<_i$" to denote the ordering in S_i. It is perfectly possible that we have $\{p, q\} \subset S_i \cap S_j, p <_i q, q <_j p$. We attach to each set S_i a nonnegative integer r_i, so that the $\{r_i\}$ satisfy a condition of "super modularity", which also involves the orderings $\{<_i\}$. First, we require some definitions.

If $p \in S_i \cap S_j$, we define

$$(i, p, j) \equiv \{q: q \in S_i, q <_i p\} \cup \{p\} \cup \{r: r \in S_j, p <_j r\}. \quad (1.1)$$

We always have at least one S_i, $i = 0, 1, \ldots, m$, such that $S_i \subset (i, p, j)$, namely S_0. We require, for all i, j, p such that $p \in S_i \cap S_j$,

$$\max_{k \mid S_k \subset (i,p,j)} r_k + \max_{l \mid S_l \subset (j,p,i)} r_l \geq r_i + r_j; \quad (1.2)$$

we also require

$$r_0 = 0. \quad (1.3)$$

Example 1.1. Let us be given a network with source and sink. Let U consist of all edges in the network, and $\mathcal{S}$ consist of the empty set and all paths from source to sink, with $r_i = 1$ for $i > 0$. Note that (1.2) is satisfied: it states that (i, p, j) contains a subset of edges forming a path from source to sink, which is manifest. If we change this example to consist of nodes in a graph forming paths joining two disjoint sets of nodes, everything applies mutatis mutandis.

Example 2.2. Another example consists of taking an acyclic directed graph G, and a subset $\mathcal{S}$ of directed paths of G such that if S_i and S_j are paths, $p \in S_i \cap S_j$, then (i, p, j), with the "natural" ordering, is also in $\mathcal{S}$. If we assign weights w_j to the elements p_j, and define $r_i = \Sigma_{p_j \in S_i} w_j$, then (1.2) and (1.3) hold.

Example 2.3. Let G be a bipartite graph, with parts U_1 and U_2. Let $\mathcal{S}$ consist of the empty set and all the edges (u_i, v_j) of G, with $u_i \in U_1$, $v_j \in U_2$, $u_i < v_j$. Let $r_{ij} \geq 0$ be an arbitrary nonnegative integer assigned to the edge (u_i, v_j). Then, if $S_a = (u_i, v_j)$, $S_b = (u_i, v_k)$, $p = u_i \in S_a \cap S_b$, it is clear that $S_b \subset (a, p, b)$, $S_a \subset (b, p, a)$ and (1.2) holds. Similarly, (1.2) holds if $S_a = (u_i, v_j)$, $S_b = (u_k, v_j)$.

            A.J. Hoffman, A generalization of max flow–min cut

Theorem 2.4. *Let* $U = \{p_1, \ldots, p_n\}$, $\mathcal{S} = \{S_0, S_1, \ldots, S_m\}$. *Let* $c = (c_1, \ldots, c_n)$ *have all its components nonnegative integers. Let* $r = (r_0, r_1, \ldots, r_m)$ *satisfy* (1.2) *and* (1.3). *Then the dual linear programming problems*

$$\text{minimize} \quad \sum_j c_j x_j \,,$$

$$\text{subject to} \quad x_j \geqslant 0, \quad \sum_{p_j \in S_i} x_j \geqslant r_i \,, \tag{1.4}$$

$$\text{maximize} \quad \sum_i r_i y_i \,,$$

$$\text{subject to} \quad y_i \geqslant 0, \quad \sum_{p_j \in S_i} y_i \leqslant c_j \tag{1.5}$$

each have integral optimal vectors.

In view of the examples cited, we assert that Theorem 2.4 fulfills the claims of the first paragraph of the introduction.

2. Proof of the theorem

Let A be a $(0,1)$ matrix with rows $0, \ldots, m$, columns $1, \ldots, n$, with $a_{ij} = 1$ if and only if $p_j \in S_i$.

Lemma 2.1. *If* $y^{\mathrm{T}} A \leqslant c^{\mathrm{T}}$, $y \geqslant 0$, *and maximize* $\Sigma\, y_i\, r_i$ *has for all* $c \geqslant 0$ *and integral always an optimal integral vector, then* $A\, x \geqslant r$, $x \geqslant 0$ *has all vertices integral.*

Proof. Suppose $\bar{x} = (\bar{x}_1, \ldots, \bar{x}_n)$ is a vertex of $P = \{x : A\, x \geqslant r, x \geqslant 0\}$, and $\bar{x}_1$ (say) not an integer. Let $T = \{j : \bar{x}_j = 0\}$, $V = \{i : \Sigma a_{ij} \bar{x}_j = r_i\}$, then the linear form $\Sigma_{i \in V} \Sigma_j a_{ij} x_j + \Sigma_{j \in T} x_j$ is of the form $\Sigma d_j x_j$, where each d_j is a nonnegative integer $(d, \bar{x}) = \Sigma_{i \in V} r_i$, and $(d, \bar{x}) < (d, x)$ for all $x \in P$, $x \neq \bar{x}$. Further, we may assume each $d_j > 0$. Now let $t = 1, 2, 3, \ldots$. For each t, let $d(t) = (t\, d_1 + 1, t\, d_2, \ldots, t\, d_n)$. Suppose that for each t there is a vector $x(t) \in P$ such that $(d(t), \bar{x}) > (d(t), x(t))$. If the set of vectors $x(1), x(2)$ is unbounded, this violates $d_j > 0$ for all j. So the set of vectors $x(1), x(2)$ has an accumulation point $\bar{\bar{x}} = (\bar{\bar{x}}_1, \ldots, \bar{\bar{x}}_2)$, and

$$t(\dot{d}, x) + \bar{x}_1 \geqslant t(d, \bar{\bar{x}}) + \bar{\bar{x}}_1 - \epsilon(t), \tag{2.1}$$

where $\epsilon(t) \to 0$ as $t \to \infty$. But (2.1) implies

$$(d, \bar{x}) + \bar{x}_1/t \geqslant (d, \bar{\bar{x}}) + \bar{\bar{x}}_1/t - \epsilon(t)/t \,.$$

Letting $t \to \infty$, we get $(d, \bar{x}) \geqslant (d, \bar{\bar{x}})$, a contradiction.

So there exists some value of t, say t_0, such that $\Sigma\, t_0\, d_j\, x_j$ and $(t_0\, d_1 + 1)\, x_1 + \Sigma_{j>1}\, t_0\, d_j\, x_j$ both attain their minimum at $\bar{x}$. Since both minima are integers (by hypothesis and the duality theorem), it follows that their difference $\bar{x}_1$ is an integer, which is a contradiction, completing the proof.

Now define a function $f(i, p, j)$, where $p \in S_i \cap S_j$, by the stipulation $r_{f(i,p,j)} = \max_{k\,|\,S_k \in (i,p,j)} r_k$. Then (1.2) can be rewritten as

$$r_{f(i,p,j)} + r_{f(j,p,i)} \geqslant r_i + r_j. \tag{2.2}$$

Lemma 2.2. *Let* $x \in P = \{x: Ax \geqslant r, x \geqslant 0\}$, *and let* $\delta_x = \{i: \Sigma\, a_{ij}\, x_j = r_i\}$. *Then* δ_x *satisfies* (1.2) *and* (1.3).

Proof. Clearly $0 \in \delta_x$. So all we need to show is that $i, j \in \delta_x$, $p \in S_i \cap S_j$ implies $f(i, p, j), f(j, p, i) \in \delta_x$.

$$S_{f(i,p,j)} \cup S_{f(j,p,i)} \subset S_i \cup S_j, \qquad S_{f(i,p,j)} \cap S_{f(j,p,i)} \subset S_i \cap S_j \,,$$

and thus:

$$r_{f(i,p,j)} + r_{f(j,p,i)} \leqslant \sum_{p_t \in S_{f(i,p,j)}} x_t + \sum_{p_t \in S_{f(j,p,i)}} x_t$$

$$\leqslant \sum_{p_t \in S_i} x_t + \sum_{p_t \in S_j} x_t = r_i + r_j. \tag{2.3}$$

Combined with (2.2), (2.3) yields Lemma 2.2. In fact, the proof of Lemma 2.2. implies the following corollary.

Corollary 2.3. *If* $x \in P$, $U_x = \{k: x_k > 0\}$, *then* $i, j \in \delta_x$, $p \in S_i \cap S_j$ *imply* $a_{ik} + a_{jk} = a_{f(i,p,j),k} + a_{f(j,p,i),k}$ *for all* $k \in U_x$.

For the remainder of the proof of Theorem 2.4, we will argue by induction on m, since the theorem is evidently true for $m = 0$. So we assume we have the smallest value of m for which the theorem is false. Let $\bar{x}$ be any optimal solution to minimize (c, x), $x \in P$. By Lemma 2.2 and the induction hypothesis, $S_{\bar{x}} = (0, 1, ..., m)$.

Lemma 2.4. *Let $Q = \{y : y^T A \leqslant c^T, y \geqslant 0\}$, and let Y_0 be the set of all vectors satisfying*

$$y_i \geqslant 0, \quad \sum_i y_i a_{ij} = c_j \quad \text{for } j \in U_{\bar{x}}, \quad \sum_i y_i a_{ij} \leqslant c_j, \quad \text{for } j \notin U_{\bar{x}}. \tag{2.4}$$

Then

$$y_0 \in Y_0 \quad \text{implies} \quad (y_0, r) \geqslant (y, r) \quad \text{for all } y \in Q. \tag{2.5}$$

Let Y_1 be the set of solutions to the linear programming problem

$$\underset{Y \in Y_0}{\text{minimize}} \sum_{j \notin U_{\bar{x}}} \sum_i y_i a_{ij}. \tag{2.6}$$

Then $y \in Y_1$, $y_i > 0$, $y_j > 0$, $p \in S_i \cap S_j$ implies

$$a_{ik} + a_{jk} = a_{f(i,p,j),k} + a_{f(j,p,i),k} \quad \text{for all } k. \tag{2.7}$$

Let $Y_2 \subset Y_1$ consist of all vectors in Y_1 in which the largest number of coordinates is positive. Then

$$y \in Y_2 \text{ implies all coordinates of } y \text{ are positive.} \tag{2.8}$$

Proof. That (2.4) implies (2.5) follows from examining the bilinear form $y^T A \bar{x}$, which shows that $y \in Y_0$ is feasible and optimal for the problem maximize (y, r), $y \in Q$. Next, suppose $y \in Y_1$. Since the right side of (2.7) is at most the left side, suppose it is strictly less than the left side. Decrease y_i and y_j by $\epsilon > 0$, where $y_i > \epsilon$, $y_j > \epsilon$, and add ϵ to $y_{f(i,p,j)}$ and $y_{f(j,p,i)}$. The new vector will satisfy (2.4) (by Corollary 2.3), and will give a smaller value for (2.6).

Next, from (2.7) and ϵ-changes, it follows that, if $y \in Y_2$, $y_i > 0$, $y_j > 0$, $p \in S_i \cap S_j$, then $y_{f(i,p,j)} > 0$ and $y_{f(j,p,i)} > 0$. Also, we can assume $y_0 > 0$. Hence by the induction hypothesis, $\{i : y_i > 0\}$ consists of all of $\{0, 1, ..., m\}$.

Lemma 2.5. *Let G be a directed graph on nodes $\{p_1, \ldots, p_n\}$ with a directed edge from p_i to p_j if and only if there is an S_k such that $p_i, p_j \in S_k$ and p_j immediately follows p_i in $<_k$. Then G is an acyclic graph. Further, if $B = \{p_i : p_i$ is a minimal element in some $S_k\}$, $E = \{p_i : p_i$ is a maximal element in some $S_k\}$, then $\mathcal{S}$ consists of the empty set and all directed paths in G with initial node in B and terminal node in E (note B and E may overlap).*

Proof. We first show that there do not exist k and l such that $p_i <_k p_j$, $p_j <_l p_i$. If they do exist, since $p_i \in S_k \cap S_l$, we have a violation of (2.7) on the column index j.

Second, if $p_i <_1 p_j$, $p_j <_2 p_k$, it follows from (2.7) that for $l = f(1, p_j, 2)$, S_l contains p_i, p_j and p_k. Further, $p_i <_l p_j$, otherwise the preceding paragraph is violated, and $p_j <_l p_k$ for the same reason. Consequently, $p_i <_l p_k$. This implies that the graph G is acyclic.

Now, to prove the second assertion of the lemma, let P be any directed path in G of the form $p_0, p_1, p_2, \ldots, p_k$, where $p_0 \in B$, $p_k \in E$. There is an S_l beginning with p_0, by the definition of B. Assume there is an S_l beginning with $p_0, p_1, \ldots, p_i$, $i < k$, we shall find some S_t beginning with $p_0, p_1, \ldots, p_i, p_{i+1}$. Since there is an S_r in which p_{i+1} follows p_i, let $t = f(l, p_i, r)$. By (2.7), S_t contains $p_0, p_1, \ldots, p_i, p_{i+1}$. Further, their order must be $p_0 <_t p_1 <_t \cdots <_t p_i <_t p_{i+1}$, otherwise we violate the preceding paragraph of the proof. Also, p_0 must be the minimal element of S_t, otherwise S_l contains an element preceding p_0 (impossible by the definition of S_l) or S_r contains an element $>_l p_i$, but $<_t p_0$, which is also impossible. Similar reasoning shows that S_t contains no element between p_j and p_{j+1} ($j = 0, \ldots, i$).

Continuing in this way, we obtain an S_l beginning $p_0, p_1, \ldots, p_k$. Since $p_k \in E$, there is an S_r whose maximal element is p_k. Then $S_{f(l, p_k, r)}$ is the desired path.

3. Completion of proof

In view of Lemmas 2.1, 2.4 and 2.5, all we need to show is that, if y satisfies (2.4), there is an integral vector y satisfying (2.4).

Let y satisfy (2.4), $a \geqslant \Sigma y_i$, a an integer. We shall construct a matrix Z with $(n + 1)$ rows and columns as follows. Temporarily ignore the diagonal entries. For $j = 1, \ldots, n$, set $z_{0j} = \Sigma' y_k$, where the sum is taken over all k such that p_j is a minimal element of S_k. Similarly, $z_{i0} = \Sigma' y_k$,

                    A.J. Hoffman, A generalization of max flow—min cut

where the sum is taken over all k such that p_i is a maximal element of S_k. For $i \neq j$, $i, j = 1, \ldots, n$, set $z_{ij} = \Sigma' y_k$, where the sum is taken over all k such that p_j immediately follows p_i in S_k. Finally, set

$$z_{00} = a - \Sigma z_{0j} = a - \Sigma z_{i0} ;$$

$$z_{jj} = c_j - \sum_i y_i a_{ij} = c_j - \sum_{\substack{i=0 \\ i \neq j}}^{n} z_{ij} = c_j - \sum_{\substack{i=0 \\ i \neq j}}^{n} z_{ji} .$$

Observe that Z is a nonnegative matrix satisfying the conditions that all row and column sums are integers. Consequently, Z is a convex sum of nonnegative integral matrices. It follows that there exists a nonnegative integral $W = (w_{ij})$ satisfying $\Sigma w_{0j} = \Sigma w_{i0} = a$, $\Sigma_j w_{ij} = \Sigma_i w_{ji} = c_i$, and $w_{ij} = 0$ whenever $z_{ij} = 0$. In particular, $w_{ii} = 0$ for all $i \in U_{\bar{x}}$, and there is no cycle $w_{i_1 i_2} > 0$, $w_{i_2 i_3} > 0, \ldots, w_{i_k i_1} > 0$ of length exceeding 1 if all $i_j > 0$.

Now let $w_{0j_1} > 0$. It follows that $c_{j_1} > 0$, and there must be a $j_2 \neq j_1$ such that $w_{j_1 j_2} > 0$. If $j_2 = 0$, stop. If not, there must be a $j_3 \neq j_2$ such that $w_{j_2 j_3} > 0$. If $j_3 = 0$, stop; otherwise, continue. Since there are no cycles not involving 0, it follows that we must eventually stop by finding some k such that $w_{j_{k-1}, j_k} > 0$ and $j_k > 0$. Let $w = \min (w_{0j_1}, w_{j_1 j_2}, \ldots, w_{j_{k-1} 0})$. Then $\{p_{j_1}, p_{j_2}, \ldots, p_{j_{k-1}}\}$ is some S_l. We assign the integer weight $w_l = w$ to S_l, subtract w from each w_{ij} in the cycle, subtract w from a and from each c_j in the cycle, and repeat the process. We continue until all off-diagonal elements in the first row and column are 0. Since the remaining matrix contains no cycle of positive elements, only diagonal entries, if any, are left. But since $w_{ii} = 0$ for all $i \in U_{\bar{x}}$, it follows that our assignment of weights w_l has produced nonnegative integers satisfying (2.4). In view of (2.5) and Lemma 2.1, the theorem is proved.

Remark 3.1. The device used in Section 3 is an adaptation of [1]. But for the case all $r_i = 1$ for $i > 1$, it is not needed, as the interested reader will discover for himself.

Acknowledgment

We are very grateful to D.R. Fulkerson and Ellis L. Johnson for useful conversations about this material.

References

[1] G.B. Dantzig and A.J. Hoffman, "Dilworth's theorem on partially ordered sets", in: *Linear inequalities and related systems*, Annals of mathematics study No. 38, Eds. H.W. Kuhn and A.W. Tucker (Princeton University Press, Princeton, N.J., 1956) pp. 207—214.
[2] J. Edmonds, "Submodular functions, matroids and certain polyhedra", in: *Combinatorial structures and their applications*, Eds. H. Guy, H. Hanani, N. Sauer and J. Schonheim (Gordon and Breach, New York, 1970) pp. 69—87.
[3] L.R. Ford, Jr. and D.R. Fulkerson, "Maximal flow through a network", *Canadian Journal of Mathematics* 8 (1956) 399—404.

Mathematical Programming Study 8 (1978) 197–207.
North-Holland Publishing Company.

ON LATTICE POLYHEDRA III: BLOCKERS AND ANTI-BLOCKERS OF LATTICE CLUTTERS

A.J. HOFFMAN*

IBM T.J. Watson Research Center, Yorktown Heights, New York, U.S.A.

Received 3 February 1977
Revised manuscript received 30 May 1977

We consider two classes (called upper and lower) of clutters satisfying postulates we have previously encountered in defining lattice polyhedra, and prove that lower clutters are maximal anti-chains in a partially ordered set, upper clutters are cuts of a family of paths closed with respect to switching.

1. Introduction

In [8], we introduced the concept of lattice polyhedron to give a unification of various theorems of Fulkerson [3], Greene [4], Johnson [9], and Greene and Kleitman [5], as well as to derive new extremal combinatorial theorems. Methods for constructing various lattice polyhedra, including the polymatroid intersection polyhedra, were given in [7]. Lattice polyhedra are defined in terms of a partially ordered set $\mathcal{L}$ admitting certain lattice-like operations, together with certain mappings from $\mathcal{L}$ into subsets of a set $\mathcal{U}$. It is desirable to have more homely descriptions of these combinatorial objects, and we have succeeded in doing this in two important special cases. The principal tool is Fulkerson's theory of blocking and anti-blocking polyhedra [3], and the Ford-Fulkerson max flow-min cut theorem [2] in the formulation given in [6].

Although the motivation for this investigation is in polyhedral combinatorics, most of our discussion can be cast in such a way that no background in linear programming is required except for citing appropriate references.

2. Lattice clutters

We shall be dealing throughout with a fixed finite set $\mathcal{U}$ and a family $\mathcal{L}$ of subsets of $\mathcal{U}$ forming a clutter. This means that $\mathcal{L}$ is not empty, and that $S, T \in \mathcal{L} \Rightarrow S \not\subseteq T$. In particular, $\emptyset \notin \mathcal{L}$. We assume that $\mathcal{L}$ is partially ordered by "$\leq$" without specifying the source of that partial order. In general, it is not set inclusion. We assume further that the partial order on $\mathcal{L}$ satisfies

$$R < S < T \Rightarrow (R \cap T) \subset S. \tag{2.1}$$

* This work was supported (in part) by the Army Research Office under contract number DAAG29-74C-0007.

Next, we assume that, for every $S, T \in \mathcal{L}$, there exist $S \wedge T$ and $S \vee T \in \mathcal{L}$ (in general, *not* set intersection and union). These operations satisfy

and

$$S \wedge T = T \wedge S; \quad S \wedge T \leq S, T; \quad S \leq T \Rightarrow S \wedge T = S; \tag{2.2}$$

$$S \vee T = T \vee S; \quad S, T \leq S \vee T; \quad S \leq T \Rightarrow S \vee T = T. \tag{2.3}$$

We are not assuming that $\mathcal{L}$ is a lattice, however: $S \vee T$ is some upper bound to S and T, not necessarily a least upper bound.

Assume that $\mathcal{L}$ satisfies also

$$(S \wedge T) \cup (S \vee T) \subset (S \cup T). \tag{2.4}$$

Then we call $\mathcal{L}$ an *upper clutter*.

If, instead of (2.4), we assume

$$(S \cup T) \subset (S \wedge T) \cup (S \vee T) \quad \text{and} \quad (S \cap T) \subset (S \wedge T) \cap (S \vee T), \tag{2.4'}$$

then $\mathcal{L}$ is called a *lower clutter*. Note, that, in the case of an upper clutter,

$$(S \wedge T) \cap (S \vee T) \subset (S \cap T)$$

follows from (2.1).

Thus, an upper clutter satisfies (2.1)–(2.4), a lower clutter satisfies (2.1)–(2.3) and (2.4').

Let $\mathcal{M}$ be any family of subsets of $\mathcal{U}$. The blocker of $\mathcal{M}$, donated by $\mathcal{B}(\mathcal{M})$, consists of all subsets $B \subset \mathcal{U}$ such that, for all $S \in \mathcal{M}$, $B \cap S \neq \emptyset$, but this statement is false for any proper subset of B. Similarly, the anti-blocker of $\mathcal{M}$, denoted by $\mathcal{A}(\mathcal{M})$, consists of all subsets $B \subset \mathcal{U}$ such that, for all $S \in M$, $|B \cap S| \leq 1$, but this statement is false for any set properly containing B. We will be considering blockers of upper clutters and anti-blockers of lower clutters.

> *Problem* 1. Describe all upper clutters and their corresponding blockers.
>
> *Problem* 2. Describe all lower clutters and their corresponding anti-blockers.

3. Solution to Problem 1

In this section, we give the required description, deferring the proof of its correctness to later. Similarly, we will give the solution to Problem 2 in Section 4, and its proof later.

Let G be a directed graph with two nodes labelled 0 and 1, which may be thought of as source and sink, such that there is no edge from 0 to 1. A path from source to sink is a sequence of distinct nodes $0 = a_0, a_1, \ldots, a_{k-1}, a_k = 1$ such that there is a directed edge from a_i to a_{i+1} for $i = 0, \ldots, k - 1$.

If p and q are source-sink paths with at least one common node $x \neq 0, 1$, the symbol (p, x, q) means the set of nodes consisting of 0 and all nodes in p preceding x on path p, together with x, together with all nodes in q following x on path q including 1.

A clutter $\mathcal{P}$ of $(0, 1)$ paths in G is said to be closed with respect to switching if

$$p, q \in \mathcal{P}, \quad x \in p \cap q \text{ implies there exists an } r \in \mathcal{P} \tag{3.1}$$

all of whose nodes are in (p, x, q). (See [6] or Section 7 for an alternate statement of these concepts.)

The following figure gives a clutter of paths in a directed graph G closed with respect to switching. We omit the initial 0 and terminal 1.

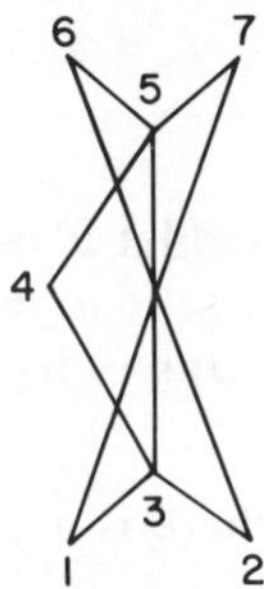

Fig. 1. A set of paths P closed with respect to switching: 1356, 17, 26, 23457.

Next, consider all minimal subsets S of the nodes of $\mathscr{P}$ other than 0 and 1 with the property that, for every $p \in P$, $S \cap p \neq \emptyset$. Call the collection of such subsets $\mathscr{L}$. We now partially order $\mathscr{L}$. For every path $p \in \mathscr{P}$,

$$p = \{0 = a_0, a_1, a_2, \ldots, a_{k-1}, a_k = 1\}, \tag{3.2}$$

and for every $S \in \mathscr{L}$, let $p(S) = \min\{i \mid a_i \in S\}$.

Because S meets every path, $p(S)$ is always defined. We say that $S \leq T$ if, for every $p \in \mathscr{P}$,

$$p(s) \leq p(T). \tag{3.3}$$

Later on we shall show that $S \leq T$ and $T \leq S$ implies $S = T$. Assuming this has been shown, it is manifest that we have a partial order of the sets in $\mathscr{L}$.

We now define $S \wedge T$. This is the set of nodes x in $S \cup T$ each of which satisfies the condition: for some $p \in \mathscr{P}$ given by (3.2), $x = a_i$, and $(S \cup T) \cap \{a_1, \ldots, a_{i-1}\} = \emptyset$. Similarly, $S \vee T$ is the set of nodes x in $S \cup T$ each of which satisfies the condition: for some $p \in \mathscr{P}$ given by (3.2), $x = a_i$ and $(S \cup T) \cap \{a_{i+1}, \ldots, a_{k-1}\} = \emptyset$. We shall show later that $S \vee T$ and $S \wedge T$ are in $\mathscr{L}$, and they are in fact the l.u.b. and g.l.b. respectively of S and T in the partial ordering of $\mathscr{L}$, i.e., $\mathscr{L}$ is a lattice. Further, all of (2.1) and (2.4) are satisfied, with the "$\vee$" and "$\wedge$" of (2.2)–(2.4) precisely the lattice operation. We illustrate this in Fig. 2 describing the lattice $\mathscr{L}$ corresponding to the paths from Fig. 1.

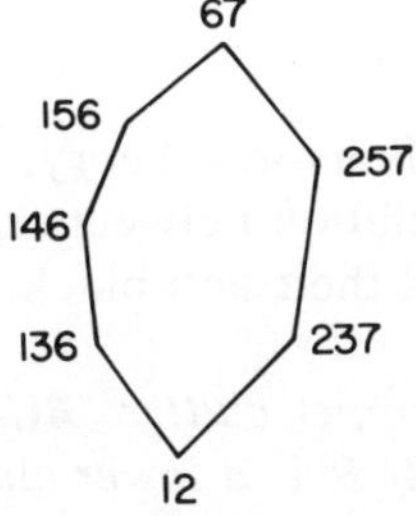

Fig. 2. Hasse diagram of L.

Theorem 3.1. *If $\mathcal{L}$ is any upper clutter on a finite set $\mathcal{U}$ and $\mathcal{B}(\mathcal{L})$ its blocker, then the elements of $\mathcal{U}$ may be identified with the nodes $\neq 0, 1$ of a directed graph G, $\mathcal{B}(\mathcal{L})$ is a clutter of $(0, 1)$ paths in G closed with respect to switching, and $\mathcal{L}$ is the blocker of $\mathcal{B}(\mathcal{L})$, with the lattice structure given by (3.3).*

4. Solution to Problem 2

Let Q be a partially ordered set, and let $\mathcal{L}$ be the set of all maximal anti-chains in Q. We partially order $\mathcal{L}$ as follows. If a and b are maximal anti-chains in Q, then each element in a is comparable to at least one element of b and conversely. We define

$$a \leq b \text{ if for each } x \in a, \text{ there is a } y \in b \text{ such that } x \leq y. \tag{4.1}$$

It is clear that this is a partial ordering, and we shall show later that, under this partial ordering, $\mathcal{L}$ is a lattice. Further, all of (2.1)–(2.3) and (2.4') are satisfied. But, in contrast with the situation described in Theorem (3.1), the "$\vee$" and "$\wedge$" of (2.2), (2.3) and (2.4') do not necessarily have to be the lattice operations. For example, Fig. 3 gives the Hasse diagram for a partially ordered set Q and for the corresponding $\mathcal{L}$. But if $b \vee c$ is taken to be e, rather than d, (2.2), (2.3) and (2.4') are also.

Fig. 3. $a = \{1, 2\}$, $b = \{2, 3\}$, $c = \{1, 5\}$, $d = \{3, 4, 5\}$, $e = \{3, 6, 5\}$.

Theorem 4.1. *If $\mathcal{L}$ is any lower clutter, and $\mathcal{A}(\mathcal{L})$ its anti-blocker, then $\mathcal{A}(\mathcal{L})$ is the set of all maximal chains of a partially ordered set Q, and $\mathcal{L}$ is the set of maximal anti-chains of Q, partially ordered by (4.1). $\mathcal{L}$ is a lattice, and the operations $\vee$ and $\wedge$ of (2.2), (2.3) and (2.4') can be the lattice operations.*

5. Blockers and anti-blockers

In this section, we invoke Fulkerson's theory, together with results of [1] or [8] to establish the fundamental relation between upper clutters and their blockers, and between lower clutters and their anti-blockers.

Proposition 5.1. (a) *If $\mathcal{L}$ is an upper clutter, $\mathcal{B}(\mathcal{L})$ its blocker, then the sets of $\mathcal{L}$ from the blocker of $\mathcal{B}(\mathcal{L})$.* (b) *If $\mathcal{L}$ is a lower clutter, $\mathcal{A}(\mathcal{L})$ its anti-blocker, then the sets of $\mathcal{L}$ from the anti-blocker of $\mathcal{A}(\mathcal{L})$.*

Proof. (a) follows from [1]. Alternatively, let $\mathscr{L}$ be an upper clutter, $\mathscr{L}$ the corresponding incidence matrix; i.e., the rows of L corresponding to the sets of $\mathscr{L}$, the columns to the elements of $\mathscr{U}$, and $L_{su} = 1$ if $u \in S$, $= 0$ if $u \notin S$. It is shown in [8] that (2.1)–(2.4) imply that the vertices of the polyhedron

$$Lx \geq \vec{1}, \quad x \geq 0$$

are the set of all $(0, 1)$ vectors y such that the elements u satisfying $y_u = 1$ meet every set of $\mathscr{L}$ but no proper subset of these elements does. Hence the vertices of the polyhedron are the sets in $\mathscr{B}(\mathscr{L})$. By [3], the sets of $\mathscr{L}$ form the blocker of $\mathscr{B}(\mathscr{L})$.

Next, let $\mathscr{L}$ be a lower clutter, and L the corresponding incidence matrix. By [7], (2.1)–(2.3) and (2.4′) imply that the vertices of

$$Lx \leq \vec{1}, \quad x \geq 0$$

are all $(0, 1)$ vectors y such that the set of elements u satisfying $y_u = 1$ meet each set of $\mathscr{L}$ in at most one element. This means that the vertices of the polyhedron consist of the sets in $\mathscr{A}(\mathscr{L})$ and all their subsets. By [3], this implies that the sets in $\mathscr{L}$ are the anti-blocker of $\mathscr{A}(\mathscr{L})$.

6. Proof of Theorem 4.2

We begin with Theorem 4.2 because its proof is easier. Suppose $\mathscr{L}$ is a lower clutter. We construct a partial ordering of the elements of $\mathscr{U}$, to produce a partially ordered set Q, as follows. Let $x \in \mathscr{U}$, and consider all the sets in $\mathscr{L}$ which contain x. Call this family of sets $\mathscr{L}(x)$. If $S, T \in \mathscr{L}(x)$, so is $S \wedge T$ by (2.4′). It follows that $\mathscr{L}(x)$ contains a least element $\underline{l}(x)$. We define the partially ordered set Q by the rule:

$$x < y \text{ in } Q \text{ if } \mathscr{L}(x) \cap \mathscr{L}(y) = \emptyset \text{ and } \underline{l}(x) < \underline{l}(y) \text{ in } \mathscr{L}. \tag{6.1}$$

We first prove

$$\text{if } \mathscr{L}(x) \cap \mathscr{L}(y) = \emptyset, \text{ then either } x < y \text{ or } y < x. \tag{6.2}$$

For assume the contrary. Then $\underline{l}(x)$ and $\underline{l}(y)$ are incomparable in $\mathscr{L}$. Consider $\underline{l}(x) \wedge \underline{l}(y)$. It cannot contain both x and y since $\mathscr{L}(x) \cap \mathscr{L}(y) = \emptyset$. Suppose it contains neither x nor y. By (2.4′), it follows that $\underline{l}(x) \vee \underline{l}(y)$ contains both x and y, a contradiction. Hence $\underline{l}(x) \wedge \underline{l}(y)$ contains, say, x. But $x \in \underline{l}(x) \wedge \underline{l}(y) < \underline{l}(x)$, contradicting the definition of $\underline{l}(x)$.

Next, define $\bar{l}(x)$ to be the maximal element in $\mathscr{L}(x)$. Then

$$\text{if } x < y \text{ in } Q, \text{ then } \bar{l}(x) < \bar{l}(y) \text{ in } \mathscr{L}. \tag{6.3}$$

The reason is that, using arguments analogous to the foregoing, either $\bar{l}(x) < \bar{l}(y)$ or $\bar{l}(x) > \bar{l}(y)$. Suppose the latter occurs. Then we have in $\mathscr{L}$

$$\underline{l}(x) < \underline{l}(y) \leq \bar{l}(y) < \bar{l}(x).$$

But $\mathscr{L}(x) \cap \mathscr{L}(y) = \emptyset$ and the preceding chain contradict (2.1).

More generally,

if $x < y$ in Q, $l(x) \in \mathcal{L}(x)$, $l(y) \in \mathcal{L}(y)$, and $l(x)$ and $l(y)$ are
comparable in $\mathcal{L}$, then $l(x) < l(y)$. $\hspace{2cm}$ (6.4)

This has the same proof as (6.3).

We now show that Q is partially ordered by proving transitivity.

If, in Q, $x < y$ $\quad$ and $\quad$ $y < z$, then $x < z$. $\hspace{2cm}$ (6.5)

We first show that $\mathcal{L}(x) \cap \mathcal{L}(z) = \emptyset$. Suppose not, and $S \in \mathcal{L}(x) \cap \mathcal{L}(z)$. Then

$$\underline{l}(y) < \underline{l}(z) \le S \le \bar{l}(x) \text{ in } \mathcal{L},$$

implying $\underline{l}(y) < \bar{l}(x)$, contradicting (6.4). It follows that $x < z$ or $z < x$. Suppose the latter occurs. Then

$$\underline{l}(z) < \underline{l}(x) < \underline{l}(y) \text{ in } \mathcal{L},$$

contradicting $y < z$ in Q.

It follows from (6.1), (6.2) and (6.5) that, if $\mathcal{L}$ is a lower clutter, then $\mathcal{A}(\mathcal{L})$ consists of all maximal chains in a partially ordered set Q. By proposition 5.1(b), the sets in $\mathcal{L}$ are all maximal anti-chains of Q. We shall give the set of these anti-chains a partial order, as explained in Section 4, and call the resulting partially ordered set $\mathcal{L}^*$. Our first task is to show that $\mathcal{L}^*$ and $\mathcal{L}$ have the same partial order. We first note the following alternate definitions of the partial order in $\mathcal{L}^*$, both equivalent to the definition given in Section 4.

$S \le T$ in $\mathcal{L}^*$ if for each element y of T then is an $x \in S$
such that $x \le y$ in Q. $\hspace{2cm}$ (6.6)

$S \le T$ in $\mathcal{L}^*$ if whenever $x \in S - S \cap T$ and $y \in T - S \cap T$
are comparable in Q, then $x < y$ in Q. $\hspace{2cm}$ (6.7)

Now assume $S < T$ in $\mathcal{L}$, and suppose (6.7) does not hold. This means that there exists $x \in S - S \cap T$ and $y \in T - S \cap T$ with $y < x$ in Q. But $S \in \mathcal{L}(x)$, $T \in \mathcal{L}(y)$ and we have violated (6.4). Next, assume $S < T$ in $\mathcal{L}^*$, and consider $S \wedge T$ in $\mathcal{L}$. But (2.4'), $S \cap T \subset S \wedge T$. If $S \wedge T$ contained all of S, then $S \wedge T = S$ (since the sets in $\mathcal{L}$ are maximal anti-chains of Q), whence we would have $S < T$ in $\mathcal{L}$. So assume $S \wedge T$ does not contain all of S. Thus there exists an element $x \in S - S \wedge T$, and we know $x \notin S \cap T$. By (2.4'), $x \in S \vee T$. Since $T < S \wedge T$ in $\mathcal{L}$, we know the conclusion of (6.7) holds for T and $S \vee T$ by the first two sentences of this paragraph. Therefore the conclusion of (6.6) holds, and there is a $y \in T$ such that $x \ge y$ in Q. We cannot have $y = x$, so $x > y$. But this contradicts (6.7) for S and T. Thus the partial order in $\mathcal{L}^*$ agrees with the partial order in $\mathcal{L}$, and the reader should note that (2.1) obviously holds.

We will be finished with the proof of Theorem 4.1 if we show that $\mathcal{L}^*$ is a lattice and that the lattice operations satisfy (2.2), (2.3) and (2.4'). Let $S, T \in \mathcal{L}^*$. We define

$$S \wedge T = \{x \in Q \mid \exists y \in S, z \in T \text{ with } x \le y,\ x \le \cdot\ \ \nexists x' \in Q,$$
$$y' \in S, z' \in T \text{ with } x' \le y',\ x' \cdot \qquad x < x'\} \hspace{1cm} (6.8)$$

Similarly, define

$$S \vee T = \{x \in Q \mid \exists y \in S, z \in T \text{ with } y \le x, z \le x; \not\exists x' \in Q,$$
$$y' \in S, z' \in T \text{ with } y' \le x', z' \le x', x' < x\}. \tag{6.9}$$

We must first show that $S \wedge T$ is a maximal anti-chain in Q, and that $S \wedge T$ is indeed the l.u.b. of all $V \in \mathcal{L}^*$ such that $V \le S$, $V \le T$. The analogous arguments will apply to $S \vee T$.

It is clear that $S \wedge T$ is an anti-chain. Let us see that $S \wedge T$ is a maximal anti-chain. Suppose $x \in Q$ is not comparable to any element of $S \wedge T$. Let us first note that all elements of $S \cup T$ which are minimal elements, with respect to the ordering in Q of $S \cup T$, are in $S \wedge T$. Next, suppose $x \in S$. Since $x \notin S \wedge T$, x must be a maximal element of $S \cup T$ which is not in $S \wedge T$. Therefore there must be some $y \in S \wedge T$ (indeed $y \in \{$minimal elements of $S \cup T\}$) such that $y < x$. This contradicts the assumption that x is not comparable to any element of $S \wedge T$. So $x \notin S$, and similarly $x \notin T$. Therefore, there exist $y \in S$, $z \in T$ such that x is comparable to y and to z. We cannot have $x < y$ and $x < z$, for then x would precede some element of $S \wedge T$. On the other hand, if $x > y$, (or z), then y is either a minimal element or maximal element of $S \cup T$, hence $y \ge w$ for some $w \in S \wedge T$ so $x > w$, a contradiction.

Next, suppose that in $\mathcal{L}^*$, $V \le S$, $V \le T$. Then for each $v \in V$, there exist $x \in S$ and $y \in T$ such that $v \le x$ and $v \le y$. But (6.8) shows that $V \le S \wedge T$. Thus $S \wedge T$ is the l.u.b. of all V such that $V \le s$ and $V \le T$.

All that remains is to verify (2.4'). But we have already done this by remarking that the maximal elements of $S \cup T$ are in $S \vee T$ and the minimal elements of $S \cap T$ are in $S \vee T$. This completes the proof of Theorem 4.1.

7. Proof of Theorem 3.1

Let $\mathcal{L}$ be an upper clutter. In order to study $\mathcal{B}(\mathcal{L})$ we begin by considering the following structure [6]. In the set $\mathcal{U}$ suppose $\mathcal{P}$ is a family of non-empty subsets of $\mathcal{U}$ each of which is linearly ordered. The ordering on one subset $p \in \mathcal{P}$ may not be consonant with another subset $q \in \mathcal{P}$; in particular, if $x, y \in p \cap q$, we may have $x < y$ in the p-ordering (written $x <_p y$) and also $y <_q x$. But we assume a certain relation among the paths and their orderings. If $p, q \in \mathcal{P}$, and $x \in p \cap q$, define

$$(p, x, q) = \{y \mid y \in p, y <_p x\} \cup \{x\} \cup \{z \mid z \in q, x <_q z\}.$$

The symbol (p, x, q) simply denotes the set of elements so defined. We postulate

If $p, q \in \mathcal{P}$, $x \in p \cap q$, then $\exists r \in \mathcal{P}$ such that all elements
of r are contained in (p, x, q). $\tag{7.1}$

Referring back to Section 3, the reader will see that what we have in mind (and shall reach eventually) is a clutter of $(0, 1)$ paths, closed with respect to switching, in a directed graph G (deleting the 0 and 1 from each path).

Return to the upper clutter $\mathcal{L}$ and its blocker $\mathcal{B}(\mathcal{L})$. We consider any

$p \in \mathcal{B}(\mathcal{L})$, and describe a linear order among the elements of p by the following rule. If x is any element of p, there is at least one $S \in \mathcal{L}$, such that $S \cap p = \{x\}$. Let $\mathcal{L}_p(x)$ be the set of all such sets. We first show

> $\mathcal{L}_p(x)$ contains a set minimal in the ordering on $\mathcal{L}$. Denote
> it by $\underline{l}_p(x)$.
> Also, $\mathcal{L}_p(x)$ contains a set maximal in the ordering a $\mathcal{L}$.
> Denote it by $\bar{l}_p(x)$. $\hspace{4cm}$ (7.2)

To prove (7.2), observe that if $S, T \in \mathcal{L}_p(x)$, $S \wedge T$ can contain no element of p other than x, by (2.4), so it must contain x or p would not be in $\mathcal{B}(\mathcal{L})$. The same proof establishes the existence of $\bar{l}_p(x)$.

$$\text{If } x, y \in p, x \neq y, \text{ then either } \underline{l}_p(x) < \underline{l}_p(y) \text{ or } \underline{l}_p(y) < \underline{l}_p(x). \hspace{1cm} (7.3)$$

For suppose $\underline{l}_p(x)$ and $\underline{l}_p(y)$ were incomparable. Then $\underline{l}_p(x) \wedge \underline{l}_p(y)$ must contain at least one of x and y, otherwise $p \notin \mathcal{B}(\mathcal{L})$. Similarly, $\underline{l}_p(x) \vee \underline{l}_p(y)$ contains at least one of x and y. Suppose $\underline{l}_p(x) \wedge \underline{l}_p(y)$ contains x and y, and (say) $x \in \underline{l}_p(x) \vee \underline{l}_p(y)$. Then $x \in \underline{l}_p(x) \wedge \underline{l}_p(y)$ and $x \in \underline{l}_p(x) \vee \underline{l}_p(y)$ implies (see (2.1)) that $x \in \underline{l}_p(y)$, which is false. Thus $\underline{l}_p(x) \wedge \underline{l}_p(y)$ contains exactly one of x and y, say x, which means $\underline{l}_p(x) \wedge \underline{l}_p(y) \in \mathcal{L}_p(x)$, and contradicts the definition of $\underline{l}_p(x)$.

Define

$$\text{If } x, y \in p, x <_p y \text{ if } \underline{l}_p(x) < \underline{l}_p(y) \text{ in } \mathcal{L}. \hspace{2cm} (7.4)$$

Definition (7.4) simply orders p in view of (7.3) and the partial ordering of $\mathcal{L}$. Further, in analogy to (6.3) and (6.4), we have

$$\text{if } x <_p y, \text{ the } \bar{l}_p(x) < \bar{l}_p(y) \text{ in } \mathcal{L}. \hspace{2cm} (7.5)$$

> if $x <_p y$ and $S \in \mathcal{L}_p(x)$, $T \in \mathcal{L}_p y$, and S and T are com-
> parable in $\mathcal{L}$, then $S < T$. $\hspace{4cm}$ (7.6)

Let $\mathcal{P}$ be the collection of sets in $\mathcal{B}(\mathcal{L})$ simply ordered by definition (7.4). We now verify that (7.1) holds. To do this, it is sufficient to show that for every $p, q \in \mathcal{P}$ and $x \in p \cap q$, and every $S \in \mathcal{L}$,

$$S \cap (p, x, q) \neq \emptyset. \hspace{3cm} (7.7)$$

For if we establish (7.7), there must be some $r \in \mathcal{P}$ such that the elements of r are contained in (p, x, q).

Assume (7.7) false. Since $p \in \mathcal{B}(\mathcal{L})$, it follows that $(S \cap p) \neq \emptyset$ and $y \in (S \cap p)$ implies $x <_p y$. We restate this as

$$(S \cap p) = \{y_1, \ldots, y_j\}, \quad x <_p y_1 <_p y_2 <_p \cdots <_p y_j. \hspace{1cm} (7.8)$$

Similarly

$$(S \cap q) = \{z_1, \ldots, z_k\}, \quad z_1 <_q z_2 <_q \cdots <_q z_{k-1} <_q z_k <_q x. \hspace{1cm} (7.9)$$

We first show that $S \leq \bar{l}_q(z_k)$. Assume otherwise. Then $\bar{l}_q(z_k) < \bar{l}_q(z_k) \vee S = T$(say). If T contains z_t, $t < k$, then $z \in \bar{l}_q(z_t) < \bar{l}_q(z_k) < T$. But $z_t \notin \bar{l}_q(z_k)$ so (2.1) would be violated. Therefore, each of $z_1, \ldots, z_{k-1} \notin T$. Since $q \in \mathcal{B}(\mathcal{L})$, q contains at least one element w of T. By (2.4), $w \in \bar{l}_q(z_k) \cup S$, so w must be z_k.

Therefore, $z_k \in T$ and $T \in \mathcal{L}_q(z_k)$, violating the definition of $\bar{l}_q(z_k)$. Consequently,

$$S \leq \bar{l}_q(z_k) \tag{7.10}$$

Similarly, using (7.8)

$$\underline{l}_p(y_1) \leq S \tag{7.11}$$

But (7.8), (7.9), (7.10) and (7.11) imply

$$\underline{l}_p(x) < \underline{l}_p(y_1) \leq S \leq \bar{l}_q(z_k) < \bar{l}_q(x), \text{ implying}$$
$$\underline{l}_p(x) < S < \bar{l}_q(x).$$

By (2.1), this implies $x \in S$, a contradiction.

Now construct a directed graph G by starting with a node 0, directing an edge from 0 to the initial element of each $p \in \mathcal{P}$, an edge from each element to its immediate successor in any path $p \in \mathcal{P}$, and from the terminal element of each $p \in \mathcal{P}$ to a node 1. It is clear from the definition of blocker that the paths in $\mathcal{P}$ are a clutter of $(0, 1)$ paths (with the 0 and 1 deleted) in G, closed with respect to switching.

By proposition 5.1(a), the sets of $\mathcal{L}$ from the blocker of $\mathcal{B}(\mathcal{L})$. We will order the sets of $\mathcal{L}$ by the rule (3.3). We shall show that this ordering is a partial ordering $\mathcal{L}^*$, that the ordering in $\mathcal{L}^*$ is the same as $\mathcal{L}$, that $\mathcal{L}^*$ is a lattice, and that the lattice operations in $\mathcal{L}^*$ are precisely the $\vee$ and $\wedge$ of $\mathcal{L}$.

We first note

$$\text{if } x \in S \in \mathcal{L}, \quad \exists p \in \mathcal{P} \text{ such that } p \cap S = \{x\}. \tag{7.12}$$

This follows from the fact that the sets of $\mathcal{P}$ are the blocker of $\mathcal{L}$.

Now suppose $S \leq T$ and $T \leq S$ in $\mathcal{L}^*$. We must show $S = T$. Suppose $x \in S - T$. Using the p of (7.12), $T \leq S$ in $\mathcal{L}^*$ implies $p(T) < p(s)$. Therefore $S \leq T$ in $\mathcal{L}^*$ is impossible. Thus (3.3), as remarked in Section 3, yields a partial order.

Next, suppose $S < T$ in $\mathcal{L}$, we wish to prove $S < T$ in $\mathcal{L}^*$. Assume otherwise. Then we have a path p such that $p(T) < p(s)$. This means that $p \cap (S \cup T) = \{Z_1, z_2, \ldots, z_k\}$ with

$$z_1 <_p z_2 <_p \cdots <_p z_k, \quad \text{and} \quad z_1 \in T - S. \tag{7.13}$$

We first show that S and $\underline{l}_p(z_1)$ are incomparable in $\mathcal{L}$. If $\underline{l}_p(z_1) < S$, then $z_1 \in \underline{l}_p(z_1) < S < T$, $z_1 \in T$ implies by (2.1) that $z_1 \in S$, a contradiction. If $S < \underline{l}_p(z_1)$, note that $p \in \mathcal{B}(\mathcal{L})$ implies that for some $t > 1$, $z_t \in S$. Then $z_t \in S < \underline{l}_p(z_1) < \underline{l}_p(z_t)$ (which we infer from (7.13)) again contradicts (2.1). And we cannot have $S = \underline{l}_p(z_1)$, since $z_1 \in \underline{l}_p(z_1)$, but $z_1 \notin S$. Hence, in $\mathcal{L}$,

$$S \wedge \underline{l}_p(z_1) < S \quad \text{and} \quad S \wedge \underline{l}_p(z_1) < \underline{l}_p(z_1). \tag{7.14}$$

Suppose $z_1 \in p \cap (S \wedge \underline{l}_p(z_1))$. Then $z_1 \in S \wedge \underline{l}_p(z_1) < S < T$ contradicts (7.14) and (2.1).

Suppose $z_1 \notin p \cap (S \wedge \underline{l}_p(z_1))$. Then by (2.4), the nonempty set $S \wedge \underline{l}_p(z_1)) \cap p$

                    A.J. Hoffman/ Lattice clutters

must contain an element of S, say z_t, $t > 1$. Then $z_t \in S \wedge \underline{l}_p(z_1) < \underline{l}_p(z_1) < \underline{l}_p(z_t)$ contradicts (7.13), (7.14) and (2.1). Thus we have shown that $S < T$ in $\mathscr{L}$ implies $S < T$ in $\mathscr{L}^*$.

Suppose $S < T$ in $\mathscr{L}^*$. We want to show $S < T$ in $\mathscr{L}$, which is equivalent to showing that, in $\mathscr{L}$, $S \wedge T = S$. Since $S \wedge T \leq S$ in $\mathscr{L}$, we know $S \wedge T \leq S$ in $\mathscr{L}^*$. This means $p(S \wedge T) \leq p(s)$ for every $p \in \mathscr{P}$. But (2.4) shows that $S \wedge T \subset S \cup T$. Together with $p(S) \leq p(T)$ this implies $p(S \wedge T) = p(S)$ for every p. As we have seen, this means $S \wedge T = S$.

We turn now to the lattice operations in $\mathscr{L}^*$. As explained in Section 3, in $\mathscr{L}^*$, $S \wedge T$ is the set of elements of $S \cup T$ each of which is the first element of $S \cup T$ encountered on same path p; similarly, in $\mathscr{L}^*$, $S \vee T$ is the set of elements of $S \cup T$ each of which is the last element of $S \cup T$ encountered in some path. To keep clear the distinction, we shall call these operations $\wedge^*$ and $\vee^*$ to distinguish from the $\wedge$ and $\vee$ of $\mathscr{L}$.

We first show

$$\text{if } S, T \in \mathscr{L}^*, \ S \wedge^* T \in \mathscr{L}^*. \quad \text{(Similarly, } S \vee^* T \in \mathscr{L}^*\text{).} \tag{7.15}$$

It is clear that $S \wedge^* T$ meets every path. What we must show (to prove $S \wedge^* T \in \mathscr{B}(\mathscr{B}(\mathscr{L}))$ is that, if any $x \in S \wedge^* T$ is deleted, then there exists a path p which does not meet $S \wedge^* T - \{x\}$. Assume $x \in S$ (we may also have $x \in T$ as well, but that does not matter). By (7.12), there exists a path p such that $p \cap S = \{x\}$. Suppose there is a $y \in S \wedge^* T$ which occurs in p "after" x. Then $y \in T$, and there exists a path q such that the first element of $S \cup T$ encountered on q is y. It follows that (q, y, p) contains no element of S, a contradiction. So we may assume that no element of $S \wedge^* T$ occurs in p after x. Since $x \in S \wedge^* T$, there exists a path r containing x such that no element of $S \cup T$ is encountered before x. It follows that $(r, x, p) \cap (S \wedge^* T) = \{x\}$. This means that $S \wedge^* T$ is in $\mathscr{B}(\mathscr{B}(\mathscr{L}))$, hence a set of $\mathscr{L}^*$.

We leave to the reader the verification that $\wedge^*$ and $\vee^*$ are indeed lattice operations in the partial ordering of $\mathscr{L}^*$ (= the partial ordering of $\mathscr{L}$). To finish our discussion, we must show that $\wedge$ is the same as $\wedge^*$ (and similarly $\vee$ is the same as $\vee^*$). It is sufficient to show $S \wedge T \subseteq S \wedge^* T$, so assume this is false; i.e., there is some element $x \in (S \wedge T) - (S \wedge^* T)$. By (2.4), $x \in S \cup T$. Hence, x is an element of $S \cup T$ encountered on p. On the other hand, since $S \wedge^* T$ is the l.u.b. of all $V \leq S, T$, it follows that $S \wedge T \leq S \wedge^* T$. Let p be a path such that $(S \wedge T) \cap p = \{x\}$. Then $S \wedge T \leq S \wedge^* T$ implies $p(S \wedge T) \leq p(S \wedge^* T)$, a contradiction.

References

[1] J. Edmonds and D.R. Fulkerson, "Bottleneck extrema", *Journal of Combinatorial Theory* 8 (1970) 299–306.

[2] L.R. Ford and D.R. Fulkerson, "Maximal flow through a network", *Canadian Journal of Mathematics* 8 (1956) 399–404.

[3] D.R. Fulkerson, "Blocking and anti-blocking pairs of polyhedra", *Mathematical Programming* 1 (1971) 168–193.

[4] C. Greene, "Some partitions associated with a partially ordered set", *Journal of Combinatorial Theory* A (1976) 69–79.

[5] C. Greene and D.J. Kleitman, "The structure of Sperner k-families", *Journal of Combinatorial Theory* A (1976) 41–68.

[6] A.J. Hoffman, "A generalization of mas flow–min cut", *Mathematical Programming* 6 (1974) 352–359.

[7] A.J. Hoffman, "On lattice polyhedra II", IBM Research Report RC 6268 (1976).

[8] A.J. Hoffman and D.E. Schwartz, "On lattice polyhedra", in: Proceedings of the 5th Hungarian Colloquium on Combinatorics, 1967 (to appear).

[9] E. Johnson, "On cut set integer polyhedra", *Cahiers du Centre de Recherches Opérationelles* 17 (1965) 235–251.

Annals of Discrete Mathematics 2 (1978) 201–209.
© North-Holland Publishing Company

LOCAL UNIMODULARITY IN THE MATCHING POLYTOPE*

A.J. HOFFMAN
IBM Thomas J. Watson Research Center, Yorktown Heights, NY 10598, U.S.A.

Rosa OPPENHEIM
Rutgers Univ. Graduate School of Business Administration, Newark, NJ 07102, U.S.A.

1. Introduction

In the first decade of linear programming, it was observed that various extremal combinatorial theorems (Dilworth, Menger, etc.), could be derived as applications of the duality principle of linear programmings. The basic idea was that the combinatorial theorem would follow from linear programming duality if optimal vertices of both primal and dual problems were integral. In all the cases treated, the linear programming matrix A was totally unimodular (i.e., every minor of A had absolute value 0 or 1), so application of Cramer's rule yielded the integrality of the vertices. A summary of that work is given in [7].

Starting with [2], Edmonds has led a development in which he and others have found several interesting classes of combinatorial problems to which the preceding argument roughly applies, even though the relevant matrix A is not totally unimodular. Nevertheless, a vestigial form of unimodularity is still present in at least some of these instances (see [3, 8] and the references cited there). (In some cases, most notably in the dual problem for the perfect graph theorem ([5, 9]), we do not yet know whether it is present at all.)

Let A be a matrix, b a vector, each with all entries integral, and let x^0 be a vertex of the polyhedron $\{x \mid Ax \le b, x \ge 0\}$. Suppose we are interested in knowing whether or not x^0 is integral. To study this question, we consider the submatrix $\tilde{A}$ of A formed by columns j such that $x_j > 0$ and rows i such that $(Ax)_i = b_i$. Defining $\tilde{b}$ in the obvious way, we know that the nonzero coordinates of x^0 are obtained from the unique solution to

$$\tilde{A}z = \tilde{b}.$$

Let $\tilde{A}$ have p rows and q columns. A sufficient condition for x^0 to be integral is [10]

* This work was supported (in part) by the Army Research Office under contract number DAHC04–74C–0007. It was part of the talk "A menu of research topics in polyhedral combinatorics", given at the Qualicum Beach Conference (1976).

A portion of this material is taken from the dissertation submitted to the Faculty of the polytechnic Institute of Brooklyn in partial fulfillment of the requirements for the degree Doctor of Philosophy (Operations Research) (1973).

A.J. Hoffman, R. Oppenheim

that the g.c.d of all determinants of order q in $\tilde{A}$ be 1. Under these circumstances, we shall say that the given polyhedron is *locally unimodular* at x^0. (This is an abuse of language: we should really speak of the pair (A, b) rather than the polyhedron, which may have many presentations, but context will make (A, b) clear.) If at least one of the determinants of order p in $\tilde{A}$ is 1, we shall say the given polyhedron is *locally strongly unimodular* at x^0.

In some cases, the primal polyhedron is locally strongly unimodular at every vertex, for the arguments establish that $\tilde{A}$ contains a nonsingular submatrix of order q which is not only unimodular, but totally unimodular. (The same arguments establish that the dual polyhedron has at least one optimal vertex locally strongly unimodular.) In this paper, we prove that the matching polytope is locally strongly unimodular, at every vertex, though not locally totally unimodular, provided one includes certain natural but possibly superfluous inequalities. We believe that the concept of local unimodularity is a useful idea in this subject: at the very least, a phenomenon whose presence or absence should be investigated. We now turn to the matching polytope.

Let G be a graph on m vertices A its associated node-edge incidence matrix; i.e.,

$$A = (a_{i,e}) = \begin{cases} 1, & \text{if node } i \text{ is on edge } e \\ 0, & \text{otherwise.} \end{cases}$$

Let $b = (b_1, \ldots, b_m)$ be a vector with nonnegative integral coordinates; and let

$$P(G, b) = \{x \mid Ax \leq b, x \geq 0\}. \tag{1.1}$$

Edmonds proved the following

Theorem 1.1. *Let $M(G, b)$ be the polyhedron given by the system of inequalities*

$$x \geq 0,$$

$$Ax \leq b, \tag{1.2}$$

$$\forall S \subset \{1, \ldots, n\} \text{ such that } |S| \geq 2, \sum_{e \in S} x_e \leq \left[\frac{1}{2} \sum_{i \in S} b_i \right]. \tag{1.3}$$

(In (1.3), $e \in S$ means that both endpoints of the edge e are in S; the symbol $[y]$ means the largest integer at most y.)

Then $M(G, b)$ is the convex hull of the integral vectors in $P(G, b)$ ([4, 1]).

Theorem 1.2. *The polyhedron $M(G, b)$ is locally strongly unimodular at every vertex.*

Our strategy will be to give a new (inductive) proof of Theorem 1.1, and then to observe that Theorem 1.2 follows from the steps in the proof of Theorem 1.1. The new proof of Theorem 1.1 may be of independent interest. It is worth noting that

Theorem 1.2 is not true if one is parsimonious in listing the inequalities in (1.3). In case $\Sigma_{i \in S} b_i$ is even, the corresponding inequality is superfluous, but Theorem 1.2 wants that inequality listed! To see this, consider the graph K_3, where each $b_i = 2$.

2. Proof of Theorem 1.1.

Assume x is a nonzero vertex of $M(G, b)$ we must prove x is integral. (It is obvious that any integral point satisfying (1.1) is in $M(G, b)$.) Let $G(x)$ be the subgraph of G formed by all its nodes $\{1, \ldots, m\}$ and all edges (i, j) such that $x_{ij} > 0$. If a node i satisfies $\Sigma a_{ie} x_e = b_i$, i is said to be *tight* (with respect to x). If $S \subset \{1, \ldots, m\}$, $|S| \geq 2$, S is said to be *tight* (with respect to x), if $\Sigma_{e \in S} x_e = [\frac{1}{2} \Sigma_{i \in S} b_i]$. For any $x \in M(G, b)$, $C(x)$ is the submatrix (of the matrix specified by (1.2) and (1.3)) whose columns correspond to $G(x)$, rows to tight sets and tight nodes.

We shall prove our theorem by induction on m, the number of nodes of G. Hence, we can assume $G(x)$ connected.

Lemma 2.1. *If S has no tight sets, or has $\{1, \ldots, m\}$ as its only tight set, x is integral.*

Proof. Since $G(x)$ is connected, it must have at least $m - 1$ edges. If it has more than $m + 1$ edges, then we must have equality in at least two of the inequalities (1.3), so there must be at least one tight set other than $\{1, \ldots, m\}$. If $G(x)$ has exactly $m - 1$ edges, it is a tree; the node-edge incidence matrix of a tree is totally unimodular, so x is integral. If $G(x)$ has exactly $m + 1$ edges, and then there is at least one tight set (which must be $\{1, \ldots, m\}$) and no other, and, for all i, $\Sigma_e a_{ie} x_e = b_i$, and $\Sigma_e x_e = \frac{1}{2} \Sigma_i b_i$. But since $G(x)$ has $m + 1$ edges, it must contain at least two cycles. Reasoning as in [1] or [6], this implies x is not a vertex.

Finally, assume $G(x)$ has exactly m edges. Then $\{1, \ldots, m\}$ must be tight, because at least m inequalities in (1.2) and (1.3) must be equations; if all such are in (1.2), then $\{1, \ldots, m\}$ is tight anyway. Next, since $G(x)$ has exactly m edges, it contains exactly one cycle (which is odd, otherwise x is not a vertex). Therefore there is either exactly one i, say i^*, such that $\Sigma_e a^*_{ie} x_e < b^*_i$ or no such i.

In either case, if we look at the submatrix of (1.2) and (1.3) corresponding to positive x, all rows from (1.2), and the single row from (1.3) corresponding to $S = \{1, \ldots, m\}$, every $m \times m$ submatrix is nonsingular. Now $C(x)$ consists of this matrix, or this matrix with one row from (1.2) deleted. But if any row from (1.2) is deleted from this $(m + 1) \times m$ matrix, the remaining $m \times m$ has determinant ± 1. The reason is that a connected graph consisting of a tree and one additional edge forming an odd cycle has the property: for each node i^*, there exists a set T of nodes, $i^* \notin T$, such that

    (i) $i, j \in T$, $i \neq j$ implies that the edges on i are distinct from the edges on j,

    (ii) the union of all edges on all nodes $i \in T$ consists of all edges of H except for one edge on the odd cycle.

A.J. Hoffman, R. Oppenheim

Hence, subtracting the rows of $C(x)$ corresponding to nodes in T from the row of T corresponding to $S = \{1, \ldots, m\}$ produces a matrix whose determinant is the same as before, and expansion by cofactors of the last row yields an $(m-1) \times (m-1)$ determinant of a matrix with at most two 1's in each column, and the columns containing two 1's form the node-edge incidence matrix of a forest. Hence, $C(x)$ is unimodular.

By virtue of Lemma 2.1, we need only consider the case where there exists a tight set $S \neq \{1, \ldots, m\}$, and no tight set $S' \subset S$, $S' \neq S$. Further, since $G(x)$ is connected, $\sum_{i \in S} b_i$ is odd. Henceforth, we assume this. Let T be the set of nodes of G not in S, so $T \neq \emptyset$. We now define a vector $x(S, S \times T)$ as follows:

$$x(S, S \times T)_e = \begin{cases} 0, & \text{if } e \in T, \\ x_e, & \text{otherwise.} \end{cases}$$

Lemma 2.2. *The vector $x(S, S \times T)$ is a convex combination of integral vectors satisfying* (1.1).

Proof. We first consider the case where T consists of one node, say $T = \{m\}$, in which case $x(S, S \times T) = x$, and the lemma coincides with the theorem. Let $x(S)$ be defined by

$$x(S)_e = \begin{cases} x_e & \text{if } e \in S \\ 0 & \text{otherwise.} \end{cases}$$

By the induction hypothesis, since $|S| = m - 1 < m$, $x(S)$ is a convex combination of nonnegative integral vectors y, each of which must satisfy

$$\sum_{e \in S} y_e(S) = \left[\frac{1}{2} \sum_{i \in S} b_i \right].$$

Since $\sum_{i \in S} b_i$ is odd, we may partition the set of these vectors into subsets $V_1, \ldots, V_{m-1}$, such that V_i consists of nonnegative integral vectors y_i^i satisfying

$$\sum_{e \in S} a_{ie}(y_t^i)_e = b_i - 1, \qquad \sum_{e \in S} a_{je}(y_t^i)_e = b_j, \qquad j = 1, \ldots, m-1, \ j \neq 1.$$

Thus we may write

$$x(S) = \sum_{k=1}^{m-1} \sum_t \lambda_{kt} y_t^k,$$

where

$$y_t^k \in V_k, \qquad \sum_k \sum_t \lambda_{kt} = 1, \qquad \lambda_{kt} \geq 0.$$

Let $\mathbf{y}_t^k$ be the vector formed by adding to y_t^k a vector with 1 for the coordinate corresponding to edge (k, m), 0 everywhere else. Clearly

$$x = \sum_{k=1}^{m-1} \sum_t \lambda_{kt} \frac{x_{km}}{\sum_u \lambda_{ku}} \mathbf{y}_t^k + \sum_{k=1}^{m-1} \sum_t \lambda_{kt} \left(1 - \frac{x_{km}}{\sum_u \lambda_{ku}} \right) y_t^k.$$

(Note $0 \le x_{km} \le \sum_u \lambda_{ku}$ follows from the definition of V_k.) But this expresses x as a convex combination of integral vectors. The theorem is completed if $|T| = 1$ by observing that, since x is a vertex, it must be one of these integral vectors.

Now assume $|T| \ge 2$. Let S^* be the graph formed by nodes in S and one additional node p (representing a collapse of T), and

$$x(S^*)_e = \begin{cases} x(S)_e, & \text{if } e \in S, \\ \sum_{j \in T} x_{ij}, & \text{if } e = (i, p). \end{cases}$$

Also, define the vector b^* by $b_i^* = b_i$ if $i \in S$, $b_p^* = \sum_{j \in T} b_j$. Then $x(S^*)$ is in $M(S^*, b^*)$, $|V(S^*)| < m$, so the induction and the case just discussed above show that

$$x(S^*) = \sum_{k \in S} \sum_t \lambda_{kt} y_t^k + \sum_{k \in S} \sum_r \mu_{kr} y_t^{kp},$$

each λ_{kt} and μ_{kr} is nonnegative,

$$\sum_{k \in S} \left(\sum_t \lambda_{kt} + \sum_r \mu_{kr} \right) = 1,$$

each y_t^k and y_t^{kp} an integral nonnegative vector,

$$(y_t^k)_{(i,p)} = 0, (y_t^{kp})_{(i,p)} = \delta_{ik}, \qquad \sum_{e \in S} a_{ke} (y_t^k)_e = \sum_{e \in S} a_{ke} (y_t^{kp})_e = b_k - 1,$$

$$\sum_{e \in S} a_{je} (y_t^k) = \sum_{e \in S} a_{je} (y_t^{kp})_e = b_j, \qquad j \ne k, j, \quad k \in S.$$

Let $y(S, S \times T)_t^k$ be obtained from y_t^k by putting 0 in the coordinate corresponding to edges e such that $e \notin S$. Let $y_t^{kj}(S, S \times T)$ be the vector formed from y_t^{kp} by putting 1 in the coordinate position corresponding to edge (k, j), $j \in T$, all other coordinates corresponding to edges $e \notin S$ are 0. Then the equation

$$x(S, S \times T) = \sum_{k \in S} \sum_t \lambda_{kt} y_t^k + \sum_{k \in S} \sum_r \mu_{kr} \sum_{j \in T} \frac{x_{kj}}{\sum_{i \in T} x_{ki}} y_t^{kj} \tag{2.1}$$

expresses $x(S, S \times T)$ as a convex combination of integral vectors. This completes the proof. Note we have not used the minimality of S (only $|S| < m$) here, but we will use it in proving Theorem 1.2.

Given x, define

$$x(S \times T, T)_e = \begin{cases} 0, & \text{if } e \in S, \\ x_e, & \text{otherwise.} \end{cases}$$

Lemma 2.3. *The vector $x(S \times T, T)$ is a convex combination of integral vectors satisfying* (1.1).

A.J. Hoffman, R. Oppenheim

Proof. Let T^* be the graph formed by nodes in T and one additional node q (representing a collapse of S), and

$$x(T^*)_e = \begin{cases} x_e, & e \in T, \\ \sum_{i \in S} x_{ej}, & \text{if } e = (q, j). \end{cases}$$

Define b^* by $b_j^* = b_j$ if $j \in T$, $b_q^* = 1$. $x(T^*)$ obviously satisfies the relevant (1.2). To see that it satisfies the appropriate (1.3), we need only consider sets $Q = \bar{Q} \cup \{q\}$, $\bar{Q} \subset T$, $\sum_{j \in \bar{Q}} b_j$ is even, say $2c$. But suppose

$$\sum_{i \in S} \sum_{j \in Q} x_{(i,j)} + \sum_{e \in \bar{Q}} x_e > c$$

Then, since $\sum_{e \in S} x_e = \frac{1}{2}(\sum_{i \in S} b_i - 1)$, we would have

$$\sum_{e \in S \cup \bar{Q}} x_e > \frac{1}{2}\left(\sum_{i \in S} b_i - 1\right) + c = \left[\frac{1}{2}\left(\sum_{i \in S} b_i + \sum_{j \in \bar{Q}} b_j\right)\right],$$

violating the original (1.3) for the vector x. Thus $x(T^*) \in M(T^*, b^*)$. Since $|V(T^*)| < m$, the induction hypothesis applies. Reasoning as in Lemma 2.2, we can write $x(S \times T, T)$ as a convex combination of integral vectors.

$$x(S \times T, T) = \sum_r \alpha_r w_r + \sum_{i \in S} \sum_{j \in T} \sum_t \beta_{ijt} w_t^{ij}, \tag{2.2}$$

where the α's and β's are nonnegative and sum to 1, each w_r and w_t^{ij} is an integral vector in $M(G, b)$, w_r has all coordinates 0 in positions corresponding to edges $e \notin T$, w_t^{ij} has the coordinate corresponding to (i, j) as 1, all other coordinates corresponding to edges $e \notin T$ are 0. This completes the proof of Lemma 2.3.

To prove the theorem, let us first rewrite (2.1) as

$$x(S, S \times T) = \sum_s \lambda_s y_s + \sum_{i \in S} \sum_{j \in T} \sum_u \nu_{iju} y_u^{ij}, \tag{2.3}$$

where the λ's and ν's are nonnegative and sum to 1.

Note that $\sum_s \lambda_s = \sum_r \alpha_r = 1 - \sum_{i \in S} \sum_{j \in T} x_{ij}$. Also, for each $i \in S$, $j \in T$, $\sum_t \beta_{ijt} = \sum_u \nu_{iju} = x_{ij}$. For each $i \in S$, $j \in T$, let z_{ut}^{ij} be the vector which agrees with y_u^{ij} on edges in S, with w_t^{ij} on edges in T, and has $z_{ut}^{ij} = 1$; all other $z_{ut}^{i'j'}$, $i' \in S$, $j' \in T$, $(i', j') \neq (i, j)$ are 0. It follows that

$$x = \frac{\sum_s \sum_r \lambda_s \alpha_r (w_r + y_s)}{1 - \sum_{i \in S} \sum_{j \in T} x_{ij}} + \sum_{i \in S} \sum_{j \in T} \sum_u \sum_t \frac{\beta_{ijt} \nu_{iju}}{x_{ij}} z_{ut}^{ij}$$

expresses x as a convex combination of integral vectors satisfying (1.1). But x is a vertex of $M(G, b)$, and each integral vector satisfying (1.1) is in $M(G, b)$. It follows that x must be one of the vectors $w_r + y_s$ or one of the vectors z_{ut}^{ij}. (Of course, since $G(x)$ is connected, x cannot be $w_r + y_s$.)

3. Proof of Theorem 1.2.

We shall use induction on m, and shall also be guided by our proof of Theorem 1.1. Clearly, we may assume that $G(x)$ is connected, and x is a vector of the type z^{ij} just discussed above. If there is no tight set, or the only tight set is $\{1,\ldots,m\}$, the discussion given in the proof of Lemma 2.1 proves the theorem. The case where the minimal tight set $S = \{1,\ldots, m-1\}$ we shall ignore, since the reader will readily see the proof from our discussion of the remaining cases. So assume $2 \le |S| \le m-2$. It is easy to see that the restriction of $G(z^{ij})$ to S, which we shall call $G_S(z^{ij})$, is connected.

There are two possibilities: $G_S(z^{ij})$ is a tree, or a tree and an additional edge forming an odd cycle. We inherit this knowledge from the proof of Lemma 2.1. Let w^{ij} (from 2.2) be the restriction of z^{ij} to T^*. By induction, since $|T^*| < m$, $C(w^{ij})$ contains a unimodular matrix $\mathcal{V}$ of rank equal to the number of positive coordinates of w^{ij}. There are two possibilities: one of the node constraints of type (1.2) that appears in $\mathcal{V}$ involves the artificial vertex q, or not. Thus we have $2 \times 2 = 4$ cases to consider.

The theorem will be proved by induction, using the fact that it holds for T^*. We will make use of the fact that S is a minimal tight set and invoke the material developed in the proof of Lemma 2.1. We will also consider what happens if edge (q, j) is in a tight node or set of T^*. If a tight node, that node is either i or j. If $q, j \in Q$, a tight set of T^*, then $Q = \bar{Q} \cup \{q\}$, $\bar{Q} \subset T$. Now

$$\sum_{e \in S \cup \bar{Q}} x_e = \sum_{e \in S} x_e + \sum_{e \in Q} x_e = \frac{1}{2}\left(\sum_{i \in S} b_i - 1\right) + \left[\frac{1}{2}\left(1 + \sum_{i \in Q} b_i\right)\right] = \left[\frac{1}{2} \sum_{i \in S \cup \bar{Q}} b_i\right].$$

This proves that, if $q, j \in Q$, a tight set for T^*, then $(i, j) \in S \cup \bar{Q}$, a tight set for G.

In what follows, we will assume that the unimodular matrix $\mathcal{V}$ mentioned above is of order $t + 1$. The last $t + 1$ columns of each of the matrices $F_1 - F_4$, discussed below correspond to positive w^{ij}; the first set of columns, including the middle one, correspond to positive y^{ij}; the middle one corresponds to edge (i, j).

Case 1. $G_S(z^{ij})$ is a tree, the node constraint involving q is in $\mathcal{V}$. Let w^{ij} have $t + 1$ positive coordinates. Consider the matrix

$$
F_1 =
\begin{array}{c}
\begin{array}{ccc} \overbrace{|S|-1}\qquad & \overbrace{1}\quad & \overbrace{t}\quad \end{array} \\
\begin{array}{c}
|S| \\
1 \\
t
\end{array}
\left[
\begin{array}{c|c|c}
L & \begin{matrix}1\\0\\ \vdots\\0\end{matrix} & 0 \\
\hline
1\,1\ldots1 & 0 & 0\,0\ldots0 \\
\hline
P & ? & N
\end{array}
\right],
\end{array}
$$

which contains some of the rows of $C(z^{ij})$. L is the node-edge incidence matrix of $G_S(z^{ij})$. The first row of F_1 and the last t rows of F_1 meet the last $t+1$ columns of F_1 in the $(t+1)\times(t+1)$ unimodular matrix. $\mathscr{V}$ (therefore N is unimodular). Note that node 1 of S is simulating q, with the middle column corresponding to edge (i, j); i.e., node 1 is node i. The column **?** contains a 1 in a given position if edge (i,j) is present in a node constraint (for $j \in T$) or a set constraint on w^{ij} in T^*. If a node constraint, then the corresponding row of P is all 0. If a set constraint then the corresponding row of P is all 1. Now F_1 is contained in $C(z^{ij})$. If we delete the middle row of F_1, which corresponds to the set constraint on S, we obtain a unimodular matrix of full rank.

Case 2. $G_S(z^{ij})$ is a tree, the node constraint involving q is not in $\mathscr{V}$. Define

$$
F_2 = \quad
\begin{array}{c}
\\ |S| \\ \\ 1 \\ \\ t+1
\end{array}
\left[
\begin{array}{c|c|c}
L & \begin{matrix}1\\0\\ \vdots \\0\end{matrix} & 0 \\
\hline
1\ldots1 & 0 & 0\ 0\ldots0 \\
\hline
P & ? & N
\end{array}
\right],
$$

(with column groups $|S|-1$, 1, t)

when the matrices have the same meaning as before, except that $\mathscr{V}$ now is formed by last $t+1$ rows and columns. Now F_2 is contained in $C(z^{ij})$. Deleting middle row and the first row of F_2, we obtain a unimodular matrix of full rank. (Note that $[?N]$ is unimodular).

Case 3. $G_S(z^{ij})$ is not a tree, the node constraint involving q is in $\mathscr{V}$. Consider

$$
F_3 = \quad
\begin{array}{c}
\\ |S| \\ \\ 1 \\ \\ t
\end{array}
\left[
\begin{array}{c|c|c}
L & \begin{matrix}1\\0\\ \vdots \\0\end{matrix} & 0 \\
\hline
1\ldots1 & 0 & 0\ 0\ldots0 \\
\hline
P & ? & N
\end{array}
\right],
$$

(with column groups $|S|$, 1, t)

F_3 is contained in $C(z^{ij})$; and the last $t+1$ columns of F_3, together with the first row and last t rows of F_3 form $\mathscr{V}$. To prove F_3 is unimodular, we use Laplace's expansion of det F_3, based on rows $2, \ldots, |S|+1$. Only one term is nonzero, and it is ± 1.

Case 4. $G_3(z^{ij})$ is not a tree, the node constraint involving q is not in $\mathcal{V}$. Consider

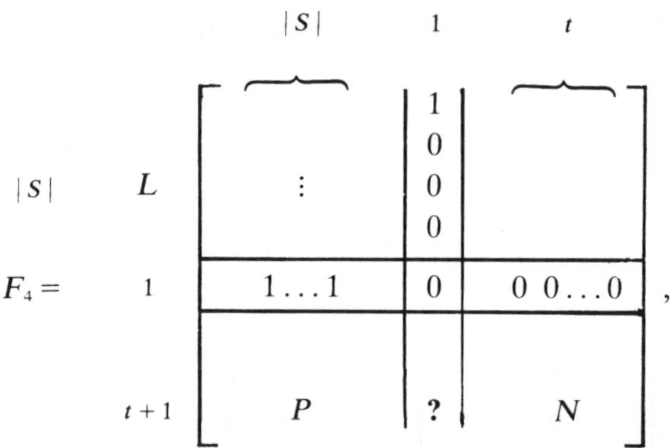

$$F_4 = ,$$

Note that $[?N]$ is unimodular. If we delete row 1 of F_4, we obtain a unimodular matrix of full rank.

References

[1] M. Balinski, Establishing the matching polytope, J. Combinatorial Theory Ser. B. 13 (1970) 1–13.

[2] J. Edmonds, Paths, trees and flowers, Canad. J. Math. 17 (1965) 449–467.

[3] J. Edmonds, Submodular functions, matroids and certain polyhedra, in Combinatorial Structures and their Applications (Gordon and Breach, 1970) 69–87.

[4] J. Edmonds and W.R. Pulleyblank, Optimum Matching (Johns Hopkins University Press) to appear.

[5] D. R. Fulkerson, Blocking and anti-blocking pairs of polyhedra, Math. Programming 1 (1971) 127–136.

[6] D.R. Fulkerson, A.J. Hoffman and M.H. McAndrew, Some properties of graphs with multiple edges, Canad. J. Math. 17 (1963) 957–969.

[7] A.J. Hoffman, Some recent applications of the theory of linear inequalities to extremal combinatorial analysis, Proc. Sympos. Appl. Math., Vol. 10, 113–127 (American Mathematical Society, Providence, RI, 1960).

[8] E.L. Johnson, On cut-set integer polyhedra, in: Journées Franco-Belgiques (9–10 May 1974).

[9] L. Lovasz, Normal hypergraphs and the perfect graph conjecture, Discrete Math. 2 (1972) 253–267.

[10] C.C. MacDuffee, The Theory of Matrices (Chelsea Publishing Co., NY, 1946).

[11] W.R. Pulleyblank, Faces of matching polyhedra, Ph.D. Thesis, University of Waterloo (1973).

A Fast Algorithm that makes Matrices Optimally Sparse

Alan J. Hoffman

IBM Thomas J. Watson Research Center
Yorktown Heights, New York 10598

S. Thomas McCormick

Department of Operations Research
Stanford University
Stanford, California 94305

Under a non-degeneracy assumption on the non-zero entries of a given sparse matrix, a polynomially-bounded algorithm is presented that performs row operations on the given matrix which reduce it to a sparsest possible matrix with the same row space. For each row of the matrix, the algorithm performs a maximum cardinality matching on the bipartite graph associated with a submatrix which is induced by that row. The dual of the optimal matching then specifies the row operations that will be performed on that row. We also describe a variant algorithm that processes the matrix in place, thus conserving storage and time. The modifications needed to apply the algorithm to matrices that do not necessarily satisfy the non-degeneracy assumption are also described. A particularly promising application of this algorithm is in the reduction of linear constraint matrices.

1. Introduction

An important factor in our present ability to solve many large-scale numerical problems is the recognition that these problems are nearly always sparse, and that taking advantage of sparsity can turn a hitherto practically unsolvable problem into a solvable one. Perhaps the best example of this is in large-scale linear programming, where highly refined sparse matrix factorization routines have allowed problems with huge coefficient matrices to be solved (see, e.g., Duff (1980) or Bunch and Rose (1976)). However, although sparsity is known to be helpful, relatively little attention seems to have been paid to techniques that economically increase sparsity (decrease density), thereby improving the efficiency of sparse algorithms. In this context, this paper considers the Sparsity Problem (SP):

185

Given a large, sparse system of linear equations

$$Ax = b, \tag{1}$$

find an equivalent system

$$\hat{A}x = \hat{b} \tag{2}$$

which has the minimum possible number of non-zero entries in $\hat{A}$.

Constraints of the form (1) are among the most common in large-scale optimization, so that it is potentially very useful to solve (SP). Under a non-degeneracy assumption, we shall present an efficient algorithm that solves (SP) using maximum cardinality matching. Sections 2-4 will assume familiarity with notions of graph theory and maximum cardinality bipartite matching (see, e.g., Lawler (1976)). Section 2 develops most of the machinery needed for the proof, and uses it to derive an algorithm that solves a subproblem of (SP). In Section 3 we use the algorithm of Section 2 to construct the full algorithm, and prove that it solves (SP). We then give a variant algorithm that uses less space and show that it also solves (SP). Section 4 discusses the modifications necessary to apply the algorithm on matrices that do not necessarily satisfy the non-degeneracy assumption. Finally, Section 5 considers further questions raised by this research.

2. Transforms and the One Row Algorithm

In this paper we shall assume that the matrix A in (1) has full row rank. We know from linear algebra that (2) is equivalent to (1) if and only if $\hat{A} = TA$ and $\hat{b} = Tb$ for some square non-singular matrix T. We are aiming for a general algorithm that makes no assumptions about any special structure in A, and thus can find T almost solely from the *sparsity pattern* of A (the positions of the non-zeros in A). What can go wrong in this aim is that we can encounter "unexpected" cancellation. To illustrate, consider the following two A's, with the same sparsity pattern, treated with the same T:

$$TA_1 = \begin{pmatrix} 1 & -1 & 1 \\ 0 & 1 & 0 \\ 0 & 0 & 1 \end{pmatrix} \begin{pmatrix} 1 & 1 & 0 & 0 & 0 \\ 0 & 1 & 1 & 1 & 1 \\ 0 & 0 & 1 & 1 & 1 \end{pmatrix} = \begin{pmatrix} 1 & 0 & 0 & 0 & 0 \\ 0 & 1 & 1 & 1 & 1 \\ 0 & 0 & 1 & 1 & 1 \end{pmatrix};$$

$$TA_2 = \begin{pmatrix} 1 & -1 & 1 \\ 0 & 1 & 0 \\ 0 & 0 & 1 \end{pmatrix} \begin{pmatrix} 1 & 1 & 0 & 0 & 0 \\ 0 & 1 & 1 & 2 & 3 \\ 0 & 0 & 1 & 1 & 1 \end{pmatrix} = \begin{pmatrix} 1 & 0 & 0 & -1 & -2 \\ 0 & 1 & 1 & 2 & 3 \\ 0 & 0 & 1 & 1 & 1 \end{pmatrix}.$$

In both cases T represents the unique linear transformation that adds the multiples of rows 2 and 3 to row 1 which makes a_{12} zero and avoids

FAST SPARSE MATRIX ALGORITHM   **187**

fill-in in a_{13}. In the first case the sparsity increased, in the second case it decreased. The difficulty is that A_1 has some dependent submatrices that are not apparent from the sparsity pattern alone. The possibility of this phenomenon makes solving (SP) too difficult in general, as shown by the following result.

THEOREM 1. (Stockmeyer (1982)) (SP) is NP-Hard in general. (See the Appendix for the proof.)

Thus, to get a polynomial algorithm for (SP), we must make some assumption about A. Suppose that A is $m \times n$, and let $R \subseteq \{1,\ldots,m\}$, $C \subseteq \{1,\ldots,n\}$. We denote the submatrix of A indexed by the rows in R and the columns in C by A_{RC}. The sparsity pattern of A_{RC} naturally induces a bipartite graph $G_{RC} = (R, C, E)$ where $E = \{(i, j) \in R \times C \mid a_{ij} \neq 0\}$. Let $M(G)$ be the number of edges in a maximum cardinality matching in the bipartite graph G. If $|R| = |C|$, then the usual expansion of det A_{RC} has at least one non-zero term precisely when $M(G_{RC}) = |R|$, and when A is "general", we expect the converse of this to be true as well. This reasoning leads us to assume henceforth that A has the

Matching Property (MP): rank $A_{RC} = M(G_{RC})$ for all R and C.

For example, A_1 above does not satisfy (MP) whereas A_2 does. Note that if the entries of A are independent algebraic indeterminates than (MP) is satisfied

Since T must be non-singular, $G(T)$ has a perfect matching which we can assume without loss of generality is the main diagonal. We can further assume that $t_{ii} = 1$, $i = 1,2,\ldots,m$ by scaling the rows of T, so that the non-zero entries in row i of T indicate the multipliers for the rows to be added to row i of A. Viewed in this way, (SP) breaks down into m one row sparsity problems $(ORSP_i)$, $i = 1,2,\ldots,m$. That is, $(ORSP_i)$ is the problem:

Find $\{\lambda_k, k \neq i\}$ so that

$$\hat{A}_{i,0} = A_{i,0} + \sum_{k \neq i} \lambda_k A_{k,0} \tag{3}$$

has the minimum possible number of non-zeros.

Not all solutions to (3) are equally good. Since we expect that the amount of arithmetic needed to do the calculations in (3) depends upon the number of rows with non-zero multipliers, ideally we would also like to solve the Strong $(ORSP_i)$:

Among all optimal solutions to (3), find one that minimizes

$$|\{k \mid \lambda_k \neq 0\}|.$$

188   HOFFMAN AND MCCORMICK

It is not clear at this point that we can solve (SP) by successively solving $(ORSP_i)$ for $i = 1,2,\ldots,m$; nevertheless we shall concentrate on $(ORSP_1)$ in the remainder of this Section.

A set of multipliers $\{\lambda_k \mid k > 1\}$ for (3) when $i = 1$ defines the following index subsets:

$$U = \{k > 1 \mid \lambda_k \neq 0\},$$

$$H = \{j \mid \hat{\alpha}_{1j} = 0 \text{ and } \alpha_{1j} \neq 0\},$$

$$S = \{j \mid \hat{\alpha}_{1j} = 0 \text{ and } \alpha_{1j} = 0 \text{ and } \alpha_{kj} \neq 0 \text{ for some } k \in U\},$$

$$G = H \cup S,$$

$$F = \{j \mid \hat{\alpha}_{1j} \neq 0 \text{ and } \alpha_{1j} = 0\},$$

$$P = F \cup S = \{j \mid \alpha_{1j} = 0 \text{ and } \alpha_{kj} \neq 0 \text{ for some } k \in U\}, \text{ and}$$

$$Z = \{j \mid \alpha_{1j} = 0\}.$$

That is, U is the set of used rows; H, the set of hit columns, where a non-zero was changed to a zero; S, the set of saved columns, where a zero that we would have expected to be filled-in (since $\alpha_{kj} \neq 0$) was not filled-in; G, the set of good columns, where the entry was actively manipulated for the better; F is the set of filled-in columns; P is the set of potential fill-in columns; and Z is the set of zero columns. The net decrease in non-zeros in row 1 is then $|H| - |F|$, which we want to maximize to solve $(ORSP_1)$. The next theorem states the intuitive result that if k columns are affected for the good, then at least k independent rows must have been used (we omit the technical proof).

THEOREM 2. For any set of multipliers, $M(G_{UG}) = |G|$, and hence rank $A_{UG} = |G|$.

Theorem 2 implies in particular that $|U| \geq |G|$ always holds. If $|U| > |G|$, we can select a $|G|$-subset of U which perfectly matches to G and use the corresponding square non-singular (by (MP)) submatrix of A to zero out A_{1G}, thus achieving the same result with less work. Conversely, if A_{RC} is a square submatrix with a perfect matching, Theorem 2 ensures that if we use A_{RC} to zero out A_{1C}, then $G = C$, i.e., only non-zeros in C are hit, and fill-in occurs in every position where it would be expected. That is, Theorem 2 shows the crucial fact that (MP) implies that there is no "unexpected cancellation".

Hence, we can assume that the canonical situation is that $|U| = |G|$ and G_{UG} has a perfect matching. Then A_{UG} is non-singular by (MP), so the $\{\lambda_k\}$ will be uniquely determined by solving

$$\lambda^{*T} A_{UG} = A_{1G}. \tag{4}$$

Equation (4) allows us to think of the $\{\lambda_k\}$ as coming from U and G rather than vice versa, thereby reducing (SP) to the more combinatorial problem of finding optimal U and G.

Thus we need only consider all possible U, and for each U consider only the G which match perfectly into U. There are potentially many possible ways to select $G \subseteq \{1,2,\ldots,n\}$ so that G perfectly matches to U. The next theorem shows that for a given U it suffices to check only one such G.

THEOREM 3. Let G_1 and G_2 be two sets of columns that perfectly match into U, and denote the set of hit columns corresponding to G_i by H_i, $i = 1,2$, etc. Then

$$|H_1| - |F_1| = |H_2| - |F_2| .$$

PROOF. Since the definition of the column subset P depends only on U and not on G_i (for a given A), we denote P by $P(U)$. Then it is easy to see that $|H_i| = |U| - |S_i|$ and $|F_i| = |P(U)| - |S_i|$, so that $|H_i| - |F_i| = |U| - |P(U)|$, $i = 1,2$.

If we fix a full-rank matching M in G_{00} , then any row subset U induces a unique matched column subset G relative to M. Any such (U, G) pair will have a perfect matching, namely M restricted to A_{UG}, so (MP) ensures that A_{UG} will be non-singular. Hence the multipliers can be found as in (4). Theorem 3 ensures that the best (U, G) pair from among this restricted class of such pairs will solve $(ORSP_1)$.

Through (MP), Theorem 2 and Theorem 3 we have reduced the apparently algebraic problem $(ORSP_1)$ into the purely combinatorial one of maximizing $|U| - |P(U)|$ over all $U \subseteq \{2,\ldots,m\}$. Define $R = \{2,\ldots,m\}$ and $\bar{U} = R \setminus U$. Then

$$\max_U (|U| - |P(U)|) = (m - 1) - \min_U(|P(U)| + |\bar{U}|). \qquad (5)$$

By definition of $\bar{U}$ and $P(U)$, every non-zero in A_{RZ} (the first zero-section of A) is contained in either a row in $\bar{U}$ or a column in $P(U)$. If we call rows and columns lines, then in this situation we say that A_{RZ} is covered by the lines in $\bar{U} \cup P(U)$. Conversely, given a minimal covering of A_{RZ} by lines in L, by letting $U = R \setminus \{$ rows in $L\}$, we then must have $P(U) = \{$ columns in $L\}$, so that L is expressible as $\bar{U} \cup P(U)$. Thus by (5), finding $\max_U (|U| - |P(U)|)$ is equivalent to finding a minimum covering of A_{RZ} by lines. But by the classical theorem of König and Egervary (see Lawler (1976) p. 190), such a minimum cover can be computed through a maximum matching in G_{RZ} :

THEOREM 4. $M(G_{RZ}) = \min_U (|P(U)| + |\bar{U}|)$, and a maximum matching and a minimum covering by lines are dual combinatorial

190   HOFFMAN AND MCCORMICK

objects.

By the duality theory of matching algorithms, if we find a maximum matching in G_{RZ} through a labelling algorithm, then an optimum U for $(ORSP_1)$ is the set of rows reachable via an alternating path from an unmatched row. That is, Theorem 4 shows that this U solves the right-hand side of (5), so it also solves the left-hand side, and so must solve $(ORSP_1)$.

Even better, the next theorem shows that the optimum U defined above also solves the Strong $(ORSP_1)$. For a network flow problem with source s and optimal flow f, define the standard minimum cut $K^* = \{i \mid$ there is an augmenting path $s \to i$ under $f\}$. Note that the optimal U defined above is a standard minimum cut for the usual way of solving a maximum cardinality bipartite matching problem by converting it to an equivalent network flow problem.

THEOREM 5. In a given network, the standard min cut is a subset of every min cut. Thus the standard min cut is the same for every optimal flow, hence it is well-defined and has minimum cardinality among all min cuts (see Ford and Fulkerson (1962) p. 13 for a proof).

Theorems 4 and 5 together imply that we can solve the Strong $(ORSP_1)$ through maximum matching, and that the optimal U is unique.

3. Two Algorithms for (SP)

Once we have found the optimal U for each row i (say, U_i^*) through matching, as noted above we can easily generate the sets G_i of columns by choosing G_i to be the set of columns that matches into U_i^* under the fixed matching M. These (U_i^*, G_i) pairs completely determine the non-zero off-diagonal entries of a matrix T^* as defined by (4). The question arises: is T^* non-singular?

To answer this question, it is necessary to investigate what the uniqueness properties of the U_i imply for the structure of T^*. Define a directed graph D with vertices $V = \{1, \ldots, m\}$ and edges $E = \{(k, i) \mid k \in U_i^*\}$; thus D represents the sparsity pattern of T^*. If the row indices of A and T^* are ordered consistently with the strong component decomposition of D, then the decomposition induces a block lower-triangular structure on T^*, where the diagonal blocks of T^* correspond to the strong components of D.

THEOREM 6. If $l \in U_k^*$ and $k \in U_i^*$ then $l \in U_i^*$.

PROOF. For ease of notation, let $U = U_i^*$, $\overline{U} = \{1, \ldots, m\} \setminus \{i\} \setminus U_i^*$, $P = P(U_i^*)$, and $\overline{P} = Z_i \setminus P(U_i^*)$. Thus U and $\overline{U}$ partition the rows of the ith zero-section of A, and P and $\overline{P}$ partition the columns.

Recall that the rows in $\overline{U}$ and the columns in P are a minimum cover of the ith zero-section of A by lines. Thus $A_{U\overline{P}} = 0$, and, since $k \in U$, $A_{k\overline{P}} = 0$. By the minimality of this cover, the submatrix $A_{\overline{U}\overline{P}}$ has a row-perfect matching, and so $|\overline{U}|$ lines are necessary to cover it. Let L^* be the standard minimum set of lines covering the kth zero-section. Since $A_{k\overline{P}} = 0$ and $k \notin \overline{U}$, the submatrix $A_{\overline{U}\overline{P}}$ is part of the kth zero-section and so must be covered by the lines in L^*. Consider the set of lines $L = L^* \cup \overline{U} \setminus \overline{P}$. Since the only non-zeros in the columns in $\overline{P}$ occur in rows in $\overline{U}$, L is a cover for the kth zero-section. The only change in lines between L^* and L is in lines passing through $A_{\overline{U}\overline{P}}$; since L has only $|\overline{U}|$ lines passing through $A_{\overline{U}\overline{P}}$, the minimum possible number L must also be a minimum cover. Finally, L contains at least as many rows as L^*, so that the U associated with L has at most as many rows as the U associated with L^*, namely U_k^*. But U_k^* has the minimum possible number of rows for any minimum cover of the kth zero-section, so $L = L^*$. But this implies that $U_k^* \subseteq U_i^*$.

The conclusion of Theorem 6 is precisely that the graph D is transitively closed. This implies that the blocks of the block lower-triangular partition of T^* are either completely dense or all zero. In particular, $U_i^* \cup \{i\} = U_k^* \cup \{k\}$ for i and k in the same strong component of D. These observations allow us to prove the following.

THEOREM 7. T^* is non-singular.

PROOF. Since T^* is block lower-triangular, it suffices to show that the diagonal blocks of T^* are non-singular. A typical diagonal block is indexed by the vertices in some strong component, say D. As shown above, the set $\tilde{D} = U_i^* \cup \{i\}$ is the same for all $i \in D$, and $D \subseteq \tilde{D}$. Assume for convenience that the fixed matching M is such that the row i matches to column i, $i = 1,\ldots,m$. If $D = \tilde{D}$, then the diagonal block associated with D is clearly just a re-scaling of $(A_{DD})^{*-1}$ (A_{DD} is non-singular by (MP)), and so is non-singular. Otherwise, let $L = \tilde{D} \setminus D$. Then the diagonal block associated with D is a re-scaling of the bottom right corner of the matrix

$$\begin{pmatrix} A_{LL} & A_{LD} \\ A_{DL} & A_{DD} \end{pmatrix}^{-1} = \begin{pmatrix} V_{LL} & V_{LD} \\ V_{DL} & V_{DD} \end{pmatrix}$$

(this matrix is non-singular by (MP)). But it is well-known (see Cottle (1974), equations (2) and (4)) that V_{DD} is non-singular if and only if A_{LL} is non-singular. But A_{LL} is indeed non-singular by (MP).

Since T^* is non-singular, we can use it to transform A into $\hat{A}$. This way of generating $\hat{A}$ processes each row in parallel, i.e. each row is solved relative to the original matrix rather than relative to a partially transformed matrix. We call this procedure the Parallel

Algorithm (PA).

THEOREM 8. (PA) solves (SP) when A satisfies (MP).

PROOF. Each row of A is made as sparse as possible in $\hat{A}$.

The "parallelism" of (PA) seems unsatisfactory for two reasons. First, it is more natural to process A *sequentially*, i.e. by solving each row's matching problem on the partially reduced A whose previous rows have already been processed. Second, by processing A sequentially we can overwrite $\hat{A}$ on A, thereby saving space, and, as we shall see later, the optimal U's can only get smaller, thus also saving time in solving equations (4). More formally, consider the **Sequential Algorithm (SA)**:

Given A.

For $i = 1, 2, \ldots, m$ do.
     Use matching in the ith zero-section of A to find U.
     Find some G so that A_{UG} is non-singular.
     Replace A_{i0} by $A_{i0} + \sum_{k \in U} \lambda_k A_{k0}$, where the λ_k are defined by (4).
     End.

A is the output. Stop.

We want to show that the output of (SA) solves (SP). The replacement step is equivalent to left-multiplying the current A by a non-singular elementary matrix, so the output A is row-equivalent to the input A. This also implies that we must be able to find a suitable G at each iteration, or else A would have lost rank at some point. Thus it remains only to show the following.

THEOREM 9. (SA) produces the same final number of non-zeros as (PA).

PROOF. Denote the optimal U for row i under (PA) by U_i^*, under (SA) by U_i, and inductively assume that $U_k \subseteq U_k^*$ for all $k < i$. Denote the original A by A^0, the A just before replacing the ith row by A^i, and the rows and columns of the ith zero-section of A^0 (which is the same for A^i) by R and Z respectively. Recall that $L^* = R \setminus U_i^* \cup P(U_i^*)$ is a minimum coverage of A_{RZ} by lines. Suppose that $k \in U_l$ for some $l \in U_i^*, l < i$. By the induction hypothesis $k \in U_l$ and $l < i$ imply that $k \in U_i^*$, so $k \in U_i^*$ by Theorem 6. Thus every row $l \in U_i^*, l \leq i$, that may have been changed in going from A^0 to A^i maintains the property that it is zero in the columns in $Z \setminus P(U_i^*)$. This means that L^* is also a covering of A_{RZ}^i by lines. Thus

$$|L^*| = M(G_{RZ}^0) = rank\ A_{RZ}^0 = rank\ A_{RZ}^i \leq M(G_{RZ}^i) \leq |L^*|, \qquad (6)$$

where the third equality holds because A_{RZ}^i is a non- singular

FAST SPARSE MATRIX ALGORITHM   **193**

transformation of A_{RZ}^0. Thus the parallel cover L^* is also minimum for A_{RZ}^i, though it may no longer be the minimum cover with the minimum cardinality U. However, Theorem 5 ensures that $U_i \subseteq U_i^*$ verifying the induction.

The improvement in non-zeros in row i under (SA) is $(m-1) - M(G_{RZ}^i)$. But (6) shows that this is equal to $(m-1) = M(G_{RZ}^0)$, which is the improvement in non-zeros in row i under (PA).

Theorem 9 together with the preceding remarks prove the final theorem.

THEOREM 10. (SA) also solves (SP) under (MP).

It is easy to get a good bound on the running time of the combinatorial part of both (PA) and (SA) Let v be the number of non-zeros in A, which we can assume is greater than n. We use the following trick to reduce the running time of both (PA) and (SA). For (SA) as well as (PA), find a fixed initial maximum matching on A; this takes $0(mv)$ operations (this can be improved to $0(\sqrt{m+n}v)$; see Papadimitriou and Stieglitz (1982)). Then, when finding a maximum matching in a zero-section, copy over the part of the fixed matching that lies in the columns of the zero- section as the starting matching; this copying takes $0(m^{*2})$ time. Since every initially unmatched row in the zero-section matches to some column outside the zero-section in the fixed matching, the number of unmatched rows in the starting solution for row i's zero-section can be at most the number of non-zeros in row i. Thus the total number of augmentations needed over all rows is $0(v)$. Since each augmentation is an $0(v)$ operation, we get an $0(v^{*2})$ overall bound for the combinatorics.

The time needed to do the numerical part of (PA) and (SA) can be bounded as follows. In the worst case we will have to solve a linear system like (4) of dimension $0(m)$ for each one of m rows. Solving one such system is bounded by $0(m^{*3})$, so the numerical part is bounded by $0(m^{*4})$ overall. However, we have assumed that A is sparse, and there are sparse equations routines that can solve a system like (4) in time more like $0(m^{*2})$. Our practical experience has been that there are only $0(1)$ rows whose linear systems are really as large as $0(m)$, and that most linear systems will be of size $0(1)$. Thus under favorable circumstances the numerical computations could take time as small as $0(m^{*2})$.

4. Practicalities

Very few real-life matrices satisfy (MP). In light of Theorem 1 we cannot hope to actually solve (SP) on all real matrices, but we can try to apply one of our algorithms or a variant as an "optimal" heuristic. Ideally, when we apply our "real" algorithms to real, full-rank matrices, they would be guaranteed to achieve at least the increase in sparsity that an "ideal" algorithm would achieve on a matrix with the same sparsity pattern that *did* satisfy (MP).

It is difficut to anticipate unexpected cancellation with real matrices. A parallel type of algorithm is therefore unsuitable, as it has to proceed without knowing where cancellation takes place. On the other hand, a sequential type of algorithm can take stock of the cancellation that arises at each step. However, guaranteeing performance becomes more subtle in the presence of cancellation. Consider the full-rank matrix

$$A = \begin{pmatrix} 1 & 3 & 0 & 5 & 5 \\ 2 & 1 & 4 & 0 & 0 \\ 0 & 3 & 0 & 5 & 5 \end{pmatrix}.$$

Any sequential algorithm will pick $U_1 = \{3\}$, and could pick $G_1 = \{2\}$. This transformation unexpectedly zeros out columns 4 and 5 of row 1. Thus if we naively process row 2 using this new row 1, we will choose $U_2 = \{1\}$. But the parallel $U_2^{\bullet} = \varnothing$, which does not contain U_2 as required by the induction hypothesis of Theorem 9, so we can no longer guarantee that our final answer will be as good as the ideal. (A close reading of that proof of Theorem 9 will reveal that the only way this difficulty can arise is when rank $A_{RZ} < M(G_{RZ})$ for some zero-section; in the second zero-section of this example, rank $A_{RZ} < 2 = M(G_{RZ})$.)

A simple trick will avoid this problem. As we perform (SA), we certainly know at each step where we expect the non-zeros to occur for subsequent steps. If we do encounter unexpected cancellation, we can merely pretend that there is still a non-zero in the cancelled position. That is, subsequent matchings are performed as if no unexpected cancellation ever took place, though we keep track of which "non-zeros" are real zeros. As long as we initially make sure that A has full row rank, the numerical operations can never create a dependence among the rows. Thus we will always be able to find a G so that A_{UG} is nonsingular even with "phantom" non-zeros. Then the proof of Theorem 9 becomes valid once again, and the modified (SA) is now guaranteed to produce an answer at least as good as the "ideal" answer.

We now make some remarks about implementing (SA). Linear

constraints are usually presented as a mixture of equalities and inequalities. If these constraints are converted to the form (1) by adding a slack variable to each inequality row, it is easy to see that there is always a maximum cardinality matching in which every inequality row is matched to its slack column. It is also easy to see that such rows can never be profitably used in the optimal U for any other row, since the slack variable will always unavoidably fill-in its column. (In fact, by this same reasoning, if A is known to have an embedded identity matrix, then A must already be optimally sparse.) Hence (SA) will still work correctly if we merely treat inequality rows as if they were permanently matched to a non-existent column without having to explicitly create a slack variable at all. This phenomenon implies that (SA) will tend to find better solutions for systems with a high proportion of equality constraints.

5. Further Questions and Conclusion

Trying to solve the Sparsity Problem as described in this paper raises some interesting questions. From an applications point of view, the chief question is: Does (SA) help in practice or not? The answer to this question must come from empirical experience with (SA) on various problems. We have implemented a preliminary version of (SA) for this purpose; our results so far are encouraging, but we have no means conclusively demonstrated the usefulness of (SA). We expect to report our computational experience with (SA) in the near future.

Although it is necessary to keep unexpected zeros as phantom non-zeros to guarantee the performance of (SA) on real matrices, it is certainly feasible to run the algorithm without this artifice. Can any guarantees be made in this case? Does this make much difference in practice? Alternatively, since "lucky" cancellation has been observed in nearly all real examples we have tried, is there some efficient heuristic for (SP) that can take advantage of this and outperform (SA), perhaps restricted to some subset of interesting problems with special structure? Finally, what happens when we try to apply these algorithms to rank- deficient matrices?

We shall continue our research on (SP) and shall try to answer some of these questions in future papers.

Appendix

PROOF. (of Theorem 1) This Theorem and its proof are due to L. Stockmeyer (1982). See Garey and Johnson (1979) for the definitions of the concepts used in this proof.

196   HOFFMAN AND MCCORMICK

The problem that we shall reduce to (SP) is

Simple Max Cut: Given an undirected graph $G = (V, E)$, partition the nodes of G into P and $V \setminus P$ so as to maximize

$$|\{\{i,j\} \in E \mid i \in P, j \in V \setminus P\}|$$

Let $n = |V|$, $m = |E|$, let $A(G)$ be the usual $(0,1)$ node-arc incidence matrix of G, and let A_i be the $n \times 2m$ matrix which is all zero except for row i, which is half $+1$ and half -1. Let e be the $2m$-vector of all ones and let f be the $(2m(n + 1) + 1)$-vector of ones. Now suppose we could solve (SP) on the matrix

$$B(G) = \begin{pmatrix} 0 & e & e & \cdots & e & f \\ A(G) & A_1 & A_2 & \cdots & A_n & 0 \end{pmatrix}$$

As before, we can assume that an optimal transform of $B(G)$, call it T^*, has unit diagonal. Because of the size of f, it will never pay to use row 1 when reducing any other row, so the first column of T^* must be $(1,0,\ldots,0)^{*T}$. Thus no choice for the first row of T^* can cause singularity problems. Because of the column size of the A_i, and since all entries are ± 1, it will pay to use every other row in reducing the first row, so the first row of T^* must be $(1, \epsilon_1, \epsilon_2, \ldots, \epsilon_n)$, where $\epsilon_i = \pm 1$ for all $i = \in V$. Let $P = \{i \mid \epsilon_i = +1\}$. Then the number of non-zeros in the first row of $\hat{B}(G)$ is clearly

$$(2m(n + 1) + mn + (m - |\{\{i,j\} \in E \mid i \in P, j \in V \setminus P\}|). \quad (7)$$

But since (7) is minimized by the optimal T, P also solves the Simple Max Cut Problem for G.

References

Bunch, J.R. and D.J. Rose, eds., "Sparse Matrix Computatins", Academic Press (New York, 1976).

Cottle, R.W. "Manifestations of the Schur Complement", *Linear Algebra and its Applications* 8 (1974), pp. 182-211.

Duff, I.S., ed., "Sparse Matrices and their Uses", Academic Press (New York, 1980).

Ford, L.R. and D.R. Fulkerson, "Flows in Networks", Princeton University Press (Princeton, 1962).

Garey, M.L. and D.S. Johnson, "Computers and Intractibility", Freeman (San Francisco, 1979).

Lawler, E.L., "Combinatorial Optimization", Holt, Rinehart and Winston (New York, 1976).

Papadimitriou, C.H. and K. Stieglitz, "Combinatorial Optimization: Algorithms and Complexity", Prentice Hall (Englewood Cliffs, 1982).

Stockmeyer, L.J., personal communication (1982).

Greedy Algorithms

1. On simple linear programming problems

All the papers in this section deal with instances where linear programming problems are solved by greedy algorithms (i.e., successive maximization of the variables in some order). I loved the name greedy algorithm, which Ray Fulkerson proposed to Jack Edmonds in lieu of the less vigorous adjective "myopic". At the time I wrote this paper, the term greedy had not been coined, and I used the anemic "simple".

During my year at ONR London, I came across a report written by a computer company congratulating itself for its splendid simplex code which had solved an important production planning problem of an automobile company. But I knew how to solve that problem simply (or greedily), and that is explained in the last part of the paper. I waited several years before publishing the paper, to avoid embarrassment to anybody. I had previously dreamed up the idea of Monge sequence (somebody told me about Monge's pamphlet, and I saw that I had a right to use the name Monge), and I found the application to production planning trying to think of ways the Monge sequence idea could be used in different contexts.

(I used the term Monge sequence on the theory that naming something after someone famous was a good way to getting that something known quickly. That theory failed totally here: it was 20 years before people notice the idea of Monge sequence (in particular, Monge array), but it is now a standard concept in combinatorial optimization.)

The part of the paper in which the warehouse problem is first transformed is an homage to Walter Jacobs. He had described in 1954 in The Caterer Problem an instance where a linear programming problem could be rewritten in simpler form (even though the polyhedron of the transformed problem properly contained the original polyhedron) if you could show that the optimum on the larger polyhedron was actually in the smaller one. I found another (less interesting, but still valid) instance of the same phenomenon here; and I have never seen another.

2. Totally balanced and greedy matrices

Kolen was kind enough to allow Sakarovich and me to join with him in publishing this paper. It turned out that we had worked on the same set of problems, but Kolen's results were more interesting. The arguments used by Sakarovich and me, now lost forever, were analogues of arguments used in the study of chordal graphs.

3. Greedy packing and series-parallel graphs

We conjectured this theorem for the case where the matrix is $(0, 1)$. Series parallel graphs are defined inductively, and we started first by using the definition in which you either make one edge two edges in series (by putting a node in the middle), or two edges in parallel (think binary fission). And we got nowhere. But there is another inductive definition, in which two series parallel graphs are connected in series or in parallel, and this worked. In fact, the completeness of the results astonished us.

4. On simple combinatorial optimization problems

The original linear programming work at the Pentagon was called Project SCOOP (scentific computation of optimum programs), and the title of this paper was intended (I swear) as a tribute to that maternity ward of linear programming. The theorem has a really ugly statement, and it took me a long time before I believed it was valid (truth is beauty ...). After a few years, I did believe it (beauty is in the eye of the beholder ...). It is inconceivable that I could have conjectured this theorem without having previously worked on paper 2 of this section.

5. Series parallel composition of greedy linear programming problems

A special case of the results here (and the general concept of combining problems in series or parallel fashion) was done by my colleagues in an earlier paper. I am particularly fond of this paper because it includes applications both of majorization (as in Hardy, Littlewood and Pólya) and Monge arrays. The main theorem, on the other hand, can be regarded as a massive generalization of the greedy solution of transportation problems with costs given by a Monge array. I like this paper because it illustrates how you can start with something deliciously simple (like Monge arrays), and create something almost hilariously complicated but still interesting: "How do I generalize thee? Let me count the ways ..."

Reprinted from
Proc. Symp. Pure Math., Vol. VII (AMS, 1963), pp. 317–327

ON SIMPLE LINEAR PROGRAMMING PROBLEMS

BY

A. J. HOFFMAN

1. Introduction. In the considerable research that has been devoted to linear programming since the subject was first formulated in 1947 by George Dantzig, there have been a number of occasions when it has been noticed that some particular classes of problems were amenable to "obvious" solution. For the most part, the source of the obvious solution has been insight into the physical or economic meaning of the problem. The purpose of this talk is to point out that almost all of the classes of problems which the author currently knows to be amenable to simple solutions, and which have the further property that the particular answers are integral when the particular data are integral, can be shown to be special cases of one simple observation. In a certain sense, therefore, this simple observation provides a unified mathematical insight as a substitute for the physical and economic insights.

Even more remarkable is the fact that the essential idea behind the observation was first noticed by G. Monge in 1781 [4]! Monge remarked that if unit quantities are to be transported from locations X and Y to Z and W (not necessarily respectively) in such a way as to minimize the total distance traveled, then the route from X and the route from Y must not intersect; and it is this idea which shall be exploited.

We first define a special class of transportation problems which can easily be solved by inspection. Next, we will take up in detail the "warehouse problem" of A. S. Cahn, which has been shown by several authors to be amenable to solution by inspection. This will be demonstrated afresh, by a succession of two transformations which result in a restatement of the problem in such a form that the Monge idea applies. While much more cumbersome than other methods of solution, our procedure has the virtue of exhibiting a variety of devices used in the trade. We close with remarks on simplifying devices and suggestions for future research.

2. The transportation problem and the Monge sequence. Let $a_1, \cdots, a_m$, $b_1, \cdots, b_n$ be given non-negative integers such that $\sum_i a_i = \sum_j b_j$. Let $C = (c_{ij})$ be an m by n matrix of real numbers. The transportation problem is to discover among all non-negative m by n matrices $X = (x_{ij})$ such that $\sum_j x_{ij} = a_i$ and $\sum_i x_{ij} = b_j$ a matrix which minimizes $\sum_i \sum_j c_{ij} x_{ij}$. For the task of actually calculating answers, several efficient iterative algorithms are known. Our concern here is to show that if the coefficients (c_{ij}) satisfy certain special conditions, then the solution can be obtained by inspection. To that end, we introduce the following definitions:

A Monge sequence is a rearrangement

                             A. J. HOFFMAN

(2.1) $$(i_1, j_1), (i_2, j_2), \cdots, (i_{mn}, j_{mn})$$

of the pairs of indices (i, j).

A Monge sequence (2.1) is said to be consonant with a matrix $C = (c_{ij})$ if, wherever

(i) $p < q, r, s,$
(ii) $(i_p, j_q) = (i_s, j_s),$
(iii) $(i_q, j_p) = (i_r, j_r),$

we have

(2.2) $$c_{i_p j_p} + c_{i_q j_q} \leq c_{i_p j_q} + c_{i_q j_p} .$$

Consider now the following procedure for choosing a matrix

$$\bar{X} = (\bar{x}_{ij}) .$$

(2.3) Step 1. Set $k = 1$.
       Step 2. Set $\bar{x}_{i_k j_k} = \min(a_{i_k}, b_{j_k})$.
       Step 3. Replace a_{i_k} by $a_{i_k} - \bar{x}_{i_k j_k}$. Replace b_{j_k} by $b_{j_k} - \bar{x}_{i_k j_k}$.
       Step 4. If $k = mn$, stop. If $k < mn$, replace k by $k + 1$ and go to Step 2.

THEOREM 1. *Let $a_1, \cdots, a_m$, $b_1, \cdots, b_n$ be given non-negative integers such that $\sum_i a_i = \sum_j b_j$, and $C = (c_{ij})$ be given m by n matrix. If the Monge sequence (2.1) is consonant with C, then the matrix $\bar{X} = (\bar{x}_{ij})$ produced by the algorithm (2.3) solves the transportation problem. If (2.1) is not consonant with C, then there exist non-negative integers $a_1, \cdots, a_m$, $b_1, \cdots, b_n$ with $\sum_i a_i = \sum_j b_j$ such that the matrix $\bar{X} = (\bar{x}_{ij})$ produced by (2.3) does not solve the transportation problem.*

PROOF. Assume (2.1) consonant with C. We shall apply induction on $m + n$. The result obviously holds if $m + n = 2$, and assume it holds for all smaller values of $m + n$ than the current one. Among all solutions of the transportation problem, let $Y = (y_{ij})$ be a solution with the largest value of $x_{i_1 j_1}$ (obvious continuity arguments establish the existence of Y). Assume that $y_{i_1 j_1} < a_{i_1}$, $y_{i_1 j_1} < b_{j_1}$. It follows that there exist indices $r \neq 1$ and $s \neq 1$ such that

$$y_{i_1 r} > 0 \quad \text{and} \quad y_{s j_1} > 0 .$$

Let $\varepsilon = \min(y_{i_1 r}, y_{s j_1})$. Consider now the matrix $Z = (z_{ij})$.

$$z_{i_1 j_1} = y_{i_1 j_1} + \varepsilon ,$$
$$z_{rs} = y_{rs} + \varepsilon ,$$
$$z_{i_1 r} = y_{i_1 r} - \varepsilon ,$$
$$z_{s j_1} = y_{s j_1} - \varepsilon ,$$
$$z_{ij} = y_{ij} \text{ for all other pairs of indices } (i, j).$$

Clearly Z satisfies the boundary conditions of the transportation problem, and by (2.2), $\sum_i \sum_j c_{ij} z_{ij}$ is not larger than $\sum_i \sum_j c_{ij} y_{ij}$. Hence Z is a solution to the transportation problem with a larger value of $x_{i_1 j_1}$ than $y_{i_1 j_1}$. This contradiction establishes that, among all solutions to the transportation problem, there is one in which $x_{i_1 j_1} = \bar{x}_{i_1 j_1}$.

By (2.3), the new value of at least one of a_{i_1}, b_{j_1} is 0. For the sake of definiteness, assume the new value of b_{j_1} is 0. The algorithm (2.3) will then compel all x_{ij_1}, $i \neq i_1$, to be zero. In fact, it is clear that our problem is now reduced to an m by $n - 1$ transportation problem obtained by deleting column j_1, and changing a_{i_1} to $a_{i_1} - \bar{x}_{i_1 j_1}$. If we consider the order in the Monge sequence obtained from (2.1) by deleting those entries corresponding to column j_1, it is clear that the conditions (2.2) are hereditary. Hence the induction hypothesis applies. This completes the proof that the consonance of the Monge sequence with C justifies (2.3).

Suppose (2.2) is not satisfied, i.e., we have

$$(2.4) \qquad c_{i_p j_p} + c_{i_q j_q} > c_{i_p j_q} + c_{i_q j_p} \, .$$

Let $a_{i_p} = a_{i_q} = b_{j_p} = b_{j_q} = 1$, all other a_i and b_j at zero. Then the algorithm (2.3) will produce $c_{i_p j_p} + c_{i_q j_q}$ for $\sum_i \sum_j c_{ij} \bar{x}_{ij}$. But this is bigger than $c_{i_p j_q} + c_{i_q j_p}$. This completes the proof of the second part of the theorem.

As a simple illustration of the application of this theorem, consider the following problem [15]. An individual has n jobs to perform. Job j takes t_j hours and is due d_j hours from now, $d_1 \leq d_2 \leq \cdots \leq d_n$, assume the t_j and the d_j are all positive integers, and the individual works in hourly units. How should he schedule his work in order to minimize the maximum tardiness? It is well known that an optimal procedure is to perform the jobs in the order of their due times. Let us now prove this as a special case of our theorem.

Let $m = \sum_j t_j$. Consider the m by n transportation problem in which $b_j = t_j$, $a_i = 1$, and the c_{ij} are defined as follows:

$$c_{ij} = 0 \text{ for } i \leq d_j, \quad c_{d_j+r, j} = M^r \, ,$$

where M is a very large number. It is clear that, for M large, the objective function is dominated by the largest r such that $x_{d_j+r, j} = 1$ for some j. Hence the solution of our transportation problem will schedule the work so as to minimize the maximum tardiness. Construct a Monge sequence $(1, 1), (2, 1), \cdots, (m, 1), (1, 2), \cdots, (m, 2), \cdots, (m, n)$. It is easy to verify that (2.2) holds.

3. The warehouse problem: first transformation. Let $p_1, \cdots, p_n, c_1, \cdots, c_{n-1}$ be given positive numbers, $0 \leq A \leq B$, a pair of constants. The problem is to choose $x_1, \cdots, x_{n-1}$ and $y_1, \cdots, y_n$ subject to

$$(3.1) \qquad x_j \geq 0 \, , \quad y_j \geq 0 \, ,$$

$$(3.2) \qquad \bar{x}_1 - y_1 \leq B - A \, ,$$

$$(x_1 + x_2) - (y_1 + y_2) \leq B - A \, ,$$

$$\cdots\cdots\cdots\cdots\cdots\cdots$$

$$(x_1 + \cdots + x_{n-1}) - (y_1 + \cdots y_{n-1}) \leq B - A \, ,$$

$$(3.3) \qquad y_1 \leq A \, ,$$

$$y_2 \leq A + x_1 - y_1 \, ,$$

$$\cdots\cdots\cdots\cdots\cdots\cdots$$

$$y_n = A + (x_1 + \cdots + x_{n+1}) - (y_1 + \cdots + y_{n-1})$$

                           A. J. HOFFMAN

in order to minimize

(3.4)
$$\sum c_j x_j - \sum p_j y_j \; .$$

Let K be the convex set of all points satisfying (3.1)–(3.3). K is not empty, since $x_j = y_j = 0$ satisfies all conditions. Further, K is bounded: (3.1) and the first line of (3.3) show that y_1 is bounded, so by the first line of (3.2) x_1 is bounded, therefore, by the second line of (3.3) y_2 is bounded, etc.

We now show that, if for some j, $p_j \geq c_j$, then the problem can be split into two problems each of the same form as the original.

Let $p_j \leq c_j$. Then there is a solution in which

(3.5)
$$y_j = A + (x_1 + \cdots + x_{j-1}) - (y_1 + \cdots + y_{j-1}) \; .$$

Otherwise, let $(\bar{x}, \bar{y}) = (\bar{x}_1, \cdots, \bar{x}_{n-1}, \bar{y}_1, \cdots, \bar{y}_n)$ be a solution in which $y_j = \bar{y}_j$ assumes its maximum value. If (3.5) does not occur, then

(3.6)
$$\bar{y}_j < A + (\bar{x}_1 + \cdots + \bar{x}_{j-1}) - (\bar{y}_1 + \cdots + \bar{y}_{j-1}) \; .$$

If we increase $\bar{x}_j$ and $\bar{y}_j$ by a small amount, it follows from (3.6) that (3.1)–(3.3) will not be violated, and since $p_j \geq c_j$, we will not have increased (3.4). This contradicts the definition of $(\bar{x}, \bar{y})$, hence we may assume (3.5).

If we substitute the value of y_j from (3.5) in (3.2) and (3.3), we obtain

(3.7)
$$x_1 - y_1 \leq B - A \; ,$$
$$\cdots\cdots\cdots\cdots\cdots$$
$$(x_1 + \cdots + x_{j-1}) - (y_1 + \cdots + y_{j-1}) \leq B - A \; ,$$

and

(3.8)
$$x_j \leq B \; ,$$
$$x_j + x_{j+1} - y_{j+1} \leq B \; ,$$
$$\cdots\cdots\cdots\cdots\cdots$$
$$x_j + (x_{j+1} + \cdots + x_{n-1}) - (y_{j+1} + \cdots + y_{n-1}) = B \; .$$

(3.9)
$$y_1 \leq A \; ,$$
$$\cdots\cdots\cdots\cdots\cdots$$
$$y_j = A + (x_1 + \cdots + x_{j-1}) - (y_1 + \cdots + y_{j-1}) \; ,$$

and

(3.10)
$$y_{j+1} \leq x_j \; ,$$
$$\cdots\cdots\cdots\cdots\cdots$$
$$y_n = x_j + (x_{j+1} + \cdots + x_{n-1}) - (y_{j+1} + \cdots + y_{n-1}) \; .$$

Now (3.7) and (3.9) involve only the variables $(x_1, \cdots, x_{j-1}, y_1, \cdots, y_j)$, (3.8) and (3.10) involve only the remaining variables. Hence our original linear programming problem has been broken into two parts. It is clear that (3.7) and (3.9) are in the same format as the original. We now work on (3.8) and (3.10). If we introduce a new variable y_j^1, and a new $p_j^1 < c_j$, and imagine $A = 0$, then it is obvious that (3.8) and (3.10) are also in the original format, with one less instance of a $p_k \geq c_k$. It follows from these considerations that we may assume the problem so posed that $p_j < c_j$ for all j and we return to consider (3.1)–(3.4) with that stipulation.

Consider now inequalities

$$(3.11) \qquad -x_1 + y_1 \leq A \, ,$$
$$-x_2 + y_2 \leq A + x_1 - y_1 \, ,$$
$$\dots\dots\dots\dots$$
$$y_n = A + (x_1 + \cdots + x_{n-1}) - (y_1 + \cdots + y_{n-1}) \, .$$

Let L be the convex set given by (3.1), (3.2) and (3.11). L is unbounded, and clearly $K \subset L$. We shall show that:

(i) (3.4) is bounded from below on L.

(ii) L has at least one vertex, and a minimum of (3.4) is attained at a vertex.

(iii) Every vertex of L is in K.

It will follow at once that minimizing (3.4) on L is equivalent to minimizing (3.4) on K.

Suppose (i) false, so that there exists a sequence of points in L on which (3.4) decreases (not necessarily monotonically) without bound. This is only possible if at least one of the y_i is unbounded. Let k be the smallest index i such that y_i is unbounded on the sequence. It follows from the kth line of (3.11) that there is some index $i \leq k$ such that x_i is unbounded on the sequence, and let $m \leq k$ be the least such index. If $m < k$, we have a contradiction of the mth line of (3.2). So $m = k < n$. Note x_k is unbounded on every subsequence on which y_k is unbounded.

Note that k depends on the sequence (x, y). Now of all sequences of vectors (x, y) such that (3.4) decreases without bound, let T be a sequence for which the corresponding $k = k(T)$ is as large as possible. Consider now a new sequence S in which each vector (x, y) is replaced by a vector (x', y') in which $x'_k = x_k - (\min x_k, y_k)$, $y'_k = y_k - (\min x_k, y_k)$, $y'_i = y_i$, $x'_i = x_i$ for all other i. Each vector of the new sequence S satisfies (3.1), (3.2), (3.11), and since $p_k < c_k$, (3.4) decreases without bound on S. Since x_k and y_k are not simultaneously positive for any vector in S, it follows from the definition of $k(T)$ that $k(S) > k(T)$. But it follows from the definition of T that $k(S) \leq k(T)$. This contradiction completes the proof of (i).

To prove (ii), observe that, since all variables are non-negative, L must contain a vertex, for it is a theorem that a closed convex set in finite dimensional space that contains no line must contain a vertex, and the first orthant contains no line. It is also a theorem that a concave function [e.g., (3.4)] bounded from below on a convex polyhedron attains its minimum, and at a vertex if the polyhedron has any vertices. (Proofs of these theorems are contained in [2].)

To prove (iii), observe that, for any k, a vertex of L cannot have both x_k and y_k positive. For we could change x_k and y_k by $\pm \varepsilon$, leaving all other coordinates unchanged, and exhibit the alleged vertex as the mid-point of a line segment contained in L. Therefore, at least one of x_k and y_k is zero. We will use this to show that each inequality in (3.3) is satisfied.

                    A. J. HOFFMAN

If $k = 1$, and $y_1 > 0$, then $x_1 = 0$, and the first inequality of (3.11) coincides
with the first inequality of (3.3). If $y_1 = 0$, then the first inequality of (3.3)
is satisfied since $A \geq 0$. If $1 < k < n$, and $y_k > 0$, then $x_k = 0$, and the kth
inequality of (3.11) coincides with the kth inequality of (3.3). If $y_k = 0$, then
the kth inequality of (3.3) coincides with the $(k-1)$th inequality of (3.11). If
$k = n$, (3.11) coincides with (3.3).

4. The warehouse problem: second transformation. It is easy from (3.2)
and (3.3) to calculate that B is an upper bound to the x_j and y_j, and that, if
one introduces the variables $B - y_j$ in lieu of the y_j's in (3.1), (3.2), (3.11) and
(3.4), then the warehouse problem takes the following form: Minimize

$$(4.1) \qquad \sum_{j=1}^{n} p_j x_j ,$$

where

$$(4.2) \qquad b_t \leq \sum_{t=1}^{j} x_t \leq c_t , \qquad\qquad j = 1, \cdots, n-1, n ,$$

with

$$(4.3) \qquad b_n = c_n$$

and

$$(4.4) \qquad 0 \leq x_j \leq a_j , \qquad\qquad j = 1, \cdots, n .$$

It is assumed that there exists at least one vector $(x_1, \cdots, x_n)$ satisfying
(4.2)–(4.4). At this point, the meaning of our symbols has shifted, and we
are dealing with a generalization of the problem.

By virtue of the non-negativeness of all x_j, it is no loss of generality to
assume

$$(4.5) \qquad 0 \leq b_1 \leq b_2 \leq \cdots \leq b_n .$$

For each b_t can clearly be replaced by $\max_{i \leq t} b_i$ without changing the problem. Henceforth we assume (4.5).

These preliminaries over (which amount to several trivial transformations
of a generalization of the problem as it appeared at the end of §3), we are
now ready to pose a transformation of the problem to the point where the
Monge algorithm applies.

First, introduce the non-negative variables $y_1, \cdots, y_n$ by the equation

$$(4.6) \qquad x_t + y_t = a_t , \qquad\qquad t = 1, \cdots, n .$$

The variables y_t may be thought of as unused capacities, if the x_t are conceived as bounded production variables. Next, introduce the non-negative
variables

$$x_{11}, x_{12}, \cdots, x_{1n} ,$$
$$x_{22}, x_{23}, \cdots, x_{2n} ,$$
$$\cdots\cdots\cdots\cdots$$
$$x_{nn}$$

by the equations

$$(4.7) \qquad
\begin{aligned}
x_1 &= x_{11} + \cdots + x_{1n} \,, \\
x_2 &= x_{22} + \cdots + x_{2n} \,, \\
&\cdots\cdots\cdots \\
x_n &= x_{nn} \,.
\end{aligned}$$

If (4.2)–(4.4) be considered as a production problem, with x_i the production in period i, then one may interpret the variable x_{ij} as that part of production in period i which is used to satisfy demand in period j. Then it is natural (and will be later justified) to impose the following conditions on the new variables:

$$(4.8) \qquad
\begin{aligned}
x_{ij} &\geqq 0 \,, \\
x_{11} &= b_1 \,, \\
x_{12} + x_{22} &= b_2 - b_1 \,, \\
x_{13} + x_{23} + x_{33} &= b_3 - b_2 \,, \\
&\cdots\cdots\cdots \\
x_{1n} + \cdots + x_{nn} &= b_n - b_{n-1} \,.
\end{aligned}$$

Note that, by (4.5), the right-hand sides are non-negative. Similarly, introduce the non-negative variables

$$
\begin{aligned}
y_{11} ,\, y_{12} ,\, &\cdots ,\, y_{1n} \,, \\
y_{22} ,\, &\cdots ,\, y_{2n} \,, \\
&\cdots\cdots\cdots \\
&y_{nn}
\end{aligned}$$

by the equations

$$(4.9) \qquad
\begin{aligned}
y_1 &= y_{11} + \cdots + y_{1n} \,, \\
y_2 &= y_{22} + \cdots + y_{2n} \,, \\
&\cdots\cdots\cdots \\
y_n &= y_{nn} \,.
\end{aligned}$$

It is not easy to interpret the y_{ij}, but one might try thinking of it as that part of the unused capacity in period i "tagged" to period j.

As a guide to the conditions analogous to (4.8), let us reconsider the right-hand inequalities of (4.2). Substituting by (4.6), we obtain

$$(a_1 - y_1) + \cdots + (a_t - y_t) \leqq c_t \,,$$

or

$$(4.10) \qquad y_1 + \cdots + y_t \geqq a_1 + \cdots + a_t - c_t \,, \qquad t = 1, \cdots, n \,.$$

Let us now define

$$(4.11) \qquad d_t = \max \left\{ 0, \max_{i \leqq t} (a_1 + \cdots + a_i - c_i) \right\} \,.$$

                          A. J. HOFFMAN

It is obvious that

$$(4.12) \qquad 0 \leqq d_1 \leqq d_2 \leqq \cdots \leqq d_n$$

and that, because the y_t are non-negative, the conditions

$$(4.13) \qquad y_1 + \cdots + y_t \geqq d_t$$

are equivalent with (4.10) and hence with the right-hand inequalities of (4.2).
We now show further that, if the inequalities (4.2)–(4.4) are consistent, then

$$(4.14) \qquad d_n = a_1 + \cdots + a_n - c_n$$

(recall (4.3)).

To prove (4.14), observe first that its right-hand side is non-negative; otherwise (4.4) and (4.3) would be inconsistent. Next, assume there is some $k < n$ such that

$$a_1 + \cdots + a_k - c_k > a_1 + \cdots + a_n - c_n = a_1 + \cdots + a_n - b_n \ ;$$

then

$$(4.15) \qquad c_k + a_{k+1} + \cdots + a_n < b_n$$

but

$$(4.16) \qquad x_1 + \cdots + x_k \leqq c_k \ ,$$

$$x_{k+1} \leqq a_{k+1} \ ,$$

$$\cdots\cdots\cdots\cdots$$

$$x_n \leqq a_n \ .$$

Adding the inequalities (4.16) and invoking the last of the inequalities of (4.2), we have

$$b_n = x_1 + \cdots + x_n \leqq c_k + a_{k+1} + \cdots + a_n \ ,$$

which violates (4.15). Thus (4.14) holds.

Now, in analogy to (4.8), we impose the following conditions on the variables (4.9):

$$(4.17) \qquad y_{ij} \geqq 0 \ ,$$

$$y_{11} = d_1 \ ,$$

$$y_{12} + y_{22} = d_2 - d_1 \ ,$$

$$y_{13} + y_{23} + y_{33} = d_3 - d_2 \ ,$$

$$\cdots\cdots\cdots\cdots$$

$$y_{1n} + \cdots + y_{nn} = d_n - d_{n-1} \ .$$

Now consider the following problem: Minimize

$$(4.18) \qquad p_1(x_{11} + \cdots + x_{1n}) + p_2(x_{22} + \cdots + x_{2n}) + \cdots + p_n x_{nn} \ ,$$

where the variables $x_{11}, \cdots, x_{nn}, y_{11}, \cdots, y_{nn}$ satisfy (4.8), (4.17) and

$$(4.19) \qquad x_{11} + \cdots + x_{1n} + y_{11} + \cdots + y_{1n} = a_1 ,$$

$$x_{22} + \cdots + x_{2n} + y_{22} + \cdots + y_{2n} = a_2 ,$$

$$\cdots \cdots \cdots$$

$$x_{nn} + y_{nn} = a_n .$$

It can be shown that:

(i) given any variables x_{ij} and y_{ij} satisfying (4.8), (4.17) and (4.19), the variables x_j obtained from (4.7), (4.9) and (4.6) satisfy (4.2) and (4.4) and yield a value for (4.1) identical with the value of (4.18);

(ii) conversely, given any variables x_j satisfying (4.2) and (4.4), one can find variables x_{ij} and y_{ij} satisfying (4.8), (4.17) and (4.19), such that (4.1) equals (4.18);

(iii) the conditions (4.8), (4.17) and (4.19) are those of a transportation problem which can be solved by inspection because an appropriate Monge sequence can be identified.

The proof of (iii) will occupy the next section. The proof of (ii) is somewhat long (see [3]) and will be omitted. The proof of (i) will now be given. Observe that the content of (i) and (ii) jointly is that our transformed problem is equivalent to the original one.

PROOF OF (i). It is clear that, using (4.7), (4.9) and (4.6), one obtains (4.4) and the equality of (4.1) and (4.18) immediately. What remains is (4.2). To prove the left side of (4.2) observe that

$$
\begin{aligned}
x_1 + \cdots + x_t &= x_{11} + \cdots + x_{1n} + \cdots + x_{tt} + \cdots + x_{tn} \\
&\geqq x_{11} + \cdots + x_{1t} + \cdots + x_{tt} \\
&= x_{11} + (x_{12} + x_{22}) + \cdots + (x_{1t} + \cdots + x_{tt}) \\
&= b_1 + (b_2 - b_1) + \cdots + (b_t - b_{t-1}) = b_t .
\end{aligned}
$$

A similar discussion shows that

$$y_1 + \cdots + y_t \geqq d_t ,$$

and ((4.11) and (4.13)) this implies (4.10) and hence the right side of (4.2).

This completes our construction of the transformed problem. Note that this construction required not only the notion of tagging production in any given period with the period whose requirements it would help satisfy, but also *the notion of tagging unused capacity in any period with some period whose requirements it would not help satisfy.*

The usefulness of this idea in the present problem will be apparent in the sequel, but it seems such a strange thought that there may very well be other opportunites for using it when its meaning has been absorbed.

5. Application of the Monge sequence. To fix our ideas, consider the case $n = 4$. All the phenomena for general n are already illustrated in this case.

Consider the four by eight transportation problem with cost coefficients, row sums and column sums given by the following tableau:

(5.1)

b_1	d_1	$b_2 - b_1$	$d_2 - d_1$	$b_3 - b_2$	$d_3 - d_2$	$b_4 - b_3$	$d_4 - d_3$	
p_1	0	p_1	0	p_1	0	p_1	0	a_1
M^2	M	p_2	0	p_2	0	p_2	0	a_2
M^4	M^3	M^2	M	p_3	0	p_3	0	a_3
M^6	M^5	M^4	M^3	M^2	M	p_4	0	a_4

M is an arbitrarily large positive number.

Notice first that this transportation problem has non-negative row and column sums, and satisfies the condition that the sum of the row sums equals the sum of the column sums, for the sum of the column sums is $b_4 + d_4$. By (4.3) and (4.14), this is

$$b_4 + a_1 + a_2 + a_3 + a_4 - b_4 = a_1 + \cdots + a_4 ,$$

which is the sum of the row sums.

The odd columns refer to variables x_{ij}, the even columns refer to variables y_{ij}. The large coefficients M^k compel certain variables to be zero. It is clear that this transportation problem is then identical with (4.18).

Next, arrange the 32 elements of the cost matrix in a sequence by the following rule:

   (i)   list first the elements of the first column in ascending order of magnitude,

   (ii)   list the elements of the second column in ascending order of magnitude,

   (iii)   list the elements of the third column in ascending order of magnitude,

   (iv)   list the elements of the fourth column by first putting the zeroes with indices whose corresponding p's are in *descending* order of magnitude, then the powers of M in ascending order,

   (v)   list the elements of the fifth column in ascending order of magnitude,

   (vi)   list the elements of the sixth column by first putting the zeroes with indices whose corresponding p's are in descending order of magnitude, then M,

   (vii)   list the elements of the seventh column in ascending order of magnitude,

   (viii)   list the elements of the eighth column in any order. Then one sees that the stipulations (2.2) have been satisfied for this sequence, so the algorithm of § 2 applies. It can be shown, of course, that if inequalities (4.2) and (4.4) are consistent, the algorithm will never choose a positive z_{ij} if c_{ij} is a power of M.

6. Remarks. The first transformation used above is a special case of a device which appears to have been used for the first time by W. Jacobs in [11]. The second transformation is based on an idea of Prager [14]. (Incidentally, the simple algorithm proposed by Beale [5] for the solution of Prager's formulation of the caterer problem can be shown to be a special case of the Monge idea; so can the algorithms presented in parts of references [6–15].)

Besides these transformations, other tricks are the use of the duality theorem [7] and various devices for standardizing the structure [1]. In the not too distant future, it should be possible to present a catalogue of devices usable in making linear programming problems "simple," whether or not the Monge idea applies; at present, they are too fragmentary to justify listing. By far, the most interesting direction of study is that initiated by Jacobs in [11]. In this instance, he gave an example of how one could minimize on a set K by minimizing on $L \supset K$, because it was possible to show that a minimum on L occurred at a point of K. A less ingenious instance of this was given in § 3 above. A comprehensive theory giving classes of cases where such transformations are possible would be very desirable.

REFERENCES

1. A. J. Goldman and A. W. Tucker, *Theory of linear programming*, Linear inequalities and related systems, pp. 53–98, Annals of Mathematics Studies, No. 38, Princeton Univ. Press, Princeton, N. J., 1956.

2. W. M. Hirsch and A. J. Hoffman, *Extreme varieties, concave functions and the fixed charge problem*, Comm. Pure Appl. Math. **14** (1961), 355–369.

3. A. J. Hoffman, *Some recent applications of the theory of linear inequalities to external combinatorial analysis*, Combinatorial analysis, pp. 95–112, Proc. Sympos. Appl. Math., Vol. X, Amer. Math. Soc., Providence, R. I., 1960.

4. G. Monge, *Déblai et remblai*, Mémoires de l'Académie des Sciences, 1781.

5. E. M. L. Beale, *Letter to the editor*, Management Sci. **4** (1957), 110.

6. E. M. L. Beale., G. Morton and A. H. Land, *Solution of a purchase-storage programme*, Operational Research Quarterly **9** (1958), 174–197.

7. A. Charnes and W. W. Cooper, *Generalizations of the warehousing model*, Operational Research Quarterly **6** (1955), 131–172.

8. C. Derman and M. Klein, *Inventory depletion management*, Management Sci. **4** (1958), 450–456.

9. M. Fréchet, *Sur les tableaux de correlation dont les marges sont données*, Ann. Univ. Lyon Sect. A (3) **14** (1951), 53–77.

10. J. W. Gaddum, A. J. Hoffman and D. Sokolowky, *On the solution to the caterer problem*, Naval Res. Logist. Quart. **1** (1954), 223–229.

11. W. Jacobs, *The caterer problem*, Naval Res. Logist. Quart. **1** (1954), 154–165.

12. S. M. Johnson, *Sequential production planning over time at minimum cost*, Management Sci. **3** (1957), 435–437.

13. H. Lighthall, Jr., *Scheduling problems for a multi-commodity production model*, Tech. Rep. 2, 1959, Contract Nonr-562(15) for Logistics Branch of Office of Naval Research, Brown University, Providence, R. I.

14. W. Prager, *On the caterer problem*, Management Sci. **3** (1946), 15–23.

15. W. E. Smith, *Various optimizers for single-stage production*, Naval Res. Logist. Quart. **3** (1956), 59–66.

INTERNATIONAL BUSINESS MACHINES CORPORATION

SIAM J. ALG. DISC. METH.
Vol. 6, No. 4, October 1985

© 1985 Society for Industrial and Applied Mathematics

TOTALLY-BALANCED AND GREEDY MATRICES*

A. J. HOFFMAN†, A. W. J. KOLEN‡ AND M. SAKAROVITCH§

Abstract. Totally-balanced and greedy matrices are $(0, 1)$-matrices defined by excluding certain submatrices. For a $n \times m$ $(0, 1)$-matrix A we show that the linear programming problem max $\{by \,|\, yA \leq c, 0 \leq y \leq d\}$ can be solved by a greedy algorithm for all $c \geq 0$, $d \geq 0$ and $b_1 \geq b_2 \geq \cdots \geq b_n \geq 0$ if and only if A is a greedy matrix. Furthermore we show constructively that if b is an integer, then the corresponding primal problem min $\{cx + dz \,|\, Ax + z \geq b, x \geq 0, z \geq 0\}$ has an integer optimal solution. A polynomial-time algorithm is presented to transform a totally-balanced matrix into a greedy matrix as well as to recognize a totally-balanced matrix. This transformation algorithm together with the result on greedy matrices enables us to solve a class of integer programming problems defined on totally-balanced matrices. Two examples arising in tree location theory are presented.

AMS(MOS) subject classifications. 05C50, 90C05, 90C10

1. Introduction. A $(0, 1)$-matrix is *balanced* if it does not contain an odd square submatrix with all row and column sums equal to two. Balanced matrices have been studied extensively by Berge [3] and Fulkerson et al. [7]. We consider a more restrictive class of matrices called totally-balanced (Lovász [11]). A $(0, 1)$-matrix is *totally-balanced* if it does not contain a square submatrix which has no identical columns and its row and column sums equal to two.

Example 1.1. Let $T = (V, E)$ be a tree with vertex set $V = \{v_1, v_2, \cdots, v_n\}$ and edge set E. Each edge $e \in E$ has a positive length $l(e)$. A *point* on the tree can be a vertex or a point anywhere along the edge. The *distance* $d(x, y)$ between the two points x and y on T is defined as the length of the path between x and y. A *neighborhood subtree* is defined as the set of all points on the tree within a given distance (called *radius*) of a given point (called *center*). Let x_i $(i = 1, 2, \cdots, m)$ be points on T and let r_i $(i = 1, 2, \cdots, m)$ be nonnegative numbers. Define the neighborhood subtrees T_i by $T_i = \{y \in T \,|\, d(y, x_i) \leq r_i\}$. Let $A = (a_{ij})$ be the $n \times m$ $(0, 1)$-matrix defined by $a_{ij} = 1$ if and only if $v_i \in T_j$. It was first proved by Giles [8] that A is totally-balanced. This result was generalized by Tamir [13]: Let Q_i $(i = 1, 2, \cdots, n)$ and R_j $(j = 1, 2, \cdots, m)$ be neighborhood subtrees·and let the $n \times m$ $(0, 1)$-matrix $B = (b_{ij})$ be defined by $b_{ij} = 1$ if and only if $Q_i \cap R_j \neq \varnothing$. Then B is totally-balanced.

Motivation for the types of problems to be studied in this paper is given by the following two examples from tree location theory stated in Example 1.2.

Example 1.2. Let $T = (V, E)$ be a tree, let T_j $(j = 1, 2, \cdots, m)$ be neighborhood subtrees and let $A = (a_{ij})$ be the $(0, 1)$-matrix as defined in Example 1.1. We interpret x_j as the possible location of a facility, and T_j as the service area of a facility at x_j, i.e., x_j can only serve clients located at T_j (we assume clients to be located at vertices). We assume there is a cost c_j associated with establishing a facility at x_j $(j = 1, 2, \cdots, m)$. The *minimum cost covering problem* is to serve all clients at minimum cost. This problem can be formulated as

$$\min \sum_{j=1}^{m} c_j x_j$$

(1.3)
$$\text{s.t. } \sum_{j=1}^{m} a_{ij} x_j \geq 1, \qquad i = 1, 2, \cdots, n,$$

$$x_j \in \{0, 1\}, \qquad j = 1, 2, \cdots, m.$$

* Received by the editors September 18, 1981, and in final revised form July 23, 1984.
† IBM T. J. Watson Research Center, Yorktown Heights, New York 10598.
‡ Econometric Institute, Erasmus University, Rotterdam, the Netherlands.
§ IMAG, Grenoble, France.

Let us relax the condition in this problem that each client has to be served by assuming that if a client located at vertex v_i is not served by a facility, then a penalty cost of d_i ($i = 1, 2, \cdots, n$) is charged. The *minimum cost operating problem* is to minimize the total cost of establishing facilities and not serving clients, i.e.,

$$\min \sum_{j=1}^{m} c_j x_j + \sum_{i=1}^{n} d_i z_i$$

(1.4)
$$\text{s.t. } \sum_{j=1}^{m} a_{ij} x_j + z_i \geqq 1, \qquad i = 1, 2, \cdots, n,$$

$$x_j \in \{0, 1\}, \qquad j = 1, 2, \cdots, m,$$

$$z_i \in \{0, 1\}, \qquad i = 1, 2, \cdots, n.$$

Let $A = (a_{ij})$ be a $(0, 1)$-matrix. We can associate a subset of rows to each column, namely those rows which have a one in this column. An $n \times m$ $(0, 1)$-matrix is called *greedy* if for all $i = 1, 2, \cdots, n$ the following holds; all columns having a one in row i can be totally ordered by inclusion when restricted to the rows $i, i+1, \cdots, n$. An equivalent definition is to say that the two 3×2 submatrices

(1.5)
$$\begin{bmatrix} 1 & 1 \\ 0 & 1 \\ 1 & 0 \end{bmatrix} \quad \text{and} \quad \begin{bmatrix} 1 & 1 \\ 1 & 0 \\ 0 & 1 \end{bmatrix}$$

do not occur. Why the name "greedy" is chosen will become clear in the next section. It is a trivial observation that each greedy matrix is totally-balanced. We will prove in § 3 that, conversely, the rows of a totally-balanced matrix can be permuted in such a way that the resulting matrix is greedy. The proof will be constructive.

Let the $n \times m$ $(0, 1)$-matrix $A = (a_{ij})$ be greedy. Consider the problem (P) given by

$$\min \sum_{j=1}^{m} c_j x_j + \sum_{i=1}^{n} d_i z_i$$

(P)
$$\text{s.t. } \sum_{j=1}^{m} a_{ij} x_j + z_i \geqq b_i, \qquad i = 1, 2, \cdots, n,$$

$$x_j \geqq 0, \qquad j = 1, 2, \cdots, m,$$

$$z_i \geqq 0, \qquad i = 1, 2, \cdots, n.$$

The dual problem D is given by

$$\max \sum_{i=1}^{n} b_i y_i$$

(D)
$$\text{s.t. } \sum_{i=1}^{n} y_i a_{ij} \leqq c_j, \qquad j = 1, 2, \cdots, m,$$

$$0 \leqq y_i \leqq d_i, \qquad i = 1, 2, \cdots, n.$$

We will show in § 2 that problem (D) can be solved by a greedy algorithm for all $c \geqq 0$, $d \geqq 0$ and $b_1 \geqq b_2 \geqq \cdots \geqq b_n \geqq 0$ if and only if the matrix A is greedy. Further we construct an optimal solution to the primal problem (P) which has the property that it is an integer solution whenever b is integer. This means that after we use the algorithm of § 3 to transform a totally-balanced matrix into a greedy matrix we can solve the two location problems using the result of § 2.

After we submitted the first version of the paper we found out about the work done by Farber. Farber [5], [6] studies strongly chordal graphs and gives polynomial-time algorithms to find a minimal weighted dominating set and minimal weighted independent dominating set. In these algorithms Farber uses the same approach as described in § 2. In another paper Anstee and Farber [1] relate strongly chordal graphs to totally-balanced matrices. This paper contains the relationship between totally-balanced and greedy matrices described in § 3 as well as a recognition algorithm for a totally-balanced matrix which, however, is less efficient than the one described here.

2. The algorithm. A greedy $(0, 1)$-matrix is in *standard greedy form* if it does not contain $\begin{bmatrix} 1 & 1 \\ 1 & 0 \end{bmatrix}$ as a submatrix. Any $n \times m$ greedy matrix can be transformed into a matrix in standard greedy form by a permutation of the columns in $O(nm)$ time as follows. Consider the columns as $(0, 1)$ vectors and sort them lexicographically reading in reverse, from bottom to top. This gives the desired permutation, for suppose $\begin{bmatrix} 1 & 1 \\ 1 & 0 \end{bmatrix}$ occurs as a submatrix with rows i_1, i_2 $(i_1 < i_2)$ and columns j_1, j_2 $(j_1 < j_2)$. Since the columns are ordered lexicographically we know that there exists a row i_3 $(i_3 > i_2)$ such that $a_{i_3 j_1} = 0$ and $a_{i_3 j_2} = 1$, but this contradicts the fact that the matrix is greedy. The algorithm of § 3 applied to a totally-balanced matrix also produces a matrix in standard greedy form. In this section we will assume that the matrix is in standard greedy form. This assumption does not affect the dual solution obtained but facilitates the description of the primal solution.

Let $A = (a_{ij})$ be an $n \times m$ $(0, 1)$-matrix in standard greedy form, let c_j $(j = 1, 2, \cdots, m)$ and d_i $(i = 1, 2, \cdots, n)$ be positive numbers (the case when one of these numbers is zero can be treated similarly) and let $b_1 \geq b_2 \geq \cdots \geq b_n \geq 0$. A feasible solution $\bar{y}$ of problem (D) is obtained by a greedy algorithm. The values of $\bar{y}_i$ are determined in order of increasing i and taken to be as large as possible. A constraint j is *tight* if $\sum_{i=1}^{n} \bar{y}_i a_{ij} = c_j$. The index $\alpha(j)$ denotes the largest index of a positive $\bar{y}$-value in the tight constraint j, J denotes a set of tight constraints. The greedy procedure is formulated in Algorithm D.

```
ALGORITHM D
begin J := ∅; ĉ := c;
        for i := 1 step 1 to n
        do y_i := min {d_i, min_{j:a_ij=1} {ĉ_j}};
            if ȳ_i > 0 then if ȳ_i = ĉ_j for some j then choose the largest j
                            and let J := J ∪ {j}; α(j) := i
                            fi;
                            ĉ_j := ĉ_j - ȳ_i for all j such that a_ij = 1
            fi
        od
end
```

For the solution $\bar{y}$ constructed by Algorithm D the following hold:

Property 2.1.   If $\bar{y}_k = d_k$, then either there is no $j \in J$ such that $a_{kj} = 1$ or there is a $j \in J$ such that $a_{kj} = 1$ and $\alpha(j) \geq k$; and

Property 2.2.   If $\bar{y}_k = 0$, then there is a $j \in J$ such that $a_{kj} = 1$ and $\alpha(j) < k$.

Property 2.1 follows immediately from the algorithm. If $\bar{y}_k = 0$, then there exists an index i, $i < k$ and a constraint $\hat{j}$ such that constraint $\hat{j}$ is tight, $a_{ij} = a_{kj} = 1$, and i is the largest index of a positive $\bar{y}$-value in constraint $\hat{j}$. During the iteration in which $\bar{y}_i$ was determined we have added an index $j \geq \hat{j}$ with $\alpha(j) = i$ to J. Since A is a standard greedy form we have $a_{kj} = 1$. This proves Property 2.2.

Example 2.3. The matrix and costs of the example as well as the results of Algorithm D are given in Fig. 2.4. We assume $d_i = 2$ $(i = 1, \cdots, 9)$ and $(b_1, b_2, b_3, b_4, b_5, b_6, b_7, b_8, b_9) = (6, 5, 4, 3, 3, 2, 2, 2, 1)$.

$$\hat{c}_1 = 3, \hat{c}_2 = 4, \hat{c}_3 = 5, \hat{c}_4 = 2, \hat{c}_5 = 3, \hat{c}_6 = 5, \hat{c}_7 = 3.$$

$$
\begin{bmatrix}
1 & 1 & 0 & 0 & 0 & 0 & 0 \\
1 & 1 & 0 & 0 & 0 & 0 & 0 \\
1 & 1 & 0 & 0 & 1 & 0 & 0 \\
0 & 0 & 1 & 0 & 0 & 1 & 0 \\
0 & 0 & 1 & 0 & 0 & 1 & 0 \\
0 & 0 & 0 & 1 & 0 & 0 & 1 \\
0 & 1 & 0 & 0 & 1 & 0 & 1 \\
0 & 0 & 1 & 0 & 0 & 1 & 1 \\
0 & 0 & 0 & 1 & 1 & 1 & 1
\end{bmatrix}
\begin{array}{l}
\bar{y}_1 = 2, \hat{c}_1 = 1, \hat{c}_2 = 2. \\
\bar{y}_2 = 1, J = \{1\}, \alpha(1) = 2, \hat{c}_1 = 0, \hat{c}_2 = 1. \\
\bar{y}_3 = 0. \\
\bar{y}_4 = 2, \hat{c}_3 = 3, \hat{c}_6 = 3. \\
\bar{y}_5 = 2, \hat{c}_3 = 1, \hat{c}_6 = 1. \\
\bar{y}_6 = 2, J = \{1, 4\}, \alpha(4) = 6, \hat{c}_4 = 0, \hat{c}_7 = 1. \\
\bar{y}_7 = 1, J = \{1, 4, 7\}, \alpha(7) = 7, \hat{c}_2 = 0, \hat{c}_5 = 2, \hat{c}_7 = 0. \\
\bar{y}_8 = 0. \\
\bar{y}_9 = 0.
\end{array}
$$

FIG. 2.4. Example of Algorithm D.

The value of the feasible dual solution $\bar{y}$ is 35.

The primal solution $\bar{x}, \bar{z}$ is constructed by Algorithm P which has as input the set of tight constraints J and the indices $\alpha(j)$ $(j \in J)$.

ALGORITHM P
begin $\hat{b} := b$; $\bar{x}_i := 0$ for all $j \notin J$;
    while $J \neq \varnothing$
    do (let k be the last column of J)
        $\bar{x}_k := \hat{b}_{\alpha(k)}$;
        $\hat{b}_i := \hat{b}_i - \bar{x}_k$ for all i such that $a_{ik} = 1$;
        $J := J \backslash \{k\}$
    od;
    for $i := 1$ **step** 1 **to** n **do** $\bar{z}_i := \max(0, \hat{b}_i)$ **od**
end

Example 2.5. Apply Algorithm P to Example 2.3.
$\bar{x}_2 = \bar{x}_3 = \bar{x}_5 = \bar{x}_6 = 0$, $\hat{b} = b$.
Iteration 1: $\bar{x}_7 = 2$, $\hat{b}_6 = \hat{b}_7 = \hat{b}_8 = 0$, $\hat{b}_9 = -1$.
Iteration 2: $\bar{x}_4 = 0$.
Iteration 3: $\bar{x}_1 = 5$, $\hat{b}_1 = 1$, $\hat{b}_2 = 0$, $\hat{b}_3 = -1$.
$\bar{z}_1 = 1$, $\bar{z}_4 = \bar{z}_5 = 3$, all other $\bar{z}_i$ values are zero.
It is easy to check that $\bar{x}, \bar{z}$ is a feasible primal solution with value 35. Since the values of the feasible primal and dual solutions are equal they are both optimal.

If we prove that $\bar{x}_j \geqq 0$ for all $j \in J$, then it is clear that $\bar{y}$ and $\bar{x}, \bar{z}$ are feasible solutions. In order to prove that they are optimal solutions we show that the complementary slackness relations of linear programming hold. These conditions are given by

$$(2.6) \qquad \bar{x}_j \left(\sum_{i=1}^{n} \bar{y}_i a_{ij} - c_j \right) = 0, \qquad j = 1, 2, \cdots, m,$$

$$(2.7) \qquad \bar{y}_i \left(\sum_{j=1}^{m} a_{ij} \bar{x}_j + \bar{z}_i - b_i \right) = 0, \qquad i = 1, 2, \cdots, n,$$

$$(2.8) \qquad \bar{z}_i (\bar{y}_i - d_i) = 0, \qquad i = 1, 2, \cdots, n.$$

Let us denote by $\hat{J}$ the set of column indices in Algorithm P which is initially equal to J and decreases by one element at each iteration. Accordingly let $b_i(\hat{J}) = b_i - \sum_{j \in J \setminus \hat{J}} a_{ij} \bar{x}_j$, $i = 1, 2, \cdots, n$. Define I by $I = \{i \mid \exists j \in J \, \alpha(j) = i\}$.

The following properties hold for Algorithm P.

Property 2.9. If $a_{ij} = a_{lj} = 1$, $i < l$, $j \in \hat{J}$, then $b_i(\hat{J}) \geqq b_l(\hat{J})$.

Proof. This is true at the start of the algorithm since $b_i \geqq b_l$, $i < l$. Let k be the last column of $\hat{J}$. Property 2.9 could be altered only if $a_{ik} = 1$ and $a_{lk} = 0$, which is ruled out by the fact that A is in standard greedy form.

Property 2.10. $b_i(\hat{J}) \geqq 0$ for all $i \in I$.

Proof. This is true at the start of the algorithm since $b_i \geqq 0$. Let k be the last column of $\hat{J}$. Using Property 2.9 we know that Property 2.10 could be altered only if $a_{ik} = 1$ and $i > \alpha(k)$, which is ruled by definition of $\alpha(k)$.

Property 2.11. $b_i(\varnothing) = 0$ for all $i \in I$.

Proof. Let $i \in I$. There exists a $j \in J$ such that $\alpha(j) = i$. At the iteration at which j was the last column of $\hat{J}$ we define $\bar{x}_j = b_i(\hat{J})$ and hence after this iteration we have $b_i(\hat{J}) = 0$. Combining this with Property 2.10 we get $b_i(\varnothing) = 0$.

Property 2.12. If $\bar{y}_k > 0$, $k \notin I$, $\sum_{j \in J} a_{kj} \bar{x}_j \leqq b_k$.

Proof. If $\bar{y}_k > 0$, $k \notin I$, then according to Property 2.1 we have to consider two cases:

1. There is no $j \in J$ such that $a_{kj} = 1$, In this case we have

$$\sum_{j \in J} a_{kj} \bar{x}_j = 0 \leqq b_k.$$

2. There is a $j \in J$ such that $a_{kj} = 1$ and $\alpha(j) > k$ (note that since $k \notin I$ we can rule out $\alpha(j) = k$). Using Properties 2.9 and 2.11 we get $b_k(\varnothing) \geqq b_{\alpha(j)}(\varnothing) = 0$.

Property 2.13. If $\bar{y}_k = 0$, then $\sum_{j \in J} a_{kj} \bar{x}_j \geqq b_k$.

Proof. If $\bar{y}_k = 0$, then according to Property 2.2 there exists a $j \in J$ such that $a_{kj} = 1$ and $\alpha(j) < k$. Using Properties 2.9 and 2.11 we get $b_k(\varnothing) \leqq b_{\alpha(j)}(\varnothing) = 0$.

It follows from Property 2.10 that $\bar{x}_j \geqq 0$ for all $j \in J$. Hence $\bar{x}, \bar{z}$ is a feasible solution. For the complementary slackness relations (2.6) follows by construction, (2.7) and (2.8) follow from Properties 2.11, 2.12 and 2.13.

THEOREM 2.14. *Problem* (D) *is solved by Algorithm* D *for all* $c \geqq 0$, $d \geqq 0$ *and* $b_1 \geqq b_2 \geqq \cdots \geqq b_n \geqq 0$ *if and only if* A *is greedy.*

Proof. If A is greedy, then we transform A into standard greedy form as indicated by a permutation of the columns. This permutation does not affect the dual solution, which was shown to be optimal.

If A is not greedy, then there exists a 3×2 submatrix of the form

$$\begin{bmatrix} 1 & 1 \\ 1 & 0 \\ 0 & 1 \end{bmatrix} \quad \text{or} \quad \begin{bmatrix} 1 & 1 \\ 0 & 1 \\ 1 & 0 \end{bmatrix}.$$

Let the rows be given by $i_1 < i_2 < i_3$ and columns by $j_1 < j_2$. Set $d_i = 0$ for all $i \notin \{i_1, i_2, i_3\}$, $d_{i_1} = d_{i_2} = d_{i_3} = 1$. $c_j = 3$ for all j except $c_{j_1} = c_{j_2} = 1$, $b_i = 1$ for all $i = 1, 2, \cdots, n$. If we apply Algorithm D we get $\bar{y}_{i_2} = 1$, all other $\bar{y}_i$ are zero. The value of this solution is 1. However $\bar{y}_{i_2} = \bar{y}_{i_3} = 1$ and all other $\bar{y}_i$ are zero is a feasible solution with value 2. This shows that Algorithm D does not solve this instance of problem (D).

3. Standard greedy form transformation. In this section we present an $O(nm^2)$ algorithm to transform an $n \times m$ totally-balanced matrix into standard greedy form. Since a matrix is in standard greedy form if and only if its transpose is in standard greedy form we may assume without loss of generality that $m \leqq n$.

726   A. J. HOFFMAN, A. W. J. KOLEN AND M. SAKAROVITCH

Let us call a $(0, 1)$-matrix *lexical* if the following two properties hold.

Property 3.1. If rows i_1, i_2 $(i_1 < i_2)$ are different, then the last column in which they differ has a zero in row i_1 and a one in row i_2.

Property 3.2. If columns j_1, j_2 $(j_1 < j_2)$ are different, then the last row in which they differ has a zero in column j_1 and a one in column j_2.

The algorithm we will present in this section transforms any $(0, 1)$-matrix into a lexical matrix by permuting the rows and by permuting the columns of the matrix. Theorem 3.3 states that a totally-balanced matrix which is lexical is in standard greedy form. Since a totally-balanced matrix is still totally-balanced after a permutation of the rows and a permutation of the columns all we have to do to transform the matrix into standard greedy form is to transform it into a lexical matrix.

THEOREM 3.3. *If a totally-balanced matrix* $A = (a_{ij})$ *is lexical, then it is in standard greedy form.*

Proof. Suppose A is not in standard greedy form. Then there exist rows i_1, i_2 $(i_1 < i_2)$ and columns j_1, j_2 $(j_1 < j_2)$ such that $a_{i_1 j_1} = a_{i_1 j_2} = a_{i_2 j_1} = 1$ and $a_{i_2 j_2} = 0$ (see Fig. 3.4).

Let i_3 be the last row in which columns j_1 and j_2 differ, and let j_3 be the last column in which rows i_1 and i_2 differ. Since A is lexical we have $a_{i_1 j_3} = 0$, $a_{i_2 j_3} = 1$ and $a_{i_3 j_1} = 0$, $a_{i_3 j_1} = 1$. Since A does not contain a 3×3 submatrix with row and column sums equal to two we know that $a_{i_3 j_3} = 0$. In general we have the submatrix of A given by Fig. 3.4 with ones on the lower and upper diagonal and the first element of the diagonal and zeros everywhere else. The rows and columns have the following properties.

	j_1	j_2	j_3	j_4	$\cdots$	$\cdots$	j_k	$\cdots$
i_1	1	1	0	0	$\cdots$	$\cdots$	0	
i_2	1	0	1	0			0	
i_3	0	1	0	1				
i_4	0	0	1	0				
$\cdot$	$\cdot$						0	
$\cdot$	$\cdot$						1	
i_k	0				0	1	0	
$\cdot$	$\cdot$							

FIG. 3.4. *Submatrix of Theorem* 3.3.

Property 3.5. i_p is the last row in which columns j_{p-2} and j_{p-1} differ $(3 \le p \le k)$.

Property 3.6. j_p is the last column in which rows i_{p-2} and i_{p-1} differ $(3 \le p \le k)$.

We shall prove that we can extend this $k \times k$ submatrix to a $(k+1) \times (k+1)$ submatrix with the same properties. So we can extend this submatrix infinitely. This contradicts the fact that A has finite dimensions. Let i_{k+1} be the last row in which j_{k-1} and j_k differ, and let j_{k+1} be the last column in which i_{k-1} and i_k differ. Since A is lexical we have $a_{i_{k+1} j_{k-1}} = 0$, $a_{i_{k+1} j_k} = 1$ and $a_{i_{k-1} j_{k+1}} = 0$, $a_{i_k j_{k+1}} = 1$. By definition of i_p and j_p $(3 \le p \le k)$ we know that $a_{i_{k+1} j_{p-2}} = a_{i_{k+1} j_{p-1}}$ and $a_{i_{p-2} j_{k+1}} = a_{i_{p-1} j_{k+1}}$. Using this for $p = k, \cdots, 3$ respectively we get $a_{i_{k+1} j_q} = a_{i_q j_{k+1}} = 0$ for $q = 1, 2, \cdots, k-1$. Since A does not contain a $(k+1) \times (k+1)$ submatrix with rows and column sums equal to two we have $a_{i_{k+1} j_{k+1}} = 0$.

Let us now describe the algorithm which transforms any $(0, 1)$-matrix into a lexical matrix. Let $A = (a_{ij})$ be any $(0, 1)$-matrix without zero rows and columns. Let us denote column j by E_j, i.e., $E_j = \{i \mid a_{ij} = 1\}$. We assume that the matrix A is given by its columns $E_1, E_2, \cdots, E_m$. The algorithm produces a 1–1 mapping $\sigma : \{1, 2, \cdots, n\} \to \{1, 2, \cdots, n\}$ corresponding to a transformation of the rows of A ($\sigma(i) = j$ indicates that row i becomes row j in the transformed matrix) and a 1–1 mapping $\tau : \{E_1, \cdots, E_m\} \to \{1, \cdots, m\}$ corresponding to a transformation of the columns of A ($\tau(E_i) = j$ indicates that column i becomes column j in the transformed matrix). We present the algorithm in an informal way and give an example to demonstrate it.

The algorithm consists of m iterations. At iteration i we determine the column E for which $\tau(E) = m - i + 1$ ($1 \leq i \leq m$). At the beginning of each iteration the rows are partitioned into a number of groups, say $G_r, \cdots, G_1$. If $i < j$, then rows in G_i will precede rows in G_j in the transformed matrix. Rows j and k belong to the same group G at the beginning of iteration i if and only if for all columns E we have determined so far, i.e. all columns E for which $\tau(E) \geq m - i + 2$, we cannot distinguish between rows j and k, i.e., $j \in E$ if and only if $k \in E$. At the beginning of iteration 1 all rows belong to the same group. Let $G_r, \cdots, G_1$ be the partitioning into groups at the beginning of iteration i ($1 \leq i \leq m$). For each column E not yet determined we calculate the vector d_E of length r, where $d_E(j) = |G_{r-j+1} \cap E|$ ($j = 1, 2, \cdots, r$). A column E for which d_E is a lexicographically largest vector is the column determined at iteration i and $\tau(E) = m - i + 1$. After we have determined E we can distinguish between some rows in the same group G if $1 \leq |G \cap E| < G$. If this is the case we shall take rows in $G \backslash E$ to precede rows in $G \cap E$ in the transformed matrix. This can be expressed by adjusting the partitioning into groups in the following way. For $j = r, r-1, \cdots, 1$ respectively we check if the intersection of G_j with E is not empty and not equal to G_j. If this is the case we increase the index of all groups with index greater than j by one and partition the group G_j into two groups called G_j and G_{j+1} where $G_{j+1} = G_j \cap E$ and $G_j = G_j \backslash E$. The algorithm ends after m interations with a partitioning into groups, say $G_r, \cdots, G_1$. The permutation σ is defined by assigning for $i = 1, 2, \cdots, r$ the values $\sum_{j=1}^{i-1} |G_j| + 1, \cdots, \sum_{j=1}^{i} |G_j|$ in an arbitrary way to the elements in group G_i. The number of computations we have to do at each iteration is $O(mn)$. Therefore the time complexity of this algorithm is $O(nm^2)$.

Example 3.7. The 9×7 $(0, 1)$-matrix A is given by its columns $E_1 = \{1, 2, 3\}$, $E_2 = (1, 2, 3, 5\}$, $E_3 = \{4, 5\}$, $E_4 = \{3, 4, 5, 9\}$, $E_5 = \{5, 8, 9\}$, $E_6 = \{6, 7, 8, 9\}$, $E_7 = \{6, 7, 8\}$.

Iteration 1: $G_1 = (1, 2, 3, 4, 5, 6, 7, 8, 9)$.
  $\quad d_{E_i} = (|E_i|)$, choose E_4, $\tau(E_4) = 7$.

Iteration 2: $G_2 = (3, 4, 5, 9)$, $G_1 = (1, 2, 6, 7, 8)$.

E	E_1	E_2	E_3	E_5	E_6	E_7
d_E	$(1, 2)$	$(2, 2)$	$(2, 0)$	$(2, 1)$	$(1, 3)$	$(0, 3)$

Choose E_2, $\tau(E_2) = 6$.

Iteration 3: $G_4 = (3, 5)$, $G_3 = (4, 9)$, $G_2 = (1, 2)$, $G_1 = (6, 7, 8)$.

E	E_1	E_3	E_5	E_6	E_7
d_E	$(1, 0, 2, 0)$	$(1, 1, 0, 0)$	$(1, 1, 0, 1)$	$(0, 1, 0, 3)$	$(0, 0, 0, 3)$

Choose E_5, $\tau(E_5) = 5$.

Iteration 4: $G_7 = (5)$, $G_6 = (3)$, $G_5 = (9)$, $G_4 = (4)$, $G_3 = (1, 2)$, $G_2 = (8)$, $G_1 = (6, 7)$.

E	d_E
E_1	$(0, 1, 0, 0, 2, 0, 0)$
E_3	$(1, 0, 0, 1, 0, 0, 0)$
E_6	$(0, 0, 1, 0, 0, 1, 2)$
E_7	$(0, 0, 0, 0, 0, 1, 2)$

Choose E_3, $\tau(E_3) = 4$.

From now on the groups do not change.

Therefore $\tau(E_1) = 3$, $\tau(E_6) = 2$, $\tau(E_7) = 1$. A mapping σ is given by $\sigma: (6, 7, 8, 1, 2, 4, 9, 3, 5) \to (1, 2, 3, 4, 5, 6, 7, 8, 9)$. The mapping τ is given by $\tau: (E_7, E_6, E_1, E_3, E_5, E_2, E_4) \to (1, 2, 3, 4, 5, 6, 7)$.

The transformed matrix is the one used in Example 2.3.

Let us now prove that a matrix transformed by the algorithm is a lexical matrix. When we say that row i is the largest row with respect to σ satisfying a property we mean that there is now row j with $\sigma(j) > \sigma(i)$ satisfying the same property. The same terminology is also used for columns with respect to τ.

LEMMA 3.8. *If rows i and j $(\sigma(i) < \sigma(j))$ are different, then for the largest column E with respect to τ in which they differ we have $i \notin E$, $j \in E$.*

Proof. Consider the last iteration in which i and j are in the same group G and let E be the column determined at this iteration. Since i and j were in the same group during all previous iterations we know that rows i and j are identical when restricted to columns which are larger than E with respect to τ. Since $\sigma(i) < \sigma(j)$ we have that after this iteration row j is in a group with larger index than the group containing row i. This implies that $j \in G \cap E$ and $i \in G \backslash E$, i.e., $i \notin E$ and $j \in E$.

LEMMA 3.9. *If columns E_k and E_l $(\tau(E_k) < \tau(E_l))$ are different, then for the largest row i with respect to σ in which they differ we have $i \notin E_k$ and $i \in E_l$.*

Proof. If E_i is strictly contained in E_j for some i, j, then we always have $\tau(E_i) < \tau(E_j)$. If $E_k \subseteq E_l$, then the lemma holds. So we may assume that $E_k \nsubseteq E_l$ and $E_l \nsubseteq E_k$. Let i be the largest row with respect to σ in $E_l \backslash E_k$, and let j be the largest row with respect to σ in $E_k \backslash E_l$. We have to prove that $\sigma(i) > \sigma(j)$. Consider the iteration in which E_l was determined. Let p be the largest index for which $G_p \cap E_k \neq G_p \cap E_l$. Since E_l was determined before E_k we know that $|G_p \cap E_l| \geq |G_p \cap E_k|$. We conclude that $i \in G_p$. If $j \in G_f$ with $f < p$, then $\sigma(j) < \sigma(i)$. If $j \in G_p$, then after this iteration G_p is partitioned into two groups $G_p \cap E_l$ and $G_p \backslash E_l$ where $G_p \backslash E_l$ precedes $G_p \cap E_l$. Since $j \in G_p \backslash E_l$ and $i \in G_p \cap E_l$ we have $\sigma(j) < \sigma(i)$.

It follows from Lemmas 3.8 and 3.9 that the transformed matrix is lexical.

In a previous paper (Brouwer and Kolen [4], see also Kolen [10]) it was shown that there exists a row of a totally-balanced matrix such that all columns covering this row can be totally ordered by inclusion. The algorithm presented gives a constructive proof that such a row exists, namely row one of the transformed matrix. As indicated by one of the referees the existence of such a row can be used to derive an $O(n^2 m)$ algorithm to transform a totally-balanced matrix into standard greedy form as compared to the $O(nm^2)$ algorithm presented (note $m \leq n$). The algorithm we gave produces a lexical matrix in standard greedy form. This is important if we consider the following result. Let A be a $n \times m$ $(0, 1)$-matrix. The row intersection matrix $B = (b_{ij})$ of A is a $n \times n$ $(0, 1)$-matrix defined by $b_{ij} = 1$ if and only if there exists a column of A which covers both row i and j. It is an easy exercise to show that if A is a lexical matrix in standard greedy form, then the row intersection matrix is in standard greedy form. This is not true for any $(0, 1)$-matrix A in standard greedy form as is shown by the following example:

$$A = \begin{bmatrix} 0 & 1 & 1 & 1 \\ 1 & 0 & 0 & 0 \\ 1 & 1 & 1 & 1 \\ 0 & 1 & 1 & 1 \end{bmatrix} \qquad B = \begin{bmatrix} 1 & 0 & 1 & 1 \\ 0 & 1 & 1 & 0 \\ 1 & 1 & 1 & 1 \\ 1 & 0 & 1 & 1 \end{bmatrix}.$$

Using the results of this section we have proved the following theorem which was first proved by Lubiw [12] by showing that the row intersection matrix of a totally-balanced matrix does not contain one of the forbidden submatrices.

THEOREM 3.10 (Lubiw [12]). *The row intersection matrix of a totally-balanced matrix is totally-balanced.*

If a matrix contains a $k \times k$ submatrix with no identical columns and row and columns sums equal to two, then the matrix transformed by the algorithm still contains such a submatrix and therefore contains $\begin{bmatrix} 1 & 1 \\ 1 & 0 \end{bmatrix}$ as a submatrix. Using Theorem 3.3 we conclude that a matrix is totally-balanced if and only if the algorithm transforms the matrix into standard greedy form. We can check in $O(nm^2)$ time whether a matrix is in standard greedy form by comparing each pair of columns.

We finish discussing the relationship between totally-balanced matrices and chordal bipartite graphs. A *chordal bipartite graph* is a bipartite graph for which every cycle of length strictly greater than four has a chord, i.e., an edge connecting two vertices which are not adjacent in the cycle. Chordal bipartite graphs were discussed by Golumbic [9] in relation with perfect Gaussian elimination for nonsymmetric matrices. Chordal bipartite graphs and totally-balanced matrices are equivalent in the following sense:

(3.11)   Given a chordal bipartite graph $H = (\{1, 2, \cdots, n\}, \{1, 2, \cdots, m\}, E)$ define the $n \times m$ $(0, 1)$-matrix $A = (a_{ij})$ by $a_{ij} = 1$ if and only if $(i, j) \in E$. Then A is totally-balanced.

Given an $n \times m$ totally-balanced matrix $A = (a_{ij})$ define the bipartite graph $H = (\{1, 2, \cdots, n\}, \{1, 2, \cdots, m\}, E)$ by $E = \{(i, j) \mid a_{ij} = 1\}$. Then H is a chordal bipartite graph.

An edge (i, j) of a bipartite graph is *bisimplicial* if the subgraph induced by all vertices adjacent to i and j is a complete bipartite graph. Let $M = (m_{ij})$ be a nonsingular nonsymmetric matrix. We can construct a bipartite graph from M equivalent to (3.11) where edges correspond to nonzero elements m_{ij}. If (i, j) is a simplicial edge in the bipartite graph, then using m_{ij} as a pivot in the matrix M to make m_{ij} to one and all other entries in the ith row and jth column equal to zero does not change any zero element into a nonzero element. This is important since sparse matrices are represented in computers by its nonzero elements. Golumbic [9] proved that a chordal bipartite graph has a bisimplicial edge. This result immediately follows from our result. The first one in the first row corresponds to a bisimplicial edge.

REFERENCES

[1] R. P. ANSTEE AND M. FARBER (1982), *Characterizations of totally balanced matrices*, Research Report CORR 82-5, Faculty of Mathematics, Univ. Waterloo, Waterloo, Ontario, Canada.

730   A. J. HOFFMAN, A. W. J. KOLEN AND M. SAKAROVITCH

[2] A. V. AHO, J. E. HOPCROFT AND J. D. ULLMAN (1974), *The Design and Analysis of Computer Algorithms*, Addison-Wesley, Reading, MA.

[3] C. BERGE (1972), *Balanced matrices*, Math. Programming, 2, pp. 19-31.

[4] A. E. BROUWER AND A. KOLEN (1980), *A super-balanced hypergraph has a nest point*, Report ZW 146/80, Mathematisch Centrum, Amsterdam.

[5] M. FARBER (1982), *Characterization of strongly chordal graphs*, Discrete Math., 43 (1983), pp. 173-189.

[6] ———, (1982), *Domination, independent domination and duality in strongly chordal graphs*, Research Report CORR 82-2, Faculty of Mathematics, Univ. Waterloo, Waterloo, Ontario, Canada.

[7] D. R. FULKERSON, A. J. HOFFMAN AND R. OPPENHEIM (1974), *On balanced matrices*, Math. Programming Study, 1, pp. 120-132.

[8] R. GILES (1978), *A balanced hypergraph defined by certain subtrees of a tree*, Ars Combinatoria, 6, pp. 179-183.

[9] M. C. GOLUMBIC (1980), *Algorithmic Graph Theory and Perfect Graphs*, Academic Press, New York.

[10] A. KOLEN (1982), *Location problems on trees and in the rectilinear plane*, Ph.D. thesis, Mathematisch Centrum, Amsterdam.

[11] L. LOVÁSZ (1979), *Combinatorial Problems and Exercises*, Akadémiai Kiadó, Budapest, p. 528.

[12] A. LUBIW (1982), *Γ-free matrices*, Master thesis, Univ. Waterloo, Waterloo, Ontario, Canada.

[13] A. TAMIR (1980), *A class of balanced matrices arising from location problems*, this Journal, 4 (1983), pp. 363-370.

Reprinted from JOURNAL OF COMBINATORIAL THEORY, Series A
All Rights Reserved by Academic Press, New York and London

Vol. 47, No. 1, January 1988

Greedy Packing and Series-Parallel Graphs

ALAN J. HOFFMAN

*Department of Mathematical Sciences, IBM Thomas J. Watson Research Center,
P.O. Box 218, Yorktown Heights, New York 10598*

AND

ALAN C. TUCKER

*SUNY at Stony Brook, Department of Applied Mathematics,
Stony Brook, New York 11794*

Communicated by the Managing Editors

Received November 5, 1986

DEDICATED TO THE MEMORY OF HERBERT J. RYSER

We characterize nonnegative matrices **A** with the following property: for every $a \geq 0$, the linear programming problem $\max(1, y)$, where $Ay \leq 0$, $y \geq 0$, is solved by successively maximizing the variables in arbitrary order. The concept of series-parallel graphs is central to the characterization.   © 1988 Academic Press, Inc.

1. INTRODUCTION

Let **A** be a nonnegative matrix in which each column and each row has at least one positive entry (which we tactily assume throughout), and let $a \geq 0$. The linear programming problem

$$\max(1, y): y \geq 0, \qquad Ay \leq a \qquad (1.1)$$

is known as the packing problem. Let σ be a permutation of the columns of **A**. The "σ-greedy algorithm," applied to the feasible region of (1.1), is

$$\max y_{\sigma 1}, \qquad \text{then } y_{\sigma 2} \cdots . \qquad (1.2)$$

We shall say that **A** is greedy if:

for every $a \geq 0$ and every σ, the σ-greedy algorithm (1.2) solves
the packing problem (1.1). $\qquad (1.3)$

6

The object of this paper is a characterization of greedy matrices. Essential to the characterization is the concept of a series-parallel graph. A series-parallel graph is a directed multigraph with single source and single sink defined as follows: if $|E(G)| > 1$, then G is obtained from a series-parallel graph G' with $|E(G')| = |E(G)| - 1$ by replacing some edge (u, v) of G' with

(i) *parallel* replacement: two copies of this edge, $(u, v)_1$, $(u, v)_2$, or

(ii) *series* replacement: two edges (u, w) and (w, v), where w is a new vertex.

An alternative definition is that a directed multigraph G with single source s and single sink t is a series-parallel graph if, when $|E(G)| > 1$, there are two edge disjoint series parallel graphs G_1 and G_2, with respective source sink pairs (s_1, t_1) and (s_2, t_2) such that G is obtained from G_1 and G_2 by

(i') *parallel* composition: s_1 and s_2 are identified and become s, t_1 and t_2 are identified and become t; or

(ii') *series* composition: t_1 and s_2 are identified, s_1 becomes s and t_2 becomes t.

For any directed multigraph G with one source and one sink, let $M(G)$ be the path-incidence matrix of G: the rows of $M(G)$ correspond to $E(G)$, the columns to the set of source-sink directed paths $\mathscr{P}(G) = \{P\}$, with $m_{ep} = 1$ if $e \in P$, 0 if $e \notin P$. It is known [1] that $M(G)$ is a greedy matrix if and only if G is a series parallel graph.

We define an *augmented* path incidence matrix $\tilde{M}(G)$ as a nonnegative matrix which can be partitioned,

$$\tilde{M}(G) = \begin{bmatrix} N & 0 \\ C & M(G) \end{bmatrix},$$

where $M(G)$ is as before, N is arbitrary, every column of C is a convex combination of the columns of $M(G)$. We do not exclude the possibility that C may be absent, or $[N\ 0]$ may be absent.

THEOREM 1.1. *A nonnegative matrix* $\mathbf{A}$ *is greedy if and only if the matrix* DA, *where* $\mathbf{D}$ *is a diagonal matrix with* $d_{ii} = (\max_i a_{ij})^{-1}$, *is an augmented path matrix* $\tilde{M}(G)$ *for some series parallel graph* $\mathbf{G}$.

In the proof, it is convenient to consider first the case where $\mathbf{A}$ is a $(0, 1)$ matrix. Then it is natural to think of the rows as elements of some finite set U, the columns of $\mathbf{A}$ as a system $\mathscr{S} = \{S_1, S_2, ...\}$ of nonempty subsets of $\mathbf{U}$ with $\bigcup S_j = U$, and $a_{uj} = 1$ if and only if $u \in U$ is contained in S_j. Let us also assume $\mathscr{S}$ is a clutter: i.e., $S_j \subset S_k$ implies $j = k$. Call $\mathscr{S}$ a greedy clutter if $\mathbf{A}$ is a greedy matrix.

THEOREM 1.2. *A clutter $\mathscr{S}$ is greedy if and only if there is a series-parallel graph $\mathbf{G}$ such that $U = E(G)$, $\mathscr{S} = \mathscr{P}(G)$.*

2. GREEDY AND SLENDER CLUTTERS

If $\mathscr{S}$ is a family of subsets of $\mathbf{U}$, a subset $T \subset V$ is a blocking set for $\mathscr{S}$ if, for every $S \in \mathscr{S}$, $S \cap T \neq \varnothing$, and a blocking set $\mathbf{T}$ is minimal if, for every $u \in T$, $T - \{u\}$ is not a blocking set. Now assume $\mathscr{S}$ is a clutter, and let $\mathscr{S}^*$ denote the clutter of all minimal blocking sets for $\mathscr{S}$. It is well known that

$$\mathscr{S}^{**} = \mathscr{S}. \tag{2.1}$$

We shall say that $\mathscr{S}$ is *slender* if, for every $T \in \mathscr{S}^*$ and every $S \in \mathscr{S}$,

$$|T \cap S| = 1. \tag{2.2}$$

Note $\mathscr{S}$ is slender if and only if $\mathscr{S}^*$ is slender.

If $\mathbf{G}$ is a series-parallel graph, let $\mathbf{G}^*$ be the graph obtained from $\mathbf{G}$ by exchanging "parallel" with "series" in either the replacement construction or the composition construction. Then it is known that

$$\mathscr{P}(G^*) = (\mathscr{P}(G))^*. \tag{2.3}$$

To establish Theorem 1.2, we prove the following lemmas.

LEMMA 2.1. *If $\mathscr{S}$ is greedy, $\mathscr{S}$ is slender.*

LEMMA 2.2. *If $\mathscr{S}$ is slender, there is a series parallel graph $\mathbf{G}$ such that $U = E(G)$, $\mathscr{S} = \mathscr{P}(G)$.*

The fact that $U = E(G)$, $\mathscr{S} = \mathscr{P}(G)$, $\mathbf{G}$ a series-parallel graph imply $\mathscr{S}$ is greedy was mentioned above.

Proof of Lemma 2.1. Assume the lemma false. Then $\exists T \in \mathscr{S}^*$, $S_0 \in \mathscr{S}$, and

$$|S_0 \cap T| > 1. \tag{2.4}$$

Define the vector a in (1.1) as follows: let $a_u = 1$ for $u \in T$, otherwise let a_u be very large. Since $T \in \mathscr{S}^*$, there exists $S_{j_1}, ..., S_{j_{|T|}} \in \mathscr{S}$ such that

$$|S_{j_k} \cap T| = 1, \qquad k = 1, ..., |T|.$$

Consequently, the optimum value in (1.1) is $|T|$. On the other hand, if σ is the permutation which considers S_0 first, the σ-greedy algorithm will produce a value for (1.1) of at most $1 + |T| - |S_0 \cap T| < |T|$, by (2.4). So lemma 2.1 is true.

Proof of Lemma 2.2.   We prove the lemma by induction on $|U|$. Suppose $\mathscr{S}$ contains a set, say S_1, which is disjoint from all other sets. Then, if there are no other sets, **G** is a path, and so a series-parallel graph. If there are other sets, they form a slender clutter in $U - S_1$, so the induction hypothesis applies. It follows that **G** is the parallel composition of a path and a series-parallel graph, with $\mathscr{S} = \mathscr{P}(G)$. So we assume that each $\mathscr{S}$ has more than one set, and any two sets in $\mathscr{S}$ have a non-empty intersection.

Now let T_1 be a set in $\mathscr{S}^*$ of maximum cardinality. If T_1 is disjoint from every other set in $\mathscr{S}^*$, then from the preceding paragraph $\mathscr{S}^* = \mathscr{P}(G)$ for some **G**. By (2.1) and (2.3), $\mathscr{S}^{**} = \mathscr{S} = \mathscr{P}(G^*)$, and we are done.

Hence there exists at least one set in $\mathscr{S}^*$ which makes a non-empty intersection with T_1. Let $T_2 \in \mathscr{S}^*$ have maximum cardinality intersection with T_1:

$$|T_2 \cap T_1| \geqq |T \cap T_1| \qquad \text{for all} \quad T \neq T_1, \ T \in \mathscr{S}^*. \tag{2.5}$$

Suppose $T_1 - T_2$ is a single element, say $T_1 - T_2 = \{u\}$. Then $T_2 - T_1$ is also a single element, say $T_2 - T_1 = \{v\}$, otherwise $|T_2| > |T_1|$. From (2.2), for each $S \in \mathscr{S}$,

$$S \text{ contains } u \qquad \text{if and only if } S \text{ contains } v. \tag{2.6}$$

Replace u and v by a single element w, and, for each j, say S_j contains w if and only if S_j contains u (and v). The new clutter $\mathscr{S}'$ on $U' = U - \{u, v\} \cup \{w\}$ is clearly slender, so by induction $\mathscr{S}' = \mathscr{P}(G')$ for some series parallel graph G'. Now if the edge w is replaced by the edges u and v in series, obtaining **G**, we reproduce the clutter $\mathscr{S}$ on U, so $\mathscr{S} = \mathscr{P}(G)$.

Therefore, we shall assume

$$|T_1 - T_2| > 1. \tag{2.7}$$

We shall show that if $u, v \in T_1 - T_2$, then for each $T \in \mathscr{S}^*$,

$$T \text{ contains } u \qquad \text{if and only if } T \text{ contains } v. \tag{2.8}$$

By the reasoning of the previous paragraph, this will show that $\mathscr{S}^* = \mathscr{P}(G)$, for some series parallel graph **G**, so $\mathscr{S} = \mathscr{P}(G^*)$.

Suppose (2.8) is false. Then there is a set $T_3 \in \mathscr{S}^*$ such that (say),

$$u, v \in T_1 - T_2, \qquad u \in T_3, \qquad v \notin T_3. \tag{2.9}$$

10   HOFFMAN AND TUCKER

For each nonempty subset $K \subset \{1, 2, 3\}$ set $T_{(K)} \equiv \{u \mid u \in \bigcap_{j \in K} T_j, u \notin T_j$ for all $j \in \{1, 2, 3\} - K\}$. Then

$$T_{(12)} \neq \varnothing; \qquad T_{(13)} \neq \varnothing; \qquad T_{(1)} \neq \varnothing, \qquad T_{(2)} \neq \varnothing, \qquad T_{(3)} \neq \varnothing. \qquad (2.10)$$

The reasons are as follows. If $T_{(12)}$ were empty, then $T_1 \cap T_2 \subset T_3$. But T_3 contains u. Hence $|T_3 \cap T_1| > |T_2 \cap T_1|$, violating (2.5). Since $u \in T_3$, $T_{(13)} \neq \varnothing$. Since every set $S \in \mathscr{S}$ meeting $T_{(12)}$ muts meet T_3, and (by (2.2)) cannot meet T_3 in an element also in T_2 or T_1, it follows that $T_{(3)} \neq \varnothing$. A similar argument shows $T_{(2)} \neq \varnothing$. The fact that $T_{(1)} \neq \varnothing$ follows from $v \in T_{(1)}$.

We now show

$$T_{(23)} = \varnothing. \qquad (2.11)$$

Otherwise, $T_{(123)} \cup T_{(12)} \cup T_{(1)} \cup T_{(2)}$, a blocking set for $\mathscr{S}$ not containing T_1 or T_2, would contain a minimal blocking set T' not containing T_1 or T_2. Clearly, $T_{(123)} \cup T_{(12)} \subset T'$, implying $|T' \cap T_1| > |T_2 \cap T_1|$, contradicting (2.5). Next,

if $w_2 \in T_{(2)}$ and $w_3 \in T_{(13)}$, there is a set $S \in \mathscr{S}$ containing w_2 and w_3.

$$(2.12)$$

If not, $T_1 \cup T_2 - \{w_2, w_3\}$ would be a blocking set for $\mathscr{S}$, containing a minimal blocking set T' violating (2.5).

$T_{(123)} \cup T_{(2)} \cup T_{(3)}$ contains a minimal blocking set $\hat{T}$, and $T_{(123)} \cup T_{(2)} \subset \hat{T}$.

$$(2.13)$$

That $T_{(123)} \cup T_{(2)} \cup T_{(3)}$ is a blocking set follows from T_1 being a blocking set, for a set $S \in \mathscr{S}$ meeting T_1 in $T_{(12)}$ or $T_{(1)}$ or $T_{(13)}$ meets $T_{(3)}$ or $T_{(2)}$ or $T_{(2)}$, respectively. Further, by (2.12), $T_{(2)} \subset \hat{T}$. The fact that $T_{(123)}$ (if not empty) is contained in $\hat{T}$ is obvious.

$$\text{If } \check{T} \equiv T_{(3)} \cap \hat{T}, \text{ then } \check{T} \neq \varnothing, \check{T} \neq T_{(3)}. \qquad (2.14)$$

A set $S \in \mathscr{S}$ meeting $T_{(12)}$ meets $T_{(3)}$ in an element of $\hat{T}$, so $\check{T} \neq \varnothing$. A set $S \in \mathscr{S}$ meeting $T_{(1)}$ meets both $T_{(2)}$ and $T_{(3)}$. But from (2.13) $T_{(2)} \subset \hat{T}$. If $\check{T} = T_{(3)}$, $\hat{T}$ would contain two elements of this S.

$$T_{(123)} \cup T_{(12)} \cup T_{(13)} \cup (T_{(3)} - \check{T}) \text{ is a blocking set.} \qquad (2.15)$$

All we need show is that a set $S \in \mathscr{S}$ meeting $T_{(1)}$ meets $T_{(3)} - \check{T}$. If not, S would meet $\check{T}$ and $T_{(2)}$, a contradiction.

But the blocking set (2.15) contains a minimal blocking set which, by

(2.14), is neither T_1 nor T_3, but must contain $T_{(123)} \cup T_{(12)} \cup T_{(13)}$, which contradicts (2.5).

So (2.8) is true, and so is Theorem 1.2.

3. GREEDY AND SLENDER MATRICES

Given a nonnegative **A**, let $Q(A) \equiv \{x \mid x'A \geqq 1', \ x \geqq 0\}$. We say **A** is *slender* if

$$\text{every vertex } \bar{x} \text{ of } Q(A) \text{ satisfies } \bar{x}'A = 1. \tag{3.1}$$

LEMMA 3.1.   *If **A** is greedy, **A** is slender.*

Proof.   Let $\bar{x}$ be a vertex of **Q(A)**. It is well known that there is an objective vector a_0 and $\varepsilon > 0$ such that the linear programming problem

$$\min(a, x): x \in Q(A), \tag{3.2}$$

for all a such that $|a - a_0| < \varepsilon$, has its unique minimum at $\bar{x}$. It follows that $a_0 > 0$, for if any coordinate of a_0 were nonpositive, there would exist a with $|a - a_0| < \varepsilon$ and at least one coordinate of a negative. But, for such an a, (3.2) would have no minimum. Hence $a_0 > 0$.

Consider the problem dual to (3.2):

$$\max(1, y): y \geqq 0, \ Ay \leqq a_0. \tag{3.3}$$

Let y_j be any coordinate of y, and choose σ so that $\sigma_1 = j$. Since $a_0 > 0$, the σ-greedy algorithm produces a $\bar{y}$ such that $\bar{y}_j > 0$. By the duality theorem of linear programming, this implies $(\bar{x}'A)_j = 1$. Hence **A** is slender.

Let us index the rows of **A** by elements of a set U, with $|U| = m$. Assume **A** has n columns $A_1, ..., A_n$. Let $S_j = \{u \mid a_{uj} > 0\}$, and let $\mathscr{S} = \{S_1, ..., S_n\}$.

LEMMA 3.2.   *If A is slender, then a subset $T \subset U$ is a minimal blocking set for $\mathscr{S}$ if and only if there is a vertex $\bar{x}$ of **Q(A)** such that $T = \text{Supp } \bar{x}$.*

Proof.   Let $\bar{x}$ be a vector of **Q(A)**. Since $\bar{x} \in Q(A)$, it follows that Supp $\bar{x}$ is a blocking set for $\mathscr{S}$, so there is a $T \in \mathscr{S}^*$ with

$$T \subset \text{Supp } \bar{x}. \tag{3.4}$$

Let $x(T)$ be (the indicator vector for **T**) defined by $x_u(T) = 1$ if $u \in T$, 0 otherwise. Then for sufficiently large t, the vector $tx(T) \in Q(A)$. It is known [2] that every vector in **Q(A)** is the sum of a nonnegative vector and a vector in the convex hull of the vertices of $\mathbf{Q}(\mathbf{A})$. This means that, for some vertex $\bar{\bar{x}}$ of $\mathbf{Q}(\mathbf{A})$,

$$\text{Supp } \bar{\bar{x}} \subset \text{Supp } tx(T) = T. \tag{3.5}$$

Now we invoke the hypothesis that $\mathbf{A}$ is slender. Since $1 = \bar{x}'A = \bar{\bar{x}}'A$, and Supp $\bar{\bar{x}} \subset$ Supp $\bar{x}$, the only way that $\bar{x}$ can be a vertex of $\mathbf{Q}(\mathbf{A})$ is if $\bar{x} = \bar{\bar{x}}$. From (3.4) and (3.5), this means $T = \text{Supp } \bar{x}$.

On the other hand, if $T \in \mathscr{S}^*$, we see from the above that there is a vertex $\bar{\bar{x}}$ of $\mathbf{Q}(\mathbf{A})$ such that (3.5) holds. But we already know that supp $\bar{\bar{x}} \in \mathscr{S}^*$, so $T = \text{Supp } \bar{\bar{x}}$.

For the remainder of the paper, we assume that the largest entry in each row of $\mathbf{A}$ is 1. Note that premultiplying $\mathbf{A}$ by a positive diagonal matrix does not affect the properties "slender" or "greedy."

LEMMA 3.3.   *If $\mathbf{A}$ is slender, every vertex of $\mathbf{Q}(\mathbf{A})$ is a $(0, 1)$ vector.*

Proof. By Lemma 3.2, $\bar{x} \in Q(A)$ is a vertex of $\mathbf{Q}(\mathbf{A})$ if and only if Supp $\bar{x} = T$, where $T \in \mathscr{S}^*$. So there exists a set $K \subset \{1, ..., n\}$ of $|T|$ columns of $\mathbf{A}$ which meet the rows of $\mathbf{T}$ in a submatrix with exactly one nonzero in each row and column. If one of these nonzero a_{uj} is less than 1, then the requirement that $\bar{x}'A_j = 1$ makes $\bar{x}_u > 1$. But some column $k \in \{1, ..., n\}$ has $a_{uk} = 1$, so $\bar{x}'A_k > 1$, violating (3.1).

LEMMA 3.4.   *Assume $\mathbf{A}$ is slender, and let $R \equiv \{j \mid every\ entry\ in\ A_j\ is\ 0$ or $1\}$. Then*

$$R \neq \varnothing, \tag{3.6}$$

and

$$\text{if } \mathscr{S}_R = \{\mathscr{S}_j : j \in R\}, \text{ then } \mathscr{S}^* = \mathscr{S}_R^*. \tag{3.7}$$

Proof. Let $\overset{\lessgtr}{S}_j = \{u \mid 0 < a_{uj} < 1\}$, $\overset{\lessgtr}{\mathscr{S}} = \{\overset{\lessgtr}{S}_j\}$. If (3.6) were false, then any $\overset{\lessgtr}{T} \in \mathscr{S}^*$ is a blocking set for $\overset{\lessgtr}{\mathscr{S}}$, so there is a $T \in \mathscr{S}^*$ contained in $\overset{\lessgtr}{T}$. Since $\overset{\lessgtr}{T} \in \overset{\lessgtr}{\mathscr{S}}^*$, there is a set K of $|\overset{\lessgtr}{T}|$ columns of $\mathbf{A}$ which meets the rows in $\overset{\lessgtr}{T}$ in a submatrix containing exactly one number strictly between 0 and 1 in each row and column. Consider now (Lemma 3.3) the vertex $x(T)$ of $\mathbf{Q}(\mathbf{A})$. Let $u \in T$, $j \in K$ with $0 < a_{uj} < 1$. Then $x(T)' A_j$ cannot be an integer, contradicting (3.1).

To prove (3.7), it is sufficient to show that $T_R \in \mathscr{S}_R^*$ implies T_R is a blocking set for $\mathscr{S}$. Assume otherwise, so $R' \equiv \{j \mid T_R \cap S_j = \varnothing\}$ is not empty. Let $\overset{\lessgtr}{\mathscr{S}}' = \{\overset{\lessgtr}{S}_j \mid j \in R'\}$, and let $T' \in \overset{\lessgtr}{\mathscr{S}}'^*$. Then there is a set

$K \subset \{1, ..., n\}$ of $|T'|$ columns of $\mathbf{A}$ which meet the rows of T' in a submatrix containing exactly one number strictly between 0 and 1 in each row and column.

Now $T_R \cup T'$ is a blocking set for $\mathscr{S}$, so there is a subset $T \subset (T_R \cup T')$, where $T \in \mathscr{S}^*$. Further, $T \cap T' \neq \varnothing$. Let $u \in T \cap T'$, then there is a j such that $0 < a_{uj} < 1$, with every other entry in A_j, in the rows of $\mathbf{T}$, 0, or 1. It follows (see Lemma 3.3) that $x'(T) A_j$ is not an integer, violating (3.1).

LEMMA 3.5. *Let y be feasible for* (1.1), *and satisfy*

$$z \text{ feasible for } (1.1), \qquad y \leq z \text{ imply } y = z. \tag{3.8}$$

Then. if $\mathbf{A}$ is slender, y solves (1.1). Thus, $\mathbf{A}$ slender implies $\mathbf{A}$ greedy.

Proof. The last sentence follows from the preceding, since the σ-greedy algorithm produces $\bar{y}$ which satisfies (3.8). Now assume y satisfies (3.8), and let $T' = \{u \mid A(y)_u = a_u\}$. Then T' is a blocking set for $\mathscr{S}$, for if $S_j \cap T = \varnothing$, we could raise y_j and violate (3.8). Hence, there is a $T \in \mathscr{S}^*$, with $T \subset T'$. Consideration of $x(T) Ay$ shows that $\sum y_j = \sum a_i : i \in T$, so y and $x(T)$ are feasible solutions to (1.1) and its dual, with the same values for the objective function. Hence y is optimum in (1.1).

Now we are ready to prove Theorem 1.1. Assume $A = \tilde{M}(G)$, where $\mathbf{G}$ is a series-parallel graph. By Lemma 3.5, to prove $\mathbf{A}$ greedy, it is sufficient to prove $\mathbf{A}$ slender. Let $\bar{x}$ be a vertex of $Q(M(G))$, and let z be the subvector of $\bar{x}$ corresponding to rows of $M(G)$. Then $\bar{z}'M(G) \geq 1$, which (since each column of C is a convex combination of columns of $M(G)$) implies $\bar{z}'C \geq 1$. This in turn implies that the coordinates of $\bar{x}$ not in $\bar{z}$ must be 0. So all we need prove is $B = [CM(G)]$ is slender. But $z'M(G) \geq 1$ implies $z'C \geq 1$ and $z'M(G) = 1$ implies $z'C = 1$. So all we need prove is that $\mathbf{M(G)}$ is slender. But $\mathbf{M(G)}$ is slender, since $\mathbf{M(G)}$ is greedy (Lemma 3.1).

Now we must prove that, if $\mathbf{A}$ is greedy, there is a $\mathbf{G}$ such that $A = \tilde{M}(G)$. By Lemma 3.1, $\mathbf{A}$ is slender. Return to Lemma 3.4, and let $K \subset R$ have the property that $\mathscr{S}_K$, the family of corresponding subsets, are a set of distinct subsets of $\mathscr{S}_R$ corresponding to minimal subsets of the family $\mathscr{S}_R$. Let $V = \bigcup S_j : j \in K$. Then $\mathscr{S}_K$ is a clutter on $\mathbf{V}$. Now $\mathbf{A}$ has the form

$$A = \begin{bmatrix} N & 0 \\ C & M \end{bmatrix},$$

where $\mathbf{M}$ is the incidence matrix of the clutter $\mathscr{S}_K$. Since $\mathbf{A}$ is slender, it follows from Lemmas 3.2 and 3.4 that the clutter $\mathscr{S}_K$ is slender, so by Theorem 1.2, $M = M(G)$ for some series-parallel graph $\mathbf{G}$.

14 HOFFMAN AND TUCKER

All that remains to be shown is that every column of **C** is a convex combination of columns of **M(G)**; i.e., of the incidence vectors of paths **P** in $\mathscr{P}(\mathbf{G})$.

We will argue by induction on $|V|$. Assume first that **G** arises from series replacement of an edge in a graph $\hat{G}$, which means that edge w in $\hat{G}$ is replaced by two edges u and v in series to produce **G**. Let $T \in \mathscr{S}^*$ contain u. Then **T** does not contain v, but $(T - \{u\}) \cup \{v\} \in \mathscr{S}^*$. It follows from Lemmas 3.1–3.3 that the rows of A corresponding to u and v are identical. If we delete one of these rows, the resulting matrix $\hat{A}$ is slender. But $\hat{A}$ is a slender matrix where $\hat{M} = M(\hat{G})$, and the induction hypothesis applies. Hence, $A = \tilde{M}(G)$.

So assume **G** is obtained from $\hat{G}$ by parallel replacement, in particular, edge w in $\hat{G}$ is replaced by parallel edges u and v to produce **G**. We will make use of the following proposition whose proof we leave to the reader:

if Q is a polyhedron with at least one vertex, and all vertices of Q lie on a hyperplane H, then $Q \cap H$ has the same vertices as Q. (3.9)

We now take a closer look at **A**:

$$
A = \begin{array}{c} \\ u \\ v \\ \\ \end{array}
\begin{bmatrix}
N & & O & \\
C'_u & 1 & 0 & 0 \\
C'_v & 0 & 1 & 0 \\
\bar{C} & \bar{M} & \bar{M} & \bar{\bar{M}}
\end{bmatrix},
$$

where the fact that u and v are in parallel implies that A has the above form. Let us now consider the matrix $\hat{A}$:

$$
\hat{A} = w \begin{bmatrix}
N & & O \\
C'_u + C'_v & 1 & 0 \\
\bar{C} & \bar{M} & \bar{\bar{M}}
\end{bmatrix}.
$$

Note that by Lemmas 3.2–3.4 every vertex of **Q(A)** has the coordinate corresponding to u and v, z_u and z_v, the same, both 0 or both 1. Hence, every vertex of **Q(A)** is in the linear space $z_u - z_v = 0$. Now $\hat{z} \in Q(\hat{A})$ if and only if $z \in Q(A)$ where $z_u = z_v = \hat{z}_w$, all other coordinates of z the same as in $\hat{z}$. By (3.9), if $\hat{\bar{z}}$ is a vertex of $Q(\hat{A})$, the corresponding $\bar{z}$ is a vertex of **Q(A)**. Hence $\hat{A}$ is slender.

By the induction hypothesis, every column $\hat{C}_j$ of $\hat{C}$ is a convex combination of the columns of $\hat{M}$, where

GREEDY PACKING AND SERIES-PARALLEL GRAPHS 15

$$\hat{C}_j = \begin{bmatrix} c_{uj} + c_{vj} \\ \bar{C}_j \end{bmatrix}, \qquad \hat{M} = \begin{matrix} b_1 \quad b_2 \\ \begin{bmatrix} 1 & 0 \\ \bar{M} & \bar{\bar{M}} \end{bmatrix} \end{matrix}$$

so $\hat{C}_j = \sum_1^{b_1 + b_2} \lambda_j \hat{M}_j$, where $\lambda_j \geqq 0$, $\sum \lambda_j = 1$.
If we write

$$C_j = \begin{bmatrix} c_{uj} \\ c_{vj} \\ \bar{c}_j \end{bmatrix}, \qquad M = \begin{matrix} b_1 \quad b_1 \quad b_2 \\ \begin{bmatrix} 1 & 0 & 0 \\ 0 & 1 & 0 \\ \bar{M} & \bar{M} & \bar{\bar{M}} \end{bmatrix} \end{matrix}$$

then (using the convention $0/0 = 0$), we have

$$C_j = \sum_{j=1}^{b_1} \frac{c_{uj}}{c_{uj} + c_{vj}} \lambda_j M_j + \sum_{j=1}^{b_1} \frac{c_{vj}}{c_{uj} + c_{vj}} \lambda_j M_{j+b_1}$$
$$+ \sum_{j=b_1+1}^{b_1+b_2} \lambda_j M_{j+b_1},$$

a convex combination of the columns of **M**.

ACKNOWLEDGMENT

We are very grateful to Louis Weinberg for stimulating conversations about series-parallel graphs.

REFERENCES

1. W. W. BEIN, P. BRUCKER, AND A. TAMIR, Minimum cost flow algorithms for series-parallel networks, *Discrete Appl. Math.* **10** (1985), 117–124.
2. A. J. GOLDMAN, Resolution and separation theorems for polyhedral convex sets, *in* "Linear Inequalities and Related Systems" (H. W. Kuhn and A. W. Tucker, Eds.), Annals of Mathematics Studies Vol. 38, pp. 41–52, Princeton University Press, Princeton, NJ, 1956.

Discrete Mathematics 106/107 (1992) 285–289
North-Holland

On simple combinatorial optimization problems

A.J. Hoffman

Department of Mathematical Sciences, IBM Research Division, T.J. Watson Research Center, P.O. Box 218, Yorktown Height, NY 10598, USA

Received 3 January 1992
Revised 4 March 1992

Abstract

Hoffman, A.J., On simple combinatorial optimization problems, Discrete Mathematics 106/107 (1992) 285–289.

We characterize $(0, 1)$ linear programming matrices for which a greedy algorithm and its dual solve certain covering and packing problems. Special cases are shortest path and minimum spanning tree algorithms.

1. Introduction

Two of the best known, conceptually simple and computationally easy combinatorial optimization problems are: to find the shortest path from a node s to a node t in a directed graph with nonnegative edge lengths; and to find a minimum spanning tree in a graph (more generally, a minimum rooted spanning arborescence in a directed graph). We announce a general theorem which includes as special cases the well-known algorithms for solving these problems. The theorem will alo include as special cases the algorithm for finding a maximum flow in a series parallel graph [4], an optimum coloring of an interval graph, and all the algorithms for the problems described in the opening sections of [3]. For a survey of related material, see [2].

2. Sequentially greedy matrices

Let A be a $(0, 1)$ matrix with m rows and n columns for which each column has at least one 1. We consider, for a given $b \geq 0$, the problem

$$\max \sum x_j : x \geq 0, \qquad Ax \leq b \tag{2.1}$$

Correspondence to: A.J. Hoffman, Department of Mathematical Sciences, IBM Research Division, T.J. Watson Research Center, P.O. Box 218, Yorktown Heights, NY 10598, USA

                                    A.J. Hoffman

and its dual

$$\min \sum y_i b_i : y \geq 0, \qquad y'A \geq \vec{1}. \tag{2.2}$$

The sequentially greedy (SG) algorithm for solving (2.1) can be informally summarized as follows. Let $b_{i(1)} = \min\{b_k : a_{k1} = 1\}$. Set $\bar{x}_1 = b_{i(1)}$. Subtract $\bar{x}_1$ from all b_k such that $a_{k1} = 1$. Delete from A row $i(1)$ and all columns j such that $a_{i(1),j} = 1$. Proceed inductively.

This process will produce a set of chosen columns $\hat{C} = \{j(1) = 1, j(2), \ldots, j(k)\}$ and chosen rows $\hat{R} = \{i(1), i(2), \ldots, i(k)\}$ such that the submatrix $A(\hat{R}, \hat{C})$ formed by them is (essentially) triangular; and such that the vector $\bar{x} = (\bar{x}_1, \ldots, \bar{x}_n)$ given by

$$\bar{x}_j = \begin{cases} \bar{x}_{j(t)} & \text{if } j = j(t),\ t = 1,\ldots,k, \\ 0 & \text{otherwise} \end{cases}$$

is feasible for (2.1). We shall characterize those A such that, for all $b \geq 0$, SG produces $\bar{x}$ which is optimum.

The dual sequential greedy algorithm (DSG) is obtained by solving $\bar{z}A(\hat{R}, \hat{C}) = \vec{1}$, and setting

$$\bar{y}_i = \begin{cases} \bar{z}_i & \text{if } i \text{ is chosen,} \\ 0 & \text{otherwise.} \end{cases}$$

We shall characterize those A such that, for all $b \geq 0$, DSG produces $\bar{y} = (\bar{y}_1, \ldots, \bar{y}_m)$ which is feasible (hence optimum for (2.2), by linear programming duality). The main theorem is the following.

Theorem. *The following conditions on A are equivalent*:
   (2.3) *for all $b \geq 0$, SG solves (2.1)*;
   (2.4) *for all $b \geq 0$, DSG solves (2.2)*;
   (2.5) *if A contains a submatrix*

	j_1	j_2	j_3			j_1	j_2	j_3
i_1	1	1	0	or	i_1	1	0	1
i_2	1	0	1		i_2	1	1	0

then at least one of the following holds:
   (2.5a) *for some j, $a_{i_1,j} = a_{i_2 j} = 0$, and for all k, $a_{kj} \leq a_{kj_2} + a_{kj_3}$*;
   (2.5b) *for some $j < j_1$, and for all k, $a_{kj} \leq a_{kj_1} + a_{kj_2} + a_{kj_3}$.*

Here is an example. A has 6 rows and 8 columns and satisfies (2.5),

	1	2	3	4	5	6	7	8	b
1	1	◇	0	0	0	0	1	0	7
2	0	1	0	0	1	0	0	◇	8
3	0	0	0	◇	0	1	1	0	4
4	◇	0	1	0	0	0	1	0	5
5	1	1	1	1	◇	0	1	0	12
6	0	1	0	0	0	0	0	0	20

The algorithm first chooses $\bar{x}_1 = 5$, because 5 is the smallest value of b_k among those k such that $a_{k1} = 1$. So $j(1) = 1$, $i(1) = 4$ (because $b_4 = 5$). Columns 3 and 7 are deleted because a_{43} and a_{47} are 1, so x_3 and x_7 can only be 0. The value of b_1 and b_5 are reduced by 5. We continue inductively. The marked entries denote $\{(i(1), j(1)), (i(2), j(2)), \ldots , (i(5), j(5))\}$. The submatrix formed by these rows and columns is

	1	2	4	5	8
4	1	0	0	0	0
1	1	1	0	0	0
3	0	0	1	0	0
5	1	1	1	1	0
2	0	1	0	1	1

The solution to primal and dual problems are

$$\bar{x} = (5, 2, 0, 4, 1, 0, 0, 5), \qquad \bar{y} = (0, 1, 1, 1, 0).$$

3. Applications

(3.1) Given a directed graph G with distinguished nodes $s \neq t$.

Let A be the following $(0, 1)$ matrix. The rows correspond to edges of G, columns to subsets $S \subseteq v(G)$, $s \in S$, $t \notin S$, with

$$a_{es} = \begin{cases} 1 & \text{if edge } e \text{ 'leaves' } s, \\ 0 & \text{otherwise.} \end{cases}$$

Then, if S are numbered by increasing cardinality of $|S|$, A satisfies (2.5). SG is max cut packing (cf. [6, p. 592]), and DSG is Dijkstra's algorithm.

(3.2) Given a directed graph G with distinguished node r. Let rows of A corresponding to edges of G, columns to subsets $S \subseteq V(G)$, $r \notin S$. Set

$$a_{es} = \begin{cases} 1 & \text{if edge } e \text{ 'enters' } s, \\ 0 & \text{otherwise.} \end{cases}$$

                    A.J. Hoffman

If the columns of A are ordered by increasing size of $|S|$, then A satisfies (2.5), and DSG is the algorithm described in [5].

(3.3) Let $A = [{}^B_I]$. Then A satisfies (2.5) if and only if B contains neither of the 2×3 matrices mentioned there. Hence, our theorem includes the problems mentioned in [3].

(3.4) It is well known that sequential greediness for any sequence of $s - t$ paths solves the max flow problem for series–parallel graphs. Consider the incidence matrix A of edges versus paths of such a graph. It is easy to see that (2.5a) applies.

(3.5) Following Ford and Fulkerson in [1], one can find an optimum coloring of an interval graph G by the following procedure. Say interval I_i precedes interval I_j if the right-hand endpoint of I_i is to the left of the left-hand endpoint of I_j. We can color G optimally by finding the smallest number of chains covering this partially ordered set, which is equivalent to finding a max match in the bipartite graph on $I_1, \ldots, I_m$ and $I'_1, \ldots, I'_m$ where I_i and I'_j are joined by an edge if I_i precedes I_j.

Observe that if

$$M = (m_{ij}) = \begin{cases} 1 & \text{if } I_i \text{ precedes } I_j, \\ 0 & \text{otherwise,} \end{cases}$$

and the numbering of rows and columns is consonant with the partial ordering, then M does not contain

$$\begin{bmatrix} 1 & 1 \\ 1 & 0 \end{bmatrix} \tag{3.4a}$$

as a submatrix.

But the non-existence of (3.4a) as a submatrix of M implies that the linear program

$$\max \sum x_{ij},$$

$x_{ij} \text{ defined only if } m_{ij} = 1, \ x_{ij} \geq 0,$

$$\sum_j x_{ij} \leq 1 \quad \text{for all } i,$$

$$\sum_i x_{ij} \leq 1 \quad \text{for all } j,$$

is solved by 'Northwest Corner' greediness, because of (2.5a).

Acknowledgement

We are very grateful to P. Krishnarao, Stanford University, for his help in clarifying this material.

References

[1] L.R. Ford and D.R. Fulkerson, Flows in Networks (Princeton Univ. Press, Princeton, 1962).

[2] A.J. Hoffman, On greedy algorithms that succeed, in: Surveys in Combinatorics (Cambridge Univ. Press, Cambridge, 1985) 97–112.

[3] A.J. Hoffman, A. Kolen and M. Sakarovitch, Totally balanced and greedy matrices, SIAM J. Algebraic Discrete Methods (1985) 721–730.

[4] A.J. Hoffman and A.C. Tucker, Greedy packing and series–parallel graphs, J. Combin. Theory Ser. A 47 (1988) 6–15.

[5] E.L. Lawler, Combinatorial Optimization: Networks and Matroids (Holt, Rinehart and Winston, New York, 1962).

[6] G.L. Nemhauser and L.A. Woolsey, Integer and Combinatorial Optimization (Wiley, New York, 1988).

Mathematical Programming 62 (1993) 1–14
North-Holland

Series parallel composition of greedy linear programming problems

Wolfgang W. Bein

American Airlines Decision Technologies, Dallas/Fort Worth, TX, USA

Peter Brucker

Fachbereich Mathematik/Informatik, Universität Osnabrück, Germany

Alan J. Hoffman

Department of Mathematical Science, IBM Thomas J. Watson Research Center, Yorktown Heights, NY, USA

Received 6 April 1992
Revised manuscript received 8 February 1993

This paper is dedicated to Phil Wolfe on the occasion of his 65th birthday.

We study the concept of series and parallel composition of linear programming problems and show that greedy properties are inherited by such compositions. Our results are inspired by earlier work on compositions of flow problems. We make use of certain Monge properties as well as convexity properties which support the greedy method in other contexts.

Key words: Greedy algorithm, Monge arrays, series parallel graphs, linear programming, network flow, transportation problem, integrality, convexity.

1. Introduction

Hoffmann [7] showed that the transportation problem is solved by a greedy algorithm if the underlying cost array is a Monge array (so named after the mathematician G. Monge, who first considered such properties [11]). Meanwhile many new results concerning the question when greedy algorithms solve linear programming problems have been obtained (see [8] for a survey), but at the same time many aspects are still not fully understood.

In [2, 4, 3] Bein, Brucker and Tamir explore the concept of series parallel compositions of network flow problems. They consider linear programming descriptions of cost network flow problems and study the programming description of the flow problems that result when two networks are combined by a series or parallel composition. They show that the greedy algorithm solves the combined problem if it solves the original problems. Based on [2] and ideas presented in [1] Hoffman [9] generalized this work further. He shows that the

Correspondence to: W.W. Bein, American Airlines Decision Technologies, P.O. Box 619616, MD 4462, Dallas/Fort Worth Airport, TX 75261, USA.

compositions preserve the greedy property not only if path costs are obtained from edge costs by summation but also if they are obtained from more general operations, if they have certain monotonicity and Monge properties.

This paper is inspired by the earlier work in [3] and [9]. In this paper we show that under certain conditions the assumption that the underlying linear programs are specific descriptions of flow problems can, in fact, be dropped entirely.

We will state our main results now and prove them in Section 2. In Section 3 we discuss how earlier work can be reinterpreted in the framework of series parallel composition of linear programs. Section 3 also contains a lemma that links a certain convexity with Monge arrays. We close with a number of technical remarks.

In what follows we will assume that all matrices are nonnegative real matrices without zero columns. Consider then the two linear programming problems I and II:

$$\text{I:} \quad \max \quad \sum_i c_i x_i$$

$$\text{s.t.} \quad \sum_i x_i A_i \leqslant a, \tag{1.1}$$

$$x_i \geqslant 0,$$

$$\text{II:} \quad \max \quad \sum_j d_j y_j$$

$$\text{s.t.} \quad \sum_j y_j B_j \leqslant b, \tag{1.2}$$

$$y_j \geqslant 0,$$

where A_i and B_j denote the columns of A and B and along with those matrices all other constants a, b, c, d are nonnegative and real. Without loss of generality we will assume that

$$c_1 \geqslant c_2 \geqslant \cdots \quad \text{and} \quad d_1 \geqslant d_2 \geqslant \cdots. \tag{1.3}$$

and introduce intervals $K := [0, c_1]$ and $L := [0, d_1]$, which contain all c_i and d_j.

Furthermore we will consider the parametric programs I' and II' where the constraints $\sum_i x_i = v_{\text{I}}$ and $\sum_j y_j = v_{\text{II}}$ are added.

The *parallel composition* of I and II is then defined as

$$\text{III:} \quad \max \quad \sum_i c_i x_i + \sum_j d_j y_j$$

$$\text{s.t.} \quad \sum_i x_i A_i \leqslant a,$$

$$\sum_j y_j B_j \leqslant b, \tag{1.4}$$

$$x_i, y_j \geqslant 0.$$

with $\sum_i x_i + \sum_j y_j = v_{\text{III}}$ added for the parametric problem III'.

We now define the *series composition* of I and II. For a given function $F : K \times L \to \mathbb{R}^+$ it is defined as

$$\text{IV:} \quad \max \quad \sum_{i,j} F(c_i, d_j) z_{ij}$$

$$\text{s.t.} \quad \sum_{ij} z_{ij} \binom{A_i}{B_j} \leqslant \binom{a}{b}, \tag{1.5}$$

$$z_{ij} \geqslant 0,$$

(the columns of IV are all possible combinations of I and II) with $\sum_{ij} z_{ij} = v_{\text{IV}}$ added for the parametric problem IV'.

In the following we will obtain results about the inheritance of certain properties under series and parallel compositions. A linear program such as I (or II) is called a *greedy* linear program if the vector $\bar{x} = (\bar{x}_1, \bar{x}_2, ...)$ found by successively maximizing x_1 then $x_2, ...$ satisfies

$$\sum_i c_i \bar{x}_i = \max \left\{ \sum_i c_i x_i : \sum_i x_i A_i \leqslant a, \, x_i \geqslant 0 \right\}. \tag{1.6}$$

For our context we introduce a somewhat stronger greedy property:
Let

$$v_1^* = \max \left\{ \sum_i x_i : \sum_i x_i A_i \leqslant a, \, x_i \geqslant 0 \right\}. \tag{1.7}$$

For any $0 \leqslant v_1 \leqslant v_1^*$ consider $\bar{x}^{v_1}$ the vector $\bar{x}$ truncated at v_1.[1] We then call I (or II respectively) a *strongly constrained parametrically greedy* (*s.c.p.g.*) linear program if

$$\sum_i c_i \bar{x}_i^{v_1} \quad \text{maximizes I' for all } 0 \leqslant v_1 \leqslant v_1^*. \tag{1.8}$$

The notion of strongly greedy linear programs is quite natural. In fact, many programs that are greedy linear programs are also strongly greedy. Examples are polymatroids or the flow problems considered in [2, 4, 3]. But we do not know of any problems where the new aspects of "greediness" described in this paper illuminate any cases where a greedy algorithm was previously sought or is now joyously welcomed.

We are now ready to state two central results, which we will prove in the next section:

Theorem 1.1. *If linear program I (1.1) and II (1.2) are s.c.p.g., so is their parallel composition III (1.4).*

Theorem 1.2. *If linear program I (1.1) and II (1.2) are s.c.p.g., so is their series composition IV (1.5) if F has the following properties:*

$$F(\cdot, v) \text{ and } F(u, \cdot) \text{ are nondecreasing,} \tag{1.9}$$

---

[1]Vector z truncated by w is defined inductively by $z_1 = \min(w, z_1)$; $z_i = \min(w - \sum_{k=1}^{i-1} z_k, z_i)$ for $i > 1$.

$$\text{for each } v \in L, \quad F(\cdot, v) \text{ is convex,}$$

$$\text{for each } u \in K, \quad F(u, \cdot) \text{ is convex,} \tag{1.10}$$

$$u_1 \geqslant u_2, \ v_1 \geqslant v_2 \quad \text{imply} \quad F(u_1, v_1) + F(u_2, v_2) \geqslant F(u_1, v_2) + F(u_2, v_1). \tag{1.11}$$

Property (1.11) is known as the *Monge property*, which we mentioned at the beginning of this section. Note that if F is differentiable (1.9) becomes the condition that the first partials are nonnegative, (1.10) and (1.11) says that all second partials are nonnegative.

Based on compositions (1.4) and (1.5) one can introduce the notion of a *series parallel linear program*. A two terminal directed graph $G = (V, E)$ is called a *series parallel graph* if it fits the following recursive definition (see [13] for a detailed treatment of series parallel graphs): A single edge from one terminal s (usually called the *source*) to the other terminal t (usually called the *sink*) is a series parallel graph. If G_1 and G_2 are series parallel graphs with respective source sink pairs s_1, t_1 and s_2, t_2, their parallel composition is the graph obtained by identifying s_1 and s_2, and also identifying t_1 and t_2. Their series composition (G_1 followed by G_2) is the graph obtained by identifying t_1 and s_2.

Given a graph G, we associate with each $e \in G$ a linear program (e) of the form of I. Denote the data of the individual problem (e) by $A^{(e)}$, $a^{(e)}$, $c^{(e)}$, where the number of columns is $n^{(e)}$ and the number of rows is $m^{(e)}$. Assume that all $c_i^{(e)}$ are contained in some convex subset C of $\mathbb{R}$ such that $F : C \times C \to C$ is associative. If we let $F(u, v)$ be written as $u \circ v$ then we can define the *G-composition* of all the $(e)_{e \in E(G)}$ problems. The number of columns of the combined problem is

$$\sum_{p \in P} \prod_{e \in p} n^{(e)}$$

where P is the set of directed s–t paths in G. The variables are

$$z_{p:i_1, i_2, \ldots, i_k} \geqslant 0,$$

where $e_1, e_2, \ldots, e_k$ are the edges in p and $1 \leqslant i_1 \leqslant n^{(e_1)}, \ldots, 1 \leqslant i_k \leqslant n^{(e_k)}$. The corresponding cost coefficient is

$$c_{i_1} \circ c_{i_2} \circ \cdots \circ c_{i_k}.$$

For edge $e \in p$ let p have the form $e_1 e_2, \ldots, e_{r-1}, e, e_{r+1}, \ldots, e_k$; i.e. r is the position of e in p. Then the inequalities of the G-composition are

$$\sum_{p \ni e} \sum_{\substack{1 \leqslant i_1 \leqslant n^{(e_1)} \\ \vdots \\ 1 \leqslant i_k \leqslant n^{(e_k)}}} z_{p:i_1, i_2, \ldots, i_k} A_{i_r}^{(e)} \leqslant a^{(e)} \quad \forall e \in E(G).$$

Thus iterating Theorems 1.1 and 1.2 we have the following result:

Theorem 1.3. *If G is a series parallel graph, and if $F : C \times C \to C$ is associative and satisfies* *properties* (1.9)–(1.11), *then the G-composition of s.c.p.g. linear programs is a s.c.p.g. linear program.* $\square$

2. Proof of theorems

We will now discuss the validity of the Theorems 1.1 and 1.2. It is clear that the series composition (Theorem 1.2) is the interesting one, whereas the parallel case (Theorem 1.1) is straightforward. All one has to do for the parallel case is to convince oneself that an optimal solution for the composed problem can be obtained by merging the two original optimal greedy solutions.

We therefore concentrate on the series composition. For the proof of Theorem 1.2 we need the following majorization lemma:

Lemma 2.1. *Assume*

$$a_1 \geqslant \cdots \geqslant a_n, \qquad b_1 \geqslant \cdots \geqslant b_n; \tag{2.1}$$

$f_1, f_2, \ldots, f_n$ *are nondecreasing convex functions on a real*

interval C containing all a_i and all b_i; $\tag{2.2}$

$$i < j \text{ and } u > v \quad imply \quad f_i(u) + f_j(v) \geqslant f_i(v) + f_j(u). \tag{2.3}$$

Then if

$$z = (z_1, \ldots, z_n) \text{ is a nonnegative vector such that}$$

$$\sum_1^k z_i a_i \leqslant \sum_1^k z_i b_i, \quad k = 1, \ldots, n, \tag{2.4}$$

we have

$$\sum_1^n z_i f_i(a_i) \leqslant \sum_1^n z_i f_i(b_i). \tag{2.5}$$

Proof. It is clear that it is sufficient to assume all $z_i > 0$, which we do. We shall prove the lemma by induction on n. It clearly holds for $n = 1$, where the only property of f_1 used is that it is nondecreasing.

For the inductive step we first consider the case where in addition to (2.1)–(2.4) we assume

$$\sum_1^n z_i a_i = \sum_1^n z_i b_i \tag{2.6}$$

and

$$\text{every } z_i \text{ is rational.} \tag{2.7}$$

From (2.7), there is a $\delta > 0$ such that

$$z_i = n_i \delta, \quad n_i \in \mathbb{N}^+, \quad i = 1,\dots,n. \tag{2.8}$$

Let $N = \sum n_i$. Consider the sequences

$$a'_1 \geqslant a'_2 \geqslant \cdots \geqslant a'_N \quad \text{and} \quad b'_1 \geqslant b'_2 \geqslant \cdots \geqslant b'_N, \tag{2.9}$$

where the sequence $a'_1, a'_2,\dots,a'_N$ consists of n_1 a_1's, n_2 a_2's,..., in descending order, and similarly for the sequence $b'_1,b'_2,\dots,b'_N$. From (2.1), (2.4) and (2.8), we have

$$a'_1 \leqslant b'_2,$$

$$a'_1 + a'_2 \leqslant b'_1 + b'_2,$$

$$\vdots$$

$$a'_1 + \cdots + a'_N = b'_1 + \cdots + b'_N. \tag{2.10}$$

It is well known (see [6]) that (2.9) and (2.10) imply that the vector $a' = (a'_1,\dots,a'_N)$ is in the polytope of all convex combinations of the vector $b' = (b'_1,\dots,b'_N)$ and its permutations. Since each f_i is convex, the function

$$f_1(a'_1) + \cdots + f_1(a'_{n_1}) + f_2(a'_{n_1+1}) + \cdots + f_2(a'_{n_1+n_2}) + \cdots + f_n(a'_N)$$

is a convex function on this polytope, so its maximum occurs at a vertex of the polytope, namely at one of the permutations of b'. But (2.3) implies that a maximizing vertex is b' itself. So

$$\sum n_i f_i(a_i) \leqslant \sum n_i f_i(b_i).$$

Multiply both sides by δ, use (2.8) and infer (2.5).

Now we must prove (2.5) without assuming (2.6) and (2.7). When $a_1 = b_1$ it is easy to see that the lemma follows from the induction hypothesis. So assume $a_1 < b_1$. It follows that, for any $\varepsilon > 0$, there exists $z'_1,\dots,z'_n > 0$,

$$z'_i \text{ is rational}, \quad i = 1,\dots,n, \tag{2.11}$$

$$|z'_i - z_i| < \varepsilon, \quad i = 1,\dots,n, \tag{2.12}$$

and

$$\sum_1^k z'_i a_i < \sum_1^k z'_i b_i, \quad k = 1,\dots,n. \tag{2.13}$$

Define $\alpha > 0$ using (2.13) by

$$\alpha z'_1 = \min_k \sum_1^k z'_i (b_i - a_i).$$

Then letting $a^* = (a^*_1,\dots,a^*_n)$ with

$$a^*_1 = a_1 + \alpha, \quad a^*_2 = a_2,\dots,a^*_n = a_n$$

we have $a_1^* \geqslant \cdots \geqslant a_n^*$, and

$$\sum_1^k z_i' a_i^* \leqslant \sum_1^k z_i' b_i, \quad k = 1,\dots,n,$$

with equality for some $k = k^*$.

If $k^* = n$, then, from our discussion of (2.6) and (2.7), we have, from (2.11),

$$\sum_1^n z_i' f(a_i^*) \leqslant \sum_1^n z_i' f(b_i). \tag{2.14}$$

If $k^* < n$, we have for the same reason

$$\sum_1^{k^*} z_i' f(a_i^*) \leqslant \sum_1^{k^*} z_i' f(b_i). \tag{2.15}$$

On the other hand, the induction hypothesis gives

$$\sum_{k^*+1}^n z_i f(a_i^*) \leqslant \sum_{k^*+1}^n z_i f(b_i). \tag{2.16}$$

But (2.12) and (2.14), or (2.12), (2.15) and (2.16) imply

$$\sum_1^n z_i f(a_i^*) \leqslant \sum_1^n z_i f(b_i). \tag{2.17}$$

But the definition of a^*, together with the fact that f_1 is nondecreasing (2.2), shows that (2.17) implies (2.5). $\square$

Before we prove Theorem 1.2 we will first rewrite problem IV$'$.
The parametric problem

$$\text{IV}': \quad \max \quad \sum_{i,j} F(c_i, d_j) z_{ij}$$

$$\text{s.t.} \quad \sum_{ij} z_{ij} \binom{A_i}{B_j} \leqslant \binom{a}{b},$$

$$\sum_{ij} z_{ij} = v, \tag{2.18}$$

$$z_{ij} \geqslant 0,$$

can be rewritten as

$$\text{IV}'': \quad \max \quad \sum_{ij} F(c_i, d_j) z_{ij}$$

$$\left. \begin{aligned} \text{s.t.} \quad & \sum_j z_{ij} = x_i, \\ & \sum_i z_{ij} = y_j, \\ & z_{ij} \geqslant 0, \end{aligned} \right\} \text{T}(x, y)$$

$$\left. \begin{aligned} & \sum_i x_i A_i \leqslant a, \\ & \sum_j y_j B_j \leqslant b, \\ & \sum_i x_i = \sum_j y_j = v, \\ & x_i, y_j \geqslant 0. \end{aligned} \right\} \text{R}$$

$$(2.19)$$

As indicated we call the top part of problem (2.19), including the objective function $T(x, y)$, and the remaining constraints R. Notice that $T(x, y)$ is a transportation problem with right side x and y.

Hoffman [7] has shown that an optimal solution for $T(x, y)$ is given by the northwest corner rule. Formally this solution can be represented in the following way[2]: On the real axis starting at 0, plot successive closed intervals $I_1, I_2, \ldots$ where

$$|I_i| \equiv \text{length of } I_i \text{ is } x_i. \tag{2.20}$$

The intervals I_i are referred to as *x-intervals*. Proceed in the same way with y_j to obtain *y-intervals* J_j. Then we have:

Remark 2.1. *An optimal solution to* $T(x, y)$ *is given by*

$$z_{ij} = |I_i \cap J_j|. \qquad \Box \tag{2.21}$$

Remark 2.2. *Let* x^* *and* y^* *be greedy solutions to* I' *and* II' *with parameter value* v. *Then* z *defined by* (2.21) *with respect to* $T(x^*, y^*)$ *is a greedy solution of* IV' (2.18).

Proof. The correctness of the remark follows from monotonicity of F and the monotonicity of the coefficients (1.3). We leave the verification to the reader.[3] $\quad \Box$

Using Remark 2.1, let $G(x, y)$ be the value of an optimal solution z defined by (2.21) with respect to $T(x, y)$. Then problem IV'' (2.19) can again be rewritten as

---

IV''' max $\quad G(x, y)$

s.t. $\quad \sum_i x_i A_i \leqslant a,$

$$\sum_j y_j B_j \leqslant b, \qquad (2.22)$$

$$\sum_i x_i = \sum_j y_j = v,$$

$$x_i, y_j \geqslant 0.$$

We are now ready to prove the following lemma, from which Theorem 1.2 follows directly:

Lemma 2.2. *Let x, y be a feasible solution to* R, *and x^* be a greedy solutionn of* I'. *Then*

$$G(x^*, y) \geqslant G(x, y).$$

Proof. Consider (in the sense of (2.20)) the x-intervals I_i of x and y-intervals J_j of y and furthermore x^*-intervals I_{i*}^*. Now consider the common refinement of all these intervals K_l, numbered successively from the left. We want to invoke Lemma 2.1. To that end, we define numbers z_l as $|K_l|$. Numbers a_l and b_l are defined as follows: Set $a_l = c_i$ if $K_l \subset I_i$ and $b_l = c_{i*}$ if $K_l \subset I_{i*}^*$. Functions f_l are given by the rule: If $K_l \subset J_j$ then $f_l(u) = F(u, d_j)$.

Since problem I is greedy we have verified (2.1). Further since problem I is s.c.p.g., we obtain (2.4) by setting the parameter v_l successively to $\sum_1^k z_l$ for $k = 1,...,n$. As for f properties (1.9) and (1.10) of F imply (2.2) and (1.11) implies (2.3). So Lemma 2.1 implies (2.5), which is $G(x^*, y) \geqslant G(x, y)$. $\quad \square$

In a similar way one shows $G(x, y^*) \geqslant G(x, y)$ and thus $G(x^*, y^*) \geqslant G(x, y)$. Therefore, by Remark 2.2, Theorem 1.2 is proved. $\quad \square$

3. Earlier results

We shall begin with netflows and the results in [2, 4, 3] and [9]. They consider cost flow problems over a series parallel graph G. Associated with each edge $e \in E(G)$ are a nonnegative (usually integer) capacity $b(e)$ as well as a nonnegative cost $c(e)$. Then the program

max $\quad \sum_{p \in P} c(p) x(p)$

s.t. $\quad \sum_{p \ni e} x(p) \leqslant b(e) \quad$ for all $e \in E(G), \qquad (3.1)$

$$x(p) \geqslant 0 \quad \text{for all } p \in P,$$

with $c(p) = c(e_1) + \cdots + c(e_k)$ and path decision variables $x(p)$, is the path-arc description of the cost flow problem on G (see [12] for a more detailed introduction to this formulation of flow problems).

Now define for each edge the trivial program

$$\max \quad c(e)x(e)$$
$$\text{s.t.} \quad 0 \leqslant x(e) \leqslant b(e),$$

which is s.c.p.g.; then (3.1) is the G-composition of these programs, where $F(u, v) = u + v$. Therefore it follows that the cost flow problem on series parallel graphs is indeed s.c.p.g. This gives the main result of [2, 4, 3]. More generally this result holds for associative operations $F(u, v) = u \circ v$, when they satisfy (1.9)–(1.11). This implies some of the results in [9] and [1].

We now turn to the transportation problem. In [7] Hoffman has shown that a greedy algorithm solves the transportation in certain cases.

Given the problem

$$\text{TP:} \quad \max \quad \sum_i \sum_j c_{ij} x_{ij}$$

$$\text{s.t.} \quad \sum_j x_{ij} \leqslant a_i,$$
$$\sum_i x_{ij} \leqslant b_j, \tag{3.2}$$
$$x_{ij} \geqslant 0,$$

with $\sum_i a_i = \sum_j b_j$ and $c_{ij}, a_i, b_j \geqslant 0$. Then the parametric problem with parameter value $v = \sum_i a_i$ is the transportation problem (TP). If the array $(c_{ij})_{n \times m}$ has the properties

$$c_i. \text{ and } c_{.j} \text{ are nondecreasing in } i \text{ and } j; \tag{3.3}$$

(c_{ij}) is a Monge array, i.e.

$$\tag{3.4}$$

$$c_{i_1 j_1} + c_{i_2 j_2} \geqslant c_{i_1 j_2} + c_{i_2 j_1} \quad \text{for all } i_1 < i_2, j_1 < j_2;$$

the problem (3.2) is greedy (and in fact s.c.p.g.).

We will now derive this result in the framework of Theorem 1.2. Although this result is used in proving Theorem 1.2, it is amusing to derive it as a corollary of Theorem 1.2. To this end, the following lemma on Monge arrays is needed; we postpone the proof of the lemma to the end of this section.

Lemma 3.1. *Given an $n \times m$ array (c_{ij}) satisfying (3.3) and (3.4) then there exist*

$$c_1 < c_2 < \cdots < c_n, \qquad d_1 < d_2 < \cdots < d_m,$$

and a function

$$F : [c_1, c_n] \times [d_1, d_n] \to \mathbb{R}^+$$

satisfying (1.9), (1.10) *and* (1.11) *such that*

$$F(c_i, d_j) = c_{ij} \quad \text{for } i = 1, \ldots, n \text{ and } j = 1, \ldots, m.$$

Consider then the linear programs

$$\max \quad \sum_i c_i x_i$$
$$\text{s.t.} \quad 0 \leqslant x_i \leqslant a_i \tag{3.5}$$

and

$$\max \quad \sum_j d_j y_j$$
$$\text{s.t.} \quad 0 \leqslant y_j \leqslant b_j \tag{3.6}$$

which are cleary s.c.p.g. as the parallel composition of trivial linear programs, where the c_i and d_j are as in Lemma 3.1. Now the transportation problem (3.2) is the series composition of (3.5) and (3.6), using the F of Lemma 3.1, which shows that (3.2) is indeed s.c.p.g.

In fact, Hoffman [7] did not make any monotonicity assumptions on c_{ij} and showed that the given Monge property, the northwest corner algorithm solves the transportation problem optimally. Hoffman's result however can also be put into our framework by observing that (c_{ij}) can be transformed into a monotone array (c'_{ij}) in such a way that $\Sigma c_{ij} z_{ij} - \Sigma c'_{ij} z_{ij}$ is a constant. To derive (c'_{ij}) from (c_{ij}) we subtract the first column from all columns and then subtract the first row from all rows. The validity of this transformation is easily verified.

Finally, we turn to the proof of Lemma 3.1:

Proof of Lemma 3.1. Given $c_1 < c_2 < \cdots < c_n$, $d_1 < d_2 < \cdots < d_m$, we define function $F : [c_1, c_n] \times [d_1, d_n] \to \mathbb{R}^+$ by

$$F(x,y) = \sum_{\mu, \nu = 0}^{1} \alpha_\nu \beta_\mu c_{i+\nu, j+\mu} \tag{3.7}$$

where $\alpha_0, \alpha_1, \beta_0, \beta_1$ are the coefficients of the unique representation of $(x, y) \in [c_i, c_{i+1}] \times [d_j, d_{j+1}]$ of the form

$$x = \alpha_0 c_i + \alpha_1 c_{i+1}, \quad \alpha_0 + \alpha_1 = 1, \quad \alpha_0, \alpha_1 \geqslant 0,$$
$$y = \beta_0 d_j + \beta_1 d_{j+1}, \quad \beta_0 + \beta_1 = 1, \quad \beta_0, \beta_1 \geqslant 0. \tag{3.8}$$

Due to (3.3) the functions $F(c_i, \cdot)$ and $F(\cdot, d_j)$ are nondecreasing. Furthermore $c_1 < c_2 < \cdots < c_n$, $d_1 < d_2 < \cdots < d_m$, can be chosen in such a way that all functions $F(c_i, \cdot)$ and $F(\cdot, d_j)$ are also convex. We have to show that F satisfies properties (1.9), (1.10) and (1.11).

First (1.9), (1.10): Let $x \in [c_i, c_{i+1}]$. Then there exists an $0 \leqslant \alpha \leqslant 1$ such that

$$F(x, \cdot) = \alpha F(c_i, \cdot) + (1 - \alpha) F(c_{i+1}, \cdot).$$

                    W.W. Bein et al. / Greedy linear programming problems

As a convex combination of nondecreasing and convex functions $F(c_i, \cdot)$ and $F(c_{i+1}, \cdot)$ the function $F(x, \cdot)$ is nondecreasing and convex. Repeat the arguments for $F(\cdot, y)$.

Now for (1.11): We first prove that

$$F(x_1, y_1) + F(x_2, y_2) \geqslant F(x_1, y_2) + F(x_2, y_1) \tag{3.9}$$

holds for points $P_1 = (x_1, y_1)$, $P_2 = (x_1, y_2)$, $P_3 = (x_2, y_2)$, $P_4 = (x_2, y_1)$ with

$$\begin{aligned} x_1 &= \alpha_0 c_i + \alpha_1 c_{i+1}, & \alpha_0 + \alpha_1 = 1, & \alpha_0, \; \alpha_1 \geqslant 0, & \alpha_0 > \beta_0, & y_1 &= d_j, \\ x_2 &= \beta_0 c_i + \beta_1 c_{i+1}, & \beta_0 + \beta_1 = 1, & \beta_0, \; \beta_1 \geqslant 0, & & y_2 &= d_{j+1}, \end{aligned} \tag{3.10}$$

(see Figure 1). Now (3.9) may be written in the form

$$\alpha_0 c_{i,j+1} + \alpha_1 c_{i+1,j+1} + \beta_0 c_{i,j} + \beta_1 c_{i+1,j}$$

$$\leqslant \alpha_0 c_{i,j} + \alpha_1 c_{i+1,j} + \beta_0 c_{i,j+1} + \beta_1 c_{i+1,j+1}$$

which is equivalent to

$$\alpha_0(c_{i,j+1} + c_{i+1,j} - c_{i+1,j+1} - c_{i,j})$$

$$\leqslant \beta_0(c_{i,j+1} + c_{i+1,j} - c_{i+1,j+1} - c_{i,j}).$$

However the last inequality holds because $\alpha_0 > \beta_0$ and

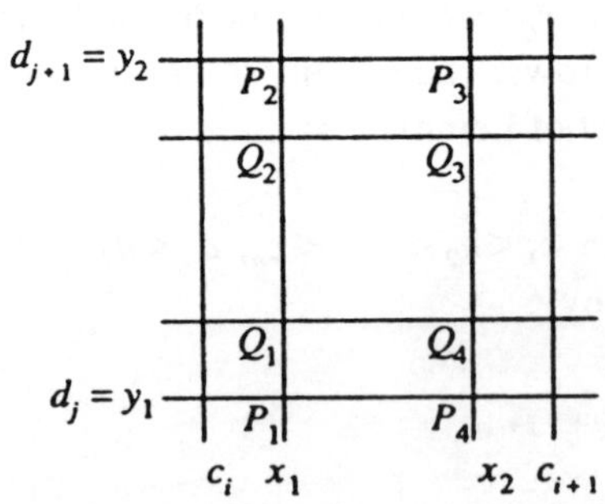

Fig. 1.

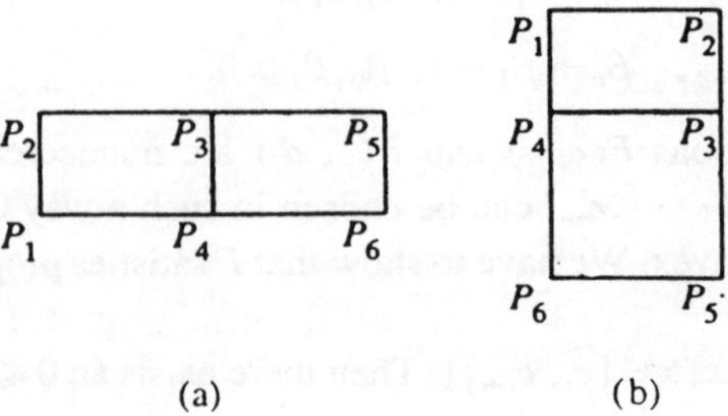

(a)   (b)

Fig. 2.

$$c_{i,j+1} + c_{i+1,j} - c_{i+1,j+1} - c_{i,j} \leqslant 0$$

due to (3.4).

Using the fact that (3.9) holds for the rectangle P_1, P_2, P_3, P_4 we derive in a similar way that (3.9) holds for the inner rectangle Q_1, Q_2, Q_3, Q_4. Next it is easy to show (see e.g. [5]) that property (3.9) is transitive in the sense that if (3.9) holds for P_1, P_2, P_3, P_4 and P_4, P_3, P_5, P_6 it also holds for P_1, P_2, P_5, P_6; see Figure 2(a) and (b). Using this transitivity and the previous result the argument can be repeated to show that (3.9) holds for arbitrary rectangles. $\square$

4. Remarks

We first note that the results of Section 1 can also be formulated for minimization problems. The corresponding results require F to be concave rather than convex and in the Monge property " $\geqslant$ " has to be replaced by " $\leqslant$ ".

The convexity assumption for Theorem 1.2 is indeed necessary. To see this, consider as program I,

$$\max \quad 12x + 2y$$

$$\text{s.t.} \quad x + y + z \leqslant 1,$$

$$2x + y \leqslant 1,$$

$$x, y, z \geqslant 0,$$

and as program II the *dummy program*

$$\max \quad 3x$$

$$\text{s.t.} \quad x \leqslant 1,$$

$$x \geqslant 0.$$

Both programs I and II are s.c.p.g., but for $F(\cdot, \cdot) = \min(\cdot, \cdot)$ the series composition is not.[4]

If we weaken the notion of greedy by replacing $\sum_i x_i = v$ by $\sum_i x_i \leqslant v$ (now called *weakly greedy*) the result on parallel compositions is still true but the result for series compositions does not hold any longer. The series composition result can be carried over if in Theorem 1.2 we require F to have the additional property that

$$F(u, 0) = F(0, v) = 0 \quad \text{for all } u \in K, \ v \in L. \tag{4.1}$$

As an example, a function satisfying those properties if $F(u, v) = u \cdot v$.

It would be interesting to characterize those F that satisfy (1.9)–(1.11) (and property

---

[4] Dummy programs are not only useful for counterexamples: If the objective function $\sum_i c_i x_i$ of a s.c.p.g. is changed to $\sum_i f(c_i) x_i$ for a monotone and convex function f, the program remains s.c.p.g. To see this, all we have to do is consider the series composition with a dummy program and $F(\cdot, 1) := f(\cdot)$.

14   *W.W. Bein et al. / Greedy linear programming problems*

(4.1) for the weak case). If F is assumed associative then there are severe restrictions. Several years ago, Jeremy Kahn [10] made significant inroads into the case where F is defined over the reals and has a fixed point $F(u, u) = u$.

Finally, what does it mean algebraically for a linear program to be s.c.p.g. Is it possible to characterize those triples (A, a, c) such that the corresponding linear program I is s.c.p.g.?

Acknowledgements

We thank Michael Shub, Don Coppersmith and Shiay Pilpel for their help on various aspects of this paper. We also thank Gene Lawler for having initiated the contacts between authors.

References

[1] Y.P. Aneja, R. Chandrasekaran and K.P.K. Nair, "Classes of linear programs with integral optimal solutions," *Mathematical Programming Study* 25 (1985) 225–237.

[2] W.W. Bein, "Netflows, polymatroids, and greedy structures," Ph.D. Thesis, Universität Osnabrück (Osnabrück, Germany, 1986).

[3] W.W. Bein and P. Brucker, "Greedy concepts for network flow problems," *Discrete Applied Mathematics* 15 (1986) 135–144.

[4] W.W. Bein, P. Brucker and A. Tamir, "Minimum cost flow algorithms for series parallel networks," *Discrete Applied Mathematics* 10 (1985) 117–124.

[5] P.C. Gilmore, E.L. Lawler and D.B. Shmoys, "Well-solved special cases," in: E.L. Lawler, J.K. Lenstra, A.H.G. Rinnooy Kan and D.B. Shmoys, eds., *The Travelling Salesman Problem – A Guided Tour of Combinatorial Optimization* (Wiley, New York, 1985) pp. 87–143.

[6] G.H. Hardy, J.E. Littlewood and G. Polya, *Inequalities* (Cambridge University Press, Cambridge, England, 1934).

[7] A.J. Hoffman, "On simple linear programming problems," in: V.L. Klee, ed., *Convexity, Proceedings of Symposia in Pure Mathematics, Vol. 7* (American Mathematical Society, Providence, RI, 1963) pp. 317–327.

[8] A.J. Hoffman, "On greedy algorithms that succeed," in: I. Anderson, ed., *Surveys in Combinatorics 1985. London Mathematical Society Lecture Note Series No. 103* (Cambridge University Press, Cambridge, England, 1985) pp. 97–112.

[9] A.J. Hoffman, "On greedy algorithms for series parallel graphs," *Mathematical Programming* 40 (1988) 197–204.

[10] J. Kahn, oral communication (1988).

[11] G. Monge, "Mémoire sur la théorie des déblai et des remblai," *Histoire de l'Academie Royale des Sciences (année 1781)* (Paris, 1784) pp. 666–704.

[12] C.H. Papadimitriou and K. Steiglitz, *Combinatorial Optimization* (Prentice-Hall, Englewood Cliffs, NJ, 1982).

[13] J. Valdes, "Parsing flowcharts and series-parallel graphs," Ph.D. Thesis, Stanford University (Stanford, CA, 1978).

Graph Spectra

1. On the uniqueness of the triangular association scheme

W. S. Connor, my colleague at the National Bureau of Standards, had proved the theorem of the title for all $n > 8$. I had always admired Bill's skill with matrices: much like Herbert Ryser, he found very ingenious ways to prove combinatorial theorems by using matrices. I realized that the problem posed could be rephrased in the language of graph theory (rather than the strange statistical description "triangular association scheme"), and I (surprise!) saw that I could exploit the spectrum of the association scheme (the least eigenvalue was an amazingly high -2) to prove the theorem held for all n except 8. While the additional information wasn't much to brag about, the method led to interesting further developments. Further, I concocted the terms "claw" and "clawfree" to use in parts of the argument, because Connor had used these concepts in places in his proof. I had previously persuaded Bose and Bruck to use these words, and their papers in the Pacific Journal brought "claws" into the mathematical literature.

If I had known about root system in Lie algebras, I would have seen much more direct routes to the answer (as Cameron, Goethals, Seidel and Shult did later), and perhaps not become involved in research on graph spectra at all.

2. On Moore graphs with diameters 2 and 3

After I discussed the preceding paper at an IBM summer workshop, E.F. Moore raised the graph theory problem described in the paper, and my GE colleague Bob Singleton and I pondered it. Moore told me the problem because he thought the eigenvalue methods I was using might find another "Moore graph" of diameter 2 besides the pentagon and the Petersen graph. Indeed, we found the Hoffman–Singleton graph with 50 nodes (and showed it was unique) and that any other Moore graph of diameter 2 had to have 3,250 nodes (and to this day, no one knows if such a graph exists). Moore declined joint authorship, so we thanked him by giving his name to the class of graphs. When it was later proved by Damerell, and also by Bannai and Ito, that there were no other Moore graphs other than the trivial odd cycles, I felt a twinge of guilt in giving Moore's name to such a small set. But I was wrong: Moore graphs, Moore geometries, etc. continue to be discussed in the profession.

At Al Tucker's suggestion, Singleton wrote a dissertation on related material (I was proud to be the de facto advisor, Tucker was the de jure advisor) for Princeton where he had been a graduate student decades earlier. He subsequently left GE and was affiliated with the Mathematics Department at Wesleyan University for many years.

3. On the polynomial of a graph

With the examples from the two preceding papers, and from various doodles at the time, I thought I discerned a general principle, which associated a polynomial with each regular connected graph in a nice way. Ernst Straus liked it, too, so I published it. This polynomial has become one of the standard tools for studying graphs with some kind of regular structure.

4. On the line graph of a symmetric balanced incomplete block design

The triangular association scheme is essentially the line graph of the complete graph. Then we studied the line graph of a finite projective plane, a finite affine plane, finally (the most interesting case) the line graph of symmetric balanced incomplete block design (the graph is the bipartite graph with blocks on one side, treatments on the other, with edges joining treatments to the blocks to which they belong). The aim was to show (or disprove) that the spectrum of the line graph characterized the graph. Our method was to look for "bad graphs" (i.e., induced subgraphs which we could exclude because of some argument involving eigenvalues and/or eigenvectors). These were fun to find and draw. Each morning Ray-Chaudhouri or I would arrive at work with another bad graph showing that the size of a counterexample had to be smaller than the bound we knew yesterday.

Eventually, the game of "can you top this?" ended with the proof that there was exactly one counterexample.

5. On eigenvalues and colorings of graphs

Herb Wilf had used the spectrum of the adjacency matrix of a graph to give an upper bound to the chromatic number. I wondered if I could use the spectrum to give a lower bound. I realized I needed a generalization of Aronzajn's inequalities (relating the eigenvalues of a symmetric matrix to the eigenvalues of the diagonal blocks in a 2×2 partition) to an $m \times m$ partition. My generalization was expanded further by Robert Thompson.

6. Eigenvalues and partitionings of the edges of a graph

If you think of graphs and eigenvalues, coloring the vertices suggests looking at the relation between the eigenvalues of the adjacency matrix and the eigenvalues of diagonal blocks. Coloring the edges (i.e., partitioning the edge set) suggests looking at the relations between the eigenvalues of a matrix A and the eigenvalues of each of a set of matrices which sum to A. At this point in my romance with investigating the connection between spectrum of a graph and "graphy" properties of a graph, I became intrigued by the question of whether certain measures on a graph were or weren't spectral functions. For example, I show that, even though you can find a lower bound on the smallest number $c(G)$ of cliques whose edges partition the edges of the graph G from the eigenvalues of G, the quantity $c(G)$ is not a "spectral function" on the set of all graphs. By this I mean: there are two sequences $G(1)$,

$G(2), \ldots$ and $H(1)$, $H(2), \ldots$ such that $c(G(i))$ goes to infinity, $c(H(i))$ remains bounded, but, for each i, $G(i)$ and $H(i)$ have the same spectrum.

7. On spectrally bounded graphs

I had done some exploring on the question of when can we know or how can we recognize that a graph has a least eigenvalue that is large (i.e., a small negative number). I knew that, if G contained a large claw or a large graph of a certain other type as an induced subgraph the least eigenvalue of G had to be a negative number large in absolute value. So I speculated that those were (in a qualitative sense) the only possibilities for G to have a least eigenvalue of large absolute value; namely, a large representative of at least one of those two families of graphs must be an induced subgraph.

I am very proud of this paper. Most theorems that a mathematician discovers are bound to be found sooner or later, and probably sooner. They are in the air, floating about in the general awareness of mathematicians working in the subspecialty and available for plucking. To put it another way (in the style of Erdös) God knows almost all theorems and chooses the particular theorems to be revealed to each mathematician. But I do not think God knew this theorem; I had to tell Him, but I still don't know if He is interested.

8. Lower bounds for the partitioning of graphs

This paper mixed together eigenvalue estimation, combinatorics and optimization, so I loved it. Together with a sequel by Cullum, Donath and Wolfe (which considered the tricky nonsmooth problem of how to choose the best diagonal of the modified Laplacian of the graph), it was an early example of using eigenvalues and semidefinite programming in an algorithm for combinatorial optimization. The paper has also been influential in suggesting various heuristics for partitioning problems.

Donath's profession is not mathematics. So it was a real pleasure for me to introduce him a few years ago to an audience of his admirers at a Rutgers conference on semidefinite programming.

9. Nearest S-matrices of given rank and the Ramsey problem for eigenvalues of bipartite S-graphs

We defined an S-matrix as any matrix whose nonzero entries are chosen from a specified set S of numbers. For a complex matrix A, the distance to the nearest complex matrix of a certain rank is governed by the singular values of the matrix. Now suppose A is an S-matrix, and we want the nearest S-matrix of a certain rank. Is this governed by the singular values of A also? In what way? We thought this question worth investigating, and the rough results are reported here.

Another theme of the paper is the concept of Ramsey function on a partially ordered set. We show that some functions on the partially ordered set of graphs are Ramsey functions and some are not. (I like the term Ramsey function because it can be shown that the celebrated theorem of Ramsey about graphs can be stated as: the number of vertices of a graph is a Ramsey function on the partially ordered set of all graphs.) We show that, for every S and k, the kth singular value is a Ramsey function. That's not true for eigenvalues of symmetric matrices, as we also show.

ON THE UNIQUENESS OF THE TRIANGULAR ASSOCIATION SCHEME

By A. J. Hoffman

General Electric Company

1. Summary. Connor [3] has shown that the relations among the parameters of the triangular association scheme themselves imply the scheme if $n \geq 9$. This result was shown by Shrikhande [6] to hold also if $n \leq 6$. (The problem has no meaning for $n < 4$.) This paper shows that the result holds if $n = 7$, but that it is false if $n = 8$.

2. Introduction. A partially balanced incomplete block design with two associate classes [1] is said to be triangular [2], [3] if the number of treatments, v, is $n(n - 1)/2$ for some integer n, and the association scheme is obtainable as follows:

Let the v treatments be regarded as all possible arcs of the graph determined by n points; let the first associates of any arc (= treatment) be all arcs each of which share exactly one end point with the given arc; let the second associates of any arc be all arcs each of which does not share an end point with the given arc and does not coincide with the given arc.

Then the following relations hold:

(2.1)   The number of first associates for any treatment is $2(n - 2)$.

(2.2)   If θ_1 and θ_2 are two treatments which are first associates, the number of treatments which are first associates of both θ_1 and θ_2 is $n - 2$.

(2.3)   If θ_1 and θ_2 are second associates, the number of treatments which are first associates of both θ_1 and θ_2 is 4.

It is natural to inquire if conditions (2.1)–(2.3) imply that the $v = n(n - 1)/2$ treatments can be represented as arcs on the graph determined by n points in the manner described above; i.e., if (2.1)–(2.3) imply the triangular association scheme. This is known ([3], [6]) to be so if $n \neq 7, 8$.

We prove the result for 7. Actually we will prove the result for all n except 8. For $n = 8$, the theorem is false, as we shall demonstrate by exhibiting a counter-example. The derivation of this counter-example and a procedure for finding all counter-examples are given in [4]. They are based on an elaboration of the devices used in Sections 3 and 4 of this paper. Other illustrations of the use of these devices are contained in [5].

Henceforth, we assume (2.1)–(2.3).

3. The Association Matrix. Number the treatments from 1 to v in any order. Define the square matrix A of order v by

$$(3.1) \qquad A = (a_{ij}) = \begin{cases} 0 & \text{if } i = j \\ 1 & \text{if } i \text{ and } j \text{ are first associates} \\ 0 & \text{if } i \text{ and } j \text{ are second associates} \end{cases}$$

Received August 31, 1959.

Note that $a_{ij} = a_{ji}$. Next let $B = AA^T = A^2$, since A is symmetric. From (2.1), we have $b_{ii} = 2(n-2)$. From (2.2), we have $b_{ij} = (n-2)$ if i and j are first associates. From (2.3), we have $b_{ij} = 4$ if i and j are second associates. If we let J be the square matrix of order v, with every entry unity, and I the identity matrix of order v, then the foregoing may be summarized by

$$(3.2) \quad A^2 = 2(n-2) I + (n-2)A + 4(J - I - A)$$
$$= (2n - 8)I + (n - 6)A + 4J.$$

All the matrices appearing in (3.2) can be simultaneously diagonalized. Imagine (3.2) in diagonal form, and one sees that the diagonal entries relate the eigenvalues of the matrices.

Now J has the eigenvalue v, corresponding to the eigenvector $(1, 1, \cdots, 1)$; all other eigenvalues of J are zero. The eigenvector $(1, 1, \cdots, 1)$ clearly corresponds to the eigenvalue $2(n-2)$ of A. Any other eigenvalue, α, of A corresponds to a zero eigenvalue of J; hence (3.2) implies that α satisfies the equation $\alpha^2 = (2n - 8) + (n - 6)\alpha$, so that $\alpha = -2$, or $\alpha = n - 4$.

The trace of A is zero, since $a_{ii} = 0$ for all i; hence the sum of the eigenvalues of A is 0. If k is the multiplicity of $n - 4$, it follows that $0 = 2n - 4 + k(n - 4) + (v - k - 1)(-2)$. So the eigenvalues of A are

$$(3.3) \quad \begin{array}{ll} \text{(a)} & 2n - 4 \text{ with multiplicity 1, eigenvector } (1, 1, \cdots, 1) \\ \text{(b)} & n - 4 \text{ with multiplicity } n - 1 \\ \text{(c)} & -2 \text{ with multiplicity } v - n. \end{array}$$

Note that $v > n$, so -2 is the least eigenvalue of A.

This is the only use we shall make of (3.3) (c) in the present paper, although it plays a major role in the analysis of the exceptional cases for $n = 8$. We shall make no use of (3.3) (b).

In what follows, we shall use two well-known properties of eigenvalues and eigenvectors of symmetric matrices, and for ease of reference, we now list them explicitly.

Let M be a (real) symmetric matrix whose least eigenvalue is β, and whose maximum eigenvalue is $\alpha > \beta$, with x an eigenvector corresponding to α. Let K be a principal submatrix of M, δ the least eigenvalue of K and y an eigenvector of K corresponding to δ. Then

$$(3.4) \qquad\qquad \delta \geqq \beta;$$

and

(3.5)   if $\delta = \beta$, then y is orthogonal to the projection of x on the subspace corresponding to K.

From (3.4) and (3.3) (c) follow the fact that a principal submatrix of A cannot have an eigenvalue less than -2. From (3.5) and (3.3) (a) and (c), if -2 is an eigenvalue of a principal submatrix of A, then the corresponding eigenvector has zero as the sum of its co-ordinates.

494   A. J. HOFFMAN

4. The Case $n \neq 8$.

LEMMA 1. *A does not contain*

(4.1)
$$\begin{matrix} 0 & 0 & 0 & 1 \\ 0 & 0 & 0 & 1 \\ 0 & 0 & 0 & 1 \\ 1 & 1 & 1 & 0 \end{matrix}$$

as a principal submatrix.

This was proved by Connor [3] for $n \geq 9$. We now prove it for all $n \neq 8$. We contend that A cannot contain any of the following three square matrices of order 5, each of which contains (4.1) as a principal submatrix:

(4.2)	(4.3)	(4.4)
$\begin{matrix} 0 & 0 & 0 & 1 & 1 \\ 0 & 0 & 0 & 1 & 1 \\ 0 & 0 & 0 & 1 & 1 \\ 1 & 1 & 1 & 0 & 0 \\ 1 & 1 & 1 & 0 & 0 \end{matrix}$	$\begin{matrix} 0 & 0 & 0 & 1 & 1 \\ 0 & 0 & 0 & 1 & 1 \\ 0 & 0 & 0 & 1 & 1 \\ 1 & 1 & 1 & 0 & 1 \\ 1 & 1 & 1 & 1 & 0 \end{matrix}$	$\begin{matrix} 0 & 0 & 0 & 1 & 1 \\ 0 & 0 & 0 & 1 & 1 \\ 0 & 0 & 0 & 1 & 0 \\ 1 & 1 & 1 & 0 & 0 \\ 1 & 1 & 0 & 0 & 0 \end{matrix}$

The impossibility of (4.2) and (4.4) follows from (3.4), since each has an eigenvalue smaller than -2. Matrix (4.3) has -2 as an eigenvalue, with $(1, 1, 1, -1, -1)$ as corresponding eigenvector, violating (3.5).

Let us denote by 1, 2, 3, 4 respectively the rows and columns of A that produced submatrix (4.1). Because (4.2) and (4.3) are impossible, it follows that 4 is the only treatment that is a first associate of 1, 2, and 3. Hence, by (2.3), there are exactly nine additional treatments, each of which is a first associate of two of the set 1, 2, 3. Since (4.4) is impossible, it follows that each of the nine is a first associate of four. Together with 1, 2, 3, this yields twelve treatments, each of which is a first associate of 4. From (2.1), we must have $12 \leq 2n - 4$, which is impossible if $n \leq 7$.

Now suppose $n \geq 9$. Treatments 1 and 4 are first associates, and, by (2.2), there are $n - 2$ first associates of each. We have previously encountered 6, three of which are first associates also of 2, and three of which are also first associates of 3. Hence there are $n - 8$ additional ones. Similary, there are $n - 8$ additional first associates of 2 and 4, and $n - 8$ additional first associates of 3 and 4. Hence, from (2.1), $2(n - 2) \geq 12 + 3(n - 8)$, which is impossible for $n \geq 9$.

Next, we prove

LEMMA 2. *If* 1 *and* 2 *are second associates,* 3, 4, 5, 6 *first associates of both* 1 *and* 2, *then (after renumbering, if necessary) the principal submatrix of A corresponding to rows and columns* 1–6 *is*

(4.5)
$$\begin{matrix} 0 & 0 & 1 & 1 & 1 & 1 \\ 0 & 0 & 1 & 1 & 1 & 1 \\ 1 & 1 & 0 & 0 & 1 & 1 \\ 1 & 1 & 0 & 0 & 1 & 1 \\ 1 & 1 & 1 & 1 & 0 & 0 \\ 1 & 1 & 1 & 1 & 0 & 0 \end{matrix}$$

PROOF: Consider the $2(n - 2)$ treatments which are first associates of 3. None of them can be second associates of both 1 and 2, for this would violate Lemma 1. Hence, if we let t be the number of first associates of 3 which are first associates of 1 and 2, we have from (2.1) and (2.2), $t + (n - 2 - t) + (n - 2 - t) = 2(n - 2) - 2$, or $t = 2$. These two must be some two of 4, 5, 6, say 5 and 6. It follows that 3 and 4 are second associates, while 3 is a first associate of both 5 and 6. The inevitability of (4.5) is now clear.

LEMMA 3. *Any matrix of form*

$$(4.6) \qquad \begin{matrix} 0 & 0 & 1 & 1 & 1 & 1 & 1 & 1 \\ 0 & 0 & 1 & 1 & 1 & 1 & 0 & 0 \\ 1 & 1 & 0 & 0 & 1 & 1 & 1 & 1 \\ 1 & 1 & 0 & 0 & 1 & 1 & 0 & 0 \\ 1 & 1 & 1 & 1 & 0 & 0 & 1 & 0 \\ 1 & 1 & 1 & 1 & 0 & 0 & 0 & 1 \\ 1 & 0 & 1 & 0 & 1 & 0 & 0 & x \\ 1 & 0 & 1 & 0 & 0 & 1 & x & 0 \end{matrix}$$

is not a principal submatrix of A.

PROOF: If (4.6) were to exist, then $x \neq 1$. For 6 and 7 would be second associates, and, if $x = 1$, then 1, 3, and 8 would mutually be first associates, but this contradicts Lemma 2. So we must take $x = 0$. But then 2, 7, and 8 are pairwise second associates; 3 is a first associate of each of 2, 7, 8, and this violates Lemma 1.

LEMMA 4. *The matrix*

$$(4.7) \qquad \begin{matrix} 0 & 0 & 1 & 1 & 1 & 1 & 1 & 1 \\ 0 & 0 & 1 & 1 & 1 & 1 & 0 & 0 \\ 1 & 1 & 0 & 0 & 1 & 1 & 1 & 1 \\ 1 & 1 & 0 & 0 & 1 & 1 & 0 & 0 \\ 1 & 1 & 1 & 1 & 0 & 0 & 1 & 1 \\ 1 & 1 & 1 & 1 & 0 & 0 & 0 & 0 \\ 1 & 0 & 1 & 0 & 1 & 0 & 0 & 0 \\ 1 & 0 & 1 & 0 & 1 & 0 & 0 & 0 \end{matrix}$$

is not a principal submatrix of A.

PROOF: All we want to show is that the other entries in (4.7) imply that 7 and 8 are first associates, not second associates as (4.7) alleges. If 7 and 8 are second associates, then using the same reasoning as in the first part of Lemma 3, some two of 1, 3, 5 are by Lemma 2 second associates. But this is not so in (4.7).

LEMMA 5. *The $2(n - 2)$ first associates of any treatment can be split into two classes so that the $n - 2$ treatments of one class are mutually first associates of each other; the $n - 2$ treatments of the other class are mutually first associates.*

PROOF: Let 1 be the treatment. Let 3 be a first associate of 1, 2 a second associate of 1 and a first associate of 3, and 4, 5, 6 chosen so that we have the submatrix of Lemma 2. In addition to 5 and 6, there are $n - 4$ other first associates of both 1 and 3. Each of these must be a first associate of at least one of

5 and 6. Otherwise it, 5 and 6, would be mutually second associates, and 1 would be a first associate of each of the three, violating Lemma 1. Further, by Lemma 3, each of these $n - 4$ treatments is a first associate of 5 or each is a first associate of 6. Without loss of generality, say it is 5. By Lemma 4, these $n - 4$ treatments are mutually first associates. Further, each is a first associate of 3 and 5, which are themselves first associates, and thus 3, 5, and these $n - 4$ treatments are altogether $n - 2$ first associates of 1, which are mutually first associates.

Of the $n - 2$ first associates of 1 and 4, 5 is in the class already described, 6 is not, and there are $n - 4$ others. These $n - 4$ are mutually first associates by the same reasoning as above; they are entirely different from the previous $n - 4$ of the first class, since each of those was a second associate of 4; each is obviously a first associate of 6 as well as 4; so 4, 6, and these $n - 4$ treatments constitute our second class.

THEOREM 1. *If $n \neq 8$, then condition (2.1)–(2.3) characterize the triangular association scheme.*

PROOF: It has been shown by Shrikhande [6] that Lemma 5 implies Theorem 1.

5. The Case $n = 8$.

THEOREM 2. *If $n = 8$, then conditions (2.1)–(2.3) do not necessarily imply the triangular association scheme.*

PROOF: Here is a counter-example. Notice that the first principal submatrix of order 5 violates the triangular association scheme.

```
0 0 1 1 1 1 0 0 0 0 0 0 1 1 1 0 1 0 1 0 1 0 1 1 0 0 0 0
0 0 1 1 1 1 0 0 0 0 0 0 0 0 0 1 0 1 0 1 0 1 0 0 1 1 1 1
1 1 0 1 1 1 1 0 0 0 0 0 0 1 1 1 0 0 0 0 0 0 1 0 1 1 0 0
1 1 1 0 1 1 0 1 0 0 0 0 0 1 0 0 1 1 0 0 0 0 1 0 0 0 1 1
1 1 1 1 0 1 0 0 1 0 0 0 0 1 0 0 0 0 1 1 0 0 0 1 1 0 1 0
1 1 1 1 1 0 0 0 0 1 0 0 0 1 0 0 0 0 0 0 1 1 0 1 0 1 0 1
0 0 1 0 0 0 0 1 1 1 1 1 0 1 1 1 0 0 0 0 0 0 1 0 1 1 0 0
0 0 0 1 0 0 1 0 1 1 1 1 0 1 0 0 1 1 0 0 0 0 1 0 0 0 1 1
0 0 0 0 1 0 1 1 0 1 1 1 0 1 0 0 0 0 1 1 0 0 0 1 1 0 1 0
0 0 0 0 0 1 1 1 1 0 1 1 0 1 0 0 0 0 0 0 1 1 0 1 0 1 0 1
0 0 0 0 0 0 1 1 1 1 0 0 0 0 1 0 1 0 1 0 1 0 0 0 1 1 1 1
0 0 0 0 0 0 1 1 1 1 0 0 1 1 0 1 0 1 0 1 0 1 1 1 0 0 0 0
1 0 0 0 0 0 0 0 0 0 0 1 0 0 1 1 1 1 1 1 1 1 1 1 0 0 0 0
1 0 1 1 1 1 1 1 1 1 0 1 0 0 0 0 0 0 0 0 0 0 1 1 0 0 0 0
1 0 1 0 0 0 1 0 0 0 1 0 1 0 0 1 1 0 1 0 1 0 1 0 1 1 0 0
0 1 1 0 0 0 1 0 0 0 0 1 1 0 1 0 0 1 0 1 0 1 1 0 1 1 0 0
1 0 0 1 0 0 0 1 0 0 1 0 1 0 1 0 0 1 1 0 1 0 1 0 0 0 1 1
0 1 0 1 0 0 0 1 0 0 0 1 1 0 0 1 1 0 0 1 0 1 1 0 0 0 1 1
1 0 0 0 1 0 0 0 1 0 1 0 1 0 1 0 1 0 0 1 1 0 0 1 1 0 1 0
0 1 0 0 1 0 0 0 1 0 0 1 1 0 0 1 0 1 1 0 0 1 0 1 1 0 1 0
1 0 0 0 0 1 0 0 0 1 1 0 1 0 1 0 1 0 1 0 0 1 0 1 0 1 0 1
0 1 0 0 0 1 0 0 0 1 0 1 1 0 0 1 0 1 0 1 1 0 0 1 0 1 0 1
1 0 1 1 0 0 1 1 0 0 0 1 1 1 1 1 1 0 0 0 0 0 0 0 0 0 0 0
1 0 0 0 1 1 0 0 1 1 0 1 1 1 0 0 0 0 1 1 1 1 0 0 0 0 0 0
0 1 1 0 1 0 1 0 1 0 1 0 0 0 1 1 0 0 1 1 0 0 0 0 0 1 1 0
0 1 1 0 0 1 1 0 0 1 1 0 0 0 1 1 0 0 0 0 1 1 0 0 1 0 0 1
0 1 0 1 1 0 0 1 1 0 1 0 0 0 0 0 1 1 1 1 0 0 0 0 1 0 0 1
0 1 0 1 0 1 0 1 0 1 1 0 0 0 0 0 1 1 0 0 1 1 0 0 0 1 1 0
```

REFERENCES

[1] R. C. Bose and K. R. Nair, "Partially balanced incomplete block designs," *Sankya*, Vol. 4 (1939), pp. 337–372.

[2] R. C. Bose and T. Shimamoto, "Classification and analysis of partially balanced designs with two associate classes," *J. Amer. Stat. Assn.*, Vol. 47 (1952), pp. 151–190.

[3] W. S. Connor, "The uniqueness of the triangular association scheme," *Ann. Math. Stat.*, Vol. 29 (1958), pp. 262–266.

[4] A. J. Hoffman, "On the exceptional case of a characterization of the arcs of a complete graph," to appear in *IBM Journal of Research*.

[5] A. J. Hoffman and R. R. Singleton, "On a graph-theoretic problem of E. F Moore," to appear in *IBM Journal of Research*.

[6] S. S. Shrikhande, "On a characterization of the triangular association scheme," *Ann. Math. Stat.*, Vol. 30 (1959), pp. 39–47.

NOTE

The results of this paper have also been obtained, using different methods, by Chang, L. C.,"The Uniqueness and Nonuniqueness of the Triangular Association Schemes," *Science Record*, Vol. III, New Series, 1959, pp. 604–613. Chang has also shown that there are exactly three counterexamples when $n = 8$ ("Association Schemes of Partially Balanced Designs with Parameters $v = 28$, $n_1 = 12$, $n_2 = 15$ and $p_{11}^2 = 4$," *Science Record*, Vol. IV, New Series, 1960, pp. 12–18).

Reprinted from IBM J. of Res. & Develop.
Vol. 4, No. 5 (1960), pp. 497–504

A. J. Hoffman*
R. R. Singleton*

On Moore Graphs with Diameters 2 and 3

Abstract: This note treats the existence of connected, undirected graphs homogeneous of degree d and of diameter k, having a number of nodes which is maximal according to a certain definition. For $k = 2$ unique graphs exist for $d = 2, 3, 7$ and possibly for $d = 57$ (which is undecided), but for no other degree. For $k = 3$ a graph exists only for $d = 2$. The proof exploits the characteristic roots and vectors of the adjacency matrix (and its principal submatrices) of the graph.

1. Introduction

In a graph of degree d and diameter k, having n nodes, let one node be distinguished. Let n_i, $i = 0$, $1, \cdots, k$ be the number of nodes at distance i from the distinguished node. Then $n_0 = 1$ and

$$n_i \leq d(d - 1)^{i-1} \quad \text{for} \quad i \geq 1. \tag{1}$$

Hence

$$\sum_{i=0}^{k} n_i = n \leq 1 + d \sum_{i=1}^{k} (d - 1)^{i-1}. \tag{2}$$

E. F. Moore has posed the problem of describing graphs for which equality holds in (2). We call such graphs "Moore graphs of type (d, k)". This note shows that for $k = 2$ the types $(2, 2)$, $(3, 2)$, $(7, 2)$ exist and there is only one graph of each of these types. Furthermore, there are no other $(d, 2)$ graphs except possibly $(57, 2)$, for which existence is undecided. For $k = 3$ only $(2, 3)$ exists; this is the 7-gon.

The results of Section 2 and Eq. (3) are due to Moore, who has also shown the nonexistence of certain isolated values of (d, k) using methods of number theory.

2. Elementary properties

Moore observed that in graphs for which equality holds in (2) every node is of degree d, since it necessitates that equality hold in (1) for each i.

Furthermore, since no node has degree exceeding d, each node counted in n_i is joined with $(d - 1)$ nodes counted in n_{i+1}, for $i = 1, \cdots, k - 1$. Hence no arc joins two nodes equally distant from some distinguished node, except when both are at distance k from the distinguished node.

Thus if arcs joining nodes at distance k from the distinguished node are deleted the residual graph is a hierarchy, as in Fig. 1. The same hierarchy results from distinguishing any node.

Figure 1

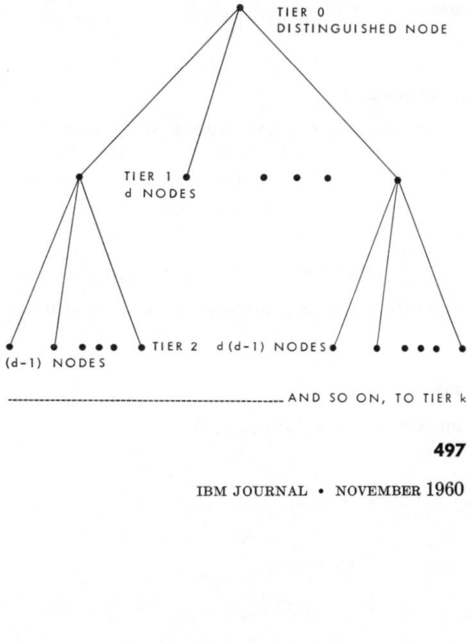

* General Electric Company, New York, N. Y

3. Notation

The discussion deals with matrices of various orders, and with some which are most clearly symbolized as arrays of matrices or blocks of lower order. We will not attempt to indicate orders by indices, but rather to indicate them explicitly or implicitly in the context.

The following symbols denote particular matrices throughout:

I is the identity matrix

0 is the zero matrix

J is the matrix all of whose elements are unity

K is a matrix of order $d(d - 1)$ which is a $d \times d$ array of diagonal blocks of J's of order $(d - 1)$. Thus

$$K = \begin{bmatrix} J & 0 & \cdots & 0 \\ 0 & J & \cdots & 0 \\ \multicolumn{4}{c}{\cdots\cdots\cdots\cdots\cdots} \\ 0 & 0 & \cdots & J \end{bmatrix}.$$

0 is used also for a vector all of whose elements are zero.

u is a vector all of whose elements are unity.

e_i is a vector whose i-th element is unity and the remainder are zero.

We use prime ($'$) to indicate the transpose of a matrix. An unprimed vector symbol is a column vector; a primed vector symbol is a row vector. Thus

$$u = \begin{bmatrix} 1 \\ 1 \\ \cdot \\ \cdot \\ \cdot \\ 1 \end{bmatrix} \qquad u' = (1, 1, \cdots, 1).$$

The subset of nodes of tier k, Fig. 1, which are joined to the ith node of tier $(k - 1)$ is designated S_i. The arcs joining nodes of tier k, which are omitted from the hierarchy, are called *re-entering arcs*.

4. Diameter 2

Consider a Moore graph with $k = 2$. Then $n = 1 + d^2$. Let A be its adjacency matrix. That is, with the nodes of the graph given any numbering,

$$a_{ij} = \begin{cases} 1 \text{ if nodes } i \text{ and } j \text{ have an} \\ \quad \text{arc in common} \\ 0 \text{ otherwise} \end{cases} i, j = 1, \cdots, n.$$

From the elementary properties, each pair of nodes is at most joined by one path of length 2. The second

order adjacencies (i.e., the pairs of nodes joined by paths of length 2 without retracing any arcs) are given by $A^2 - dI$. Using the 0th, 1st and 2nd order adjacencies,

$$A^2 + A - (d - 1)I = J. \tag{3}$$

Since J is a polynomial in A, A and J have a common set of eigenvectors. One of these is u, and

$$Ju = nu, \qquad Au = du.$$

For this eigenvector, (3) supplies the relation which is already known,

$$(1 + d^2)u = nu.$$

Let v be any other eigenvector of A corresponding to eigenvalue r. Then

$$Jv = 0, \qquad Av = rv.$$

Using (3),

$$r^2 + r - (d - 1) = 0.$$

Hence A has two other distinct eigenvalues:

$$r_1 = (-1 + \sqrt{4d - 3})/2,$$
$$r_2 = (-1 - \sqrt{4d - 3})/2. \tag{4}$$

If d is such that r_1 and r_2 are not rational then each has multiplicity $(n - 1)/2$ as eigenvalue of A, since A is rational. Since the diagonal elements of A are 0, the sum of the eigenvalues of A is 0. Hence

$$d + \frac{(n - 1)}{2}(r_1 + r_2) = d - \frac{d^2}{2} = 0.$$

The values of d which satisfy this equation are:

$d = 0$, for which $n = 1$. This is a single node, which does not have diameter 2.

$d = 2$, for which $n = 5$. This is the pentagon, clearly a Moore graph of type (2, 2), and clearly the only one of that type.

The values of d for which the r's of (4) are rational are those for which $4d - 3$ is a square integer, s^2, since any rational eigenvalues of A are also integral. Let m be the multiplicity of r_1. Then the sum of the eigenvalues is

$$d + m \frac{s - 1}{2} + (n - 1 - m) \frac{-s - 1}{2} = 0.$$

Using $n - 1 = d^2$ and $d = (s^2 + 3)/4$,

$$s^5 + s^4 + 6s^3 - 2s^2 + (9 - 32m)s - 15 = 0. \tag{5}$$

Since (5) requires solutions in integers the only candidates for s are the factors of 15. The solutions are:

$$s = 1, \quad m = 0, \quad d = 1, \quad n = 2$$
$$s = 3, \quad m = 5, \quad d = 3, \quad n = 10$$
$$s = 5, \quad m = 28, \quad d = 7, \quad n = 50$$
$$s = 15, \quad m = 1729, \quad d = 57, \quad n = 3250.$$

There is no graph of degree 1 and diameter 2.

The case $d = 3$ is the Petersen graph, which may be drawn:

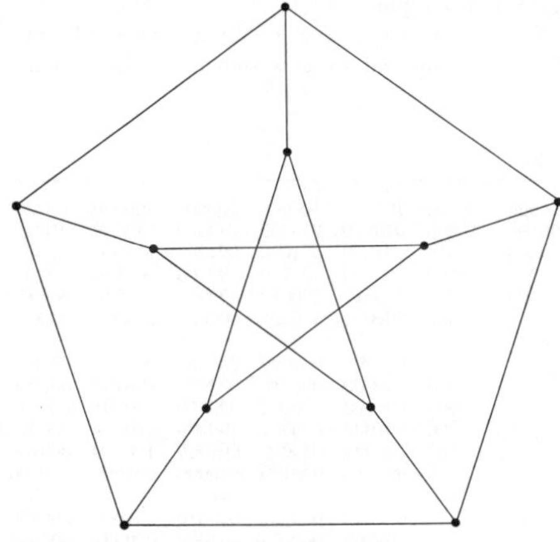

The case $d = 7$ has an exemplar which is shown later.

The case $d = 57$ is undecided.

The uniqueness of Moore graphs (3, 2) and (7, 2) is shown in the next section.

5. Uniqueness

Let the nodes be numbered as follows:

No. 0: any node,

Nos. 1 to d: the nodes adjacent to No. 0 in arbitrary order,

Nos. $(d + 1)$ to $(2d - 1)$: the nodes of S_1 in arbitrary order,

. . . .

Nos. $(i(d - 1) + 2)$ to $(i(d - 1) + d)$: the nodes of S_i in arbitrary order.

The adjacency matrix A then has the form of Fig. 2.

The P_{ij} are matrices of order $(d - 1)$, as indicated by the tabulation of the number of rows in each block. The argument will concern several of the principal submatrices of A. Consider first the principal submatrix of order $d(d - 1)$ in the lower right, outlined in heavy rules, which shows the adjacency

relations between the tier 2 nodes through the re-entering arcs. Let it be designated B. We give further form to B in the following theorems, which are rather obvious consequences of the hierarchy of Fig. 1.

• *Theorem 1*

No cycle of length less than 5 exists in the graph.

If there were such a cycle, designate one of its nodes as the distinguished node. Then equality would not hold in (1) for some i.

• *Theorem 2*

The diagonal blocks of B are 0.

Let two nodes of tier 2, a and b, be members of the same subset S_i. If they were adjacent, then a, b and the i-th node of tier one would form a cycle of length 3.

• *Theorem 3*

The blocks P_{ij} of B are permutation matrices.

Let node a be a member of S_i and b and c be members of S_j. If a were adjacent to both b and c, then a, b, c and the jth node of tier 1 would form a cycle of length 4. Hence any node of tier 2 is adjacent to at most one node in any of the subsets designated S_i. Since such a node is adjacent to $(d - 1)$ other nodes of tier 2 through the re-entering arcs, and

						No. of Rows
0	11...1	00...0	00...0	......	00...0	1
1	00...0	11...1	00...0	......	00...0	
1	00...0	00...0	11...1	......	00...0	
.	......	......	......	......	......	d
1	00...0	00...0	00...0	......	11...1	
0	10...0					
0	10...0					
.	......	0	P_{12}	......	P_{1d}	d-1
0	10...0					
0	01...0					
0	01...0					
..	......	P_{21}	0	......	P_{2d}	d-1
0	01...0					
.	......	......	......	......	......	
0	00...1					
0	00...1					
.	......	P_{d1}	P_{d2}	......	0	d-1
0	00...1					

Figure 2

since there are $(d - 1)$ subsets S_i other than the one of which it is itself a member, each node of tier 2 is adjacent to exactly one node in each of the other subsets. Hence each row and each column of each P_{ij} in B contains exactly one 1.

• *Theorem 4*

The nodes may be so numbered that $P_{1j} = P_{j1} = I$.

In arriving at the form for A shown in Fig. 2 no order was prescribed for the nodes within each S_i of tier 2. Let any order be given to the nodes of S_1. Each node of S_1 is adjacent to one node of each other subset. If each node of S_i is given the order number of its adjacent node in S_1, then $P_{1i} = I$.

Note that the orders of nodes in tier 1 and in S_1 are still arbitrary. This fact will be used later.

When the nodes are numbered so that A has the form of Fig. 2 with the further arrangement of Theorem 4, A is said to be in canonical form. By using the canonical form of A in (3) one finds that B satisfies

$$B^2 + B - (d - 1)I = J - K. \tag{6}$$

Then from Eq. (6),

$$\text{For} \quad j \neq 1, \qquad \sum_i P_{ij} = J_1;$$

$$\text{and} \quad P_{ik} + \sum_i P_{ii}P_{ik} = J \quad \text{if} \quad i \neq k. \tag{7}$$

An analysis similar to that given for A shows that the eigenvalues of B and their multiplicities are:

	eigenvalue	multiplicity
	2	1
$d = 3$	-1	2
	1	2
	-2	1
	6	1
$d = 7$	-1	6
	2	21
	-3	14
	56	1
$d = 57$	-1	56
	7	1672
	-8	1463

• *Theorem 5*

The Moore graph (3, 2) is unique.

In the canonical form

$$B = \begin{bmatrix} 0 & I & I \\ I & 0 & P \\ I & P' & 0 \end{bmatrix},$$

P cannot be I, for this would mean that $\sum_i P_{2i} = 2I$, violating (7). Hence

$$P = \begin{bmatrix} 0 & 1 \\ 1 & 0 \end{bmatrix}$$

and this is unique.

The submatrix B for a Moore graph of type (7, 2) in canonical form is shown in Fig. 3. Only

Figure 3

```
000000 100000 100000 100000 100000 100000 100000
 00000 010000 010000 010000 010000 010000 010000
  0000 001000 001000 001000 001000 001000 001000
   000 000100 000100 000100 000100 000100 000100
    00 000010 000010 000010 000010 000010 000010
     0 000001 000001 000001 000001 000001 000001

       000000 010000 001000 000100 000010 000001
        00000 100000 000010 000001 000100 001000
         0000 000100 100000 000010 000001 010000
          000 001000 000001 100000 010000 000010
           00 000001 010000 001000 100000 000100
            0 000010 000100 010000 001000 100000

              000000 000001 000010 001000 000100
               00000 000100 001000 000001 000010
                0000 000010 010000 100000 000001
                 000 010000 000001 000010 100000
                  00 001000 100000 000100 010000
                   0 100000 000100 010000 001000

                     000000 010000 000100 000010
                      00000 100000 001000 000001
                       0000 000001 010000 000100
                        000 000010 100000 001000
                         00 000100 000001 100000
                          0 001000 000010 010000

                            000000 000001 001000
                             00000 000010 000100
                              0000 000100 100000
                               000 001000 010000
                                00 010000 000001
                                 0 100000 000010

                                   000000 010000
                                    00000 100000
                                     0000 000010
                                      000 000001
                                       00 001000
                                        0 000100

                                          000000
                                           00000
                                            0000
                                             000
                                              00
                                               0
```


IBM JOURNAL • NOVEMBER 1960

the upper triangle is represented in the Figure, since B is symmetric. To show that by appropriate numbering of the nodes the adjacency matrix for any graph $(7, 2)$ may be made to correspond with that shown, and hence that there is only one such graph, requires several steps. We first show that all P_{ij} are involutions. As a preliminary:

• *Theorem 6*

The principal submatrix of A for type $(7, 2)$,

$$M = \begin{bmatrix} 0 & I & I \\ I & 0 & P_{23} \\ I & P'_{23} & 0 \end{bmatrix}$$

has an eigenvalue 2 of multiplicity 3.

The argument involves the invariant vector spaces corresponding to eigenvalue 2 of A, and some of the other principal submatrices of A. A set of vectors forming a basis for the invariant vector space of A corresponding to the characteristic root 2 is shown below. For notation, the components are segregated according to the blocks shown for A in Fig. 2. The first 8 components are written out and the last 42 components are shown as 7 vectors of dimension 6. The vectors are numbered at the left for ease of reference later.

The last 42 components of (I) form an eigenvector for B for eigenvalue 6, and the last 42 components of numbers (II) through (VII) are eigenvectors for B for eigenvalue -1. Hence there remain 21 independent vectors whose first eight components are 0 and whose last 42 components form a basis for the eigenspace corresponding to 2 of B. We symbolize these as

(VIII) $\quad 0\ 0\ 0\ 0\ 0\ 0\ 0\ 0\ v_1'\ v_2'\ v_3'\ v_4'\ v_5'\ v_6'\ v_7'.$

Because as eigenvectors of B they correspond to different eigenvalues,

$$u'v_i = 0.$$

We now consider the upper left principal submatrix L of A, of order 26, and the submatrix L^* of order 27 obtained through augmenting L by one column and the corresponding row,

$$L = \begin{array}{|c|ccccccc|c|c|c|}
\hline
0 & 1 & 1 & 1 & 1 & 1 & 1 & 1 & 0' & 0' & 0' \\
\hline
1 & & & & & & & & u' & 0' & 0' \\
1 & & & & & & & & 0' & u' & 0' \\
1 & & & & & & & & 0' & 0' & u' \\
1 & & & & 0 & & & & 0' & 0' & 0' \\
1 & & & & & & & & 0' & 0' & 0' \\
1 & & & & & & & & 0' & 0' & 0' \\
1 & & & & & & & & 0' & 0' & 0' \\
\hline
0 & u & 0 & 0 & 0 & 0 & 0 & 0 & 0 & I & I \\
\hline
0 & 0 & u & 0 & 0 & 0 & 0 & 0 & I & 0 & P_{23} \\
\hline
0 & 0 & 0 & u & 0 & 0 & 0 & 0 & I & P'_{23} & 0 \\
\hline
\end{array},$$

$$L^* = L + \begin{bmatrix} 0 \\ — \\ 0 \\ 0 \\ 0 \\ 0 \\ 1 \\ 0 \\ 0 \\ 0 \\ — \\ e_h \\ — \\ e_i \\ — \\ e_j \\ — \\ 0 \end{bmatrix}$$

where the augmenting column for L^* comes from the fourth block, and h, i and j are unspecified.

Number															
(I)	-14	-4	-4	-4	-4	-4	-4	-4	u'	u'	u'	u'	u'	u'	u'
(II)	0	3	-3	0	0	0	0	0	u'	$-u'$	$0'$	$0'$	$0'$	$0'$	$0'$
(III)	0	0	3	-3	0	0	0	0	$0'$	u'	$-u'$	$0'$	$0'$	$0'$	$0'$
(IV)	0	0	0	3	-3	0	0	0	$0'$	$0'$	u'	$-u'$	$0'$	$0'$	$0'$
(V)	0	0	0	0	3	-3	0	0	$0'$	$0'$	$0'$	u'	$-u'$	$0'$	$0'$
(VI)	0	0	0	0	0	3	-3	0	$0'$	$0'$	$0'$	$0'$	u'	$-u'$	$0'$
(VII)	0	0	0	0	0	0	3	-3	$0'$	$0'$	$0'$	$0'$	$0'$	u'	$-u'$

Figure 4

-14	8	-4	-4	-7	-7	-7	-7	$5u'$	u'	u'	$0'$	$0'$	$0'$	$0'$
0	3	-3	0	0	0	0	0	u'	$-u'$	$0'$	$0'$	$0'$	$0'$	$0'$
0	0	3	-3	0	0	0	0	$0'$	u'	$-u'$	$0'$	$0'$	$0'$	$0'$
0	0	0	0	0	0	0	0	v_1'	v_2'	v_3'	$0'$	$0'$	$0'$	$0'$

Since the eigenspace for eigenvalue 2 of A has dimension 28 (see the solutions of (5)) a subspace of this eigenspace of dimension at least 4 lies in the subspace corresponding to L. By inspection of the exhibited vectors a basis for such a 4-space is given above in Fig. 4 for some unspecified v_i.

If L be augmented by one column and row, as shown in L^* above, then a subspace of dimension at least 5 of eigenvectors for eigenvalue 2 of A lies in the subspace of L^*. The four vectors above, being characteristic vectors for A, are characteristic vectors for L^*.

A fifth vector for the basis of this 5-space is independent of the eigenvectors (IV), (V), (VI) and (VII) of A exhibited earlier, since any such dependence would introduce a component proportional to u in at least one of the last four blocks (last 24 components). But in the block containing the augmenting column the vector may have at most one nonzero component, and in the other blocks all its components are zero. Hence the fifth vector is of the form of (VIII)

$$0\ 0\ 0\ 0\ 0\ 0\ 0\ 0\ w_1'\ w_2'\ w_3'\ w_4'\ 0'\ 0'\ 0'.$$

But $u'w_i = 0$, and w_4 has at most one nonzero component. Hence $w_4 = 0$.

Of the five eigenvectors for L^* exhibited above, the two containing v's and w's are zero in all the components not corresponding to the principal submatrix M in the statement of the theorem. Hence they are eigenvectors for an eigenvalue 2 of M. They are mutually independent, and are also independent of (being orthogonal to) a vector

$$(u'\ u'\ u')$$

for M. The latter is, by inspection, an eigenvector for eigenvalue 2 of M. Hence the eigenvalue 2 of M has multiplicity at least 3. We can now show that P_{23} is an involution. For if we rewrite P_{23} as P, then M becomes

$$\begin{bmatrix} (0 & I & I) \\ (I & 0 & P) \\ (I & P' & 0) \end{bmatrix}.$$

Let us denote by x, y, z the three parts of a characteristic vector of M corresponding to 2. Then

$$y + z = 2x$$
$$x + Pz = 2y$$
$$x + P'y = 2z.$$

Substituting for x in the last two equations, we obtain

$$-3y + (I + 2P)z = 0$$
$$(I + 2P')y - 3z = 0.$$

So the multiplicity of 2 as an eigenvalue of M equals the multiplicity of 3 as an eigenvalue of

$$\begin{bmatrix} 0 & I + 2P \\ I + 2P' & 0 \end{bmatrix}.$$

Now any real matrix of the form

$$\begin{bmatrix} 0 & T \\ T' & 0 \end{bmatrix},$$

where all submatrices are square, has for its eigenvalues the square roots of the eigenvalues of TT' and their negatives. Hence the multiplicity of 2 as an eigenvalue of M is the multiplicity of 9 as an eigenvalue of

$$(I + 2P)(I + 2P') = 5I + 2(P + P').$$

Thus the multiplicity of 2 as an eigenvalue of M is the multiplicity of 2 as an eigenvalue of

$$P + P',$$

and it is clear that this is in turn equal to the number of disjoint cycles in P. So $P = P_{23}$ is composed of three disjoint cycles. Thus we have

- *Theorem 7*

In the canonical form for Moore graphs of type (7, 2) all P_{ij}, $i, j \neq 1$ and $i \neq j$, are involutions.

We adopt the notation

$$P_{ii} = 0.$$

- *Theorem 8*

In a Moore graph of type (7, 2) in canonical form

$$P_{ij}P_{jk}P_{ki} = P_{ik} \quad \text{if} \quad i \neq j, \quad i \neq k, \quad j,k \neq 1.$$

If $i = 1$ the theorem is trivial. We consider $i \neq 1$ and write the involutions as three transpositions.

Let $P_{ij} = (ab)(cd)(ef)$. In P_{ik} the companion of a must come from one of (cd) or (ef), and the companion of b from the other, because of (7). Let $P_{jk} = (ac)(be)(df)$, which is completely general. Then $P_{ij}P_{jk} = (aed)(bcf)$.

Since P_{ik} is in a row with P_{ij} and in a column with P_{jk} it may have no substitution of terms, which is the same as any substitution appearing in any of P_{ij}, P_{jk} or $P_{ij}P_{jk}$. The only involution with this property is $P_{ik} = (af)(bd)(ce)$. Evaluating the product $P_{ij}P_{jk}P_{ki}$ proves the theorem.

If all of i, j and k are different the expression in Theorem 8 may be multiplied on the left by P_{ik} and on the right by P_{ik}, and we have

• *Theorem 9*

$P_{ik}P_{ij}P_{ik} = P_{jk}$ *if* i, j, k *are all different and* j, $k \neq 1$.

• *Theorem 10*

$P_{ij} \neq P_{kl}$ *if* i, j, k *and* l *are all different.*

If any subscript is unity the theorem is trivial. For ease of presentation, we prove the theorem for a particular set of subscripts, none unity, but it obviously extends to the general case.
Suppose $P_{23} = P_{45}$. Then

$$P_{23}P_{34}P_{45} = P_{23}P_{34}P_{23} = P_{24}$$

by Theorem 9. Also,

$$P_{23}P_{34}P_{45} = P_{45}P_{34}P_{45} = P_{35}.$$

Hence $P_{24} = P_{35}$.

Similarly $P_{25} = P_{34}$.

Hence

$$P_{23} + P_{24} + P_{25} = P_{32} + P_{34} + P_{35}$$
$$= P_{42} + P_{43} + P_{45}.$$

Hence by Eq. (7),

$$P_{26} + P_{27} = P_{36} + P_{37} = P_{46} + P_{47}.$$

But $P_{26} + P_{27} = P_{36} + P_{37}$ implies $P_{36} = P_{27}$. Similarly, $P_{26} + P_{27} = P_{46} + P_{47}$ implies $P_{46} = P_{27}$. Therefore, $P_{36} = P_{46}$, violating (7).

• *Theorem 11*

The Moore graph (7, 2) *is unique.*

There are 15 different P_{ij}, $2 \leq i < j$. There are fifteen different involutions of order 6 without fixed points. Hence, for any Moore graph (7, 2) in canoni-

cal form, the involution (12)(34)(56) appears once. By an appropriate numbering of the nodes of tier 1 it may be brought to the P_{23} position.

Because of Eq. (7), in the remaining P_{2j}, $j > 2$, the first row of each is one of e'_3, e'_4, e'_5, e'_6, and each of these appears once. By an appropriate numbering of nodes 4, 5, 6, 7 of tier 1 the P_{2j} may be brought to the sequence of Fig. 3.

With $P_{23} = (12)(34)(56)$, because of Eq. (7), and the ordering of nodes of tier 1 already assigned, P_{24} might be only (13)(25)(46) or (13)(26)(45). The order of the fourth and fifth nodes of S_1 may be transposed, if necessary, to achieve (13)(25)(46). The remaining P_{2i} are then uniquely determined. The argument is similar to that used in Theorem 8.

The second row of B having been determined, all other P_{ij} are uniquely determined by the relation of Theorem 9, with $j = 2$. Hence, any (7, 2) graph may be numbered to have the adjacency matrix of Fig. 4.

6. Diameter 3

• *Theorem 12*

If the polynomial which is characteristic of Moore graphs of type (d, k), $k \geq 2$, *is irreducible in the field of rational numbers, then no such graphs exist unless* $d = 2$.

The polynomials $F_i(x)$ satisfy the difference equation

$$F_{i+1} = xF_i - (d - 1)F_{i-1}$$
$$F_1 = x + 1, \qquad F_2 = x^2 + x - (d - 1)$$

and the equation for the adjacency matrix for diameter k is

$$F_k(A) = J,$$

similar to (3). An adjacency matrix satisfying this equation has the number d as one of its eigenvalues, and it has exactly k distinct other eigenvalues which are the roots of the irreducible $F_k(x)$. Let those roots be r_i, $i = 1, \cdots, k$.

The first and second coefficients of F_k, $k \geq 2$, are both unity. Hence

$$\sum_{i=1}^{k} r_i = -1.$$

If F_k is irreducible its roots have equal multiplicity as eigenvalues of A. The number of nodes in a Moore graph of diameter k, if $d > 2$, is

$$n = 1 + d \frac{(d - 1)^k - 1}{d - 2}$$

and hence the multiplicity of each r_i is

$$m = d\,\frac{(d-1)^k - 1}{k(d-2)}.$$

Since the trace of A is 0,

$$d + m \sum_{i=1}^{k} r_i = 0.$$

Substituting for m, this reduces to

$$(d-1)^k - k(d-1) + (k-1) = 0.$$

Considering this as a polynomial in $(d-1)$, and remarking $k \geq 2$, by the rule of signs it has at most two positive roots. Since it has a double root at $d-1 = 1$, no $d > 2$ satisfies it.

Of course, $d = 2$ corresponds to the $(2k+1)$-gon, which is a Moore graph.

• *Theorem 13*

The only Moore graph of diameter 3 is $(2, 3)$.

The polynomial equation for $k = 3$ is

$$x^3 + x^2 - 2(d-1)x - (d-1) = 0.$$

If a graph $(d, 3)$ exists, $d > 2$, where d of course is an integer, then the above equation has at least one root which is an integer. Let r be such a root. Then

$$d - 1 = \frac{r^2(r+1)}{2r+1}.$$

Now $2r + 1$ is relatively prime to both r and $r + 1$. Hence the denominator is 1 or -1, and for both of these $d = 1$, but the type $(1, 3)$ does not exist.

Received April 12, 1960.

Reprinted from the
Amer. Math. Monthly, Vol. 70, No. 1 (1963), pp. 30–36

ON THE POLYNOMIAL OF A GRAPH

A. J. HOFFMAN, IBM Research Center, Yorktown Heights, N. Y.

1. Introduction. Several recent investigations in graph theory and studies of "association schemes" arising in the design of experiments (all references other than [1], [5], and [10]) have been variations on the following theme. For each pair of (not necessarily distinct) vertices i, j of a graph G, let $p_t(i, j)$ be the number of different paths in G from i to j of length t (we allow paths to be reentrant and cross themselves without restriction, and we also stipulate that $p_0(i, j) = \delta_{ij}$). The question pursued in these investigations is: given a positive integer n, and rational coefficients $a_0, a_1, \cdots, a_k$, to find all graphs G with n vertices such that

$$(1.1) \qquad a_0 p_0(i, j) + a_1 p_1(i, j) + \cdots + a_k p_k(i, j) = 1$$

for all pairs of vertices i, j of G.

This paper:

(i) points out that a graph satisfies (1.1) for some set of coefficients if and only if it is regular (the number of edges meeting each vertex is a constant) and connected, and suggests some appropriate terminology for considering (1.1);

(ii) characterizes bicolored (every cycle is of even length) regular and connected graphs by properties of the coefficients;

(iii) applies this characterization to the study of the graphs formed by the vertices and edges of the m-dimensional cube, for $m \leq 4$.

The intent of (ii) and (iii) is to exhibit simple instances of the methods used to investigate particular instances of (1.1), namely an interplay among properties of matrices, polynomials, and graphs.

2. Notation and Main Theorem. Let G be an unoriented* graph with vertices $1, \cdots, n$ with no edges from i to i, and at most one edge joining i and $j(i \neq j)$. Let A be the square matrix of order n, called the adjacency matrix of G, given by

$$A = (a_{ij}) = \begin{cases} 1 & \text{if } i \text{ and } j \text{ are joined by an edge} \\ 0 & \text{otherwise.} \end{cases}$$

Let u be the vector of order n, every entry of which is unity, and let J be the square matrix of order n, every entry of which is unity. Let d_i (called the degree of i) be the number of edges meeting vertex i. G is said to be regular (of degree d) if $d = d_i$ for each i.

THEOREM 1. *There exists a polynomial $P(x)$ such that*

$$(2.1) \qquad\qquad\qquad J = P(A)$$

if and only if G is regular and connected.

Note the equivalence of (2.1) with (1.1). E. C. Dade and K. Goldberg (and perhaps others) have known Theorem 1 for some time, but it does not seem to be in the literature. It would be worthwhile to find a proof which does not use the concept of characteristic root.

Proof. Assume (2.1). Then A commutes with J, hence $d_i = (i, j)$th entry of $AJ = (i, j)$th entry of $JA = d_j$, so G is regular. Further, if i and j are any vertices of G, there is, for some t, a nonzero number as the (i, j)th entry of A^t; otherwise, no linear combination of the powers of A could have 1 as the (i, j)th entry, and

---

* This hypothesis is not needed, but we use it to keep the discussion simple.

(2.1) would be false. Thus, for some t, there is at least one path of length t from i to j. But this means G is connected.

Conversely, assume G regular (of degree d) and connected. As we saw in the proof of necessity, because G is regular, A commutes with J. Thus, since A and J are symmetric commuting matrices, there exists an orthogonal matrix U such that

$$(2.2) \qquad J = UJ_0U^T, \qquad A = UA_0U^T,$$

where J_0 is a diagonal matrix whose diagonal entries are the eigenvalues of J, namely $(n, 0, 0, \cdots, 0)$, and A_0 is a diagonal matrix whose diagonal entries are the eigenvalues of A, namely $(\alpha_1, \cdots, \alpha_n)$. Now u is an eigenvector of both A and J, with d and n the corresponding eigenvalues, a consequence of the fact that G is regular of degree d. It is a classic result from the theory of matrices with nonnegative entries that, because G is connected, d is an eigenvalue of A of multiplicity 1 (also, an eigenvalue of largest absolute value; see [10]). Let $d, \beta_1, \cdots, \beta_k$ be the distinct eigenvalues of A, and let

$$(2.3) \qquad P(x) = \frac{n \prod_{i=1}^{k} (x - \beta_i)}{\prod_{i=1}^{k} (d - \beta_i)}.$$

Then $P(A_0) = J_0$, so (2.2) implies (2.1).

Let us call (2.3) the polynomial of the graph G, and say that the polynomial and graph belong to each other. It is clear that (2.3) is the polynomial of smallest degree for which (2.1) holds. Further, the distinct eigenvalues of A, other than d, are the roots of $P(x)$.

3. A lemma on bicolored graphs. A graph G is bicolored [5] if and only if it is possible so to number its nodes that the adjacency matrix is

$$(3.1) \qquad A = \begin{bmatrix} 0 & B \\ B^T & 0 \end{bmatrix},$$

where the 0's are square blocks along the diagonal. Of course, if G is regular, then the squares are of the same order, and B is a square matrix also of that order. To see this, it is sufficient to show that B is square. Suppose G regular, of degree d, and that B has p rows and q columns, then the number of 1's in B is pd (adding by rows) or qd (adding by columns). Hence, $p = q$.

LEMMA 1. *If G is a regular connected graph of degree d, and $P(x)$ is the polynomial of G, then G is a bicolored graph if and only if $P(-d) = 0$.*

Proof. It is clear that if A has the form (3.1), then the vector $(u; -u)$ is an eigenvector of A, corresponding to the eigenvalue $-d$. Conversely, suppose that $-d$ is an eigenvalue of A, where A is the adjacency matrix of a regular connected

 ON THE POLYNOMIAL OF A GRAPH

graph of degree d, and that $v = (v_1, \cdots, v_n)$ is a corresponding eigenvector. There is no loss of generality in assuming that the largest absolute value of the components of v is 1, and that $v_{i_0} = 1$. Since $\sum a_{i_0 j} v_j = -d$, it follows that $v_{i_1} = -1$ for all vertices i_1 joined to i_0 by an edge. Similarly, if i_2 is a vertex joined to an i_1 by an edge, then $v_{i_2} = +1$, and so on. Because G is connected, every coordinate of v is ± 1, and every edge of G joins two vertices such that the corresponding coordinates of v are different. But this is equivalent to saying that G is bicolored.

It is worth remarking, that if G is bicolored, then $(x-d)P(x)$ is an even function. This follows from Wielandt's observation [1] that the eigenvalues of A in (3.1) are the nonnegative square roots of the eigenvalues of BB' and their negatives.

4. The polynomials belonging to the graphs of low-dimensional cubes. Let Q_m be the graph of 2^m nodes and $m2^{m-1}$ arcs, $m \leq 4$, formed by the vertices and edges of the m-dimensional cube. The adjacency matrix of Q_m is of the form (3.1), with

$$
(4.1) \quad
\begin{cases}
\text{(a)} \quad B = \begin{bmatrix} 1 & 1 \\ 1 & 1 \end{bmatrix} & \text{in case } m = 2 \\[2em]
\text{(b)} \quad B = \begin{bmatrix} 1 & 1 & 1 & 0 \\ 1 & 1 & 0 & 1 \\ 1 & 0 & 1 & 1 \\ 0 & 1 & 1 & 1 \end{bmatrix} & \text{in case } m = 3 \\[4em]
\text{(c)} \quad B = \begin{bmatrix} 1 & 1 & 1 & 1 & 0 & 0 & 0 & 0 \\ 0 & 0 & 0 & 0 & 1 & 1 & 1 & 1 \\ 1 & 1 & 0 & 0 & 1 & 1 & 0 & 0 \\ 0 & 0 & 1 & 1 & 0 & 0 & 1 & 1 \\ 1 & 0 & 1 & 0 & 1 & 0 & 1 & 0 \\ 0 & 1 & 0 & 1 & 0 & 1 & 0 & 1 \\ 1 & 0 & 0 & 1 & 0 & 1 & 1 & 0 \\ 0 & 1 & 1 & 0 & 1 & 0 & 0 & 1 \end{bmatrix} & \text{in case } m = 4.
\end{cases}
$$

It is also easy to calculate that the polynomial $P(x)$ of smallest degree such that $J = P(A)$, i.e., the polynomial of the graph, is

$$
(4.2) \quad
\begin{cases}
\text{(a)} \quad P_2(x) = \dfrac{x^2}{2} + x & \text{in case } m = 2 \\[2em]
\text{(b)} \quad P_3(x) = \dfrac{x^3}{6} + \dfrac{x^2}{2} - \dfrac{x}{6} - \dfrac{1}{2} & \text{in case } m = 3 \\[2em]
\text{(c)} \quad P_4(x) = \dfrac{x^4}{24} + \dfrac{x^3}{6} - \dfrac{x^2}{6} - \dfrac{2x}{3} & \text{in case } m = 4.
\end{cases}
$$

The question we now raise for consideration is: for $m = 2$, 3, 4 respectively, do 2^m (the number of vertices of Q_m) and $P_m(x)$, as given by (4.2), characterize Q_m? Is there possibly another graph with 2^m vertices belonging to $P_m(x)$?

5. The graphs with 2^m vertices belonging to $P_m(x)$ $(m = 2, 3, 4)$.

THEOREM 2. *For $m = 2$ or 3, Q_m is the only graph with 2^m vertices belonging to $P_m(x)$. But Q_4 and exactly one other graph of 16 vertices (given by (5.1) and (5.5) below) belong to $P_4(x)$.*

Proof. We begin with some general considerations. Let H_m be any graph with 2^m nodes belonging to $P_m(x)$. Let d_m be the degree of H_m. Then, by (2.3), d_m is the unique positive x satisfying $P_m(x) = 2^m$, namely (as can be seen from (4.2)), $d_m = m$. In particular, $d_2 = 2$ and it follows at once from (4.2) (a) that $H_2 = G_2$.

Secondly, it also follows from (4.2) that $P_m(-m) = 0$, hence (Lemma 1) H_m is a bicolored graph. It follows that if C is the adjacency matrix of H_m, we may assume

$$(5.1) \qquad\qquad C = \begin{bmatrix} 0 & D \\ D^T & 0 \end{bmatrix}.$$

It is clear that the even powers of C can be nonzero only in the two blocks along the diagonal, the odd powers of C can be nonzero only in the two off-diagonal blocks.

If we let

$$(5.2) \qquad\qquad F = DD^T,$$

it follows from (4.2) (b) and (c) that

$$(5.3) \qquad \begin{array}{lll} \text{(b)} & 2J = F - 1 & \text{in case } m = 3 \\ \text{(c)} & 24J = F^2 - 4F & \text{in case } m = 4 \end{array}$$

when F and J are of order 4 in case $m = 3$, and 8 in case $m = 4$.

Now it is obvious from (5.2) and (5.3) (b) that the only possibility for D in case $m = 3$ is a matrix which, under arbitrary permutations of rows and columns, is of the form (4.1) (b). Hence, $H_3 = G_3$.

There remains the case $m = 4$. Because $d_4 = 4$, we know that the diagonal entries of F are 4, and the sum of the entries in any row of F is 16. From (5.3) (c), we know that each diagonal element of F^2 is 40. Hence, since F is symmetric, the sum of the squares of the elements in any row of F is 40. It follows from these facts that the off-diagonal elements of any row of F are nonnegative integers whose sum is 12, and the sum of whose squares is 24. There are only two possibilities:

$$(5.4) \qquad \begin{array}{lll} \text{(a)} & 2,\ 2,\ 2,\ 2,\ 2,\ 2,\ 0, & \text{and} \\ \text{(b)} & 3,\ 2,\ 2,\ 2,\ 1,\ 1,\ 1. \end{array}$$

We first show that (5.4) (a) and (5.4) (b) cannot occur in the same F. It is no loss of generality to assume that the first row of F is

$$4 \quad 2 \quad 2 \quad 2 \quad 2 \quad 2 \quad 2 \quad 0$$

and the second row contains a 3. If the 3 were in the last place, then the inner product of the two rows would be 30, whereas (5.3) (c) requires that it be 32. Hence, there is no loss of generality in having the second row begin

$$2, \ 4, \ 3.$$

Then the only way the remaining elements of the second row can be placed so that the inner product of the two rows will be 32 is so the first two rows look like

$$4 \quad 2 \quad 2 \quad 2 \quad 2 \quad 2 \quad 2 \quad 0$$
$$2 \quad 4 \quad 3 \quad 1 \quad 1 \quad 1 \quad 2 \quad 2.$$

Then it is easy to see that the first three rows are

$$4 \quad 2 \quad 2 \quad 2 \quad 2 \quad 2 \quad 2 \quad 0 \qquad\qquad 4 \quad 2 \quad 2 \quad 2 \quad 2 \quad 2 \quad 2 \quad 0$$
$$2 \quad 4 \quad 3 \quad 1 \quad 1 \quad 1 \quad 2 \quad 2 \quad \text{or} \quad 2 \quad 4 \quad 3 \quad 1 \quad 1 \quad 1 \quad 2 \quad 2$$
$$2 \quad 3 \quad 4 \quad 1 \quad 1 \quad 1 \quad 2 \quad 2 \qquad\qquad 2 \quad 3 \quad 4 \quad 1 \quad 1 \quad 2 \quad 1 \quad 2.$$

But then the inner product of the second and third row is 39 or 38, whereas (5.3) (c) requires that it be 36.

Hence, F is composed entirely of rows whose off-diagonal elements are all of the form (5.4) (a) or all of the form (5.4) (b). It is then possible to show by an extensive use of (5.3) (c) that, apart from renumbering of the vertices, either $D = (4.1)$ (c) or:

$$(5.5) \qquad D = \begin{bmatrix} 1 & 1 & 1 & 1 & 0 & 0 & 0 & 0 \\ 1 & 1 & 1 & 0 & 1 & 0 & 0 & 0 \\ 1 & 0 & 0 & 1 & 0 & 1 & 1 & 0 \\ 0 & 1 & 0 & 1 & 0 & 1 & 0 & 1 \\ 0 & 0 & 1 & 1 & 0 & 0 & 1 & 1 \\ 1 & 0 & 0 & 0 & 1 & 1 & 1 & 0 \\ 0 & 1 & 0 & 0 & 1 & 1 & 0 & 1 \\ 0 & 0 & 1 & 0 & 1 & 0 & 1 & 1 \end{bmatrix}$$

In the former case, $H_4 = Q_4$. In the latter case, we can verify that the graph H_4 does satisfy (4.2) (c), and it is obviously different from Q_4.

We are grateful for stimulating conversations with J. H. Griesmer, E. F. Moore and R. R. Singleton.

References

1. A. R. Amir-Moez, and A. L. Fass, Elements of Linear Spaces, Edwards Brothers, Ann Arbor, Mich., 1961, p. 139.

2. L. C. Chang, The Uniqueness and Nonuniqueness of the Triangular Association Scheme, Science Record 3 (1959) 604–613.

3. ———, Association Schemes of Partially Balanced Block Designs with Parameters $v = 28$, $n_1 = 12$, $n_2 = 15$ and $p_{11}^2 = 4''$, Science Record 4 (1960) 12–18.

4. W. S. Connor, The Uniqueness of the Triangular Association Scheme, Annals of Mathematical Statistics 29 (1958) 262–266.

5. F. Harary, Structural Duality, Behavioral Science 2 (1957) 255–265.

6. A. J. Hoffman, On the Uniqueness of the Triangular Association Scheme, Ann. Math. Statis. 31 (1960) 492–497.

7. ———, On the Exceptional Case in a Characterization of the Arcs of a Complete Graph, IBM J. Res. Develop. 4 (1961) 487–496.

8. A. J. Hoffman and R. R. Singleton, On Moore Graphs With Diameters 2 and 3, IBM J. Res. Develop. 4 (1960) 497–504.

9. D. M. Mesner, An Investigation of Certain Combinatorial Properties of Partially Balanced Incomplete Block Designs and Association Schemes, With a Detailed Study of Designs of Latin Squares and Related Types, unpublished thesis, Michigan State University, 1956.

10. O. Perron, Zur Theorie der Matrizen, Math. Ann. 64 (1907) 248–263.

11. S. S. Shrikhande, On a Characterization of the Triangular Association Scheme, Ann. Math. Statis. 30 (1959) 39–47.

12. ———, The Uniqueness of the L_2 Association Scheme, *ibid.* 30 (1959) 781–798.

Reprinted from the
Trans. Amer. Math. Soc., Vol. 116, issue 4 (1965), pp. 238–252

ON THE LINE GRAPH OF A SYMMETRIC BALANCED INCOMPLETE BLOCK DESIGN

BY

A. J. HOFFMAN[1] AND D. K. RAY-CHAUDHURI

1. **Introduction.** We shall study the relations between an infinite family of finite graphs and the eigenvalues of the corresponding adjacency matrices. All graphs we consider are undirected, finite, with at most one edge joining a pair of vertices, and with no edge joining a vertex to itself. Also, they are all connected and regular (every vertex has the same valence). If G is a graph, its adjacency matrix $A = A(G)$ is given by

$$a_{ij} = \begin{cases} 1 \text{ if } i \text{ and } j \text{ are adjacent vertices,} \\ 0 \text{ otherwise.} \end{cases}$$

The line graph $L(G)$ (also called the interchange graph, and the adjoint graph) of a graph G is the graph whose vertices are the edges of G. With two vertices of $L(G)$ adjacent if and only if the corresponding edges of G are adjacent.

There have been several investigations in recent years of the extent to which a regular connected graph is characterized by the eigenvalues of its adjacency matrix, especially in the case of line graphs (see [4] for a bibliography, and [2]). Most germane to the present investigation is the result of [4], which we now briefly describe.

Let Π be a finite projective plane with $n + 1$ points on a line. We regard Π as a bipartite graph with $2(n^2 + n + 1)$ vertices, which are all points and lines of Π, with two vertices adjacent if and only if one is a point, the other is a line, and the point is on the line. Let $L(\Pi)$ be the line graph of Π. A useful way of visualizing $L(\Pi)$ is to imagine its vertices as the 1's in the incidence matrix of Π (see [4]), with two 1's corresponding to adjacent vertices if and only if they are in the same row or column of the incidence matrix. Then $L(\Pi)$ is a regular connected graph with $(n + 1) \cdot (n^2 + n + 1)$ vertices whose adjacency matrix has

(1.1) $$2n, \, -2, \, n - 1 \pm \sqrt{n}$$

as its distinct eigenvalues. It is shown in [4] that any regular connected graph on $(n + 1)(n^2 + n + 1)$ vertices whose distinct eigenvalues are given by (1.1) must be isomorphic to the line graph of a plane Π with $n + 1$ points on a line. (It is, of course, impossible for (1.1) to distinguish

---

Received by the editors May 13, 1964.

[1] This research was supported in part by the Office of Naval Research under Contract No. Nonr 3775(00), NR 047070.

between nonisomorphic planes of the same order n.)

In this paper we generalize this result to symmetric balanced incomplete block designs (also called λ-planes). An SBIB $\Pi(v, k, \lambda)$ can be conceived as a bipartite graph on $v + v$ vertices, each vertex having valence k, with any two vertices in the same part adjacent to exactly λ vertices of the other part. It is assumed that $0 < \lambda < k < v$, and it is well known that

$$(1.2) \qquad \lambda = \frac{k(k-1)}{v-1}.$$

Just as in $[4]$, one readily shows (see §4) that $L(\Pi)$ is a regular connected graph on vk vertices, and its adjacency matrix has

$$(1.3) \qquad 2k - 2, \, -2, \, k - 2 \pm \sqrt{(k - \lambda)}$$

as its distinct eigenvalues. We then raise the question: if H is a regular connected graph on vk vertices, with (1.3) as the distinct eigenvalues of its adjacency matrix, is H isomorphic to some $L(\Pi(v, k, \lambda))$?

The answer is yes, unless $v = 4$, $k = 3$, $\lambda = 2$, in which case there is exactly one exception.

2. Outline of proof.

A (three-fingered) *claw* is a graph consisting of four vertices $0, 1, 2, 3$ such that 0 is adjacent to $1, 2, 3$ but i is not adjacent to j $(i, j = 1, 2, 3)$. We shall denote such a claw by the notation $(0; 1, 2, 3)$. It is clear that a line graph contains no claw, and, conversely, if we can show under suitable hypotheses that H contains no claw, then the remainder of the proof that $H \cong L(\Pi)$ will be quite straightforward. Our central problem then is to prove H contains no claw.

Let $A = A(H)$, and consider the matrix

$$(2.1) \qquad B = A^2 - (2k - 2)I - (k - 2)A.$$

We shall show below in §4 that, for each i,

$$(2.2) \qquad \sum_j b_{ij}(b_{ij} - 1) = 2(\lambda - 1)(k - 1).$$

Consider also

$$(2.3) \qquad C = A^2 - (2k - 2)I - (k - 2)A - (J - I - A),$$

where J is a matrix of all 1's.

We shall show in §4 that, for each i,

$$(2.4) \qquad \sum_j c_{ij}(c_{ij} - 1) = 2(v - k)(k - \lambda).$$

After further preliminaries, we consider the case when we assume that H is *edge regular* (i.e., every edge is contained in the same number of triangles). With this additional hypothesis, the nonexistence of claws is readily established, the only case requiring any effort being $k = 4$. Next, we consider the case when H is *not* edge regular. Then claws must exist satisfying certain properties. But we show that, apart from the exception cited in the introduction, these claws would produce violations of (2.2) or (2.4). These violations are the result of a counting process, and the counting is facilitated by showing that certain graphs cannot be subgraphs of H. (The discussion of the edge regular case also uses the nonexistence of some subgraphs.) A list of the "impossible" subgraphs is given in §3, and we now explain the principles used in proving these subgraphs impossible. They are based on elementary facts about eigenvalues and eigenvectors of symmetric matrices.

The first principle is: if K is a subgraph of H, if $M = M(K)$ is the adjacency matrix of K, if -2 is an eigenvalue of M, and x the corresponding eigenvector, then the sum of the coordinates of x must be zero.

The reason is as follows. Let y be the row vector with vk components obtained by adjoining to the vector x additional coordinates all zero. It easily follows that $yAy^T = -2yy^T$. Since -2 is the minimum eigenvalue of A, y is an eigenvector of A corresponding to the eigenvalue -2. Now $2k - 2$ is also an eigenvalue of A, corresponding to the eigenvector $(1, 1, \cdots, 1)$ (see [3] for a brief justification). In a symmetric matrix, two eigenvectors corresponding to different eigenvalues must be orthogonal. Hence, y must be orthogonal to $(1, 1, \cdots, 1)$, i.e., the sum of the coordinates of x is 0.

Thus, the graph

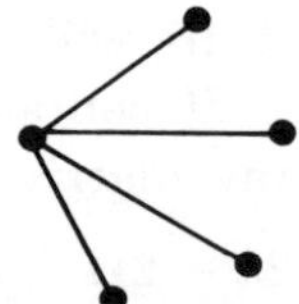

whose corresponding adjacency matrix is

$$(2.5) \qquad \begin{pmatrix} 0 & 1 & 1 & 1 & 1 \\ 1 & 0 & 0 & 0 & 0 \\ 1 & 0 & 0 & 0 & 0 \\ 1 & 0 & 0 & 0 & 0 \\ 1 & 0 & 0 & 0 & 0 \end{pmatrix},$$

cannot be a subgraph of H, since $(-2, 1, 1, 1, 1)$ is an eigenvector of (2.5), with -2 the corresponding eigenvalue.

Our second principle is that, if the sum of the coordinates of x is 0, then, if a is any vertex of H not in K, the sum of the coordinates of x corresponding to vertices of K adjacent to a must be 0. The proof is a direct application of the minimum characterization of the least eigenvalue.

This principle makes the following graph impossible:

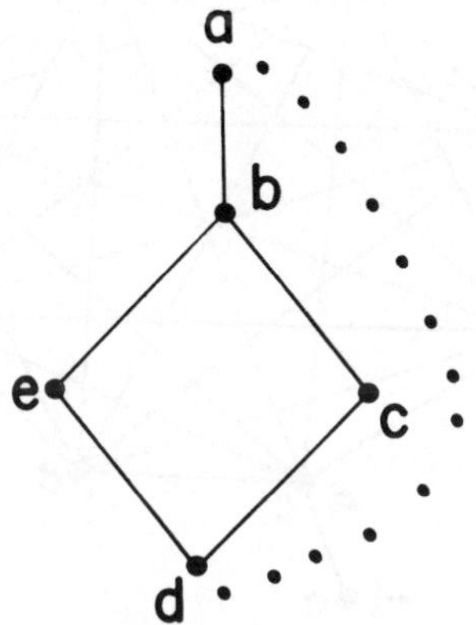

where the dotted line indicates that there may or may not be an edge of H. For the graph K whose vertices are b, c, d, e has $(1, -1, 1, -1)$ as eigenvector of the corresponding adjacency matrix.

3. **Impossible subgraphs.** We first list some subgraphs impossible because of the first principle. Accompanying each vertex will be a latin letter (for later reference) and a number giving the coordinate of the corresponding eigenvector.

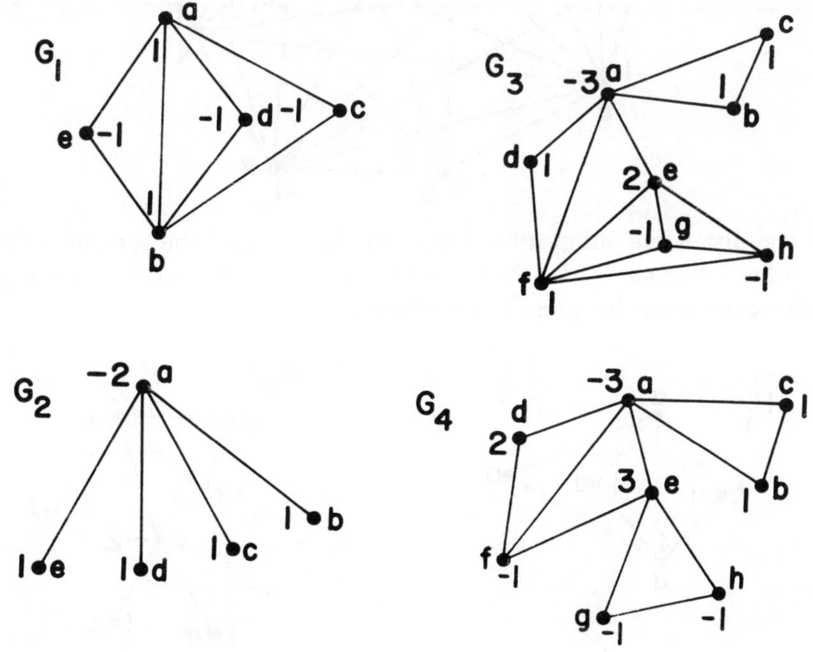

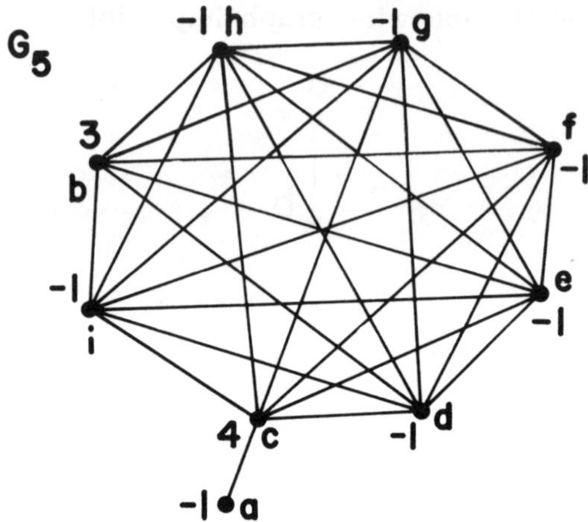

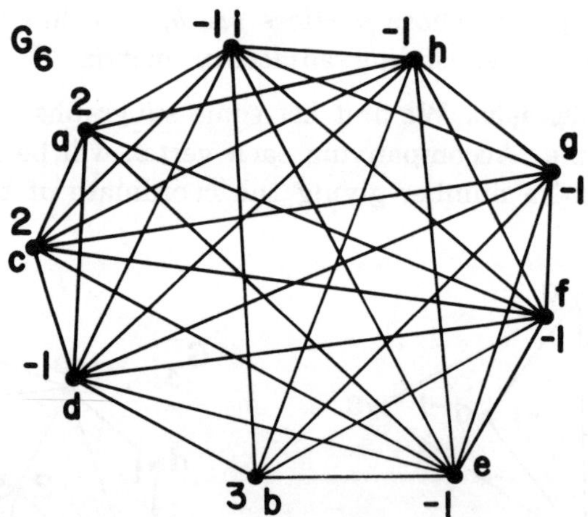

We now list some subgraphs impossible because of the second principle. The "other" vertex is denoted by the letter α. A dotted line signifies that the edge may be present or absent.

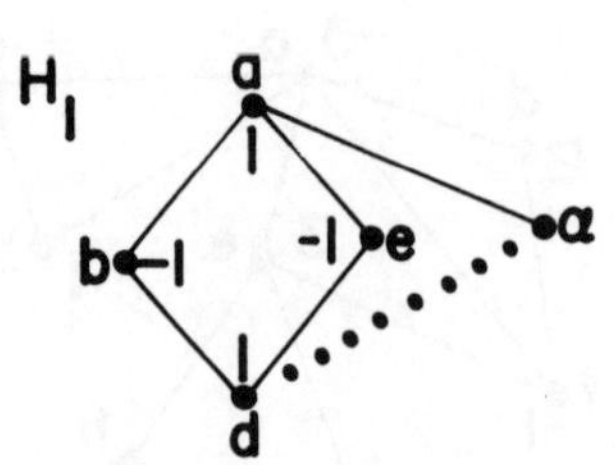

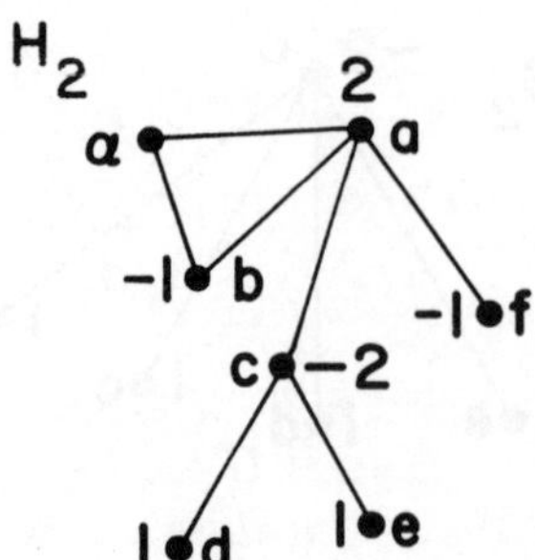

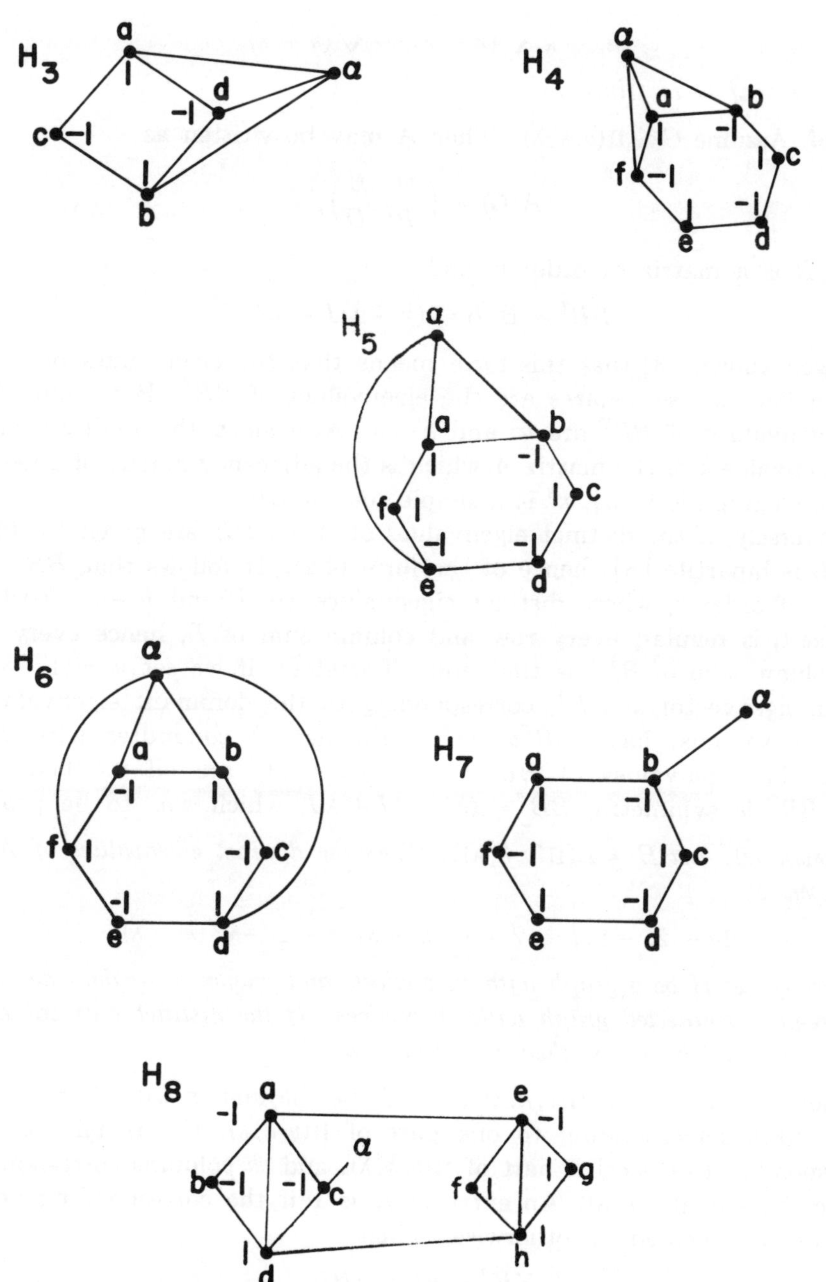

4. Some preliminaries on matrices. We begin with two lemmas.

LEMMA 4.1. *Let G be a regular connected graph on $2v$ vertices, $A = A(G)$. The distinct eigenvalues of A are given by*

$$(4.1) \qquad k, \; -k, \; \sqrt{(k-\lambda)}, \; -\sqrt{(k-\lambda)}$$

if and only if G is $\Pi(v, k, \lambda)$.

Proof. Assume $G \cong \Pi(v, k, \lambda)$. Then A may be written as

$$(4.2) \qquad A(G) = \begin{pmatrix} O & B \\ B^T & O \end{pmatrix},$$

where B is a matrix of order v, and

$$(4.3) \qquad BB^T = B^T B = (k - \lambda)I + \lambda J.$$

It is well known [3] that this form means that the eigenvalues of A are the numbers whose squares are the eigenvalues of BB^T. But from (4.3), the eigenvalues of BB^T are k^2 and $k - \lambda$. Also since the multiplicity of the eigenvalue k of the matrix A which is the adjacency matrix of a regular connected graph is 1 [3], k^2 is a simple root of BB^T.

Conversely, if the distinct eigenvalues of $A = A(G)$ are given by (4.1), then G is bipartite [3], hence of the form (4.2). It follows that BB^T is a matrix of order v, whose distinct eigenvalues are k^2 and $k - \lambda$. Further, because G is regular, every row and column sum of B, hence every row and column sum of BB^T is the same. Therefore, if we set $u = (1, \cdots, 1)$, u is an eigenvector of BB^T, corresponding to the dominant eigenvalue of BB^T, so we must have $BB^T u = k^2 u$. Further, J commutes with BB^T. Hence the eigenvalues of $BB^T - ((k - \lambda)I + \lambda J)$ are all 0. Therefore, since BB^T is symmetric $BB^T = (k - \lambda)I + \lambda J$, which was to be proven.

LEMMA 4.2. *Let $H = L(\Pi(v, k, \lambda))$. Then the distinct eigenvalues of $A(H)$ are given by*

$$(4.4) \qquad 2k - 2, \; -2, \; k - 2 + \sqrt{(k - \lambda)}, \; k - 2 - \sqrt{(k - \lambda)}.$$

Conversely, let H be a graph with vk vertices, and known to be the line graph of a regular connected graph with $2v$ vertices. If the distinct eigenvalues of $A(H)$ are given by (4.4), then $H \cong L(\Pi(v, k, \lambda))$.

Proof. Assume $H = L(\Pi(v, k, \lambda))$. Let K be the matrix with $2v$ rows (the first v rows corresponding to one part of $\Pi(v, k, \lambda)$, the remaining rows corresponding to the other part of $\Pi(v, k, \lambda)$, and vk columns corresponding to the edges of $\Pi(v, k, \lambda)$. An entry in K is 1 if the corresponding vertex and edge are incident, 0 otherwise. Then

$$(4.5) \qquad KK^T = kI + A(G),$$

where $A(G)$ is as in (4.2).

$$(4.6) \qquad K^T K = 2I + A(H).$$

The nonzero eigenvalues of KK^T and $K^T K$ are the same. Further, 0 is an eigenvalue of $K^T K$, since K has more columns than rows, and 0 is an eigenvalue of KK^T since the sum of the first v rows of K minus the sum of the last v rows is 0. Hence KK^T and K^TK have exactly the same set of distinct eigenvalues. From (4.5), (4.6) and (4.1), we infer (4.4).

Conversely, if H is the line graph with vk vertices, if $H = L(G)$, where G has $2v$ vertices, and if the distinct eigenvalues $A(H)$ are given by (4.4), then the above discussion is completely reversible unless the rows of K are *not* linearly dependent, (i.e., G is not bipartite). This would mean that $A(G)$ has for its distinct eigenvalues k, $\sqrt{(k-\lambda)}$ and $-\sqrt{(k-\lambda)}$. Then the polynomial of G (see [3]) would be

$$\frac{2}{\lambda}(x^2 - (k - \lambda)),$$

so

$$(A(G))^2 - (k - \lambda)I = \frac{\lambda}{2}J.$$

In other words, the diagonal element of $(A(G))^2$ would be $k - \lambda + \lambda/2$. But since G is regular and k is the dominant eigenvalue of $A(G)$, every row sum of $A(G)$ must be k. Therefore, every diagonal element of $(A(G))^2$ must be k. This is a contradiction.

Henceforth, we assume H is a regular connected graph with vk vertices, and $A(H)$ has (4.4) as its distinct eigenvalues. We also write $A = A(H)$.

LEMMA 4.3. *The matrix A satisfies the equation*

$$(4.7)\,(A + 2I)(A - (k - 2 + \sqrt{(k - \lambda)})I)(A - (k - 2 - \sqrt{(k - \lambda)})I) = 2\lambda J.$$

Proof. See [3].

LEMMA 4.4. *If B is defined by (2.1), then (2.2) holds. If C is defined by (2.3), then (2.4) holds.*

Proof. It is clear that (2.2) and (2.4) can be established if we can calculate the diagonal element of any power of A, and of any power of A multiplied by J. Since the row sums of A are all $(2k - 2)$, it follows that $A^t J = (2k - 2)^t J$. Also, we know the diagonal entries of I, A, and A^2. Since the left side of (4.7) is a third degree polynomial in A, we can calculate the diagonal entires of A^3. Multiplying (4.7) by A, we can then calculate the diagonal elements of A^4.

5. **Some preliminaries on claws.** In this section, we assume a claw $0, 1, 2, 3$, in the form

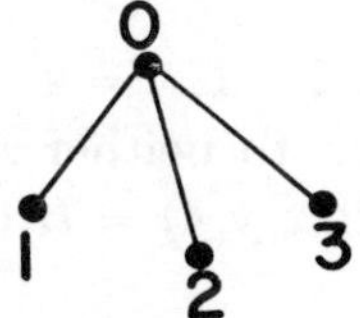

The subgraph of H determined by the vertices $x_1, \cdots, x_n$ will be written as $H(x_1, \cdots, x_n)$. We define

$$S_i = \{x \mid x \text{ is adjacent to } 0 \text{ and } i, \text{ but not to } j \text{ or } k\}.$$

$$S_{ij} = \{x \mid x \text{ is adjacent to } 0, i \text{ and } j, \text{ but not to } k\}.$$

Note that no vertex is adjacent to $0,1,2,3$. If a vertex (say 4) is adjacent to $0,1,2,3$, then $H(0,1,2,3,4) = G_1(a,c,d,e,b)$. (The equality $H(0,1,2,3,4) = G_1(a,c,d,e,b)$ means that the graph G_1 is the same as the graph $H(0,1,2,3,4)$ and the vertices $(0,1,2,3,4)$ are respectively identified with the vertices (a,c,d,e,b).) Also, every vertex other than $1,2,3$ adjacent to 0 must be in some S_i or some S_{ij}. Otherwise, we would have graph G_2. Thus, using $|S|$ to denote the number of elements in S, we have.

$$(5.1) \qquad\qquad \sum |S_i| + \sum |S_{ij}| = 2k - 5,$$

since the valence of 0 is $2k - 2$.

We also define

$$\widetilde{S}_i = \{x \mid x \text{ is adjacent to } i, \text{ but not to } 0\}.$$

Note that $x \in \widetilde{S}_i$ implies x is not adjacent to either j or k, for if x is adjacent to both i and j, then

$$H(0, x, i, j, k) = H_1(a, d, b, c, \alpha).$$

We define a clique to be a graph of which each pair of vertices is adjacent.

LEMMA 5.1. *If $S_i \neq \emptyset$, and $j \neq i$, then $\widetilde{S}_j$ is a clique.*

Proof. Assume otherwise. Then there are two vertices, say 4 and 5, in $\widetilde{S}_j$, which are not adjacent. Let 6 be any vertex in S_i. Since $\{0; 6, j, k\}$ is a claw, by a previous remark, neither 4 nor 5 is adjacent to 6. It follows that $H\{0,i,j,k,4,5,6\} = H_2\{a,b,c,f,d,e,\alpha\}$.

LEMMA 5.2. *If $S_{ij} \neq \emptyset$, and $k \neq i,j$, then $S_k = \emptyset$.*

Proof. To fix ideas assume $k = 3$, and let $4 \in S_3$, $5 \in S_{12}$. Since there are at least three vertices (namely $1,3,4$) adjacent to 0 but not to 2, there must be at least 3 vertices $6, 7$, and 8 in $\widetilde{S}_2$. They form a clique, by Lemma 5.1. Either 5 is adjacent to at least two vertices in $\widetilde{S}_2$ (say 6 and 7), or 5 is not adjacent to at least two vertices in $\widetilde{S}_2$ (say $6'$ and $7'$). In the former case, the graph $H(0,1,2,3,4,5,6,7) = G_3(a,d,e,b,c,f,g,h)$. In the latter case, the graph $H(0,1,2,3,4,5,6',7') = G_4(a,d,e,b,c,f,g,h)$.

LEMMA 5.3. $|S_{ij}| \leq 10$.

Proof. To fix ideas, assume $i = 1$, $j = 2$, and let $4 \in \widetilde{S}_2$. Now let $x, y, \in S_{12}$, and assume 4 is adjacent to neither x nor y. Then x must be adjacent to y, otherwise $H(1,2,x,y,4) = H_1(a,d,b,c,\alpha)$. Next, assume 4 is adjacent to both x and y. Then x must be adjacent to y, otherwise $H(1,2,x,y,4) = H_3(a,b,c,d,\alpha)$. To summarize, the subset S_{12}^* of S_{12} consisting of vertices each of which is not adjacent to 4 must be a clique; the subset S_{12}^{**} consisting of vertices each of which is adjacent to 4 is also a clique. If $|S_{12}| > 10$, then, since $S_{12} = S_{12}^* \cup S_{12}^{**}$, either $|S_{12}^*| \geqq 6$ or $|S_{12}^{**}| \geqq 6$.

In the former case $H(4,1,2,5,6,7,8,9,10) = G_5(a,b,c,d,e,f,g,h,i)$ where $5,6,7,8,9$, and 10 are 6 vertices of S_{12}^*. In the latter case

$$H(4,1,2,5,6,7,8,9,10) = G_6(a,b,c,d,e,f,g,h,i)$$

where $5,6,7,8,9$ and 10 are 6 vertices of S_{12}^{**}.

6. **The nonexistence of claws in the edge regular case.** The graph is called edge regular if every edge is contained in the same number of triangles. Since H is edge regular, and each diagonal entry of A^3 is $2(k-1)(k-2)$ (from 4.7), it follows that every edge of H is contained in exactly $k-2$ triangles. Assume a claw as in §4.

LEMMA 6.1. *For each i, $\widetilde{S}_i$ contains two nonadjacent vertices.*

Proof. To fix ideas, take $i = 3$. Since the valence of 3 is $2k-2$, $\widetilde{S}_3$ must contain $k-1$ vertices. We assume they form a clique, and will establish a contradiction. Because $\widetilde{S}_3$ is a clique, each edge joining 3 to a vertex in $\widetilde{S}_3$ is contained in $k-2$ triangles where the third vertex is in $\widetilde{S}_3$. Consequently, by the edge regularity, $0,3$, and all vertices adjacent to 0 and 3 form a clique. In turn, this implies that all vertices adjacent to 0 but not to 3 form a clique. But 1 and 2 are adjacent to 0 and not to 3, yet 1 and 2 are not adjacent to each other.

LEMMA 6.2. *If $k \neq 4$, then H contains no claw.*

Proof. By Lemma 6.1, each $\widetilde{S}_i$ contains two vertices not adjacent to each other. By Lemma 5.1, this means each S_i is empty. Using the edge regularity condition on the edges $(0,1)$ $(0,2)$ and $(0,3)$, and adding, we have

$$2 \sum |S_{ij}| = 3(k-2) = 3k - 6.$$

By (5.1),

$$2 \sum |S_{ij}| = 4k - 10.$$

Therefore, $k = 4$.

LEMMA 6.3. *If $k = 4$, then H contains no claw.*

Proof. Assume $k = 4$, and we have the claw $0,1,2,3$. By edge regularity, $|S_{12}| = |S_{13}| = |S_{23}| = 1$, and let $\{4\} = S_{12}$, $\{5\} = S_{13}$, $\{6\} = S_{23}$. We then have

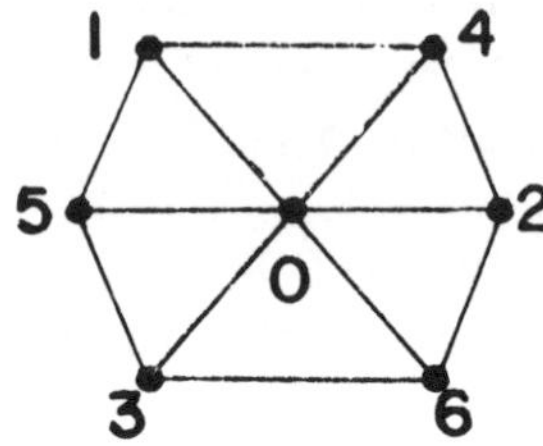

The valence of each vertex of H is $2k - 2 + 6$. There must be a vertex (say 7) adjacent to both 1 and 4, since $k - 2 = 2$, and 7 cannot be adjacent to 0. If 7 is adjacent to any vertex in $\{2, 3, 5, 6\}$, it must be adjacent to at least one other, otherwise $H(1, 2, 3, 4, 5, 6, 7)$ would be H_4 or H_5.

Suppose 7 is adjacent to 5. Then, in order to avoid both H_4 and H_6, 7 must be adjacent to 2 or 3. Without loss of generality, we can take 7 adjacent to 2 and 5. $H\{0, 2, 5, 7, 1\}$ is $H_3(a, c, d, b, \alpha)$. Hence 7 cannot be adjacent to 5. In a like manner we can show that 7 is not adjacent to 2. Suppose 7 is adjacent to neither 2 nor 5, but is adjacent to 6. Then $H\{4, 6, 0, 7, 2\} = H_3(a, b, d, c, \alpha)$. Hence 7 is not adjacent to 6. Similarly 7 cannot be adjacent to 3.

Therefore, 7 is not adjacent to 2, 3, 5, or 6. In like manner we can find distinct vertices $8, 9, 10, 11, 12$ all distinct, with 8 adjacent to 2 and 4, 9 adjacent to 2 and 6, 10 adjacent to 6 and 3, 11 adjacent to 3 and 5, 12 adjacent to 5 and 1. Referring back to (2.1) and (2.2), we have

$$\sum_{j=7}^{12} b_{0j}(b_{0j} - 1) = 6 \times 2 \cdot 1 = 12 \leqq 2(\lambda - 1) \cdot 3.$$

Therefore, $\lambda - 1 \geqq 2$ or $\lambda \geqq 3$. Since $k = 4$, this means $\lambda = 3$. $v = 5$, $vk = 20$. Now we shall use (2.3) and (2.4). If a vertex j is not connected to 0 by a path of length 2, then $c_{0j} = -1$. This means that the number of vertices at distance greater than two from 0 must be at most one.

Now each of $\{1, \cdots, 6\}$ has valence 6, and we have already identified, for each, 5 adjacent vertices. Therefore, there are at most 6 more vertices at distance two from 0. If there were exactly 6 such vertices, we would have identified 18 vertices at distance one or two from 0, and not yet have a violation of (2.4). But if there are fewer than 6 such vertices, we would have a violation of (2.4).

Let 13 be adjacent to 4, but not to 1 or 2. If 13 were adjacent to 5, then $\{13, 4, 0, 5, 1\}$ would form a graph H_3. Similarly, 13 cannot be adjacent to 6. If 13 were not adjacent to 3, we would have $H\{1, 2, 3, 4, 5, 6, 13\}$

$= H_7(a,c,e,b,f,d,\alpha)$. Therefore, 13 must be adjacent to 3, so we do not have six additional vertices at distance two from 0, and hence have a violation of (2.4).

7. Proof that H is edge regular if $k > 3$. In this section, we assume H is not edge regular and $k > 3$, and show this leads to a contradiction.

If H is not edge regular, then there is some edge (say $(0,1)$) contained in $k - 3 - a$ triangles ($a \geq 0$), and every edge of H is contained in at least $k - 3 - a$ triangles.

LEMMA 7.1. *There exists a claw* $\{0,1,2,3\}$.

Proof. If no such claw existed, then we would have $k + a$ vertices adjacent to 0 and not adjacent to 1 all forming a clique. For each such vertex j, $(A^2)_{0j}$ would be at least $k + a - 1$. For each other vertex j, adjacent to 0, we have $(A^2)_{0j}$ is at least $k - 3 - a$. Therefore, $(A^3)_{00}$ would be at least $(k - 2 - a)(k - 3 - a) + (k + a)(k + a - 1)$, which exceeds $2(k - 1) \cdot (k - 2)$.

LEMMA 7.2. $S_1 = S_2 = S_3 = \emptyset$. $|S_{23}| = k - 2 + a$.

Proof. The same reasoning which established the claw proves that $\widetilde{S_1}$ is not a clique. Therefore, $S_2 = S_3 = \emptyset$, by Lemma 5.1. Since the number of vertices adjacent to 0 and 1 is $k - 3 - a$, the number of vertices adjacent to 0 but not adjacent to 1 is $k + a$, which means $|S_{23}| = k - 2 + a$. But $k - 2 + a > 0$, which (by Lemma 5.2), implies $S_1 = \emptyset$.

LEMMA 7.3. *If $k > 3$, H is edge regular.*

Proof. If H is not edge regular, the previous lemmas of this section apply, and we have a claw $\{0; 1, 2, 3\}$ with $|S_{23}| = k - 2 + a$, $|S_{12}| + |S_{13}| = k - 3 - a$. Without loss of generality, we can assume

$$|S_{12}| \geq \frac{k - 3 - a}{2}.$$

By Lemma 5.3, $|S_{23}| = k - 2 + a \leq 10$. Therefore, $k \leq 12$.

Now let us make the tentative assumption that $\lambda < k - 1$. By (1.2), this means that, in case $k = 12$, for example, $\lambda \leq 6$. Therefore, the right side of (2.2) is at most 110. But, in (2.2), the left side is at least $b_{23}(b_{23} - 1) + b_{21}(b_{21} - 1)$, which is

$$(11 + a)(10 + a) + \left[\frac{12 - a}{2} \right]\left(\left[\frac{12 - a}{2} \right] - 1 \right).$$

Since $a \geq 0$, this is a contradiction. This line of reasoning eliminates all possible values of k, $4 \leq k \leq 12$, with $\lambda < k - 1$, except $k = 9$.

If $k = 9$, and $\lambda < k - 1$, then $\lambda = 6$ or $\lambda \leq 4$, $\lambda \leq 4$, the above reasoning applies. If $\lambda = 6$, then, from (1.2), $v = 13$. By (2.3) and (2.4), $c_{23}(c_{23} - 1)$ $= |S_{23}| \cdot (|S_{23}| - 1) \leq 24$. But $|S_{23}| \geq 7$, a contradiction.

Therefore, all we need consider is the case $\lambda = k - 1$, so $v = k + 1$. Since the right side of (2.4) is then 2, it follows that $|S_{23}| \leq 2$. Since $k - 2 \leq |S_{23}|$, we have only to consider the case $k = 4$. When $k = 4$, $|S_{12}| = 1$, $|S_{23}| = 2$. Therefore, in (2.4), $c_{23} = 2$. Since, in general $(A^3)_{00} = 2(k - 1)(k - 2)$, the number of edges in the graph subtended by vertices adjacent to 0 must be $(k - 1)(k - 2)$, or be 6 for the case $k = 4$. But $|S_{12}| = 1$, $|S_{23}| = 2$ already picks out six edges, so that, if $S_{23} = \{4, 5\}$, and $S_{12} = \{6\}$, we have, in the graph subtended by $\{2, 3, 4, 5, 6\}$ the graph H_1.

8. The main theorem.

LEMMA 8.1. *Let H be edge regular and contain no claw. Then*
(8.1) *every edge of H is contained in exactly one clique of order k,*
(8.2) *every maximal clique of H contains k vertices,*
(8.3) *every vertex is contained in two cliques of order k,*
(8.4) *there are $2v$ cliques of order k.*

Observe first that if $0, 1$ are adjacent vertices, then the $k - 1$ vertices adjacent to 0 and not to 1, together with 0 must form a clique of order k. Clearly every edge of H is accounted for in a clique exactly once this way, which proves (8.1), (8.2) and (8.3) are equally obvious. Let T denote the total number of cliques of order k. Since every vertex is contained in two cliques, $kT = 2vk$, or $T = 2v$, which is (8.4).

THEOREM. *Let H be a regular connected graph on vk vertices, such that the distinct eigenvalues of its adjacency matrix $A = A(H)$ are*

$$2k - 2 - 2, k - 2 \pm \sqrt{(k - \lambda)}.$$

Then $H \cong L(\Pi(v, k, \lambda))$ unless $k = 3$, $\lambda = 2$, when there is exactly one exception.

Proof. If $k > 3$ H is edge regular and contains no claw and Lemma 8.1 applies. Let $\widetilde{H}$ be the graph with $2v$ vertices corresponding to the cliques of order k in H, and two vertices of $\widetilde{H}$ adjacent if the corresponding cliques of H have a common vertex. By Lemma 8.1, $\widetilde{H}$ is a regular connected graph on $2v$ vertices, and H is its line graph. The therorem will then follow from Lemma 4.2.

By Lemmas 6.2, 6.3 and 7.3, the theorem holds if $k > 3$. If $k = 3$, and $\lambda = 1$, [4] applies. If $k = 3$, $\lambda = 2$, and the theorem does not hold, H is not edge regular, and $\lambda = 2$. In this case $k - 2 = 1$. Since H is not edge regular, there exists an edge (a, b) which is not contained in a triangle. Since the number of triangles containing a given vertex is $(k - 1)(k - 2) = 2$, we must have the following subgraph.

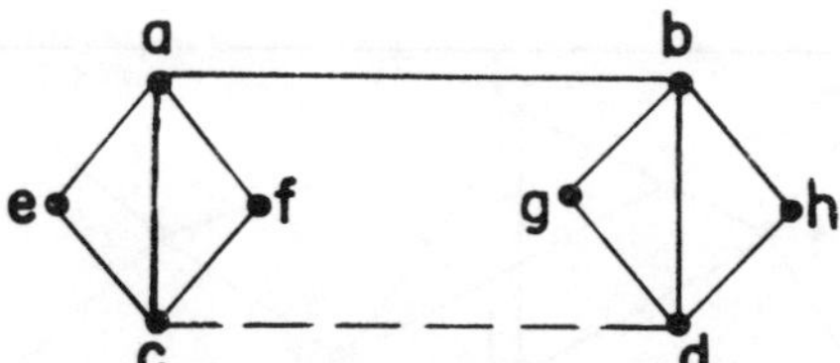

In case $k = 3, \lambda = 2, v = 4, vk = 12$, the polynomial of the graph is $(x^3 - 4x)/4$. Therefore there must be exactly eight paths of length 3 from a to b. This, however, is impossible unless c and d are adjacent, so our subgraph becomes:

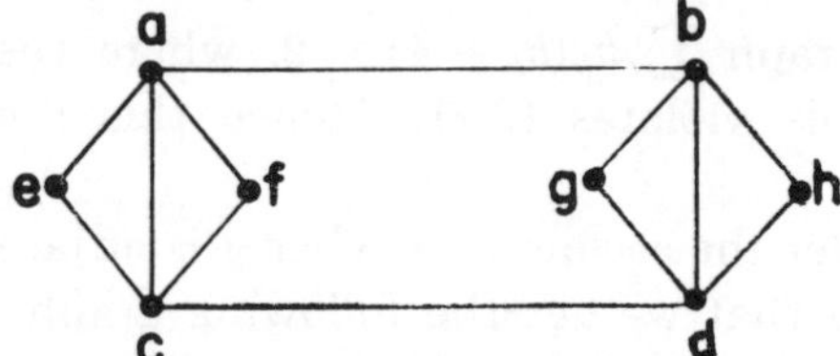

Now in order that $(A^3)_{ff} = 4$, it is necessary that there exist vertices i and j such that f, i, j form a triangle. Since the valence of every vertex is 4, vertex i cannot be adjacent to a, c, b, and d. Vertex i must be adjacent to at least one of e, g, h otherwise the vertices a, b, c, e, f, g, h, i would subtend a graph H_8. But i cannot be adjacent to e, otherwise $\{a, e, f, c, i\}$ would subtend H_3. Vertex i cannot be adjacent to both g and h, otherwise vertices i, g, h, d, f will subtend H_1. So i is adjacent to exactly one of g and h, say g, and we have

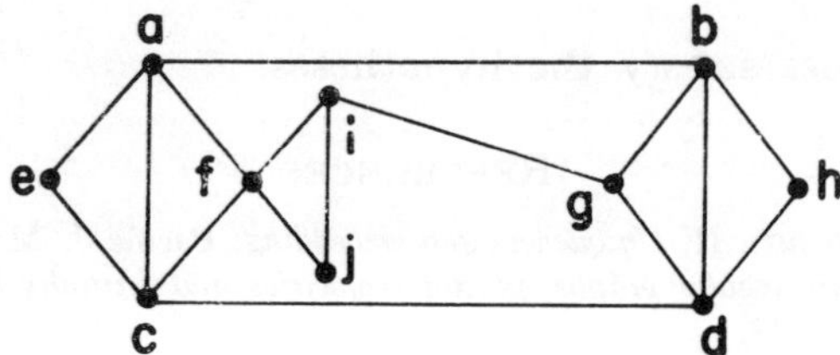

By the same argument, j cannot be adjacent to e and must be adjacent to one of g and h. Now there are two possible cases. In the first case we assume j to be adjacent to h. In the second case j is adjacent to g. Let us consider the first case. In this case vertex h cannot be adjacent to e, otherwise vertices $\{e, c, d, h, j\}$ subtend a graph H_1. Hence there is a new vertex k adjacent to h and to fulfill $A^4_{hh} = 4$, j and k must be adjacent. In the subgraph $\{a, b, c, d, e, f, g, h, i, j, k\}$ valence of every vertex other than i, g and e is 4. Vertices i and g are already adjacent. We have shown that vertices i and e cannot be adjacent. Hence the twelfth vertex l is adjacent to i. It is easily checked that we get the following graph:

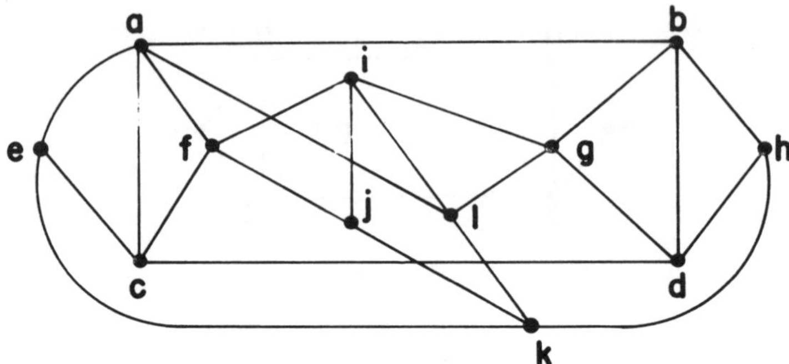

However for this graph $\sum_r b_{fr}(b_{fr} - 1) = 2$, where the summation is over all the vertices. This violates (2.2). Hence this graph does not satisfy the hypothesis.

Now let us consider the second case when j is adjacent to g. In this case it is readily checked that we get the following graph

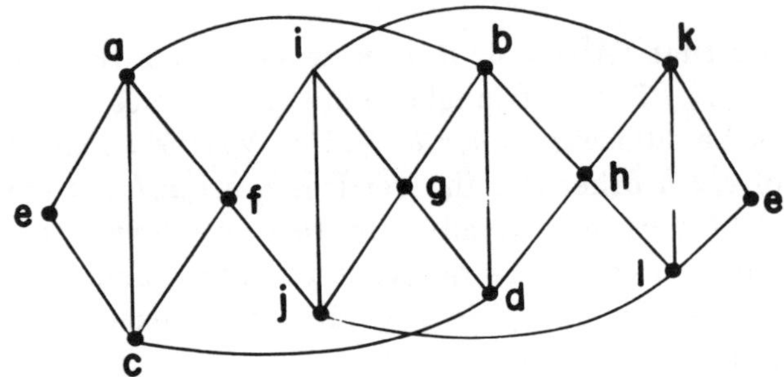

and this graph does satisfy the hypotheses.

REFERENCES

1. R. H. Bruck, *Finite nets. II. Uniqueness and embedding*, Pacific J. Math. **13** (1963), 421-457.

2. R. C. Bose, *Strongly regular graphs, partial geometries and partially balanced designs*, Pacific J. Math. **13** (1963), 389-419.

3. A. J. Hoffman, *On the polynomial of a graph*, Amer. Math. Monthly **70** (1963), 30-36.

4. ______, *On the line graph of a projective plane*, Proc. Amer. Math. Soc. **16** (1965), 297-302.

5. A. J. Hoffman and D. K. Ray-Chaudhuri, *On the line graph of a finite affine plane*, Canad. J. Math. (to appear).

INTERNATIONAL BUSINESS MACHINES CORPORATION,
  YORKTOWN HEIGHTS, NEW YORK

On Eigenvalues and Colorings of Graphs

ALAN J. HOFFMAN

1. Introduction.

Let G be a graph (undirected, on a finite number of vertices, and no edge joining a vertex to itself). Let $A = A(G) = (a_{ij})$ be the adjacency matrix of G , i. e.,

$$a_{ij} = \begin{cases} 1 & \text{if } i \text{ and } j \text{ are adjacent vertices} \\ 0 & \text{if } i \text{ and } j \text{ are not adjacent vertices} \end{cases}.$$

For any real symmetric matrix A of order n , we will denote its eigenvalues arranged in descending order by

$$\lambda_1(A) \geq \cdots \geq \lambda_n(A) ,$$

and the eigenvalues in ascending order by

$$\lambda^1(A) \leq \cdots \leq \lambda^n(A).$$

(Thus $\lambda_i(A) = \lambda^{n-i+1}(A)$): If $A = A(G)$, we shall write $\lambda_i(G)$ and $\lambda^i(G)$ for $\lambda_i(A(G))$ and $\lambda^i(A(G))$ respectively.
Over the past ten years, there has been much work done on the question of relating geometric properties of G to the eigenvalues of $\lambda_i(G)$, and it had been my original intention to devote this talk to summarizing what has been accomplished since the survey given in the spring of 1967 [6]. Unfortunately, I could find no pedagogically sound way to organize this material. Instead, I will describe some

ALAN J. HOFFMAN

observations I made during the summer connecting the eigen-values of a graph with its coloring number and with some related concepts.

By this tactic, I hope that those who have never been previously exposed to any of the work relating eigenvalues to graphs will become convinced that there is some relevance.

We require some definitions. If G is a graph, its set of vertices is denoted by $V(G)$, its set of edges by $E(G)$. If $S \subset V(G)$, $S \neq \emptyset$, $G(S)$ is the graph such that $V(G(S)) = S$, $E(G(S)) =$ the subset of $E(G)$ consisting of all edges both of whose vertices are in S. If G and H are graphs, $G \subset H$ if there is a subset $S \subset V(H)$ such that $G = H(S)$. If G is a graph on n vertices, and $|E(G)| = \binom{n}{2}$, then $V(G)$ is called a clique on n vertices and G is denoted by K. If G is a graph, $\bar{G}$ is the graph with $V(\bar{G}) = V(G)$, $E(\bar{G}) \cap E(G) = \emptyset$, $|E(\bar{G})| + |E(G)| = \binom{n}{2}$. The graph $\bar{G}$ is called the complementary graph of G, and satisfies $A(G) + A(\bar{G}) = J-I$, where J is the $n \times n$ matrix all of whose elements are unity. If $E(G) = \emptyset$, then $V(G)$ is called an independent set.

If i is a vertex of G, $d_i(G)$ is the number of ver-tices of G adjacent to i, and is called the valence of i. We also define

$$D(G) = \max_i d_i(G); \quad d(G) = \min_i d_i(G).$$

Note that $d_i(G) = \sum_j (A(G))_{ij}$.

A coloring of a graph G is a partitioning of $V(G)$ into independent sets. The coloring number of G is the smallest number of independent sets arising in such a parti-tion and is denoted by $\gamma(G)$. More generally, let

$$(1.1) \qquad V(G) = K^1 \cup \ldots \cup K^r \cup I^1 \cup \ldots \cup I^s$$

be a partition of $V(G)$ into subsets such that

> (i)   each K^i is a clique with at least two vertices, $i = 1, \ldots, r$;

> (ii)   each I^i is an independent set;

if such a decomposition (1.1) is possible, then G will be

ON EIGENVALUES AND COLORINGS OF GRAPHS

said to admit an (r, s) decomposition.

The inspiration for the present investigation is an elegant result of Wilf [11]:

$$(1.2) \qquad \gamma(G) \leq 1 + \lambda_1(G) \ .$$

This upper bound is an improvement of part of a theorem of Brooks [3]

$$(1.3) \qquad \gamma(G) \leq 1 + D(G) \ .$$

In §2, (1.2) is proved and shown to imply (1.3).

To describe the results of §3 we need further definitions. Let $S_1 \cup \ldots \cup S_t$, $t \geq 2$ be a partition of $\{1, \ldots, n\}$ into non-empty subsets. For any real symmetric matrix A of order n, let $A_{ij}(i, j = 1, \ldots, t)$ be the submatrix of A with rows in S_i and columns in S_j. Aronzajn [1] has proved that if $t = 2$, $i_1 < |S_1|$, $i_2 < |S_2|$, then

$$(1.4) \qquad \lambda^1(A) + \lambda_{i_1+i_2+1}(A) \leq \lambda_{i_1+1}(A_{11}) + \lambda_{i_2+1}(A_{22}) \ .$$

With the help of a theorem of Wielandt [10] we prove that, for all $t \geq 2$, if $i_1 < |S_1|, \ldots, i_t < |S_t|$, then

$$(1.5) \qquad \lambda_{i_1+\ldots+i_t+1}(A) + \sum_{i=1}^{t-1} \lambda^i(A) \leq \sum_{k=1}^{t} \lambda_{i_k+1}(A_{kk}).$$

In §4, (1.5) is used to derive lower bounds on $\gamma(G)$ (more generally, on r and s in an (r, s) decomposition of G) in terms of the eigenvalues of G. For example,

$$(1.6) \qquad \frac{\lambda_1(G)}{|\lambda^1(G)|} + 1 \leq \gamma(G) \ .$$

We show in §5 that (1.6) is sharp in a number of interesting cases, and we also attempt to compare (1.6) with a lower bound for $\gamma(G)$ given in terms of $\{d_i(G)\}$ due to Bondy [2]. There does not seem to be an easy way to compare (1.6) with

[2], except in the case of regular graphs $(D(G) = d(G))$, where the comparison is favorable to (1.6).

In §6, we consider the cliquomatic number $\kappa(G)$ of a graph T, defined by $\kappa(G) = \gamma(\bar{G})$. Using a theorem of Lidskii [9], we derive from §4, lower bounds on $\kappa(G)$ in terms of the eigenvalues of G.

Finally, we mention in §7, results of a different kind: upper bounds for $\gamma(G)$ and $\kappa(G)$ as functions of (respectively) the number of eigenvalues of G each of which is at most -1, and the number of nonnegative eigenvalues of G. Also, we state that for each k, there exists upper bounds for r and s in an (r, s) decomposition of G, where the respective upper bounds depend on max $(\lambda_k(G), -\lambda^1(G))$.

2. Wilf's Theorem

For the sake of amusement, we will give a proof slightly different from Wilf's, using the maximum characterization of eigenvalues of a real symmetric matrix in lieu of his appeal to the Perron-Frobenius theory.

Let A be a real symmetric matrix of order n, B a principal submatrix of W. Then from the maximum principle we infer

$$(2.1) \qquad \lambda_1(A) \geq \lambda_1(B),$$

$$(2.2) \qquad \lambda^1(B) \geq \lambda^1(A)$$

(we will use (2.2) in later sections). Further

$$(2.3) \qquad \lambda_1(A) \geq \frac{1}{n} \sum_{i,j} a_{ij} .$$

For, let $u = (1, \ldots, 1)$. Then the right-hand side of (2.3) is the Rayleigh quotient $(Au, u)/(u, u)$, and every Rayleigh quotient formed from A is at most $\lambda_1(A)$. This argument is contained in [4].

If $A = A(G)$, we infer from (2.3) that

ON EIGENVALUES AND COLORINGS OF GRAPHS

$$(2.4) \qquad \lambda_1(G) \geq \frac{1}{n} \sum_i d_i(G) \geq d(G) \ .$$

Further

$$(2.5) \qquad \lambda_1(G) \leq D(G) \ ,$$

since by Gersgorin's theorem, $\lambda_1(G) \leq \max_i \sum_i a_{ij} = D(G)$.
Comment: This also follows from the Perron–Frobenius Theory– viz $\max \lambda_1 \leq \min(\max_j \Sigma \, a_{ij}, \ \max_i \Sigma \, a_{ij})$ which reduced to the same.

To prove (1.2) we first observe that there must exist a subgraph $H \subset G$ such that $d(H) \geq \gamma(G) - 1$ (otherwise we would contradict that $\gamma(G)$ is the coloring number of G). From (2.4), we have

$$(2.6) \qquad \lambda_1(H) \geq d(H) \geq \gamma(G) - 1.$$

From (2.1), $\lambda_1(H) \leq \lambda_1(G)$. Combining this with (2.6), we infer (1.2). Next, inequality (1.3) follows from (1.2) and (2.5)

3. A Lemma on Partitioned Matrices.

Let A be a real symmetric matrix of order n, and let $S_1 \cup \ldots \cup S_t$, $t \geq 2$, be a partition of $\{1, \ldots, n\}$ into nonempty subsets. If $i_k < |S_k|$, $k = 1, \ldots, t$, then

$$(3.1) \qquad \lambda_{1 + \sum_{k=1}^{t} i_k}(A) + \sum_{i=1}^{t-1} \lambda^i(A) \leq \sum_{k=1}^{t} \lambda_{i_k + 1}(A_{kk}).$$

Proof: We shall prove the lemma by induction on t. In case $t = 2$, the lemma is Aronzajn's inequality [1] (see (1.4)). Assume therefore that the lemma has been proved for $t - 1$. Let $T = S_1 \cup \ldots \cup S_{t-1}$ and let $A[T]$ be the principal submatrix of A formed by rows and columns in T. By the induction hypothesis

$$(3.2) \qquad \lambda_{1 + \sum_{k=1}^{t-1} i_k}(A[T]) + \sum_{i=1}^{t-2} \lambda^i(A[T]) \leq \sum_{k=1}^{t-1} \lambda_{i_k + 1}(A_{kk}) \ .$$

Let $\tilde{x}^1, \tilde{x}^2, \ldots, \tilde{x}^{|T|}$ be an orthonormal set of eigenvectors of $A[T]$, so that

$$(3.3) \qquad A[T]\tilde{x}^j = \lambda^j(A[T])\tilde{x}^j, \quad j = 1, \ldots, |T|.$$

Let $x^1, \ldots, x^{|T|}$ be the vectors with n coordinates obtained respectively from $\tilde{x}^1, \ldots, \tilde{x}^{|T|}$ by putting 0 for all coordinates x_ℓ^j, $\ell \in T$. Let B be the matrix representing the same linear operator as A, with respect to the orthonormal basis $x^1, \ldots, x^{|T|}$ and unit vectors v^ℓ, $\ell \in T$. Then B has the same eigenvalues as A, and $B[T]$ is diagonal.

Let U be the set of indices $j \in T$ such that $\{B[T]_{jj}\}_{j \in U}$ consists of $\{\lambda^{t-1}, \ldots, \lambda^{|T|}\}$. Let $N = U \cup S_t$. Thus for any $j_1 < |U|$ and $j_2 < |S_t|$, by Aronzajn's inequality (1.4)

$$\lambda^1(B[N]) + \lambda_{j_1+j_2+1}(B[N]) \leq \lambda_{j_1+1}(B[U]) + \lambda_{j_2+1}(A_{tt}).$$

Since $|U| = |T| - t + 2$ and $i_k \leq |S_k| - 1$, $k = 1, 2, \ldots, t-1$,
$$\sum_{k=1}^{t-1} i_k \leq \sum_{k=1}^{t-1} |S_k| - t+1 = |T| - t+1 < |U|, \quad \text{setting } j_1 = \sum_{k=1}^{t-1} i_k$$
and $j_2 = i_t$, we obtain

$$(3.4) \quad \lambda_{1+\sum_{k=1}^{t} i_k}(B[N]) + \lambda^1(B[N]) \leq \lambda_{1+\sum_{k=1}^{t-1} i_k}(B[U]) + \lambda_{i_t+1}(A_{tt}).$$

Note that by our construction of U,

$$\lambda_{1+\sum_{k=1}^{t-1} i_k}(B[U]) = \lambda_1 + \sum_{k=1}^{t-1} i_k(A[T]),$$

and

$$\lambda^i(A[T]) = \lambda^i(B[T]) \quad i = 1, \ldots, |T|.$$

Hence, adding (3.2) and (3.4), we obtain

$$(3.5) \quad \lambda_{1+\sum_{k=1}^{t} i_k}(B[N]) + \lambda^1(B[N]) + \sum_{i=1}^{t-2} \lambda^i(B[T] \leq \sum_{k=1}^{t} \lambda_{i_k+1}(A_{kk}).$$

ON EIGENVALUES AND COLORINGS OF GRAPHS

We now invoke a lemma due to Wielandt [10]: Let C be a real symmetric matrix of order n, and let $1 \le j_1 < j_2 < \ldots < j_r \le n$ then

$$(3.6) \qquad \min_{\substack{S_{j_1} \subset \ldots \subset S_{j_r}}} \; \max_{\substack{x_\ell \in S_{j_\ell} \\ (x_\ell, x_m) = \delta_{\ell m}}} \; \sum_{\ell=1}^{r} (C x_\ell, x_\ell) = \sum_{\ell=1}^{r} \lambda^{j_\ell}(C)$$

(In the left side of (3.6), S_ℓ stands for a linear subspace of dimensional ℓ). Let $\{\tilde{y}^i\}_{i \in N}$ be an orthonormal set of eigenvectors of $B[N]$ so that

$$(3.7) \qquad B[N]\, \tilde{y}^i = \lambda^i(B[N]) , \quad i = 1, \ldots, |N| ,$$

and let y^i be the vector with n coordinates obtained from $\tilde{y}^i$ by putting 0 for all coordinates y^i_ℓ, $\ell \notin N$. Then $\{x^1, \ldots, x^{t-2}, y^1, \ldots, y^{n-t+2}\}$ are an orthonormal set of vectors. For $i = 1, \ldots, t-2$, let T_i be the vector space spanned by $\{x^1, \ldots, x^i\}$. For $i = t-1, \ldots, n$, let T_i be the vector space spanned by $\{x^1, \ldots, x^{t-2}, y^1, y^2, \ldots, y^{i-t+2}\}$. Note dim $T_i = i$ in all cases.

Let $V = \{1, \ldots, t-2, t-1, n - \sum_{k=1}^{t} i_k\}$. Then

$$(3.8) \qquad \max_{\substack{x_\ell \in T_\ell \\ (x_\ell, x_m) = \delta_{\ell m}}} \; \sum_{\ell \in V} (B x_\ell, x_\ell) = \sum_{i=1}^{t-2} \lambda^i(B[T]) + \lambda^1(B[N]) + \lambda_{1+\sum_{t=1}^{t} i_k}(B[N]),$$

from (3.3), (3.7) and the construction of the $\{T_i\}_{i \in V}$. By (3.6),

$$(3.9) \qquad \lambda_{1+\sum_{k=1}^{t} i_k}(B) + \sum_{i=1}^{t-1} \lambda^i(B) \le \max_{\substack{x_\ell \in T_\ell \\ \{x_\ell\}_{\ell \in V} \\ \text{orthonormal}}} \; \sum_{\ell \in V} (B x_\ell, x_\ell).$$

Combining (3.8), (3.9) and (3.5) yields (3.1).

85

4. Lower Bounds for Coloring and (r, s) Decompositions.

It is now a simple matter to apply (3.1) and derive lower bounds for $\gamma(G)$ and (r, s) decompositions. We first prove: if $\gamma = \gamma(G) \geq 2$, then

$$(4.1) \qquad \lambda_1(G) + \sum_{i=1}^{\gamma-1} \lambda^i(G) \leq 0 .$$

Proof: By hypothesis, $V(G)$ can be partitioned into non-empty subsets $S_1 \cup \ldots \cup S_\gamma$ such that S_k is an independent set, $k = 1, 2, \ldots, \gamma$. Consequently, $\lambda_1(A(G)_{kk}) = 0$, $k = 1, \ldots, \gamma$. Apply (3.1) with each $i_k = 0$, and (4.1) follows.

Before proceeding further, note that for $t \geq 2$, $\lambda_1(K_t) = t - 1$, $\lambda^1(K_t) = \ldots = \lambda^{t-1}(K_t) = -1$. In particular, it follows from (2.2) that if G has at least one edge (so $\gamma(G) \geq 2$, and $K_2 \subset G$), $\lambda^1(G) \leq -1$. We now infer from (4.1)

$$(4.2) \qquad \gamma(G) \geq \frac{\lambda(G)}{-\lambda^1(G)} + 1, \quad \text{if } \gamma(G) \geq 2 .$$

Proof: By (4.1), we have

$$\lambda_1(G) + (\gamma-1)\lambda^1(G) \leq 0 .$$

Using the fact that $\lambda^1(G) < 0$, (4.2) follows. We next prove: If G has an (r, s) decomposition, then

$$(4.3) \qquad \lambda_{r+1}(G) + \sum_{i=1}^{r+s-1} \lambda^i(G) \leq -r .$$

Proof: Recalling (1.1), we see that, if we use $i_k = 1$ for the indices k referring to cliques, and $i_\ell = 0$ for the indices ℓ referring to independent sets, and use $\lambda_2(K_t) = -1$ if $t \geq 2$, then (3.1) becomes (4.3).

In particular, we infer from (4.3) that

$$\lambda_{r+1}(G) + (r+s-1)\lambda^1(G) \leq -r , \qquad \text{or}$$

$$(4.4) \qquad \frac{\lambda_{r+1}(G) + r}{-\lambda^1(G)} + 1 - r \leq s .$$

ON EIGENVALUES AND COLORINGS OF GRAPHS

5.   Further Comments on the Lower Bound for $\gamma(G)$.

The lower bound for $\gamma(G)$ given by (4.2) is sharp in a number of interesting cases. For example, it is known [5] that $\gamma(G) = 2$ if and only if $\lambda^1(G) + \lambda_1(G) = 0$. In this case, the lower bound given by (4.2) is exact. If G is an odd polygon, then $\gamma(G) = 3$, $\lambda_1(G) = 2$, $-\lambda^1(G) = 2 - \epsilon$, where $0 < \epsilon \leq 1$. Thus the right hand side of (4.2) becomes $1 + \alpha$, when $1 < \alpha \leq 2$. But $\gamma(G) \geq 1 + \alpha$ implies $\gamma(G) \geq -[-(1+\alpha)]$, where $[x]$ is the largest integer not exceeding x . Thus, in this case, (4.2) is also sharp. If G has n independent sets, each consisting of m vertices, such that any pair of vertices of different independent sets are adjacent, then $\gamma(G) = n$, $\lambda_1(G) = m(n-1)$, $\lambda^1(G) = -m$, and again (4.2) is sharp.

If M is a $(0, 1)$ matrix such that every row sum is k , every column sum is k, $k \geq 2$ let $G = G(M)$ be the graph whose vertices are the 1's in M , with two vertices adjacent if the corresponding 1's are in the same row or column. Then it has been shown [6] that $\lambda_1(G) = 2k-2$, $\lambda^1(G) = -2$, so (4.2) shows that $\gamma(G) \geq k$. On the other hand, by a theorem of Konig [8] there exist permutation matrices $P_1, \ldots, P_k$ such that $M = P_1 + \ldots + P_k$. The 1's occurring in each P_i form an independent set, so $\gamma(G) \leq k$. Thus, (4.2) is sharp in this case.

It is also interesting to observe the implication of (4.2) when one knows an upper bound for $\gamma(G)$. For instance if G is planar then $\gamma(G) \leq 5$. By (4.2) this implies

$$-\lambda^1(G) \geq \frac{1}{4}\lambda_1(G) \quad \text{if G is planar} .$$

Further $-\lambda^1(G) \geq \frac{1}{3}\lambda_1(G)$ if G is planar and the four color hypothesis is true.   Similar remarks can be made, of course, for graphs imbedded on surfaces of higher genus.

Bondy [2] has given a lower bound for $\gamma(G)$ in terms of the $\{d_i(G)\}$. It is difficult to compare his lower bound with (4.2) except in the case where G is regular (of valence d). Then

$$(5.1) \qquad\qquad \gamma(G) \geq \frac{n}{n-d} .$$

ALAN J. HOFFMAN

We will show (4.2) is a better bound, by proving

$$(5.2) \qquad \frac{\lambda_1(G)}{-\lambda^1(G)} + 1 \geq \frac{n}{n-d} \quad .$$

To prove (4.2) we recall from §2 that $\lambda_1(G) = d$, so (5.2) holds if and only if

$$(5.3) \qquad 0 \leq (n-d) + \lambda^1(G) .$$

Write $J = (J - A) + A$, where $A = A(G)$, and observe that since G is regular (and therefore A commutes with J), the eigenvalues of J are the sums of corresponding eigenvalues of J $J-A$ and A. Clearly (see §1), $\lambda_1(J) = n$, $\lambda_1(J-A) = n-d$, $\lambda_1(A) = d$. Also,

$$0 = \lambda_1(J) = \lambda_i(J-A) + \lambda^{i-1}(A) \qquad i = 2, \ldots, n .$$

In particular, setting $i = 2$, we have

$$(5.4) \qquad 0 = \lambda_2(J-A) + \lambda^1(A) .$$

But $\lambda_2(J-A) \leq \lambda_1(J-A) = n-d$. Combined with (5.4), this yields (5.3). In fact, if $\bar{G}$ is connected, then $\lambda_2(J-A) < \lambda_1(J-A)$, so (4.2) is always at least as good a bound as (5.1) for regular graphs, and is a better bound if the complementary graph is connected.

6. Lower Bounds on $\kappa(G)$.

For any graph G, define $\kappa(G)$ to be the smallest number of cliques whose union contains all vertices of G. Then $\kappa(G) = \gamma(\bar{G})$. We shall prove that, if $\kappa = \kappa(G) \geq 2$, and $|V(G)| = n$ then

$$(6.1) \qquad n - \kappa \leq \sum_{i=1}^{\kappa} \lambda_i(G) .$$

Note that if $\kappa(G) \geq 2$, G contains two non-adjacent vertices- thus $A(G)$ contains a 2×2 principal submatrix which is 0. By the interlacing theorem, $\lambda_2(G) \geq 0$, hence $1 + \lambda_2(G) > 0$. Since the right-hand side of (6.1) is bounded from above by $\lambda_1(G) + (\kappa-1)\lambda_2(G)$, we infer from (6.1) that

88

ON EIGENVALUES AND COLORINGS OF GRAPHS

$$(6.2) \qquad \kappa(G) \geq \frac{n+\lambda_2(G) - \lambda_1(G)}{1 + \lambda_2(G)} \; .$$

To prove (6.1), we recall a theorem of Lidskii [9]: if A, B and C are real symmetric matrices of order n , A = B + C . $1 \leq i_1 < i_2 < \ldots < i_r \leq n$, then

$$(6.3) \qquad \sum_{\ell=1}^{r} \lambda_{i_\ell}(A) \leq \sum_{\ell=1}^{r} \lambda_{i_\ell}(B) + \sum_{\ell=1}^{r} \lambda_i(C) \; .$$

Let $\kappa = \kappa(G) = \gamma(\bar{G})$, and write $J-I = A(G) + A(\bar{G})$. Let r = k, $t_1 = 1$, $t_k = n-1 \ldots, t_2 = n-k+2$. From (6.3), we infer

$$(6.4) \qquad (n-1) - (\kappa-1) \leq \lambda_1(\bar{G}) + \sum_{i=1}^{k-1} \lambda^i(\bar{G}) + \sum_{i=1}^{\kappa} \lambda_i(G) \; .$$

But from (4.1)

$$(6.5) \qquad \lambda_1(\bar{G}) + \sum_{i=1}^{\kappa-1} \lambda^i(\bar{G}) \leq 0 \; .$$

Combining (6.4) and (6.5), we infer (6.1) .

7. Further Upper Bounds on $\gamma(G)$, $\kappa(G)$ and (r, s) Decompositions of G .

Let M(G) = the number of eigenvalues of G , each of which is at most -1, and let m(G) = the number of non-negative eigenvalues of G . Then one can prove there exist functions f and g such that

$$(7.1) \qquad \gamma(G) \leq f(M(G)) \; ,$$

$$(7.2) \qquad \kappa(G) \leq g(m(G)).$$

We conjecture that f(G) = 1 + M(G), but have not succeeded in proving this.

Let k > 1 . Using the results of [7], we can prove there exists functions f_k and g_k such that G has an (r, s) decomposition, where

$$(7.3) \qquad r \leq f_k (\max(\lambda_k(G), \; -\lambda^1(G)) \; ,$$

ALAN J. HOFFMAN

$$(7.4) \qquad s \leq g_k \, (\max \, (\lambda_k(G), \, - \lambda^1(G))).$$

We conjecture that f_k can be made a function of $\lambda^1(G)$ alone, g_k a function of λ^k alone, but have not yet succeeded in proving this.

We are grateful to Donald Newman, Robert Thompson and Herbert Wilf for useful conversations about this material.

REFERENCES

1. Aronzajn, N. "Rayleigh-Ritz and A. Weinstein Methods for Approximation of Eigenvalues. I. Operators in a Hilbert Space," Proc. Nat. Acad, Sci., U. S. A., 34(1948), 474-480.

2. Bondy, J. A. "Bounds for the Chromatic Number of a Graph," J. Combinatorial Theory, 7(1969), 96-98.

3. Brooks, R. L.. "On Colouring the Nodes of a Network, Proc. Cambridge Philos. Soc., 3791941), 194-197.

4. Collatz, L. and Sinogowitz, U., "Spektren endlicher Graphen," Abh. Math. Sem. Univ. Hamburg, 21(1957), 63-77.

5. Hoffman, A. J., "On the Polynomial of a Graph," Amer. Math. Monthly, 70 (1963), 30-36.

6. Hoffman, A. J. , "The Eigenvalues of the Adjacency Matrix of a Graph," in Combinatorial Mathematics and its Applications, edited by R. C. Bose and T. C. Dowling, University of North Carolina Press, Chapel Hill, 1969, 578-584.

7. Hoffman, A. J., "-1-$\sqrt{2}$?" to appear in Proceedings of the Calgary International Conference on Combinatorial Structures and their Applications, Canada in June 1969.

8. König, D., "Theorie der Endlichen und Unendlichen Graphen," Chelsea, New York, 1950, 170-178.

90

ON EIGENVALUES AND COLORINGS OF GRAPHS

9.   Lidskii, V. B. , "The Proper Values of the Sum and
       Product of Symmetric Matrices, " Doklady Akad. Nauk.
       SSSR (N. S.) 75(1950, 769-772.   Translation by C. D.
       Benster, National Bureau of Standards (Washington)
       Report 2248, 1953.

10.   Wielandt, H., "An Extremum Property of Sums of
       Eigenvalues, " Pacific J. Math. 5(1955), 633-638.

11.   Wilf, H. S., "The Eigenvalues of a Graph and its
       Chromatic Number, " J. London Math. Soc. 42(1967),
       330-332.

Work supported in part by the office of Naval Research
under Contract No. Nonr-3775(00).

LINEAR ALGEBRA AND ITS APPLICATIONS 5, 137–146 (1972)   137

Eigenvalues and Partitionings of the Edges of a Graph*

A. J. HOFFMAN
IBM Watson Research Center
Yorktown Heights, New York

ABSTRACT

This paper is concerned with the relationship between geometric properties of a graph and the spectrum of its adjacency matrix. For a given graph G, let $\alpha(G)$ be the smallest partition of the set of edges such that each corresponding subgraph is a clique, $\beta(G)$ the smallest partition such that each corresponding graph is a complete multipartite graph, and $\gamma(G)$ the smallest partition such that each corresponding subgraph is a complete bipartite graph. Lower bounds for α, β, γ are given in terms of the spectrum of the adjacency matrix of G. Despite these bounds, it is shown that there can exist two graphs, G_1 and G_2, with identical spectra such that $\alpha(G_1)$ is small, $\alpha(G_2)$ is enormous. A similar phenomenon holds for $\beta(G)$. By contrast, $\gamma(G)$ is essentially relevant to the spectrum of G, for it is shown that $\gamma(G)$ is bounded by and bounds a function of the number of eigenvalues each of which is at most -1.

. It is also shown that the chromatic number $\chi(G)$ is spectrally irrelevant in the sense of the results for α and β described above.

1. INTRODUCTION

Let G be a graph with $V(G)$ the set of its vertices, $E(G)$ the set of its edges. We will assume G has at least one edge. If $F \subset E(G)$, $F \neq \phi$, we will denote by G_F the subgraph of G such that $E(G_F) = F$, $V(G_F) =$ the set of vertices of G each of which is on at least one edge in F. In this paper we shall consider partitions $F_1 \cup \cdots \cup F_k$ of $E(G)$ in which each G_{F_i} is a graph of a particular class, and consider the relation of the smallest k for which such a partition of $E(G)$ exists to the eigenvalues of the adjacency matrix $A(G)$ of G. This matrix is a square symmetric $(0, 1)$ matrix of order $(V(G))$, defined by the rule: $a_{ij} = 1$ if and only if vertices i and j are adjacent (note $a_{ij} = 0$ for all i). The eigenvalues of a real symmetric matrix A of order n will be denoted by $\lambda_1(A) \geqslant \cdots \geqslant \lambda_n(A)$ or by $\lambda^1(A) \leqslant \cdots \leqslant \lambda^n(A)$ as convenience dictates. If $A = A(G)$, we may write $\lambda^i(G)$ for $\lambda^i(A(G))$ or $\lambda_i(G)$ for $\lambda_i(A(G))$.

A graph G is called a clique if every pair of its vertices is adjacent. If $V(G)$ can be partitioned into two or more subsets so that each pair of

* This work was supported in part by the Office of Naval Research under Contract Nonr 3775(00).

                                       A. J. HOFFMAN

vertices in the same subset is not adjacent, but each pair of vertices in different subsets is adjacent, G is called a complete multipartite graph. Thus a clique is a complete multipartite graph in which each subset of the vertices contains exactly one element. If, for a complete multipartite graph, the number of subsets (parts) of $V(G)$ is exactly two, the graph is called a complete bipartite graph.

Let $\alpha(G)$ be the smallest integer k such that there exists a partition

$$F_1 \cup \cdots \cup F_k = E(G) \tag{1.1}$$

and each G_{F_i} is a clique. Let $\beta(G)$ be the smallest integer h such that (1.1) holds and each G_{F_i} is a complete multipartite graph. Let $\gamma(G)$ be the smallest integer h such that (1.1) holds and each G_{F_i} is a complete bipartite graph.

In Section 2 we shall derive lower bounds for $\alpha(G)$, $\beta(G)$, $\gamma(G)$ from the eigenvalues of $A(G)$. (The bound for $\beta(G)$ was previously derived by Ronald Graham and H. S. Witsenhausen.) Somewhat related questions for partitions of $V(G)$, especially bounds on the chromatic number $\chi(G)$ were given in [2] and [3]. In Section 3 we shall show that, despite these bounds, there is no intimate relation between $\alpha(G)$ or $\beta(G)$ and the spectrum of $A(G)$. Specifically, we shall show that, given any number V, there exist two graphs G_1 and G_2 such that $A(G_1)$ and $A(G_2)$ have the same spectrum, $\alpha(G_1) = 2$, $\alpha(G_2) > N$. Similarly, we shall show that, given any number N, there exist two graphs G_1 and G_2 whose adjacency matrices have the same spectrum, $\beta(G_1) = 3$, $\alpha(G_2) > N$.

Our most interesting result, presented in Section 4, is that such a phenomenon cannot hold for $\gamma(G)$. We will show that each of the following functions of G is bound by a function of each of the others: $\gamma(G)$, the number of eigenvalues of $A(G)$ each of which is at most -1, the number of nonzero eigenvalues of $A(G)$, the number of different rows of $A(G)$.

Finally, in Section 5, we use this opportunity to point out that the phenomenon described in Section 2 holds also for the chromatic number $\chi(G)$, and mixed chromatic number $\tilde{\chi}(G)$ (to be defined later).

2. LOWER BOUNDS FOR $\alpha(G)$, $\beta(G)$, $\gamma(G)$

We shall need the Courant-Weyl inequalities: Let A and B be real symmetric matrices of order n. $C = A + B$, $0 \leqslant i, j, i + j + 1 \leqslant n$.

$$\lambda_{i+j+1}(C) \leqslant \lambda_{i+1}(A) + \lambda_{j+1}(B)$$

EIGENVALUES AND PARTITIONINGS   139

and

$$\lambda^{i+j+1}(C) \geqslant \lambda^{i+1}(A) + \lambda^{j+1}(B).$$

By applying induction, it is easy to derive from the above: If $A_1, \ldots, A_k$ are real symmetric matrices of order n, $k + 1 \leqslant n$, then

$$\lambda_{k+1}\left(\sum_1^k A_i\right) \leqslant \sum_1^k \lambda_2(A_i), \tag{2.1}$$

$$\lambda^{k+1}\left(\sum_1^k A_i\right) \geqslant \sum_1^k \lambda^2(A_i), \tag{2.2}$$

$$\lambda^1\left(\sum_1^k A_i\right) \geqslant \sum_1^k \lambda^1(A_i). \tag{2.3}$$

We next note

$$\beta(G) = 1 \quad \text{if and only if} \quad \lambda_2(G) \leqslant 0, \tag{2.4}$$

$$\gamma(G) = 1 \quad \text{if and only if} \quad \lambda^2(G) \geqslant 0, \tag{2.5}$$

$$\alpha(G) = 1 \quad \text{if and only if} \quad \lambda_2(G) \leqslant 0, \quad \lambda^1(G) = -1. \tag{2.6}$$

The necessity part of (2.4)–(2.6) is obvious. The sufficiency part of (2.4) has been shown by Smith [5]. To prove the sufficiency part of (2.6), observe that $\lambda_2(G) = 0$ implies (by (2.4)) that $\beta(G) = 1$. Let H be the complete multipartite graph found by the edges of G. All we need show is that, if one of the parts has more than one vertex, $\lambda^1(H) < -1$. But, if one of the parts has more than one vertex, $K_{1,2} \subseteq H$ implies $\lambda^1(K_{1,2}) = -\sqrt{2}$. By the interlacing theorem, $K_{1,2} \subseteq H$ implies $\lambda^1(K_{1,2}) \geqslant \lambda^1(H)$.

To prove the sufficiency statement of (2.5), first observe that, if C is any odd polygon, $\lambda^2(C) < 0$. It follows from $\lambda^2(G) \geqslant 0$ that C cannot be an induced subgraph of G; hence G is a bipartite graph (see [1]). We must show that the subgraph H of G induced by the nonisolated vertices is a complete bipartite graph. Since H is bipartite, we have

$$A(H) = \begin{bmatrix} 0 & B \\ B^T & 0 \end{bmatrix},$$

and the eigenvalues of $A(H)$ are, apart from 0, the singular values of BB^T and their negatives. Hence, $\lambda^2(H) \geqslant 0$ implies that BB^T has rank one. Since B is a $(0, 1)$ matrix, BB^T is a rank one positive semidefinite matrix in which each diagonal entry is at least as large as any off-diagonal entry in its row. It follows that B consists entirely of 1's, so H is a complete bipartite graph.

140   A. J. HOFFMAN

THEOREM 2.1. *For every graph G,*

$$\beta(G) \geq \text{number of positive eigenvalues of } A(G); \tag{2.7}$$

$$\gamma(G) \geq \text{number of negative eigenvalues of } A(G); \tag{2.8}$$

$$\lambda^1(G) \geq -\alpha(G); \tag{2.9}$$

$$\binom{1 + \alpha\left(\genfrac{}{}{0pt}{}{G}{\text{or}}{H}\right)}{2} \geq \text{number of eigenvalues of } A(G) \text{ each of which is neither } 0 \text{ nor } -1. \tag{2.10}$$

If $\beta(G) = \beta$, there exist graphs $H_1, \ldots, H_\beta$ such that

$$\beta(H_i) = 1, \qquad i = 1, \ldots, \beta, \tag{2.11}$$

and

$$A(G) = A(H_1) + \cdots + A(H_\beta). \tag{2.12}$$

By (2.4), (2.11) implies $\lambda_2(H_i) \leq 0$, $i = 1, \ldots, \beta$. Hence (2.1), applied to (2.12), shows

$$\lambda_{1+\beta}(G) \leq 0,$$

which means that β is at least the number of positive eigenvalues of $A(G)$. This proves (2.7).

The proof of (2.8) is similar, using (2.5) and (2.2). Inequality (2.9) follows from (2.6) and (2.3).

To prove (2.10), let H be the subgraph of G formed by the nonisolated vertices of G. It is clearly sufficient to prove

$$\binom{1 + \alpha\left(\genfrac{}{}{0pt}{}{G}{\text{or}}{H}\right)}{2} \geq \text{number of eigenvalues of } A(H) \text{ which are not } -1. \tag{2.13}$$

Let $\alpha(H) = \alpha$. Then there exist cliques $K^1, \ldots, K^\alpha$ which partition the edges of H. Let S_i be the set of vertices which are in K^i, but not in K^j, $j = 1, \ldots, \alpha$, $j \neq i$. If x, y are two vertices in S_i, then the vector which is $+1$ at x, -1 at y, and 0 at all other vertices of H is an eigenvector of $A(H)$ with -1 as corresponding eigenvalue. It follows that -1 is an eigenvalue of $A(H)$ of multiplicity at least $\sum_{i=1}^{\alpha} (|S_i| - 1)$. From the fact that $|V(K^i) \cap V(K^j)| \leq 1$ for $i \neq j$, we have

$$\sum_{i=1}^{\alpha} |S_i| \geqslant |V(H)| - \binom{\alpha}{2}.$$

Thus, letting m be the multiplicity of -1 as an eigenvalue of $A(H)$, we have

$$m \geqslant \sum_{i=1}^{\alpha} (|S_i| - 1) = -\alpha + \sum_{i=1}^{\alpha} |S_i| \geqslant |V(H)| - \binom{1+\alpha}{2},$$

which is a restatement of (2.13).

3. ESSENTIAL IRRELEVANCE OF SPECTRUM OF $A(G)$ TO $\alpha(G)$ AND $\beta(G)$

To construct the examples required in this section, we first note without proof the following facts:

If K_n is a clique on n vertices,

$$\lambda_1(K_n) = n - 1, \qquad \lambda_2(K_n) = \cdots = \lambda_n(K_n) = -1. \tag{3.1}$$

Let H_{2n} be a graph on $2n$ vertices, $n > 2$ constructed as follows: Take two disjoint cliques on n vertices K and K', and join vertex i of K to vertex i' of K', and to no other vertices of K'.

$$\lambda_1(H_{2n}) = n, \qquad \lambda_2(H_{2n}) = n - 2, \qquad \lambda_3(H_{2n}) = \cdots = \lambda_{n+1}(H_{2n}) = 0,$$

$$\lambda_{n+2}(H_{2n}) = \cdots = \lambda_{2n}(H_{2n}) = -2. \tag{3.2}$$

Let L_{2n+1} be the graph on $2n + 1$ vertices constructed as follows: Take two disjoint cliques on n vertices, adjoin one other vertex adjacent to each vertex of each clique.

$$\lambda_2(L_{2n+1}) = n - 1, \qquad \lambda_3(L_{2n+1}) = \cdots = \lambda_{2n}(L_{2n+1}) = -1, \tag{3.3}$$

$\lambda_1(L_{2n+1})$ and $\lambda_{2n+1}(L_{2n+1})$ are the larger and smaller of the eigenvalues of

$$\begin{bmatrix} 0 & 2n \\ 1 & n-1 \end{bmatrix}. \tag{3.4}$$

Let M_{n+2} be the graph formed as follows: Take a clique on n vertices and adjoin two additional vertices, each adjacent to each vertex of the clique, but not adjacent to each other.

$$\lambda_2(M_{n+2}) = 0, \qquad \lambda_3(M_{n+2}) = \cdots = \lambda_{n+1}(M_{n+2}) = -1. \tag{3.5}$$

$\lambda_1(M_{n+2})$ and $\lambda_{n+2}(M_{n+2})$ are the larger and smaller of the eigenvalues of

A. J. HOFFMAN

$$\begin{bmatrix} 0 & n \\ 2 & n-1 \end{bmatrix}. \tag{3.6}$$

Let P_{2n} be the complete multipartite graph on $2n$ vertices, in which each part has two vertices.

$$\lambda_1(P_{2n}) = 2n - 2, \qquad \lambda_2(P_{2n}) = \cdots = \lambda_{n+1}(P_{2n}) = 0,$$

$$\lambda_{n+2}(P_{2n}) = \cdots = \lambda_{2n}(P_{2n}) = -2. \tag{3.7}$$

THEOREM 3.1. *Let $N > 0$ be given. Then there exist graphs G_1 and G_2 such that $A(G_1)$ and $A(G_2)$ have the same spectrum, $\alpha(G_1) = 2$, $\alpha(G_2) > N$.*

Proof. Let G_1 be the graph on $2n + 2$ vertices formed by the union of L_{2n+1} and one additional vertex. Let G_2 be the graph on $2n + 2$ vertices formed by M_{n+2} and K_n. Observe that (3.4) and (3.6) have the same eigenvalues. It follows from (3.1), (3.3), and (3.5) that $A(G_1)$ and $A(G_2)$ have the same eigenvalues. Now $\alpha(G_1) = 2$ and $\alpha(G_2) > \alpha(M_{n+2})$. Let x and y be the adjoined vertices in M_{n+2}. Let the edges on x be partitioned so they are contained in r cliques of size $m_1, \ldots, m_r$. Then the edges on y will require at least $\max_i m_i$ cliques. Since $\sum M_i = n$, it follows that $\alpha(M_{n+2}) > r + n/r \geqslant 2\sqrt{n}$. Thus $\alpha(G_2) > N$ for n sufficiently large.

THEOREM 3.2. *Let $N > 0$ be given. Then there exist graphs G_1 and G_2 such that $A(G_1)$ and $A(G_2)$ have the same spectrum, $\beta(G_1) = 3$, $\beta(G_2) > N$.*

Proof. Let G_1 be the graph on $4n$ vertices formed by the disjoint union of K_{n+1}, K_{n-1}, and P_{2n}. Let G_2 be the graph on $4n$ vertices formed by the disjoint union of H_{2n}, K_{2n-1}, and one additional vertex. By (3.1), (3.2), and (3.7), $A(G_1)$ and $A(G_2)$ have the same spectrum. Now $\beta(G_1) = 3$. It is easy to see that, in H_{2n}, if i, j, k are different indices, the three edges joining i and i', j and j', k and k' cannot be present in the same complete multipartite graph. Hence $\beta(G_2) > \beta(H_{2n}) > n/2 > N$ for n sufficiently large.

4. ESSENTIAL RELEVANCE OF SPECTRUM OF $A(G)$ TO $\gamma(G)$

THEOREM 4.1. *Each of the following functions of G is bounded by a function of each of the others: $\gamma(G)$, $a(G) \equiv$ the number of different rows of $A(G)$, the number of eigenvalues of $A(G)$ each of which is at most -1, $\mathrm{rank}(A(G))$.*

LEMMA 4.1.

$$\gamma(G) \leqslant a(G) - 1, \qquad a(G) \leqslant 3^{\gamma(G)}. \tag{4.1}$$

Proof. Let $a(G) = a$, $\gamma(G) = \gamma$. Let $S_1, \ldots, S_a$ be a partition of the vertices of G such that i and j are in the same S if and only if row i and row j of $A(G)$ are identical. Note that, if $i, j \in S_k$, then i and j are not adjacent; further, for each k, l, either every vertex of S_k is adjacent to every vertex of S_l, or no vertex of S_k is adjacent to any vertex of S_l. Clearly, $\gamma \leqslant a - 1$.

Next, for the kth complete bipartite graph in the decomposition of $E(G)$ into γ parts, label one part L_k, the other part R_k, and the set of remaining vertices (if any), P_k. Clearly, row i and row j of $A(G)$ are identical if, for each k, row i and row j have the same label for each k. It follows that $a \leqslant 3^\gamma$.

Remark. Using Lemma 4.1, it is easy to prove the spectral relevance of $\gamma(G)$. Let $r = \operatorname{rank} A(G) = $ number of nonzero eigenvalues of $A(G)$. Clearly, $r \leqslant a$. Further, since $A(G)$ is a $(0, 1)$ matrix, it follows that $a \leqslant 2^r$. This proof of relevance is, however, less interesting than Theorem 4.1, whose demonstration we now resume.

LEMMA 4.2. *Let B be a square $(0, 1)$ matrix of order $3n$, $b(B)$ the number of singular values of B each of which is at least 1. If*

$$B = I, \qquad b(B) = 3n, \tag{4.2}$$

$$B = J - I, \qquad b(B) = 3n, \tag{4.3}$$

$$b_{ij} = 0 \quad \text{for} \quad i < j, \qquad b_{ij} = 1 \quad \text{for} \quad i \geqslant j, \qquad b(B) \geqslant n, \tag{4.4}$$

$$b_{ij} = 0 \quad \text{for} \quad i > j, \qquad b_{ij} = 1 \quad \text{for} \quad i \leqslant j, \qquad b(B) \geqslant n. \tag{4.5}$$

Proof. The lemma is obvious in cases (4.2) and (4.3), and clearly (4.4) and (4.5) are essentially the same case. So all we need show is that, if B is lower triangular in the form (4.4), then at least one-third of the eigenvalues of BB' are at least 1. To show this, we will prove that at least one-third of the eigenvalues of $(B^{-1})(B^{-1})'$ are at most 1 (they are positive, since $(B^{-1})(B^{-1})'$ is positive definite).

Let B have the form (4.4), and let $C = (B^{-1})(B^{-1})'$. Then $c_{ii} = 1$ if $i = 1$, $c_{ii} = 2$ if $i > 1$, $c_{i,i+1} = -1$ for $1 \leqslant i \leqslant 3n - 1$, $c_{i-1,i} = 1$ for $2 \leqslant i \leqslant 3n$, all other $c_{ij} = 0$. If we consider the principal submatrix D of order n of C formed by rows $3i$ and $3i - 1$ $(i = 1, \ldots, n)$ and the corresponding columns, then D is the direct sum of n 2×2 matrices

144   A. J. HOFFMAN

$$\begin{bmatrix} 2 & -1 \\ -1 & 2 \end{bmatrix},$$

each of which has 1 as an eigenvalue. By the interlacing theorem, at least n of the eigenvalues of C are at most 1. (One could also have appealed to the relation between C and the vibrating string problem.)

LEMMA 4.3.   *Let $r > 0$ be a given integer. Let E be a rectangular $(0, 1)$ matrix with $3 \cdot 2^{r-1} - 1$ rows, all different. Then it is possible to permute rows and columns of E so that the first $2r$ rows and r columns form a matrix $F = (f_{ij})$ such that*

$$f_{2i-1,i} = 1, \qquad f_{2i,i} = 0, \qquad i = 1,\ldots,r,$$

$$f_{2i-1,j} = f_{2i,j}, \qquad i = 1,\ldots,r, \qquad j = 1,\ldots,r, \qquad j \neq i. \qquad (4.6)$$

Proof.   We shall argue by induction. The lemma holds in case $r = 1$. If any column of E can be discarded, yet all rows of the remaining matrix are different, discard that column. Hence we may assume that, if any column of E is discarded, then at least two rows of the remaining rows of the matrix are the same. In particular, if the first column of E is removed, then (say) rows 1 and 2 of the remaining matrix are the same. Hence, after permuting, we may assume $e_{1,1} = 1$, $e_{2,1} = 0$, $e_{1j} = e_{2j}$ for all $j > 1$. Also, the number of remaining rows of E is $3 \cdot 2^{r-1} - 1 - 2 = 2(3 \cdot 2^{r-2} - 1) - 1$. Hence at least $3 \cdot 2^{r-2} - 1$ remaining rows of E have the same entry in the first column. Application of the induction hypothesis to these rows completes the proof of the lemma.

LEMMA 4.4.   *Let m, n be given. Then there exists an integer $R(n, m)$ such that, if the edges of a clique on at least $R(n, m)$ vertices are colored in m colors, a subclique on n vertices has all edges the same color.*

This is Ramsey's theorem.   See [4].

LEMMA 4.5.   *Let E be a rectangular $(0, 1)$ matrix with $b(E) =$ number of singular values of E which are at least 1. Then the number of different rows of E is less than*

$$3 \cdot 2^{R(3b(E)+3,4)-1} - 1. \qquad (4.7)$$

Proof.   Assume the contrary. By Lemma 4.3, E contains a submatrix F with $2r$ rows and r columns of the form (4.6), with $r = R(3b(E) + 3, 4)$. By Lemma 4.4 (see [4, p. 45], where precisely this application of Ramsey's theorem is made), F contains a submatrix of $3b(E) + 3$ columns and twice that many rows from which one extracts a square submatrix B of order $3b(E) + 3$ of one of the forms (4.2)–(4.5). By Lemma 4.2, the number of singular values of B each of which is at least one is at least $b(E) + 1$. But, if B is a submatrix of E, $b(B) \leqslant b(E)$. Thus we have the contradiction $b(E) \geqslant b(B) \geqslant b(E) + 1$.

If G is a graph, its chromatic number $\chi(G)$ is the smallest number of subsets which form a partition of $V(G)$ such that any two vertices in the same subset are not adjacent.

LEMMA 4.6. *Let $e(G) =$ the number of eigenvalues of $A(G)$ each of which is at most -1. There exists a function h such that*

$$\chi(G) \leqslant h(e(G)). \tag{4.8}$$

This lemma is proved in [2].

LEMMA 4.7. *If h and $e(G)$ are defined as in (4.8), then*

$$a(G) < h(e(G)) R(3 \cdot 2^{R(3e(G)+3,4)-1} - 1, h[e(G) - 1]. \tag{4.9}$$

Proof. Let G be given. Delete a row of G (and corresponding column) if it is identical with another row, and repeat. Since $a(G)$ stays the same and $e(G)$ does not increase, it is clear that it is sufficient to show (4.9) in the case where all rows of $A(G)$ are different, which we now assume. By Lemma 4.6, $V(G)$ can be partitioned into at most $h(e(G))$ subsets, so that no two in the same subset are adjacent. One of these subsets S must contain at least $v = a/h$ vertices, where $a = a(G)$, $h = h(e(G))$. Suppose

$$\frac{a}{h} \geqslant R(3 \cdot 2^{R(3e(G)+3,4)-1} - 1, h - 1). \tag{4.10}$$

We shall show that (4.10) leads to a contradiction, thus proving (4.9).

If (4.10) holds, let us color the edges of the complete graph on v vertices in at most $h - 1$ colors as follows: Label the subsets other than S by $1, \ldots, h - 1$. If $i, j \in S$, $i \neq j$, let k be the label of the subset of vertices which contains the vertex of smallest index t such that $a_{it} \neq a_{jt}$. By Lemma 4.4, A contains a submatrix

$$C = \begin{bmatrix} 0 & B \\ B^T & 0 \end{bmatrix}$$

where B has at least

$$3 \cdot 2^{R(3e(G)+3,4)-1} - 1 \tag{4.11}$$

rows which are all different. Now the nonzero eigenvalues of C are the singular values of B and their negatives, as is well known. By the interlacing theorem, $e(G) = b(B)$. Thus (4.11) contradicts Lemma 4.5.

The proof of the theorem now follows from Lemma 4.7, and the inequalities given in Theorem 2.1 (Eq. 2.8) and Lemma 4.1,

$$e(G) \leqslant \gamma(G) < a(G).$$

5. ESSENTIAL IRRELEVANCE OF SPECTRUM OF $A(G)$ TO $\chi(G)$ AND $\tilde{\chi}(G)$

Let $K_{(r)}m$ stand for the complete m-partite graph in which each part has exactly m vertices (if $r = 1$, we may write instead K_m). Let k be a positive integer. Let H_1 be the graph which is the union of

$$K_{(2^k-1)3} \cup 2K_{(2^k-2)3} \cup \cdots \cup 2^{k-1}K_{(2^0)3}.$$

Let H_2 be the graph which is the union of K_{2^k+1} and

$$2K_{(2^k-1)2} \cup 2^2K_{(2^k-2)2} \cup \cdots \cup 2^{k-1}K_{2^2}.$$

One can calculate that H_1 and H_2 have the same nonzero eigenvalues. Hence, by adding isolated vertices, we obtain cospectral graphs G_1 and G_2, $\chi(G_1) = 3$, $\chi(G_2) = 2^k + 1$.

The mixed chromatic number $\tilde{\chi}(G)$ is the smallest number of subsets in a partition of $V(G)$ such that, for each subset, either every pair of vertices is adjacent or every pair of vertices is not adjacent. Clearly, by taking n very large, nG and nG_2 are cospectral, $\tilde{\chi}(G_1) = 3$, $\tilde{\chi}(G_2) = 2^{k+1}$.

REFERENCES

1 Frank Harary, *Graph Theory*, Addison-Wesley Publishing Company, Reading, Massachusetts, 1969, p. 18.
2 A. J. Hoffman, On eigenvalues and colorings of graphs, to appear in *Proc. Advanced Seminar on Graph Theory*, Army Mathematical Center, University of Wisconsin, Madison, Wisconsin.
3 A. J. Hoffman and Leonard Howes, On eigenvalues and colorings of graphs, II, to appear in *Proc. Internat. Symp. on Combinatorial Math.*, New York Academy of Sciences.
4 H. J. Ryser, Combinatorial mathematics, Carus Monograph 14, *Math. Assoc. Amer.*, 1963.
5 J. H. Smith, in Combinatorial Structures and Their Applications, *Proc. Calgary Internat. Conf. on Combinatorial Structures*, Gordon and Breach, New York, 1970, pp. 403–405.

Received September, 1970

N. Srivastava et al., eds., *A Survey of Combinatorial Theory*
© North-Holland Publishing Company, 1973

CHAPTER 22

On Spectrally Bounded Graphs

A. J. HOFFMAN

IBM Inc., New York, N.Y., U.S.A.

1. Introduction

Let G be a graph (undirected, on a finite number of vertices, with at most one edge joining a pair of vertices, and no edge joining a vertex to itself). Two vertices are said to be adjacent if they are joined by an edge. The valence of a vertex is the number of vertices adjacent to it. The set of vertices of G is denoted by $V(G)$, the set of edges by $E(G)$.

If G is a graph, the adjacency matrix of $G \equiv A(G)$ is defined by

$$A(G) = A = (a_{ij}) = \begin{cases} 1 & \text{if vertex } i \text{ and vertex } j \text{ are adjacent,} \\ 0 & \text{otherwise.} \end{cases}$$

Thus, $A(G)$ is a real symmetric $(0, 1)$ matrix of order $|V(G)|$, with diagonal all 0. If A is any real symmetric matrix, we denote its eigenvalues in descending order by $\lambda_1(A) \geq \lambda_2(A) \geq \ldots$, in ascending order by $\lambda^1(A) \leq \lambda^2(A) \leq \ldots$. If $A = A(G)$, we will write $\lambda_i(G)$ and $\lambda^i(G)$ for $\lambda_i(A(G))$, $\lambda^i(A(G))$.

Let $\mathscr{G}$ be an infinite set of graphs. In the investigations given in the bibliography, a lower bound on $\lambda^1(G)$, as G varies in $\mathscr{G}$, played a key role; sometimes a specific lower bound (most especially -2), sometimes the existence of some unspecified bound. It is this latter situation we will explore in detail. For given $\mathscr{G}$, it may be true or false that there exists some λ such that $\lambda^1(G) \geq \lambda$ for all $G \in \mathscr{G}$. We shall give two graph theoretic characterizations of those families $\mathscr{G}$ for which a uniform lower bound exists, one "local" in terms of excluded subgraphs, one "global" in describing how each graph in $\mathscr{G}$ is constructed.

2. Statement of characterizations

We shall need further definitions. If G is a graph, $\bar{G}$ is the graph with $V(G) = V(\bar{G})$, and two distinct vertices are adjacent in $\bar{G}$ if and only if they are not adjacent in G. If H and G are graphs, we say $H \subset G$ (H is a subgraph of G) if $V(H) \subset V(G)$, and two vertices in $V(H)$ are adjacent in H if and only if they were adjacent in G. A clique on t vertices, denoted by K_t, is a graph in which each pair of distinct vertices is adjacent. $\bar{K}_t$ is called an independent set on t

vertices. The symbol C_t denotes a claw on t vertices (i.e., a set of $t+1$ vertices, one of which is adjacent to all the others, of which no two are adjacent; this graph is also sometimes denoted $K_{1,t}$). The symbol H_t denotes a graph on $2t+1$ vertices, $2t$ of which form a K_{2t}, while the remaining vertex is adjacent to exactly t of these $2t$ vertices.

If G and H are graphs with $V(G) = V(H)$, we shall define a distance $d(G, H)$. Write

$$A(G) + A(\tilde{G}) = A(H) + A(\tilde{H}),$$

where

$$(A(\tilde{G}))_{ij} = 1 \quad \text{if and only if } (A(G))_{ij} = 0, \ (A(H))_{ij} = 1$$
$$(A(\tilde{H}))_{ij} = 1 \quad \text{if and only if } (A(G))_{ij} = 1, \ (A(H))_{ij} = 0.$$

Then $d(G, H) =$ the largest of the valences of the vertices in $\tilde{G}$ and $\tilde{H}$.

Theorem. *Let $\mathcal{G}$ be an infinite set of graphs. Then the following statements about $\mathcal{G}$ are all true or all false:*

(i) there exists a number λ such that, for all $G \in \mathcal{G}$, $\lambda^1(G) \geq \lambda$;

(ii) there exists a positive integer l such that, for all $G \in \mathcal{G}$, $C_l \nsubseteq G$ and $H_l \nsubseteq G$;

(iii) there exists a positive integer L such that, for each $G \in \mathcal{G}$, there exists a graph H with $d(G, H) \leq L$, and H contains a distinguished family of cliques $K^1, K^2, \ldots$ such that

(iiia) each edge of H is in at least one K^i,

(iiib) each vertex of H is in at most L of the cliques $K^1, K^2, \ldots$,

(iiic) $|V(K^i) \cap V(K^j)| \leq L$ for $i \neq j$.

This theorem was first announced by Hoffman [1970a]. Some consequences of the theorem are reported in Hoffman [1970a,b, 1971]. In addition, a portion of these results have been used by Howes [1970] to characterize the families $\mathcal{G}$ for which there is a uniform upper bound on $\lambda_2(G)$ for all $G \in \mathcal{G}$.

Before proceeding to the proof, which occupies the remainder of the paper, some remarks are in order. Let $\mathcal{G} = \{G_1, G_2, \ldots\}$. If $\mathcal{H} = \{H_1, H_2, \ldots\}$ is an infinite sequence of graphs such that there exist two infinite sequences of indices $i_1 < i_2 \ldots$ and $j_1 < j_2 \ldots$ with $G_{i_k} \subset H_{j_k}$, $k = 1, 2, \ldots$, then we will say $\mathcal{G} \subset \mathcal{H}$. The significance of (i) $\Leftrightarrow$ (ii) in the theorem can now be restated: if $\mathcal{G} = \{G_1, G_2, \ldots\}$ is any sequence of graphs such that $\lambda^1(G_i) \to -\infty$, then $\{C_1, H_1, C_2, H_2, C_3, H_3, \ldots\} \subset \mathcal{G}$.

The second remark concerns (iii). It would be desirable, if true, to avoid the intervention of H in (iii) and assert (iiia,b,c) for G rather than H. But the intervention of H cannot be avoided. Let G_n be the cocktail party graph of order n (i.e., G_n is a graph on $2n$ vertices, in which each vertex is adjacent to all but one of the remaining vertices). Note $\lambda^1(G_n) = -2$ for all $n \geq 2$.

Let k be the largest number of vertices in a clique $\{K^i\}$, say $k = |V(K^1)|$. Let v be a vertex not adjacent to a vertex, say w, in K^1. Then v and each of the $k-1$ vertices in K^1 other than w is contained in at least one distinguished clique by (iiia). There are at most L such cliques, by (iiib), and each contains less than L vertices other than v, by (iiic). It follows that $k-1 \leq L^2$. Since $k = \max |V(K^i)|$, it follows that w is adjacent to at most $L(k-1) \leq L^3$ vertices of G_n. But this cannot be true for $2n-1 \geq L^3$.

3. Proof of Theorem

In this section, we show that (i) $\Rightarrow$ (ii) and (iii) $\Rightarrow$ (i). In the next section, we will show (ii) $\Rightarrow$ (iii).

Throughout the proof, we shall use the fact that $G \subset H$ implies $\lambda^1(G) \geq \lambda^1(H)$, which follows from the fact that $G \subset H$ means that $A(G)$ is a principal submatrix of $A(H)$.

To prove (i) $\Rightarrow$ (ii), it is sufficient to record that $\lambda^1(C_t)$, $\lambda^1(H_t)$ both tend to $-\infty$ for t large, which we have proved elsewhere (Hoffmann [1971]), or which the reader can easily establish himself. This and the preceding paragraph complete the proof.

To prove that (iii) implies (i), assume a graph G satisfies (iii) for some L. We will prove

$$\lambda^1(G) \geq -3L - \binom{L}{2}(L-1), \tag{3.1}$$

which will prove (i).

Let M be the (0, 1) incidence matrix of vertices of H versus distinguished cliques $K^1, K^2, \ldots$. Thus $m_{ij} = 1$ if $v_i \in K^j$, and 0 otherwise.

Lemma 3.1. $MM^{\mathsf{T}} = A(H) + S$, where

$$S = (s_{ij}) \tag{3.2}$$

is a matrix with all entries nonnegative, and

$$\lambda_1(S) \leq L + \binom{L}{2}(L-1). \tag{3.3}$$

Proof. To prove (3.2), note first that MM^{T} has all entries nonnegative, $A(H)_{ii} = 0$ for all i, hence $s_{ii} \geq 0$ for all i. If two vertices in H are adjacent, then there is at least one distinguished clique containing both by (iiia), so $(MM^{\mathsf{T}})_{ij} \geq 1$, $(A(H))_{ij} = 1$, so $s_{ij} \geq 0$. If two different vertices i and j of H are not adjacent, $(MM^{\mathsf{T}})_{ij} = 0$, $A(H)_{ij} = 0$, $s_{ij} = 0$. Thus, in all cases $s_{ij} \geq 0$.

To prove (3.3), we note that, by the Perron-Frobenius theorem, $\lambda_1(S) \leq \max_i \sum_j s_{ij}$. Also, by (iiib), $s_{ii} \leq L$. Hence, to prove (3.3), it is sufficient to prove that, for each i,

$$\sum_{j \neq i} s_{ij} \leq (L-1)\binom{L}{2}. \tag{3.4}$$

But, the left side of (3.4) is the number of 2×2 matrices in M which have one row i and consist entirely of 1's. But the number of 1's in row i is at most L by (iiib), so the number of pairs of columns which are candidates for a 2×2 "box" is at most $\binom{L}{2}$. Two such columns can produce at most $L-1$ boxes, by (iiic). Thus (3.4) is proved, and hence the lemma.

To complete the proof of (3.1), we invoke the theorem that, if A and B are real symmetric matrices, and $C = A + B$, then

$$\lambda^1(A) + \lambda^1(B) \leqq \lambda^1(C) \leqq \lambda^1(A) + \lambda_1(B). \tag{3.5}$$

By the Perron-Frobenius theorem, if G is a graph in which each vertex has valence at most L, then

$$-L \leqq \lambda^1(G) \leqq \lambda_1(G) \leqq L. \tag{3.6}$$

From $A(G) + A(\tilde{G}) = A(H) + A(\tilde{H})$, (3.5) and (3.6), we conclude

$$\lambda^1(G) \geqq \lambda^1(H) - 2L. \tag{3.7}$$

From Lemma 3.1 and (3.5), we conclude

$$0 \leqq \lambda^1(MM^T) \leqq \lambda^1(H) + \lambda_1(S) \leqq \lambda^1(H) + L + (L-1)\binom{L}{2}. \tag{3.8}$$

But (3.7) and (3.8) imply (3.1).

4. Proof of the Theorem (continued)

In this section, we prove (ii) $\Rightarrow$ (iii). Our reasoning here is entirely graph theoretical, since the concept of eigenvalue plays no role in the statement of (ii) or (iii). We shall prove that there exists a function $L(l)$ such that, if G satisfies (ii) for some l, then G satisfies (iii) for $L = L(l)$.

The strategy of the proof is as follows: We shall first look for large cliques in G ("large" depends on l), and define an equivalence relation on large cliques. The equivalence classes of large cliques will be shown to have properties (iiib) and (iiic) and, if additional edges (forming a graph $\tilde{G}$ in which each vertex has "small" valence) are added, the equivalence classes will be cliques. It will also turn out that the edges in G not contained in any large clique form a graph $\tilde{H}$ in which each vertex has small valence. Thus the distinguished cliques in H will be the equivalence classes of large cliques in G.

To define large cliques, we need first the Ramsey function $R(l)$ which satisfies: if $|V(G)| \geqq R(l)$, then $K_l \subset G$ or $\overline{K}_l \subset G$. We also need a function $f(m, r, l)$ defined recursively on triples of positive integers:

$$f(1, r, l) = r + 1,$$
$$f(m+1, r, l) = \max \{r + mr(l-2) + 1, f(m, r+l-1, l)\}.$$

Let

$$N = N(l) = \max \{l^2 + l + 2, l + R(l), f(l, 1, l)\}.$$

Define $\mathcal{W}$ to be the set of all cliques $K \subset G$ such that $|V(K)| \geqq N$.

Lemma 4.1. *If K, $K' \in \mathscr{W}$, define*

$$K \sim K'$$

if each vertex of K is adjacent to all but at most $l-1$ vertices of K'. Then $\sim$ is an equivalence relation.

Proof. Reflexivity is clear, since $|V(K)| \geqq l$. To prove symmetry, assume there is a vertex v in K' not adjacent to at least l vertices in K, and let A denote that set of l vertices in K. Each vertex in A is not adjacent to at most $l-2$ vertices in K' other than v, since $K \sim K'$. Hence, the set of vertices in K' each not adjacent to at least one vertex in A consists of v and at most $l(l-2)$ other vertices. Since $N > l + l(l-2) + 1$, it follows that K' contains at least l vertices each of which is adjacent to each vertex in A. Call that set of l vertices B. Then v, A, B generate an H_l, contrary to hypothesis. This contradiction proves that $\sim$ is symmetric.

To prove transitivity, assume $K^1 \sim K^2$, $K^2 \sim K^3$, $K^1 \not\sim K^3$. Then K^3 contains a vertex v not adjacent to a set C of l vertices in K^1. Since $N > 2l + l(l-1) - 1$, and $K^1 \sim K^2$, it follows that K^2 contains a subset D of $2l-1$ vertices each of which is adjacent to all vertices in C. But since $K^3 \sim K^2$, D contains some subset F of l vertices adjacent to v. Then C, F, v generate an $H_l \subset G$, which is a contradiction.

Henceforth, the letter E will denote any equivalence class of cliques in $\mathscr{W}$, and $V(E)$ will be the union of all vertices of all cliques in E.

Lemma 4.2. *Let E be an equivalence class, $v \in V(E)$. Then v is adjacent to all but at most $R(l)-1$ other vertices in $V(E)$.*

Proof. Let $K^v \in E$ be a clique containing v. By Ramsey's theorem, if $F \subset V(E), |F| \geqslant R(l)$, and every vertex in F not adjacent to v, then F contains a K_l or a $\overline{K}_l$. If $\overline{K}_l \subset F$, then since $|V(K^v)| > l^2 - 2l + 1$, there exists a vertex $w \in K^v$ adjacent to all vertices in $\overline{K}_l$. Thus $C_l \subset G$, a contradiction. If $K_l \subset F$, then $|V(K^v)| > l + l(l-2) + 1$ implies that K^v contains a set of l vertices each adjacent to all the vertices in K_l, thus generating an H_l.

Lemma 4.3. *If K, $K' \in \mathscr{W}$, $K \in E$, and $V(K') \subset V(E)$, then $K \sim K'$.*

Proof. Assume the contrary. Then there exists a vertex $v \in K'$ not adjacent to as many as l vertices in K, thus adjacent to at most $l-1$ vertices in K, thus not adjacent to more than $N-l$ vertices in K. But since $N \geq l + R(l)$, v is not adjacent to more than $R(l)$ vertices of K, contradicting Lemma 4.2.

Lemma 4.4. *If E_1 and E_2 are different equivalence classes,*

$$|V(E_1) \cap V(E_2)| \leq R(N) - 1.$$

Proof. If $|V(E_1) \cap V(E_2)| \geqslant R(N)$, then by Ramsey's theorem, $V(E_1) \cap V(E_2)$ contains K_N or $\overline{K}_N$ as a subgraph. It is impossible for $\overline{K}_N$ to be a subgraph, by Lemma 4.2. If K_N were a subgraph, let $K \in E_1$. By Lemma 4.3,

$K_N \sim K$, so $K_N \in E_1$. Similarly, $K_N \in E_2$. Therefore, $E_1 = E_2$, contrary to hypothesis.

Lemma 4.5. *If $K, K' \in \mathscr{W}$, $Z \subset V(K')$, $|Z| = l$, each vertex of Z adjacent to all but at most $l-1$ vertices of K, then $K \sim K'$.*

Proof. Assume the contrary; so there is a vertex $v \in V(K')$ adjacent to fewer than l vertices in K. Since $|V(K)| \geq N$, there are at least $(2l-1)$ vertices in K each of which is adjacent to all vertices in Z. Vertex v is not adjacent to at least l of them, thus an H_l would be generated.

Lemma 4.6. *Let $f(m, r, l)$ be the function defined at the beginning of Section 4. Let $n \geq f(m, r, l)$, and let $K^1, \ldots, K^m \in \mathscr{W}$ be inequivalent cliques, v a vertex in each of $V(K^1), \ldots, V(K^m)$, $|V(K^i)| \geq n$, $i = 1, \ldots, m$. Then there exist sets $S_i \subset V(K^i) - v$, $i = 1, \ldots, m$, such that $|S_i| = r$, and $i \neq j$ implies that each vertex in S_i is adjacent to no vertices in S_j.*

Proof (by induction on m). If $m = 1$, then $n \geq r+1$; the lemma holds. Assume the lemma to be true for some m and all r; we shall show that it holds for $m+1$ and all r.

Since $n \geq f(m+1, r, l) \geq f(m, r+l-1, l)$, it follows from the induction hypothesis that there exist subsets $S_i' \subset V(K')$, $i = 1, \ldots, m$, $|S_i'| = r+l-1$, and each vertex in S_i' adjacent to no vertices in S_j' for $i \neq j$. By Lemma 4.5, at most $l-1$ vertices in S_i' $(i = 1, \ldots, m)$ are each adjacent to at least l vertices in K^{m+1}. Consequently, S_i' contains a subset S_i, $|S_i| = r$, such that each vertex in S_i is adjacent to at most $l-2$ vertices in K^{m+1} other than v. Since

$$|V(K^{m+1})| \geq r+mr(l-2)+1,$$

there exists a subset $S_{m+1} \subset V(K^{m+1})$, $|S_{m+1}| = r$, such that each vertex in S_{m+1} is adjacent to no vertex of any S_i, $i = 1, \ldots, m$. This completes the induction.

Lemma 4.7. *Each vertex is contained in fewer than l equivalence classes.*

Proof. Since $N \geq f(l, 1, l)$, a contradiction of Lemma 4.7 would produce, by Lemma 4.6, $aC_l \subset G$.

Lemma 4.8. *Let $\tilde{H}$ be the graph formed by edges of G not in any clique in W. Then every vertice in $\tilde{H}$ has valence at most $R(\max(N-1, l))$.*

Proof. If not, then by Ramsey's theorem we would have $C_l \subset G$, or the edges in $\tilde{H}$ adjacent to v would be in a clique in $\mathscr{W}$, contradicting the definition of $\tilde{H}$.

It is clear that, following the strategy outlined at the beginning of this section, we have proved (ii) $\Rightarrow$ (iii), with

$$L(l) = \max\{R(l)-1, R(N)-1, R(\max(N-1, l), l-1\} = \max\{R(N)-1\},$$

where N is defined at the beginning of this section.

We are very grateful to Leonard Howes for his valuable help in the development and exposition of this material.

References

M. Doob, 1970, On characterizing certain graphs with few eigenvalues by their spectra, *Linear Algebra Appl.* **3**, 461–482.

A. J. Hoffman, 1960a, On the uniqueness of the triangular association scheme, *Ann. Math. Statist.* **31**, 492–497.

A. J. Hoffman, 1960b, On the exceptional case in a characterization of the arcs of a complete graph, *IBM J. Res. Develop.* **4**, 497–504.

A. J. Hoffman, 1965, On the line graph of a projective plane, *Proc. Am. Math. Soc.* **16**, 297–302.

A. J. Hoffman, 1968, Some recent results on spectral properties of graphs, *Beiträge zur Graphentheorie* (H. Sachs, H. J. Voss and H. Walther, eds.; B. G. Teubner Verlagsgesellschaft, Leipzig), pp. 75–80.

A. J. Hoffman, 1969a, The eigenvalues of the adjacency matrix of a graph, *Combinatorial Mathematics and its Applications* (R. C. Bose and T. C. Dowling, eds.; Univ. of North Carolina Press, Chapel Hill, N. Car.), pp. 578–584.

A. J. Hoffman, 1969b, The change in the least eigenvalue of the adjacency matrix of a graph under imbedding, *SIAM J. Appl. Math.*, 664–671.

A. J. Hoffman, 1970a, $-1-\sqrt{2}$?, *Combinatorial Structures and Their Applications* (R. Guy, H. Hanani, N. Taner, J. Schonheim, eds.; Gordon and Breach, New York), pp. 173–176.

A. J. Hoffman, 1970b, On eigenvalues and colorings of graphs, *Graph Theory and its Applications* (B. Harris, ed.; Academic Press, New York), pp. 79–91.

A. J. Hoffman, 1971, On vertices near a given vertex of a graph, *Studies in Pure Mathematics, papers presented to Richard Rado* (L. Mirsky, ed.; Academic Press, London), pp. 131–136.

A. J. Hoffman and Leonard Howes, 1970, On eigenvalues and colorings of graphs, II, *Intern. Conf. on Combinatorial Mathematics* (A. Gewirtz and L. Quintas, eds.), *Ann. N.Y. Acad. Sci.* **175**, 238–242.

A. J. Hoffman and A. M. Ostrowski, On the least eigenvalue of a graph of large minimum valence containing a given graph, *Linear Algebra Appl.* (to appear).

A. J. Hoffman and D. K. Ray-Chaudhuri, 1965a, On the line graph of a finite affine plane, *Canad. J. Math.* **17**, 687–694.

A. J. Hoffman and D. K. Ray-Chaudhuri, 1965b, On the line graph of a symmetric balanced incomplete block design, *Trans. Am. Math. Soc.* **116**, 238–252.

A. J. Hoffman and D. K. Ray-Chaudhuri, On a spectral characterization of regular line graphs (unpublished).

Leonard Howes, 1970, On subdominantly bounded graphs, Doctoral Dissertation, City Univ. of New York.

D. K. Ray-Chaudhuri, 1967, Characterization of line graphs, *J. Combin. Theory* **3**, 461–482.

J. J. Seidel, 1968, Strongly regular graphs with $(-1, 1, 0)$ adjacency matrix having eigenvalue 3, *Linear Algebra Appl.* **1**, 281–298.

Reprinted from IBM J. of Res. & Develop.
Vol. 17, No. 5 (1973), pp. 420–425

W. E. Donath
A. J. Hoffman

Lower Bounds for the Partitioning of Graphs

Abstract: Let a k-partition of a graph be a division of the vertices into k disjoint subsets containing $m_1 \geq m_2, \cdots, \geq m_k$ vertices. Let E_c be the number of edges whose two vertices belong to different subsets. Let $\lambda_1 \geq \lambda_2, \cdots, \geq \lambda_k$ be the k largest eigenvalues of a matrix, which is the sum of the adjacency matrix of the graph plus any diagonal matrix U such that the sum of all the elements of the sum matrix is zero. Then

$$E_c \geq \tfrac{1}{2} \sum_{r=1}^{k} - m_r \lambda_r.$$

A theorem is given that shows the effect of the maximum degree of any node being limited, and it is also shown that the right-hand side is a concave function of U. Computational studies are made of the ratio of upper bound to lower bound for the two-partition of a number of random graphs having up to 100 nodes.

Introduction

Partitioning of graphs occurs in computer logic partitioning [1, 2], paging of computer programs [3, 4], and may also find application in the area of classification [5]. Graph partitioning is the problem of dividing the vertices of a graph into a given number of disjoint subsets such that the number of nodes in each subset is less than a given number, while the number of *cut* edges, i.e., edges connecting nodes in different subsets, is a minimum. The problem of computer logic partitioning is actually somewhat different; for a thorough description of that problem, see Ref. 1. The partitioning of graphs is a simplified version of that problem.

In this paper, we assume that the number of vertices in each subset is prescribed. Let $A = A(G)$ be the adjacency matrix of the graph G, which will be defined later, and U any diagonal matrix with the property that trace (U) is the negative of the sum of the valences of the vertices. We derive in Theorem 1 a lower bound for the number of cut edges in terms of the eigenvalues of $A + U$. For the case of division into two subsets, we present, using a different method of derivation, another bound that is stricter for this special case. The bound given in Theorem 1 turns out to be a concave function of U, a fact that suggests exploitation by means of mathematical programming. Computational results are presented, in which the bound is compared with the results of actual, but not necessarily minimal, partitioning. We also compare experimentally the results when $U_{ii} = -d_i$ (the valence of vertex i), and when the $\{U_{ii}\}$ vary.

We believe that, in combinatorial problems whose complexity suggests the use of heuristic methods, such as the partitioning of graphs, it is worthwhile to have a lower bound on what can be achieved, regardless of the algorithm, provided the calculation of the bound is itself not too onerous and the bounds derived are not too far from the correct value. The results presented here may satisfy these criteria. The calculation of the bound may itself suggest new approaches to the original problem. Also, the fact that different methods are used to derive the bounds of Theorems 1 and 2 suggests that a more comprehensive approach to the problem may be possible. We should also mention that a different use of eigenvalues and eigenvectors on a related problem is discussed in Ref. 6.

This paper does not present details of our experiments on the algorithm for varying U, because a new method which converges to the maximum value of the bound has since been found by Jane Cullum. We are grateful to Jane Cullum and Philip Wolfe, of the Watson Research Center, for useful conversations about the present work.

420

Derivation of lower bound

Let G be a graph, with edge set E, and vertex set V. For any set S, $|S|$ denotes the number of elements in S. Let $A(G) = (a_{ij})$ be a square matrix of order $|V|$ and be defined by:

$$a_{ij} = \begin{cases} 1 \text{ if vertices } i \text{ and } j \text{ are joined by an edge,} \\ 0 \text{ otherwise.} \end{cases}$$

Thus, $A(G)$ — the adjacency matrix of G — is a square symmetric $(0, 1)$ matrix with 0 diagonal.

Let the eigenvalues of any real symmetric matrix M be denoted by $\lambda_1(M) \geq \lambda_2(M) \geq \cdots$; let U be any diagonal matrix such that $\Sigma_i U_{ii} = -2|E|$; let $m_1 \geq m_2 \geq \cdots \geq m_k$ be given positive integers such that $\Sigma m_i = |V|$; and let $V_1, \cdots, V_k$ be disjoint subsets of V such that $|V_i| = m_i$, $i = 1, \cdots, k$. Finally, let E_c be the set of edges of G, each of which has its two endpoints in different V_i.

Theorem 1. Given the notation above,

$$|E_c| \geq -\frac{1}{2} \sum_1^k m_i \lambda_i (A + U).$$

The right-hand side is a concave function of U.

Proof. It is easy to see that the main theorem of Hoffman and Wielandt [7], when applied to real symmetric matrices M and N of order n, yields

$$\text{Trace } MN^{\mathrm{T}} \leq \sum_1^n \lambda_i(M)\lambda_i(N), \tag{1}$$

in which the Trace of a matrix denotes the sum of all the elements of the diagonal. We note that, if N^{T} is the transpose of N, then $\text{Trace } MN^{\mathrm{T}} = \Sigma_{ij} M_{ij} N_{ij}$. Let $M = A + U$ and N be the direct sum of k matrices, each of which consists entirely of 1's, and is defined on the rows and columns corresponding to $V_i (i = 1, \cdots, k)$. Then

$$\lambda_1(N) = m_1, \cdots, \lambda_k(N) = m_k,$$
$$\lambda_{k+1}(N) = \cdots = \lambda_{|V|}(N) = 0. \tag{2}$$

It is clear that

$$\sum_1^{|V|} \lambda_i(M)\lambda_i(N) = \sum m_i \lambda_i(A + U). \tag{3}$$

On the other hand,

$$\text{Trace } MN^{\mathrm{T}} = -2|E| + 2(|E| - |E_c|) = -2|E_c|. \tag{4}$$

Inserting (3) and (4) into (1) proves the first sentence of Theorem 1.

To prove the second sentence, it is sufficient to show that $\Sigma m_i \lambda_i (A + U)$ is a convex function of U. If R, S are real symmetric matrices of order n, and if $l \leq n$, then [8]

$$\sum_1^l \lambda_i(R + S) \leq \sum_1^l \lambda_i(R) + \sum_1^l \lambda_i(S).$$

Hence

$$\sum_1^l \lambda_i \left[A + \frac{1}{2}(U_1 + U_2) \right] = \sum_1^l \lambda_i \left[\frac{1}{2}(A + U_1) \right.$$
$$\left. + \frac{1}{2}(A + U_2) \right]$$
$$\leq \sum_1^l \frac{1}{2}(A + U_1)$$
$$+ \sum_1^l \frac{1}{2} \lambda_i(A + U_2).$$

$\sum_1^l \lambda_i(A + U)$ is a convex function of U.

Next

$$\sum_1^k m_i \lambda_i(A + U) = (m_1 - m_2)\lambda_1(A + U) \tag{5}$$
$$+ (m_2 - m_3)[\lambda_1(A + U) + \lambda_2(A + U)]$$
$$+ \cdots + m_k[\lambda_1(A + U)$$
$$+ \cdots + \lambda_k(A + U)].$$

Since $m_i \geq m_{i+1}$, $i = 1, \cdots, k - 1$, and $m_k > 0$, it follows that the right-hand side of (5) is a nonnegative sum of convex functions of U and hence a convex function of U.

The next theorem is concerned with a partition into two equal groups when the maximum degree of any vertex of G is less than d.

Theorem 2. Given 1) a graph with an even number of vertices, 2) that $m_1 = m_2 = |V|/2$, 3) that the degree of any vertex does not exceed some value d, and 4) that $0 \leq \delta_1 \leq \pi/4$, $0 \leq \delta_2 \leq \pi/4$, and 5) that $x \leq 1$ represents a simultaneous solution of the equations

$$x \sin 2\delta_1 = (1 - x) \sin 2\delta_2, \tag{6}$$
$$-[\lambda_1(A + U) + \lambda_2(A + U)]/2 = x\{1 - \sin 2\delta_1$$
$$+ 2(d - 1)[1 - \cos(\delta_1 + \delta_2)]\}, \tag{7}$$

then

$$|E_c| \geq x|V|/2. \tag{8}$$

Note: Setting $\delta_1 = \delta_2 = 0$ causes this theorem to be a special case of Theorem 1, namely, the case in which $m_1 = m_2 = |V|/2$.

Proof. We first show that if there exists any partition into two equalized groups with $e < |V|/2$ edges cut, then

$$-\frac{1}{2}[\lambda_1(A + U) + \lambda_2(A + U)] \leq Z\{1 - \sin 2\alpha_1$$
$$+ 2(d - 1)[1 - \cos(\alpha_1 + \alpha_2)]\}, \tag{9}$$

where $Z = 2e/|V|$ and α_1 and α_2 are any numbers satisfying

421

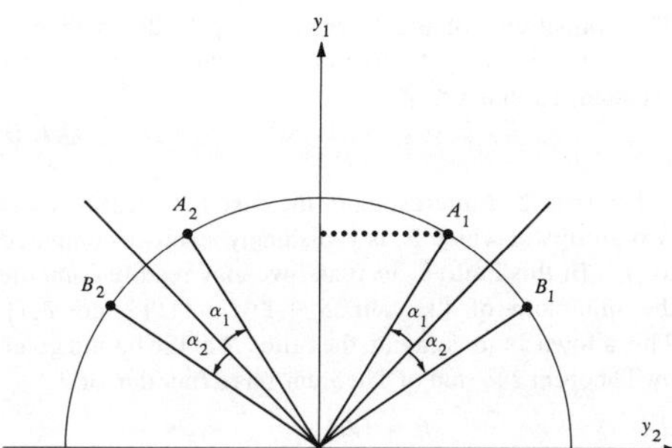

Figure 1 Locations of the groups A_1, A_2, B_1, B_2 in a coordinate system defined by y_1, y_2. These four groups are defined in the proof of Theorem 2.

$$Z \sin 2\alpha_1 = (1 - Z) \sin 2\alpha_2, \quad 0 \le \alpha_1, \alpha_2 \le \pi/4. \tag{10}$$

Later we show that the result given above is sufficient to prove the theorem.

It has been shown [9] that, if y_1 and y_2 are any two orthonormal vectors, then

$$\lambda_1(A + U) + \lambda_2(A + U) \ge$$
$$y_1^{\mathrm{T}}(A + U)y_1 + y_2^{\mathrm{T}}(A + U)y_2, \tag{11}$$

since A and U are symmetric matrices. We can furthermore see that

$$y_k^{\mathrm{T}}(A + U)y_k = \sum_i \sum_j A_{ij} y_{ki} y_{kj} + \sum_i U_{ii} y_{ki}^2$$
$$= -\frac{1}{2} \sum_i \sum_j A_{ij}(y_{ki} - y_{kj})^2 + \sum_i (U_{ii}$$
$$+ \sum_j A_{ij})y_{ki}^2, \tag{12}$$

where y_{ki} are the components of y_k. Let us define

$$U_{ii}' = U_{ii} + \sum_j A_{ij}. \tag{13}$$

Since $\sum_i \sum_j A_{ij} + \sum_i U_{ii} = 0$, we have

$$\sum_i U_{ii}' = 0. \tag{14}$$

We now divide further each of the two groups 1, 2 into which the corresponding V that was partitioned; subgroup $A_k (k = 1,2)$ is a set of exactly e vertices, which includes *all* the vertices that have connections to vertices not belonging to group k. As long as $e < |V|/2$, such a set can always be generated. The other subgroup B_k has, of course, only connections to group A_k. There are $|V|/2 - e$ vertices in B_k, and N_k connections between A_k and B_k. We now set values for $y_{mi}(m = 1,2)$, as indicated in Fig. 1.

422

W. E. DONATH AND A. J. HOFFMAN

$$i \in B_1 \qquad y_{1i} = \sqrt{2/|V|} \cos (\pi/4 + \alpha_2)$$
$$y_{2i} = \sqrt{2/|V|} \sin (\pi/4 + \alpha_2)$$
$$i \in A_1 \qquad y_{1i} = \sqrt{2/|V|} \cos (\pi/4 - \alpha_1)$$
$$y_{2i} = \sqrt{2/|V|} \sin (\pi/4 - \alpha_1)$$
$$i \in A_2 \qquad y_{1i} = \sqrt{2/|V|} \cos (\pi/4 - \alpha_1)$$
$$y_{2i} = - \sqrt{2/|V|} \sin (\pi/4 - \alpha_1)$$
$$i \in B_2 \qquad y_{1i} = \sqrt{2/|V|} \cos (\pi/4 + \alpha_2)$$
$$y_{2i} = - \sqrt{2/|V|} \sin (\pi/4 + \alpha_2). \tag{15}$$

Since $|A_1| = |A_2|$ and $|B_1| = |B_2|$, then y_1 and y_2 are clearly orthogonal. We now show that condition (10) proves $\|y_k\| = 1$, $k = 1, 2$. It can be seen that

$$y_1^{\mathrm{T}} y_1 = (2/|V|)[(|V| - 2e) \sin^2 (\pi/4 + \alpha_2)$$
$$+ 2e \sin^2 (\pi/4 - \alpha_1)],$$
$$y_2^{\mathrm{T}} y_2 = (2/|V|)[(|V| - 2e) \cos^2 (\pi/4 + \alpha_2)$$
$$+ 2e \cos^2 (\pi/4 - \alpha_1)],$$

so that $y_1^{\mathrm{T}} y_1 + y_2^{\mathrm{T}} y_2 = 2$. However, we also require for normality that $y_1^{\mathrm{T}} y_1 - y_2^{\mathrm{T}} y_2 = 0$ so that

$$0 = (2/|V|)[(|V| - 2e) \cos (\pi/2 + 2\alpha_2)$$
$$+ 2e \cos (\pi/2 - 2\alpha_1)]$$

or, dividing by two and using $Z = 2e/|V|$, we have

$$0 = (1 - Z) \cos (\pi/2 + 2\alpha_2) + Z \cos (\pi/2 - 2\alpha_1).$$

Since $\cos (\pi/2 + x) = - \sin x$, we find the above to be equivalent to $0 = -(1 - Z) \sin 2\alpha_2 + Z \sin 2\alpha_1$, which is condition (10).

Inserting Eq. (12) into (11) we have

$$-\lambda_1(A + U) - \lambda_2(A + U) \le \frac{1}{2} \sum_i \sum_j A_{ij}[(y_{1i} - y_{1j})^2$$
$$+ (y_{2i} - y_{2j})^2]$$
$$- \sum_i U_{ii}'(y_{1i}^2 + y_{2i}^2).$$

However, from Eq. (15) it follows that $y_{1i}^2 + y_{2i}^2 = 2/|V|$ and, when Eq. (14) is used, the term in U_{ii}' falls out. The other part becomes, on substitution of Eq. (15),

$$-\lambda_1 - \lambda_2 \le \sum_{i \in A_1} \sum_{j \in A_2} A_{ij}(8/|V|) \sin^2 (\pi/4 - \alpha_1)$$
$$+ \left(\sum_{i \in A_1} \sum_{j \in B_1} + \sum_{i \in A_2} \sum_{j \in B_2} \right)(2A_{ij}/|V|)[\sin (\pi/4 - \alpha_1)$$
$$- \sin (\pi/4 + \alpha_2)]^2 + [\cos (\pi/4 - \alpha_1)$$
$$- \cos (\pi/4 + \alpha_2)]^2,$$

which simplifies to

$$-\lambda_1 - \lambda_2 \leq (8e/|V|)[\sin^2 (\pi/4 - \alpha_1)]$$
$$+ (2/|V|)(N_1 + N_2)[\sin^2 (\pi/4 - \alpha_1)$$
$$+ \sin^2 (\pi/4 + \alpha_2)$$
$$- 2 \sin (\pi/4 - \alpha_1) \sin (\pi/4 + \alpha_2)$$
$$+ \cos^2 (\pi/4 - \alpha_1)$$
$$- 2 \cos (\pi/4 - \alpha_1) \cos (\pi/4 + \alpha_2)$$
$$+ \cos^2 (\pi/4 + \alpha_2)].$$

Upon using standard geometric identities, we have

$$-\lambda_1 - \lambda_2 \leq (4e/|V|)(1 - \sin 2\alpha_1) + [4(N_1 + N_2)/|V|]$$
$$\times [1 - \cos (\alpha_1 + \alpha_2)].$$

Because each node has a maximum degree of d, we have

$$e + N_1 \leq ed$$
$$e + N_2 \leq ed$$

so that $N_k \leq e(d - l)$ and using $Z = 2e/|V|$, we have after also dividing both sides by two

$$\frac{1}{2}(-\lambda_1 - \lambda_2) \leq Z(1 - \sin 2\alpha_1) + 2Z(d - 1)$$
$$\times [1 - \cos (\alpha_1 + \alpha_2)],$$

which is the inequality (9).

The second part of the proof consists in showing that any possible value of x that solves Eqs. (6) and (7) must be less than Z. The value of x is then used in the inequality (8).

Let us assume that we have found x, δ_1, δ_2 satisfying Eqs. (6) and (7) and that x exceeds the minimum possible value of Z, which is a characteristic of the graph. We fix $\alpha_1 = \delta_1$, and with $x > Z$, it turns out that α_2 exists if δ_2 exists, and furthermore, that $\alpha_2 < \delta_2$, which can be verified by inspecting conditions (6) and (10). This leads to

$$\alpha_1 + \alpha_2 < \delta_1 + \delta_2 \leq \pi/2$$

and

$$-\cos (\alpha_1 + \alpha_2) < -\cos (\delta_1 + \delta_2),$$

so that, with $d \geq 1$,

$$1 - \sin 2\delta_1 + 2(d - 1) [1 - \cos (\delta_1 + \delta_2)$$
$$> 1 - \sin 2\alpha_1 + 2(d - 1)[1 - \cos (\alpha_1 + \alpha_2)].$$

Using Eq. (9) we find

$$x\{1 - \sin 2\delta_1 + 2(d - 1)[1 - \cos (\delta_1 + \delta_2)]\}$$
$$> Z\{1 - \sin 2\alpha_1 + 2(d - 1)[1 - \cos (\alpha_1 + \alpha_2)]\}$$
$$\geq -\frac{1}{2}(\lambda_1 + \lambda_2).$$

The transitivity of the $>$ relationship leads us then to conclude, contrary to hypothesis, that Eq. (7) is not satisfied, so that $x \leq Z$.

$$Q.E.D.$$

Theorem 2 is interesting in the case for partitions into two groups in which E_c is vanishingly small as compared to $|V|$. In this limit, $\delta_2 \to 0$ and we may readily compute the minimum of $[1 - \sin 2\delta_1 + 2(d - 1)(1 - \cos \delta_1)]$. This allows us to compute the ratios R of the bound given by Theorem 2 to that of Theorem 1 as a function of d:

$d =$		$R =$
	3	1.68
	4	1.42
	5	1.30
	10	1.12
	20	1.06
	50	1.02

This shows that, for small d, the bound given by Theorem 1 will be off by a significant amount. While a factor of two to four between actual result and theoretical bound may be tolerable, since one may be able to develop heuristic rules for such a ratio, much larger factors would make the present work useless. Accordingly, some results are presented in the next section showing that the ratio R is, at least in certain cases, not excessive.

Let $B = A + D$, where D is a diagonal matrix chosen so that each row sum of B is 0; i.e., d_{ii} is the negative of the valence of vertex i. For this choice, $D = U$, we obtain another improvement of Theorem 1. This theorem yields a better estimate if the $\{m_i\}$ are different.

Theorem 3. Let B be defined as above, and $\alpha_2 \geq \cdots \geq \alpha_k$ be the roots of

$$f(x) = \left(\sum m_i\right)x^{k-1} - 2\sum_{i<j} m_i m_j x^{k-2}$$
$$+ 3\sum_{i<j<p} m_i m_j m_p x^{k-3} \mp \cdots = 0. \tag{16}$$

Then

$$E_c \geq -\frac{1}{2}\sum_{j=1}^{k} \alpha_j \lambda_j(B). \tag{17}$$

Proof. Let J be the matrix of 1's, and N be as defined in the proof of Theorem 1. By the methods used in Theorem 1, it follows that

$$E_c \geq -\frac{1}{2}\sum_{j=1}^{n} \lambda_j(tJ + N)\lambda_j(B) \text{ [because Tr } (tJ + N)B$$

$$= \text{Tr } NB] = -\frac{1}{2}\sum_{j=2}^{n} \lambda_j(tJ + N)\lambda_j(B),$$

since the hypotheses on B show that $\lambda_1(B) = 0$. Inequality (17) is valid for all $t \to \infty$. But it is easy to see that as $t \to \infty$, $\lambda_1(tJ + N) \to \infty$, $\lambda_2(tJ + N) \to \alpha_2, \cdots$,

Table 1 Computed bounds with partitioning of results into two groups.

Graph number	No. of nodes	No. of edges	B_L $U_{ii}' = 0$	Best U'	B_U (Heuristic partition)	B_U/B_L
A1	20	54	7	11	13	1.18
A2	20	51	5	11	13	1.18
A3	20	45	4	7	10	1.43
A4	20	46	6	10	15	1.50
A5	19	45	7	9	12	1.33
A6	20	40	3	7	9	1.29
A7	20	48	5	9	13	1.44
A8	20	34	2	5	7	1.40
A9	20	51	8	13	16	1.23
A10	20	51	5	10	14	1.40
A11	20	46	5	8	11	1.38
A16	20	42	4	9	11	1.22
A13	20	52	4	11	15	1.36
A14	20	43	3	8	11	1.38
A15	20	40	4	6	10	1.67
A16	20	44	4	9	12	1.33
A17	20	54	8	12	17	1.42
A18	20	35	2	5	8	1.80
A19	20	45	6	9	14	1.56
A20	20	42	6	8	13	1.63
					Average $B_U/B_L = $	1.41

Graph number	No. of nodes	No. of edges	B_L $U_{ii}' = 0$	Best U'	B_U (Heuristic partition)	B_U/B_L
B1	40	100	12	15	27	1.80
B2	40	92	8	13	23	1.77
B3	40	104	9	17	25	1.47
B4	50	80	6	9	17	1.89
B5	38	78	5	11	16	1.45
B6	40	91	9	13	21	1.62
B7	39	118	18	22	31	1.41
					Average $B_U/B_L = $	1.63
C1	59	162	13	26	41	1.58
C2	58	153	10	25	40	1.60
C3	60	152	11	24	37	1.54
C4	59	142	13	21	32	1.52
C5	59	147	9	20	33	1.65
					Average $B_U/B_L = $	1.58
D1	99	232	14	28	47	1.68
D2	100	264	21	36	54	1.50
D3	100	252	12	34	58	1.71
D4	100	238	13	30	49	1.63
D5	97	272	19	40	62	1.55
					Average $B_U/B_L = $	1.61

$\lambda_k(tJ + N) \to \alpha_k$, $\lambda_{k+1}(tJ + N) = \cdots = \lambda_n(tJ + N) = 0$. The reason is as follows. The matrix $tJ + N$ is positive semidefinite, and if x is any vector such that, for each $V_i(i = 1, \cdots, k)$, $j \in V_i \sum x_j = 0$, then $(tJ + N)x = 0$. It follows that the eigenvectors x corresponding to positive eigenvalues of $tJ + N$ have $x_k = x_\ell$ if $k, \ell \in V_i$. Accordingly, the nonzero eigenvalues of $tJ + M$ are the eigenvalues of the $k \times k$ matrix $N(t)$, where

$$[(N(t)]_{r,s} = \begin{cases} (t + 1)m_r & \text{if } r = s \\ tm_s & \text{if } r \neq s \end{cases} \quad r, s = 1, \cdots, h.$$

Clearly $\lambda_1[N(t)] \to \infty$. The other eigenvalues of $N(t)$ approach limits that are the roots of the polynomial that is the coefficient of the highest power of t present in the characteristic polynomial of $N(t)$. The characteristic polynomial of $N(t)$ is $\prod_1^k (x - m_i) - tf(x)$.

To prove that (17) is a better bound than that provided by Theorem 1 in the case $D = U$, it is sufficient to show $\alpha_i \geq m_i$ for $i = 2, \cdots, k$. But $N(t)$ is similar to diag $(m_1, \cdots, m_k) + t(\sqrt{m_i m_j})$. Since $t\sqrt{m_i m_j}$ is positive semidefinite for $t \geq 0$, we have completed the proof.

Computational results
Graphs were generated by connecting a preset number of vertices with some probability p, and removing unconnected vertices from the graph. The lower bound B_L on the number of edges cut by a partition into two equal-sized groups was first computed with $U_{ii} = -\Sigma_j A_{ij}$, and then U was varied using the procedure of the two preceding sections to obtain a "best" U with maximum B_L. A heuristic procedure was then used to obtain a partition into groups with B_U edges cut, which is an upper bound on the minimum number of edges cut by such a partition. Results are given in Table 1 for graphs having 20, 40, 60, and 100 nodes; the ratio B_U/B_L is computed for each graph and averaged over all graphs of each of the various sets of equal size. It can be seen that this ratio which is about 1.6 for many of the cases, gives a reasonable range in view of our Theorem 2.

From the results one can also see that variation of U improves B_L significantly — a factor of two improvement is the rule for the larger graphs.

Lastly, two graphs are given in detail in Tables 2 and 3, together with a partition.

Acknowledgment
Some of the work reported here by one of the authors, A. J. Hoffman, was partially supported by contract DAHC 04-72-C-0023 with the U.S. Army Research office.

References
1. R. L. Russo, P. H. Oden and P. K. Wolff, Sr., "A Heuristic Procedure for the Partitioning and Mapping of Computer Logic Blocks to Modules," to be published in the *IEEE Trans. Computers*.

Table 2 The connections and the partition of Graph A1 (see Table 1).

Node	Connections to
1	2, 3, 4, 7, 8, 17
2	3, 10, 14, 15, 16
3	8, 12, 16
4	7, 9, 11, 17
5	6, 9, 11, 15, 16, 20
6	7
7	9, 15, 16
8	10, 12, 14, 16, 18
9	12, 20
10	12, 14, 16, 19
11	18, 19, 20
12	13, 15
13	14, 16, 18, 19
14	16, 18, 19
15	16, 17, 19
17	18

Partition into two groups, where 13 edges are cut.
Group 1: 1, 4, 5, 6, 7, 9, 11, 17, 18, 20
Group 2: 2, 3, 8, 10, 12, 13, 14, 15, 16

Table 3 The connections and the partition of Graph A2 (see Table 1).

Node	Connections to
1	7, 12, 13, 14, 15, 16, 17
2	12, 17, 18, 20
3	5, 11, 13, 14, 18, 19, 20
4	6, 9
5	7, 9, 10, 12, 16, 19
6	16, 18, 20
7	8, 9, 11, 16
8	15, 18
9	11, 15, 19
11	14, 17, 18, 20
12	14
13	18, 20
14	16, 18, 20
16	18
17	18
18	20

Partition into two groups, where 13 edges are cut.
Group 1: 1, 2, 3, 11, 12, 13, 14, 17, 18, 20
Group 2: 4, 5, 6, 7, 8, 9, 10, 15, 16

2. H. R. Charney and D. L. Plato, "Efficient Partitioning of Components," Share/ACM/IEEE Design Automation Workshop, Washington, D. C., July, 1968.
3. L. W. Comeau, "A Study of User Program Optimization in a Paging System," ACM Symposium on Operating System Principles, Gatlinburg, Tennessee, October, 1967.
4. P. J. Denning, "Virtual Memory," *Computing Surveys* **2**, 153 (September 1970).
5. C. T. Zahn, "Graph Theoretical Methods for Detecting and Describing Gestalt Clusters," *IEEE Trans. Computers* **C-20**, 68 (1971).
6. K. M. Hall, "r-Dimension Quadratic Placement Algorithm," *Management Science* **17**, 219 (1971).
7. A. J. Hoffman and H. W. Wielandt, "The Variation of the Spectrum of a Normal Matrix," *Duke Math. J.* **20**, 37 (1953).
8. M. Marcus and H. Minc, *A Survey of Matrix Theory and Matrix Inequalities*, Allyn and Bacon, Inc., Boston, 1964 p. 120, Chapter II. 4.4.14.
9. K. Fan, "On a Theorem of Weyl Concerning Eigenvalues of Linear Transformations, I," *Proc. National Academy Science* USA **35**, 652 (1949).

Received March 20, 1973

The authors are located at the IBM Thomas J. Watson Research Center, Yorktown Heights, New York 10598.

425

Colloques internationaux C.N.R.S.
N° 260 - Problèmes combinatoires et théorie des graphes

NEAREST 8-MATRICES OF GIVEN RANK AND THE RAMSEY PROBLEM FOR EIGENVALUES OF BIPARTITE 8-GRAPHS

Alan J. HOFFMAN ([1])

IBM Thomas J. Watson Research Center, Yorktown Heights, New York 10598

Peter JOFFE

City University of New York

Résumé. — Soit S un ensemble de nombres réels non nuls. On désigne par $A(S)$ l'ensemble des matrices dont tous les coefficients non nuls appartiennent à S. Nous démontrons qu'il existe pour tout $k \geqq 0$, un entier $r_{k,S}$ et une fonction $f_{k,S}$ possédant la propriété (*) : pour toute matrice A de $A(S)$, il existe une matrice B de $A(S)$ telle que l'on ait :

(i) rang $B \leqq r_{k,S}$,

(ii) $\| A - B \| \leqq f_{k,S}(\mu_k(A))$

($\mu_1(A) \leqq \mu_2(A) \leqq \dots$ sont les valeurs singulières de la matrice A).

Nous donnons aussi une formule pour le plus petit entier $r_{k,S}$ tel qu'il existe une fonction $f_{k,S}$ vérifiant (*).

Nous introduisons ensuite le concept de fonction de Ramsey définie sur un ensemble partiellement ordonné, ainsi que le concept de S-graphe. Certaines valeurs propres des matrices associées de certaines classes de S-graphes sont des fonctions de Ramsey. En particulier, pour l'ensemble des S-graphes bipartis, et pour tout i, $\lambda_i(A(G))$ est une fonction de Ramsey.

1. **Introduction.** — In this note, we describe some concepts and problems, announce some results and demonstrate a counter-example.

Let S be a finite set of nonzero real numbers. A (rectangular) matrix is an S-matrix if all of its nonzero entries are selected from S. A well-known theorem of matrix theory asserts : if A is a real matrix with singular values $\mu_1(A) \geqq \mu_2(A) \geqq \cdots \geqq 0$, then there exists a matrix B such that

$$\text{rank } B \leqq k - 1 , \qquad (1.1)$$

and

$$\| A - B \| \equiv \mu_1(A - B) \leqq \mu_k(A) . \qquad (1.2)$$

Our first problem is to see if some result like the foregoing can hold if A and B are both required to be S-matrices. This will be given in paragraph 2. In paragraph 3, we introduce the notion of Ramsey function on a partially ordered set, and show by example that the notion captures the essence of Ramsey's theorem and its variations in many (not all) cases. In paragraph 4, we define the concept of S-graph and show that, in certain cases, eigenvalues of

the adjacency matrices of (classes of) S-graphs are Ramsey functions. In particular, we point out in paragraph 5 that the mechanics of the proof used in paragraph 2 show that, for all S, each eigenvalue is a Ramsey function on the class of bipartite S-graphs. In paragraph 6, we show that, for the class of all S-graphs, there is at least one S for which the least eigenvalue is not a Ramsey function.

2. **Approximation of S-matrices. — Theorem 2.1.** — *For each S, and each positive integer k, there is a function of one variable $f_{k,S}(x)$ and an integer $r_{k,S}$ such that if A is any S-matrix, there exists an S-matrix B such that*

$$\text{rank } B \leqq r_{k,S} , \qquad (2.1)$$

and

$$\| A - B \| \equiv \mu_1(A - B) \leqq f_{k,S}(\mu_k(A)) . \qquad (2.2)$$

We emphasize that A can have any number of rows and columns.

Next, define $r_{k,S}^*$ to be the smallest integer $r_{k,S}$ such that there exists a function $f_{k,S}(x)$ so that, for each S-matrix A, there exists an S-matrix B so that (2.1) and (2.2) hold. We seek a finite expression for $r_{k,S}^*$.

([1]) Part of the work in this paper was supported by Army Research Office under contract DAAG 29-74-C-0007.

Let B, C, D be $\mathcal{S}$-matrices. We say (B, C, D) is *k-allowable* if

rows of B are different, columns of B are different, and the same holds for C,   (2.3)

$$\text{rank } B \leq k - 1, \text{ rank } C \leq k - 1, \text{ rank } B + \text{rank } C - \text{rank } D \leq k - 1 ,   (2.4)$$

each row of D is linear combination of rows of B; each column D is a linear combination of columns of C.
$$(2.5)$$

Theorem 2.2

$$r^*_{k,\mathcal{S}} = \max_{\substack{(B,C,D) \\ k\text{-allowable } \mathcal{S}\text{-matrix}}} \min_{E \text{ an}} \text{ rank } \left[\begin{array}{c|c} E & B \\ \hline C & D \end{array}\right] .$$

For example, if $\mathcal{S} = \{ 1, 2 \}$, the theorem proves that $r^*_{k,\mathcal{S}} = 2k - 2$.

If $\mathcal{S} = \{ 1 \}$, the theorem enables us to prove $r_1 = 0$, and for $k \geq 1$,

$$k - 1 + \frac{1}{3}(k - 1) \ \leq r^*_{k,\mathcal{S}} \leq 2k - 3 .$$

We conjecture that, in this case, $r^*_{k,\mathcal{S}} = 2k - 3$ for all $k \geq 1$, and this has been proved by A. Lempel for $1 \leq k \leq 5$.

3. Ramsey functions on partially ordered sets. — Let P be a partially ordered set with an element 0 preceding all others, and let f be a nonnegative function defined on P such that

$$a \leq b \text{ implies } f(a) \leq f(b) .   (3.1)$$

An *f-chain* in P is a chain

$$0 = a_0 \prec a_1 \prec a_2 \prec a_3 \prec \cdots \text{ with } f(a_n) \to \infty .$$
$$(3.2)$$

Let M be a finite set of indices, and let

$$F = \{ \{ a^i_j \}, i \in M, j = 1, 2, \dots \}   (3.3)$$

be a finite collection of f-chains. For any $a \in P$, define

$$t_F(a) = \max \{ j \mid \exists i \in M \text{ with } a^i_j \leq a \} .$$

It is clear from (3.1) and (3.2) that $t_F(a)$ is properly defined, and is a nonnegative function on P.

Definition. — A nonnegative function f defined on a partially ordered set P which satisfies (3.1) is a *Ramsey function on P* if there exists a finite collection F of f-chains (3.3) such that for every $S \subset P$,

$$\sup_{a \in P} f(a) = \infty \quad \text{implies} \quad \sup_{a \in P} t_F(a) = \infty .$$

To show the relevance of the concept to certain kinds of Ramsey theorems, consider first the statement that, for each $m > 0$, $\exists R(m)$ such that, for every graph G with

$$| V(G) | \geq R(m) , \quad K_m \subset G \text{ or } \overline{K}_m \subset G .$$

(Here, and throughout, $H \subset G$ means the graph H is (isomorphic to) an induced subgraph of G.) We express this by saying that, if $\mathcal{G}$ is the partially ordered set of all graphs, ordered by « $\subset$ », then $| V(G) |$ is a Ramsey function on $\mathcal{G}$, with F consisting of the two chains

$$K_1 \subset K_2 \subset K_3 \subset \cdots \quad \text{and} \quad \overline{K}_1 \subset \overline{K}_2 \subset \overline{K}_3 \subset \cdots .$$

Let us consider the finite case of van der Waerden's theorem that, given any m, there exists an $n(m)$ such that if $N = (1, \dots, n(m))$ is partitioned into two parts, one contains an arithmetic progression of length m. We consider the partially ordered set arising as follows : for every n, take any ordered pair of sets of integers (S_0, S_1), where $S_0 \cap S_1 = \varnothing$ and

$$S_0 \cup S_1 = \{ 1, \dots, n \} .$$

Some S_i can be empty. P has as elements all such ordered pairs arising for all n. Next, let $g_{(a,b)}$ be a function defined on any set S of distinct integers, where $a \geq 1$ and b are integers :

$$g_{(a,b)}(\{ \varnothing \}) = \{ \varnothing \} ,$$

and, if S is not empty

$$g_{(a,b)}(S) = \bigcup_{x \in S} \{ ax + b \} .$$

Let (S_0, S_1) and $(T_0, T_1) \in P$. We say

$$(S_0, S_1) \leq (T_0, T_1)$$

if $\exists$ integer $a \geq 1$ and b such that

$$g_{(a,b)} S_i \subset T_i , \quad i = 0, 1$$

or

$$g_{(a,b)} S_i \subset T_{i+1} \pmod 2 , \quad i = 0, 1 .$$

Then van der Waerden's theorem asserts that $| S_0 | + | S_1 |$ is a Ramsey function on P, with F consisting of the single chain

$$(\{ 0 \}, \varnothing) \prec (\{ 0, 1 \}, \varnothing) \prec (\{ 0, 1, 2 \}, \varnothing) \prec \cdots .$$

One can use a similar idea on partitions of the m-sets of an n-set into r classes.

It is worth remarking that, if f is a Ramsey function on P, then F is essentially unique, by the following discussion. If $a_1 \prec a_2 \prec \cdots$ and $b_1 \prec b_2 \prec \cdots$ are f-chains such that, for each j there is an n_j with $a_j \prec b_{n_j}$, then clearly the f-chain of b's can be omitted from F. Assume this has been done for both F and F', which have indexing sets M and M' respectively in (3.3), with F and F' both verifying that f is a Ramsey function on P. Then, if the chains on F' are denoted by $\{ b^i_j \}$, there is a surjection $\sigma : M \to M'$ such that, for each $i \in M$, and each j, there exist n_{ij} with

$$a^i_j \prec b^{\sigma(i)}_{n_{ij}} ,$$

and conversely. In other words, the chains $\{ a^i_j \}$ and $\{ b^{\sigma(i)} \}$ are essentially equivalent.

This concept of Ramsey function on a partially ordered set emerged from conversations with Nicholas Pippenger, and we are extremely grateful to him. Calvin Elgot and Alex Heller also contributed to this formulation. Our motivation in seeking such a concept was the desire for a convenient language for stating some results and problems about eigenvalues of the adjacency matrices of graphs, but we believe it will suggest or be useful for other combinatorial investigations.

4. **Eigenvalues of S-graphs.** — As in paragraphs 1 and 2, let S be a finite set of nonzero real numbers. An S-graph G is a graph together with an assignment τ of an element of S to each edge of the graph, $\tau : E(G) \to S$. If G is an S-graph, its adjacency matrix $A(G)$ is defined by

$$A = A(G) = (a_{ij}) =$$
$$= \begin{cases} 0 \text{ if } i = j \\ 0 \text{ if } i \text{ and } j \text{ are not adjacent} \\ \tau(i, j) \text{ if } i \text{ and } j \text{ are adjacent} \end{cases}.$$

Thus, $A(G)$ is a symmetric matrix with 0 diagonal and all nonzero off diagonal entries belonging to S. For any G, we denote the eigenvalues of $A(G)$ by

$$\lambda_1(G) \geqq \lambda_2(G) \geqq \cdots, \text{ or } \lambda^1(G) \leqq \lambda^2(G) \leqq \cdots.$$

Let $S(S)$ be the set of all S-graphs. We partially order $S(S)$ by « $\subset$ ». Next, let $H \subset S(S)$. We say λ_k is a Ramsey function on H if $\max(0, \lambda_k(G))$ is a Ramsey function on H. Similarly, we say λ^k is a Ramsey function on H if $\max(0, -\lambda^k(G))$ is a Ramsey function on H. Then we advertise the following problem.

Given S, $H \subset S(S)$, and k, is λ^k or λ_k a Ramsey function on H ? The previous results on this problem are :

$S = \{ 1 \}, H = S(S), \lambda_1, \lambda_2$

    and λ^1 are Ramsey functions ([1], [6]) .   (4.1)

$S = \{ 1, -1 \}, H = S(S), \lambda_1$

    and λ^1 are Ramsey functions ([3], [4]) .   (4.2)

$S = \{ 1, -1 \}, H = $ complete graphs in $S(S), \lambda_2$

    and λ^2 are Ramsey functions ([5]) .   (4.3)

The new results are

S arbitrary, $H = $ bipartite graphs in $S(S)$, every λ_k

    and λ^k is a Ramsey function, and we will describe the F for each k in paragraph 5 .   (4.4)

On the other side, if

$S = \{ 1, \tau^{1/2} + \tau^{-1/2} \}, \tau = (\sqrt{5} + 1)/2 ,$

    $H = S(S), \lambda^1$ is not a Ramsey function .   (4.5)

This will be proved in paragraph 6.

5. **Bipartite S-graphs.** — If G is bipartite, then

$$A(G) = \begin{bmatrix} 0 & M \\ M^T & 0 \end{bmatrix},$$

where M is an S-matrix. It follows that the eigenvalues of $A(G)$ are the singular values of M and their negatives. This is the reason for the connection between this section and paragraph 2. We now proceed to describe F for λ_k, which also serves for λ^k.

The S-graphs are described by the matrix M. If M is an S-matrix with r rows and s columns, and if a and b are positive integral vectors with r and s coordinates respectively, $M(a, b)$ is the matrix with $\sum a_i$ rows and $\sum b_j$ columns obtained by duplicating the ith row of M a_i times for each i and then duplicating the jth column b_j times for each j. This is of course also the matrix of a bipartite graph in $S(S)$.

We can now describe F for λ_k (some of the chains in F may be equivalent, but we do not pause to consider this. It is sufficient to show a finite number of f-chains so that t_F and $\max(0, \lambda^k)$ are cofinal on H). We first define type I chains :

Let M be a nonsingular matrix of order k, and let e_k be the vector all of whose coordinates are 1. Then $\{ M(e_k, ne_k) \}, n = 1, 2, \ldots$ is a λ_k-chain, and the set of all such chains, arising from all nonsingular S-matrices M of order k, are all the type I chains.

The remaining chains in F are type II chains. Let

$$M = \begin{array}{c} \\ r \\ s \end{array} \begin{bmatrix} \overset{t}{E} & \overset{u}{B} \\ C & D \end{bmatrix}$$

be an S-matrix satisfying

rank $B = r \leqq k - 1$, all columns of B are different ;

rank $C = t \leqq k - 1$, all rows of C are different;   (5.1)

rows of D spanned by rows of B ;

    columns of D spanned by columns of C .   (5.2)

    $r + t - $ rank $D = k$.   (5.3)

Then, for each such M,

   $M((e_r ; ne_s), (e_t ; ne_u)) , n = 1, 2, \ldots$ is a λ_k-chain .

The set of all such chains arising from all S-matrices M satisfying (4.1)-(4.3) is the set of all type II chains. One can prove that λ_k is Ramsey on the bipartite graphs of $S(S)$ with these type I and II chains serving as F.

6. **The case $S = \{ 1, \tau^{1/2} + \tau^{-1/2} \}$.** — Recall that $\tau = (\sqrt{5} + 1)/2$ is the golden mean.

Proposition 6.1. — *For* $S = \{ 1, \tau^{1/2} + \tau^{-1/2} \}, \lambda^1$ *is not a Ramsey function on* $S(S)$.

To prove this result, we use material from [2]. Let $m \geqq 4$ and let $\hat{C}_m$ be the graph formed by a circuit of m

vertices together with an additional vertex adjacent to exactly one of the vertices of the circuit.

$$\hat{C}_m \, (m = 6) =$$

Let $A(\hat{C}_m) = (a_{ij})$ be the $(0, 1)$ adjacency matrix of $\hat{C}_m$. This means

$$a_{ij} = \begin{cases} 0 \text{ if } i = j \\ 1 \text{ if } i \neq j, i \text{ and } j \text{ are adjacent vertices} \\ 0 \text{ if } i \neq j, i \text{ and } j \text{ are not adjacent vertices} \end{cases}.$$

We use the notation $B \subset A$ to mean B is a principal submatrix of A.

Lemma 6.2. — *Let* $\alpha = \tau^{1/2} + \tau^{-1/2}$, *and assume m even. Then*

$$if \, B \subsetneq A(\hat{C}_m) + \alpha I, \lambda^1(B) > 0, \qquad (6.1)$$

$$\lambda^1(A(\hat{C}_m) + \alpha I) < 0. \qquad (6.2)$$

Proof. — If m is even, $\hat{C}_m$ is a bipartite graph, because the only cycle is of even order. Hence, $\lambda_1(A) = - \lambda^1(A)$. It is proved in [2] that $\lambda_1(\hat{C}_m) > \alpha$, whence (6.2) follows. It is also shown in [2] that, if $D \subsetneq A(\hat{C}_m)$, then $\lambda_1(D) < \alpha$, whence (6.1) follows.

Henceforth, assume m even. Now fix m, and let J_n be the matrix of order n every entry of which is 1.

Since $\lambda_1(J_n) = n$, and all other eigenvalues of J_n are 0, it follows that

$$F_n^m \equiv (A(\hat{C}_m) + \alpha I) \otimes J_n - \alpha I \qquad (6.3)$$

is the adjacency matrix of some $G \in \mathcal{G}(8)$,

$$\lim_{n \to \infty} \lambda^1(F_n^m) =$$
$$= \lim_{n \to \infty} (n\lambda^1(A(\hat{C}_m) + \alpha I) - \alpha) = - \infty, \qquad (6.4)$$

by (6.2).

Now suppose λ^1 is a Ramsey function on $\mathcal{G}(S)$. Let M be a finite set of indices, and

$$\{ G_n^i \mid i \in M, n = 1, 2, \dots \}$$

be a finite collection of λ^1-chains (i.e., for each $i \in M$, $\lambda^1(G_n^i) \to - \infty$). Let m be a fixed even integer. By (6.4), there exists an index $i(m) \in M$ and sequences $t_1 < t_2 < \cdots$ and $n_1 < n_2 < \cdots$ with

$$G_{t_j}^{i(m)} \subset F_{n_j}^m, j = 1, 2, \dots . \qquad (6.5)$$

Recall that F_n^m has each vertex of $\hat{C}_m$ « duplicated » n times. Then it follows at once from (6.1) and (3.2) that, for an infinite number of values of j, each vertex of $\hat{C}_m$ is represented at least once in the vertices of $G_{t_j}^{i(m)}$. In particular, $C_m \subset G_{t_j}^{i(m)}$ for at least one j. Note also that, if $m \neq m'$, $C_{m'} \not\subset F_n^m$, whence it follows that $i(m) \neq i(m')$ if $m \neq m'$. This last remark, together with the fact that there are an infinite number of even integers, contradicts the finiteness of M.

References

[1] A. J. HOFFMAN, On spectrally bounded graphs, in : *A Survey of Combinatorial Theory*, North-Holland (1973) 277-283.

[2] A. J. HOFFMAN, On limit points of spectral radii of nonnegative integral matrices, in : *Graph Theory and its Applications*, Springer-Verlag, Berlin (1972).

[3] A. J. HOFFMAN, On eigenvalues of symmetric $(+1, -1)$ matrices, *Israel Journal of Mathematics* **17** (1974) 69-75.

[4] A. J. HOFFMAN, On spectrally bounded signed graphs, in : *Transactions of the 21st Conference of Army Mathematics*, U.S. Army Research Office, Durham (abstract) 1-5.

[5] A. J. HOFFMAN and J. J. SEIDEL, On the second eigenvalue of $(- 1, 1)$ adjacency matrices (in preparation).

[6] L. HOWES, On subdominantly bounded graphs — summary of results, in : *Recent Trends in Graph Theory*, Springer-Verlag, Berlin (1971) 181-183.